ALLE · ZEIT · WACH
SJ
1842

Jürgen Fuchs

Oxidative Injury in Dermatopathology

With 122 Figures and 17 Tables

Springer-Verlag
Berlin Heidelberg New York
London Paris Tokyo
Hong Kong Barcelona
Budapest

Dr. phil. nat., Dr. med. Jürgen Fuchs
Klinikum der Johann Wolfgang Goethe-Universität
Zentrum der Dermatologie und Venerologie
Abteilung II
Theodor-Stern-Kai 7
W-6000 Frankfurt 70, FRG

ISBN-13:978-3-642-76825-5 e-ISBN-13:978-3-642-76823-1
DOI: 10.1007/978-3-642-76823-1

Library of Congress Cataloging-in-Publication Data
Fuchs, J. (Jürgen), 1957 – Oxidative injury in dermatopathology / J. Fuchs. p. cm. Includes bibliographical references and index.
ISBN-13:978-3-642-76825-5
1. Skin – Pathophysiology. 2. Active oxygen – Pathophysiology. 3. Free radicals (Chemistry) – Pathophysiology. I. Title. [DNLM: 1. Oxygen – adverse effects. 2. Oxygen – metabolism. 3. Skin Diseases. WR 140 F951o] RL96.F83 1992 616.5'07 – dc20 DNLM/DLC 91-5192

Typesetting: Appl, Wemding
27/3145–5 4 3 2 1 0-Printed on acid-free paper

Preface

Dermatology is a complex and puzzling world of itching bumps, pimples, and rashes. The multitude of clinically distinct skin diseases, their frequently unresolved pathogenesis, and the exponentially increasing amount of scientific information add to the confusion about skin diseases. The great prevalence of skin diseases makes them an urgent priority for intensive research effort, and although many scientists and academic clinicians are vigorously trying to uncover their secrets, we are only at the very brink of understanding the etiology of most dermatoses.

The principle mechanisms of general organ pathology (physical, chemical, microbial, ischemic, degenerative, and neoplastic disturbances) are believed to be relatively well understood. In contrast to skin pathomorphology, however little is known regarding the biochemistry and physiology of dermatoses. The difficulty in understanding skin diseases may be overcome partially by finding biomedical simplifications, and the concept of "oxidative injury in dermatopathology" is just such a simplification. It should, of course, always be kept in mind that no single mechanism alone can explain the pathogenesis of a disease and that there may be a danger of overlooking other important biological determinants.

One major mechanism involved in organ pathology is oxygen metabolism and the formation of reactive oxygen species, such as the superoxide anion and hydroxyl radical. The continuous exposure of aerobic organisms to oxygen leads to tissue destruction when the flux of reactive oxygen species is high, or when protective antioxidant mechanisms fail: Humans do not usually become rancid while they are alive, even when they consume increased amounts of oxygen, such as occurs during strenuous physical exercise, as their polyunsaturated fatty acids are well protected by efficient antioxidant mechanisms. However, these antioxidant defenses break down after death, and the human body takes up much more oxygen within the first 3 days post mortem than during the same time period when alive and breathing.

In the past decade, increasing attention has been paid to the role of reactive oxygen species and free radicals in dermatopathology, although there is still controversy over how and to what extent free radicals and reactive oxygen species participate in organ pathology, particularly in skin. This skepticism reflects both our inability to measure and analyze exactly biochemical reactions in skin induced by reactive oxygen species, and some scientific ignorance regarding what is occurring in skin at a biochemical level.

Despite intensive and worldwide research efforts on oxidative skin injury, a comprehensive analysis of the existing experimental and clinical data is not available. This work is the result of my efforts to produce a state-of-the-art review of the topic of oxidative injury in dermatopathology, with objective of summarizing and discussing, from the perspective of an academic clinician, developments in our understanding of the role of free radicals and reactive oxygen species.

Most of the opinions and hypotheses presented in this book were extracted from the references cited, and a few results and statements were literally translated from the biomedical literature, including my own studies. The review of the literature carried out reveals that publications sometimes contain differences in results, and that controversy exists about interpretation of other results. Occasionally, authors make speculations based on a limited amount of solid and direct experimental evidence.

This book should serve as a valuable source of information for dermatologists and biomedical scientists interested in the field of free radicals and reactive oxygen species in cutaneous biology and medicine. A comprehensive list of references is also provided to enable the interested reader to obtain more detailed background information. I am aware that the physiological limitations of the human mind only enable me as an individual to give a subjective perspective on this topic, and hope that this does not detract from the value of the work.

Jürgen Fuchs

Contents

Chapter 1
History of a Concept

The free radicals which provided the impetus for life's beginning also damage the same biomolecules that they helped to originate. Biological life is a combustion process which needs oxygen, but it is also endangered by reactive oxygen metabolites, so-called reactive oxygen species. It was suggested by Harman that life originated as a result of free radical reactions, selected free radical reactions to play a major part in metabolism, and assured evolution by employing them to provide for mutation and death (Harman 1984). The historical roots of the concept of oxidative injury in biological systems, and particularly in dermatopathology, are heterogeneous.

Oxygen. On earth about 3 billion years ago, primitive aerobic organisms were created under high doses of ionizing and nonionizing irradiation in a reducing chemical environment mainly composed of water, hydrogen, methane, and ammonia. The evolution of photosynthetic blue-green algae and photodissociation of water led to increasing formation of molecular oxygen. Oxygen is toxic to many life processes, and biological defense mechanisms evolved to counteract the toxic mechanism of this biradical gas (Möhn 1984; Prosser 1986).

Priestley (1771, 1775) and Scheele (1773, 1777) are credited with independently discovering oxygen. Priestley isolated oxygen by liberating it from oxides of mercury, and later discovered that it is consumed by animals and produced by plants. Ingenhousz (1779–1796) showed that light is required for oxygen production by green plants. Lavoisier (1780–1789) demonstrated that animals require oxygen, recognized that respiration is oxidation, and measured the oxygen consumption of a human subject. In 1912, Warburg postulated a respiratory enzyme for the activation of oxygen.

In 1775 Priestley had predicted oxygen's toxicity, and in 1785 Lavoisier was very likely the first to demonstrate the pathological effects of oxygen in guinea pigs. Animal studies on oxygen toxicity were continued by Bert, who documented the toxicity of oxygen in mammals in 1874 but failed to discover the phenomenon of pulmonary oxygen toxicity. A few years later, pulmonary oxygen toxicity was described by Lorrain Smith (Smith 1897, 1899). Oxygen was recognized as a radiation sensitizer in the 1920s (Holthusen 1921). The suggestion that free radicals may be involved in oxygen toxicity and radiation damage (Risse 1929; Gerschmann 1954) can be considered as the beginning in our understanding of free radical biology in human disease. The superoxide anion radical, a reactive oxygen intermediate, was discovered in 1933 (Neu-

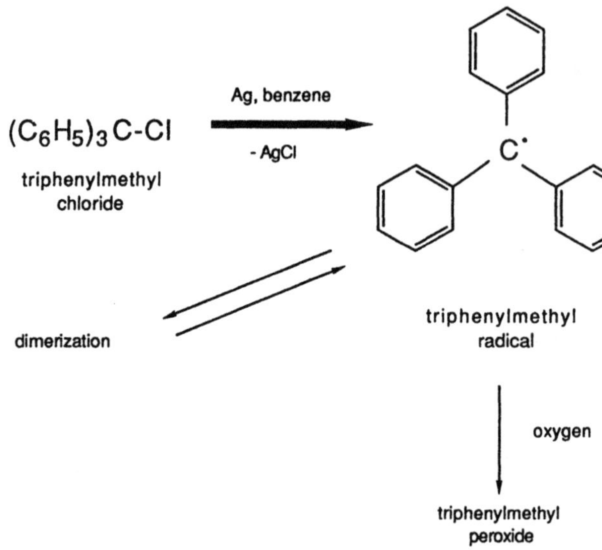

Fig. 1.1. Production of triphenylmethyl radicals

man 1934), and subsequently the chemical reaction between hydrogen peroxide and superoxide anion radical, resulting in the formation of hydroxyl radicals, was characterized (Haber 1932, 1934; Weiss 1944). This provided key information for unravelling the molecular mechanisms of oxygen toxicity. The first experiments in humans demonstrating the effects of oxygen toxicity were carried out by Behnke in 1936.

Radicals. Free radicals on carbon have probably the oldest history of all reaction intermediates in chemistry. Although the existence of free radical species as chemical reaction intermediates or even as stable compounds was disputed by many chemists in the nineteenth century, Gomberg made history in 1900 by observing the transient appearance of triphenylmethyl radicals on warming a benzene solution of triphenylmethylchloride over silver powder (Fig. 1.1).

Nitroxide Radicals. The stable inorganic nitroxide radical Fremy's salt (nitrosodisulfonate) had already been discovered in 1845. Organic nitroxide type radicals were synthesized first by Piloty and Graf Schwerin in 1901. The free radical properties of these substances were only hypothetical at that time. Lebedev proved the free radical properties of a synthesized organic nitroxide by elemental analysis, molecular weight determination, and electron spin resonance spectroscopy (Lebedev 1959). Soon, other types of persistent free radicals were synthesized, and free radical chemistry became commonplace (Forrester 1968). Free radical chemistry is now an extensively studied field (Nonhebel 1979).

Nitroxides are used for various analytical purposes in biomedicine and also in dermatologic research (Fuchs and Milbradt 1990). The chemistry of ni-

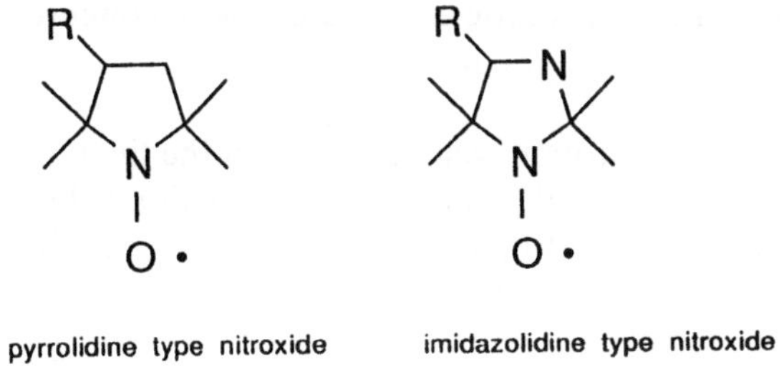

Fig. 1.2. Pyrrolidine and imidazolidine type nitroxides

troxide radicals was investigated by Rozantsev (1970), Rassat (1971), and Volodarsky (1988, 1989). Nitroxides were introduced as spin labels in molecular biology by McConnell (1970) and Likhtenstein (1976). Nitrones, nitroxide precursurs, are used for the detection of extremely short-lived free radicals (spin trapping) (Kalyanaraman 1982). Various types of nitroxides are known, which have different stabilities in biological material. Imidazolidine and pyrrolidine type nitroxides, which are quite inert concerning bioreduction, are frequently used for biomedical studies (Fig. 1.2).

Electron Paramagnetic Resonance Spectroscopy. The basic theories of electron parmagnetic resonance (EPR) spectroscopy were developed in the 1930s, and in 1944 the Russian student Zavoisky made history by measuring for the first time a paramagnetic substance by means of EPR spectroscopy with a self-built spectrometer (Zavoisky 1945). His discovery enabled scientists to study free radical reactions in complex and heterogeneous biological material for the first time. The application of EPR spectroscopy to biological materials in the 1950s provided a specific and sensitive method for monitoring free radicals and other paramagnetic species in tissues. In the late 1950s it was possible for the first time to detect free radicals in biological material by this new method. High free radical content was found in tissues with raised metabolic activity. An extraordinarily high free radical concentration was detected in unfertilized, pigmented frog eggs and identified as melanin radicals (Commoner et al. 1954).

EPR spectroscopy is becoming an increasingly important analytical method in basic dermatologic research (Fuchs and Milbradt 1990). Recently, an EPR tomography technique was developed (Herrling et al. 1982) which allows paramagnetic imaging in different skin planes (Fuchs et al. 1990b). It can be used for spatially resolved skin studies: (a) measuring free radical reactions, (b) monitoring penetration of spin labeled drugs, (c) analyzing biophysical membrane properties, and (d) visualizing biochemical reactions of spin labeled, reactive substances (Fuchs et al. 1990f).

Antioxidants in Polymer Chemistry. In 1861 Hoffman showed that atmospheric oxygen was the primary cause of the aging of rubber. Soon it was found that this slow deterioration could be prevented by vulcanization, a process that involves heating rubber with sulfur and accelerators, many of which have since been shown to possess antioxidant activity. Subsequently, study of the chemical mechanisms of natural and synthetic antioxidants was pioneered by Moureau and Dufraisse during the 1920s (Scott 1985).

Antioxidants in Biomedicine. In 1939 a protein was described in the liver and blood of mammals that contained copper and had no obvious function (Mann and Keilin 1939). Later, it was revealed that this protein fraction had antiinflammatory activity (Huber 1981), and in 1969 an enzymic function was attributed to it, namely dismutation of superoxide radicals to oxygen and hydrogen peroxide (McCord and Fridovich 1969). Glutathione peroxidase was discovered by Mills in 1957. Evans and Bishop reported in 1922 the existence of a new vitamin, which they designated vitamin E. Its mode of action remained unknown until its antioxidant properties were described by Olcot and Emerson.

Lipid Peroxidation. In 1820 Saussurer measured the oxygen uptake of walnut oil exposed to air and noticed that the oil subsequently became rancid. Berzelius suggested a few years later that oxygen uptake of oils when exposed to air is involved in their autoxidation reactions. The presence of lipid peroxidation products in skin and their pathophysiological role were suspected by the dermatologist Unna in 1910. The biological significance of lipid peroxidation for human pathology remained unclear for several more decades. In 1969 the concept of "biological rancidification" was introduced by Dormandy, suggesting that lipid peroxidation is significantly involved in key pathophysiological events. Its vital contribution in dermatopathology was subsequently investigated (Meffert and Lohrisch 1971 a, b, Meffert and Reichmann 1972, Meffert et al. 1972).

Oxidative Stress in Biomedicine. The crucial role of phagocytes in host defense reactions was discovered by the Russian zoologist Metschnikoff in 1883. Involvement of oxygen in this reaction was described in 1933 (Baldrige and Gerard 1933), and later on hydrogen peroxide (Iyer et al. 1961) and superoxide anion radical (Babior et al. 1973) were identified as reactive mediators. Circumstantial experimental and clinical evidence suggested a significant role for free radicals in a variety of pathological processes such as inflammation, carcinogenesis, aging, and degenerative diseases (Floyd 1980; Tso et al. 1980; Halliwell and Gutteridge 1985; Yagi 1982; Sohal 1986). Oxidative stress is considered a key pathophysiological event in the chronic struggle of organisms with oxygen (Sies 1985). Following Unna's work on the cellular localization of oxygen metabolism in the skin, Gans (1923) measured skin oxygen uptake using a manometric method developed by Warburg. Steigleder (1952) and Glasenapp and Leonhardi (1953) continued his work on biological oxidation in the skin.

Basic dermatologic studies have recently focused on the role of reactive oxidants and autoxidative tissue injury in skin diseases. The most comprehensive studies were done in Japan by the groups of Yagi, Ogura, Niwa, and Miyachi. Meffert and Fuchs in Germany, Roschchupkin and Potapenko in the Soviet Union, and Packer, Pathak, Mukhtar, and Black in the USA are doing extensive research in this area.

Chapter 2

The Skin and Oxidative Stress

A. Introduction

I. Skin and Environmental Stress

Skin is a biological interface with the external environment and functions as the first line of defense against noxious stimuli. As such, it is often the first organ to come into contact with harmful stresses. Skin primarily provides protection from external insults to the body's interior. From an evolutionary point of view, the oldest function of skin is resistance to and separation from the environment. Other basic functions include heat regulation, communication and involvement in hormone metabolism and in immunological surveillance. Skin is also challenged by intrinsic aggressive stimuli such as autoimmune reactions, e.g., autoaggressive inflammatory cells, and specific metabolic dysfunctions. The integument mainly functions as a immunological, microbial, and physicochemical barrier. The physicochemical barrier function resides principally in the stratum corneum. The epidermodermal interface is a second barrier due to a change from a lipophilic to a hydrophilic skin compartment. The protective mechanisms can break down under stresses caused by external and internal insults. There are many different forms of harmful environmental stress conditions, such as extremes brought about by oxygen depletion or excess, extremes of temperature, water stress, nutrient depletion or excess, and exposure to environmental pollutants. Environmentally related disorders of the skin are responsible for about one third of all cases of occupational diseases in the USA. It has been claimed that there may be as many as 2000 airborne chemicals to which persons could be exposed in the community and workplace. At the same time, there is a widespread lack of information on the toxicity of most chemical substances, and the mechanism of it. The involvement of free radicals and reactive oxygen species (reactive oxidants) in environmental stress conditions such as hyperoxia (Nohl et al. 1981; Frank and Massaro 1980; Housset et al. 1983), ischemia and reperfusion (McCord 1985, 1986; Zweier et al. 1987), heat shock (Loven 1988), and metabolism of environmental pollutants and toxic chemicals (Mason 1982, Mason and Chignell 1982; Cavalier and Rogan 1985; O'Brien 1984; Pryor 1982) is well documented.

II. Oxidative Stress

Oxidative stress refers to a condition of imbalanced prooxidant/antioxidant equilibrium in favor of the former leading to potential damage (Sies 1991). Reactive oxidants affect virtually all aspects of biological life by reaction with and modification of structural and functional material. They are common intermediates in cell physiology (Williams 1985; Yamazaki et al. 1985) but also play an important role in various pathological situations in humans. The same type of a reactive species can exert molecular damage in a highly susceptible cellular microenvironment if the steady state concentration is high enough and may be innocuous in a well protected environment, especially at a low steady state concentration. To prevent uncontrolled and deleterious reactions these reactive species are either of controlled reactivity or confined to a restricted microenvironment. In addition, in biological systems free radical reactions are well controlled by a complex system of enzymatic and nonenzymatic defense and repair systems. Oxidative injury may evolve when antioxidant defense functions such as free radical prevention, interception, and repair processes are impaired, and free radical flux from endogenous or exogenous sources is increased. Free radical pathology is thought to qualify as a basic mechanism of molecular pathology (Balentine 1982) based upon the findings that key molecular structures are highly susceptible to free radical reactions.

III. Skin as a Target Organ of Oxidative Injury

Skin is a potential target organ of oxidative injury because it is continuously exposed to visible and ultraviolet irradiation and high oxygen concentrations and contains a variety of oxidizable structures critical for the maintenance of cellular homeostasis. It is one of the largest body organs and serves as a major portal of entry for many airborne environmental pollutants, some of them free-radical-generating agents. Skin harbors a large population of saprophytic microbes and is frequently in contact with pathogenic species. Skin infection and subsequent inflammatory reactions can result in autoxidative tissue injury. Skin is frequently anaerobic through pressure; it has no oxygen reserve and has a very low quotient of oxygen uptake. However, the circulation through the skin provides mainly for thermoregulation, and only a fraction of the blood (about 20%) is needed for metabolic exchanges. The luxurious vascularization of the integument may account for the strong inflammatory skin reactions, frequently observed as a nonspecific reaction to different etiopathogenetic factors. Skin is the most peripheral organ of the immune system, and Langerhans cells recognize allergens. These cells are involved in the complex regulation process of a balanced skin immune system. They are very susceptible to oxidative stress factors such as solar irradiation and oxidizing compounds.

B. Biological Oxidants

Oxygen Redox Chemistry. Various chemically different species are of interest in "free radical biology." These species may themselves be free radicals or may be nonradical compounds which have a great potential to induce radical or radical chain reactions in biological systems. However, some of these species do not always react via free radical pathways. A free radical is a chemical species with an unpaired electron that has the tendency to react very rapidly with other free radicals or nonradicals (Pryor 1986).

Oxygen-centered radicals are formed during one-electron activation of molecular oxygen. The standard redox potential of H_2O/O_2 $E_0' = 0.85$ V indicates that oxygen is thermodynamically a good oxidant; however, its reactivity is kinetically restricted (spin forbidden). Molecular oxygen (ground state) is a diradical gas, possessing two unpaired free electrons which have the same spin quantum number. Usually an electron pair in an orbital has antiparallel spin. Due to their parallel electron spin arrangement, the one-electron reduction of oxygen predominates over the two-electron reduction pathway. The univalent pathway is outlined in Fig. 2.1.

Free Radical Half-lifes. Reactive oxygen products and other biologically important free radical species are usually very unstable in biological material due to their high reactivity. High chemical reactivity results in low specificity of the reaction in biological material; biological oxidants with low reactivity have a high chemical specificity. Free radicals have characteristic half-lives due to their chemical reactivity and are distinguished from radicals bound to, e.g., enzymes (ribonucleotide reductase), which are not considered "free" radicals. Some radicals are stable enough to diffuse across biological membranes; others are so reactive that they react in the chemical microenvironment at their site of formation. Half-lives of radicals can be determined by measuring their rate constant with a given substrate, e.g., an unsaturated fatty acid (linoleic acid). Estimated average half-lives and half-life ways of biologically relevant free radicals and reactive oxygen species are listed in Table 2.1. For comparison, the thickness of the phospholipid bilayer is 4 nm, and the cell diameter of an erythrocyte is 7 μm.

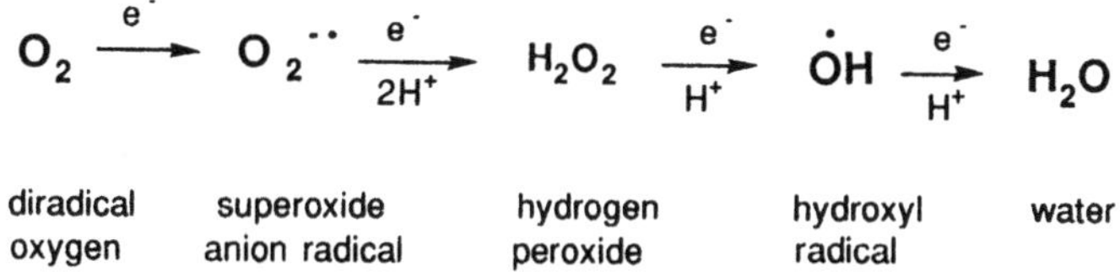

Fig. 2.1. Univalent oxygen reduction pathway

Table 2.1. Half-lives and half-life ways of biologically relevant, free radicals and reactive oxygen species

Reactive species	Half-lives	Half-life way
Hydroxyl radical ($OH\cdot$)	0.3 ns	1.8 nm
Alkyl radical ($R\cdot$)		10 ns
Alkoxy radical ($RO\cdot$)		1 μs
Alkylperoxy radical ($ROO\cdot$)		1–10 s
Hydrated electron ($e\cdot$)	3.7 ns	9.2 nm
Hydrogen atom ($H\cdot$)	9 ns	19 nm
Superoxide anion radical ($O_2^{-\cdot}$)	0.4 μs–1 ms	55 nm–3 μm
Singlet oxygen ($^1O_2$)	μs–ms	
Nitrous oxide radical ($NO\cdot$ = endothelial derived relaxing factor)	seconds	
Ascorbyl anion radical	seconds	
α-Tocopheroxyl radical	seconds	
Ubisemiquinone-type radicals	hours	
Hydrogen peroxide (H_2O_2)	persistent	
Lipid hydroperoxide (ROOH)	persistent	
Radical chelates ($Me \ldots X\cdot$)	persistent	
Melanin-semiquinone radicals	persistent	
Anthralin brown radicals	persistent	

Data according to Kasha and Kahn (1970) Pryor (1986), and others

Very reactive free radicals (hydroxyl radical) cause biological damage only if generated in close proximity to a potential target molecule (e.g., DNA), because they are immediately scavenged by the high concentration of organic molecules in the cell. If they are to cause cell damage directly, they need to be generated directly at the critical cellular target site.

Free radicals of intermediate reactivity are able to diffuse over significant distances and may then react with some specificity and selectivity with target molecules. They are the most likely species to lead to direct biological damage. Other types are physiological reaction intermediates which do not primarily cause cell damage. Examples of such endogeneous radicals are tocopheroxyl, ascorbyl, and certain semiquinone-type radicals. Only under special circumstances may these species lead to pathophysiological reactions.

Persistent free radicals are rather biologically unimportant with respect to direct cell damage. However, they may be damaging by indirect means. Nitroxide-type radicals are one of the best characterized group of persistent radicals. Organic nitroxides are used as contrast-enhancing agents in magnetic resonance imaging (MRI) (Brasch et al. 1983), and the toxicology of these compounds has been studied in detail (Couet et al. 1985). Under special conditions nitroxides may cause cell mutations (Sies and Mehlhorn 1986). Otherwise, organic nitroxide radicals do not have physiological or pathophysiological relevance, with the exception of the inorganic NO· radical (endothelial derived relaxing factor, EDRF). NO· is involved in the regulation of vasotonus and is a product of arginine metabolism. Other persistent free radicals of physiological relevance are melanin radicals, which can be detected directly in skin.

Biological Oxidants. Radicals of interest in biological systems can have their unpaired electron on different atoms. Oxygen-centered radicals are to be distinguished from carbon-, sulfur-, and nitrogen-centered radicals. The latter can be products of drug metabolism. Carbon-centered radicals are also formed during lipid peroxidation, and sulfur-centered radicals result from the oxidation of endogenous thiol compounds. Free radicals can be formed by homolysis, photolysis, or radiolysis of molecules or be redox reactions, and they may be uncharged (OH·, hydroxyl radical) or charged (O_2–·, superoxide anion radical).

Reactive nonradical oxygen compounds of interest in oxidative injury are ozone, hydroperoxides, hypobromic and hypochloric acid, hydrogen peroxide, singlet oxygen, and excited carbonyls. Since these species do not have an odd electron on their oxygen atom, the term "reactive oxygen species" is used to described them. "Reactive oxidants" refers to all free radical or nonradical compounds which may cause "oxidative injury". The label "Free radical pathology" is potentially misleading because only some of the key intermediates of pathobiological redox reactions are free radicals. Other important nonradical intermediates are transition metals, hydroperoxides, singlet oxygen, and excited carbonyls. A more appropriate term may be "oxidative injury." This implies that reactive oxidants are involved in key pathophysiological processes.

In general, these reactions are oxidative processes: A reactive oxidant abstracts an electron from a biological donor molecule. However, the term "reactive oxidants" should not be interpreted too closely. The superoxide anion radical can be an oxidizing or reducing species in a free-radical-type reaction, depending on the redox potential of its reactant. Hydroxyl radicals may generate the powerful reductant carbonate anion radical $CO_3^{-\cdot}$, which may be damaging through reductive mechanisms. Hydrogen atoms (H·) and hydrated electrons $[e^{-}] \times H_2O$, which are formed during water radiolysis, are primarily reducing radicals.

Detection of Reactive Oxidants in Skin and Skin Cells. Evidence for the existence of reactive oxidants in skin tissue and skin cells is difficult to establish because of the high reactivity of these species, their low steady state concentration, and the heterogeneity of the organ. Skin is composed of three morphologically distinct layers (epidermis, dermis, subcutis). The epidermis has a complicated substructure, and all three skin layers consist of a variety of distinct cell populations. In addition to its high compartmentation, an extensive database on skin biochemistry is not available because this was rather a neglected field in the past.

Direct evidence of the formation of free radicals in skin can be obtained by low temperature EPR spectroscopy of highly reactive free radicals or by ambient temperature spectroscopy of persistent free radicals.

1. Low temperature EPR. Ultraviolet irradiation of epidermis induces the formation of EPR signals in human skin (Norins 1962; Stratton and Pathak 1968; Ogura 1981; Nishi et al. 1991 a), which are highly unstable at temperatures above −70°C.

2. Ambient temperature EPR: In the affected skin of psoriasis patients increased concentrations of persistent free radicals have been detected (Zadorozhny et al. 1973; Verbenko et al. 1986). The ascorbyl radical is found in mouse skin; its level rises further after UV irradiation (Büttner et al. 1986). Epidermal application of free-radical-generating compounds, such as 4-hydroxyanisol (Riley 1970) and anthralin (Schroot and Brown 1986; Fuchs and Packer 1989), results in the generation of persistent organic radicals in skin.

Indirect experimental evidence for the formation of reactive oxidants in epidermal and dermal cells as well as in intact skin can be obtained by the detection of reaction products, spin trapping, or measurement of chemiluminescence.

1. Detection of free radical reaction products: Lipid peroxidation products such as thiobarbituric acid reactive substances (TBRS) and volatile alkanes are used as indicators of free-radical-mediated lipid peroxidation. Lipid peroxidation products were noted in UV-exposed skin (Meffert and Reich 1969 a) and in inflammatory dermatoses (Niwa et al. 1987 b). Ethane/pentane exhalation is increased in animals topically treated with the prooxidant anthralin (Müller et al. 1987).

2. Spin trapping highly reactive free radicals: Spin trapping is a method applying a diamagnetic free radical trap (e.g., nitrone), which is converted after reaction with a highly reactive radical (e.g., hydroxyl radical) into a persistent free radical adduct (e.g., nitroxide), which has a characteristic EPR spectrum that allows chemical identification (Janzen 1980). Glutathione thiyl radicals were spin trapped in keratinocytes with 5,5-dimethyl-1-pyrroline-*N*-oxide (DMPO) (Schreiber et al. 1989). Alkyl and alkoxyl radicals were also detected in keratinocytes with DMPO (Taffe et al. 1987), and lipid radicals were

spin trapped by DMPO in epidermis exposed to UV (Nishi et al. 1990). Unidentified thiyl radicals arising after photosensitization in intact skin of hairless mice were also spin trapped by DMPO (Li, personal communication).

3. Measurement of chemiluminescence: The detection of light emission from biological samples is a useful method for studying oxidative reactions in intact biological systems (Boveris et al. 1979; Cadenas and Sies 1984). The tumor promotor 12-*O*-tetradecanoylphorbol-13-acetate is a stimulator of luminol-enhanced chemiluminescence in mouse keratinocytes, indicating formation of reactive oxygen species in epidermal cells (Fischer and Adams 1985; Fischer et al. 1986). Chemiluminescence is augmented after stimulating primary cultures of human fibroblasts with lipopolysaccharides and culture supernatants of bacteria causing polyarthritis (Meier et al. 1988), joint aspirations of patients suffering from rheumatic arthritis (Meier et al. 1990), and interleukin-1 (Il-1) or tumor necrosis factor (TNF) (Meier et al. 1989).

I. Superoxide Anion Radical

The principal mammalian source of superoxide anion radical ($O_2^{-\cdot}$) under normal conditions is the mitochondrion (Chance et al. 1979). Increased formation may occur in skin via photosensitization, invasion of the epidermis and dermis by leukocytes with their respiratory burst, redox cycling of xenobiotics, autoxidation of endogenous metabolites, and altered electron transport chain activity of mitochondria, microsomes, and the plasma membrane. In addition, hydrogen peroxide is photochemically converted by near UV into superoxide anion radical (Ahmad 1981). Superoxide anion radical may react as a reducing reagent (standard redox potential of $O_2^{-\cdot}/O_2$ $E_0' = -0.16$ V) or as an oxidizing agent (standard redox potential $H_2O_2/O_2^{-\cdot}$ $E_0' = 0.94$ V), depending on the redox properties of its reactant.

It is generally not considered to be very reactive with lipids, proteins, polysaccharides, or nucleic acids. Since it is a charged species, it usually does not penetrate membranes. It is poorly reactive in aqueous solution but exhibits a higher reactivity in hydrophobic environments. In the protic microenvironment of the polyanionic cell surface, formation of the hydroperoxy from the superoxide anion radical is favored, which is a stronger oxidant and more lipid soluble and has a longer half-life. The hydroperoxy radical ($HO_2^{-\cdot}$) might be the predominant form of superoxide in phospholipid membranes due to their lower dielectric and acidic microenvironment. It oxidizes polyunsaturated fatty acids and reduced nicotinamide adenine dinucleotide (NADH) at a rate of 1×10^3 M^{-1} s^{-1} and 1.8×10^5 M^{-1} s^{-1}, respectively, whereas superoxide does not detectably react with them.

The low chemical reactivity of superoxide cannot explain the toxic effects produced by systems generating aqueous superoxide anion radical. Many of the biological effects of superoxide may be ascribed to products derived from its dismutation (Borg et al. 1978) or via formation of thiyl radicals by reaction

$$O_2^{-\cdot} + H_2O_2 \xrightarrow{Me^{n+}} O_2 + OH^- + OH\cdot$$

Fig. 2.2. Haber-Weiss reaction

of superoxide with endogeneous thiol groups (Asada and Kanematsu et al. 1976; Ross 1985). Superoxide spontaneously or enzymatically dismutates to hydrogen peroxide and oxygen. Superoxide can further react with its own dismutation product, hydrogen peroxide, in a metal-ion-catalyzed (Me^{n+}) mechanism to yield the highly reactive hydroxyl radical (Haber-Weiss reaction; Fig. 2.2).

This reaction occurs only very slowly under most physiological conditions and is thought to be an unlikely source of high fluxes of hydroxyl radicals in vivo (Sutton 1985).

II. Hydrogen Peroxide

Hydrogen peroxide is the most stable intermediate of oxygen reduction products, and was suspected to be formed in skin following treatment for psoriasis with cignolin (Unna 1916). Based on data from isolated cell fractions, hydrogen peroxide is known to be generated in rat liver by microsomes (45%), peroxisomes (35%), mitochondria (15%), and cytosolic sources (5%). It can result from the univalent reduction of oxygen followed by dismutation of the superoxide anion radical, and is considered the major cytotoxic product formed by the xanthine/xanthine oxidase system (Link and Riley 1988). In addition, it is the product of several oxidases localized in peroxisomes and can be created by various inflammatory cells and by near-UV irradiation via endogenous photosensitizers. It arises intra- and extracellularly and crosses biological membranes nearly as freely as water.

Hydrogen peroxide is cytotoxic and a weak oxidizing agent. It can oxidize sulfhydryl groups in enzymes and polyunsaturated fatty acids. However, these reactions proceed at reasonable rates only at concentrations of hydrogen peroxide which exceed those found in vivo by about four orders of magnitude. It is believed that hydrogen peroxide is not the ultimate oxidant. Its cytotoxicity may be mediated through extremely reactive hydroxyl radicals formed by the reaction of hydrogen peroxide with transition metal ions or chelates (Fenton reaction). Fenton observed in 1894 that a mixture of hydrogen peroxide and iron (II) salts reacts readily with many organic compounds.

In the Fenton reaction, the reduced form of a transition metal ion reacts with hydrogen peroxide, yielding the oxidized metal ion and the hydroxyl radical. Superoxide anion radical may then react with the oxidized form of the metal ion, resulting in reduction of the metal ion and the generation of molecular oxygen. The metal ion via this cycle participates in numerous redox reactions (Fig. 2.3).

$$Me^{(n-1)+} \text{ chelate} + H_2O_2 \longrightarrow Me^{n+} \text{ chelate} + OH^- + OH\cdot$$

$$Me^{n+} \text{ chelate} + O_2^{-\cdot} \longrightarrow Me^{(n-1)+} \text{ chelate} + O_2$$

Fig. 2.3. Fenton reaction

Fenton Chemistry in Biological Membranes. In homogeneous solutions, the effectiveness of chelated iron compounds as catalysts of the Haber-Weiss reaction is probably low (Sutton 1985). However, the intramembranous formation of hydroxyl radicals and initiation of lipid peroxidation via the Fenton reaction in the lipid phases of membranes in vivo seem feasible (Schaich and Borg 1988). Under in vivo conditions, injection of Fe^{2+} into rat brain cortex results in lipid peroxidation and induction of seizures (Triggs and Willmore 1984). In mouse skin, inflammation can be induced by intradermal injection of Fe^{2+}/H_2O_2. Injection of H_2O_2 or Fe^{2+} alone causes only minor inflammatory reactions (Fuchs et al. 1990d).

The effects resulting from Fenton chemistry in biological membranes depend on the microenvironment in which the redox-active transition metal ion is bound. The presence of other target molecules, such as proteins or DNA, and the specific site to which the metal catalyst binds are considered important factors for the molecular manifestation of cellular damage. Besides hydroxyl radicals, other highly reactive oxidizing species such as the ferryl radical ($[Fe\text{-}OH]^{3+}$ or $[Fe=O]^{2+}$) and the perferryl radical ($[Fe=O]^{3+}$ or $[Fe^{3+}–O_2^{-\cdot}]$) may be involved in transition-metal-ion-mediated damage in biological membranes.

III. Hydroxyl Radical

The hydroxyl radical is the most reactive oxygen intermediate and is a strong oxidizing species with the standard redox potential of $H_2O/HO\cdot$ $E_0' = 2.0$ V. It reacts with almost every type of molecule found in living cells with an extremely high rate constant. Its reactivity is so great that if are is formed in a biological system, it will react immediately with whatever biological molecule is in the vicinity.

Thus, the hydroxyl radical usually does not escape from its site of generation, and it is not membrane permeable. Molecular damage to DNA can only occur if the hydroxyl radical is generated in close proximity to its target, for example, in a site-specific formation by transition metal ions bound ot the DNA molecule. Reactions of the hydroxyl radical are of three main types: (a) hydrogen abstraction, (b) addition to, e.g., an aromatic ring structure, and (c) electron transfer from, e.g., chlorine anion to yield the chlor radical.

The toxicity of the hydroxyl radical may also be mediated through other $HO\cdot$-dependent compounds, such as the powerful reducing agent carbonate anion radical $CO_3^{-\cdot}$ (standard redox potential $CO_3^{-\cdot}/CO_2$ $E_0' = -2.0$ V).

Carbonate anion radical is formed by the reaction of the hydroxyl radical with the carbonate anion.

Hydroxyl radicals are presumably generated during the respiratory burst of neutrophils and macrophages and are one of the primary products of ionizing irradiation in skin. It is probably the cytotoxic principle of the relatively nontoxic superoxide anion radical and hydrogen peroxide. Degradation of lipid hydroperoxides by reaction with superoxide anion radical or transition metal ions is a further source of hydroxyl radicals in vivo.

IV. Singlet Oxygen

Singlet oxygen is not a free radical but an electronically excited form of oxygen. It is a strongly electrophilic molecule generated in many photochemical reactions and as a result of the reaction of hydrogen peroxide with hypochlorite. Dismutation of superoxide anion radical in water can lead to the formation of singlet oxygen (Corey et al. 1987). If one of the two unpaired electrons of molecular oxygen absorbs energy and is transformed into an orbital position of higher energy, with inversion of spin, then singlet oxygen arises (Fig. 2.3). Delta singlet oxygen is formed ($\Delta\,^1O_2$) when the excited electron of molecular oxygen forms an electron pair and occupies the same orbital. Delta singlet oxygen has a half-life of microseconds in water and can react both as an electrophile and a nucleophile. When the excited electron remains unpaired, occupying a different orbital, sigma singlet oxygen is formed ($\Sigma\,^1O_2$), which is even less stable than delta singlet oxygen and usually decays to delta singlet oxygen before it has time to react with a substrate (Fig. 2.4).

Singlet oxygen may be the major oxidant generated during ultraviolet exposure of mice topically photosensitized by rose bengal, causing lipid peroxidation of epidermal lipids (Müller and Bonne 1988). Singlet oxygen has a high reactivity with a variety of biomolecules and readily combines chemically with them. It can also transfer its electronic energy, forming the excited state of the donor molecule and then returning to ground state (quenching). Singlet

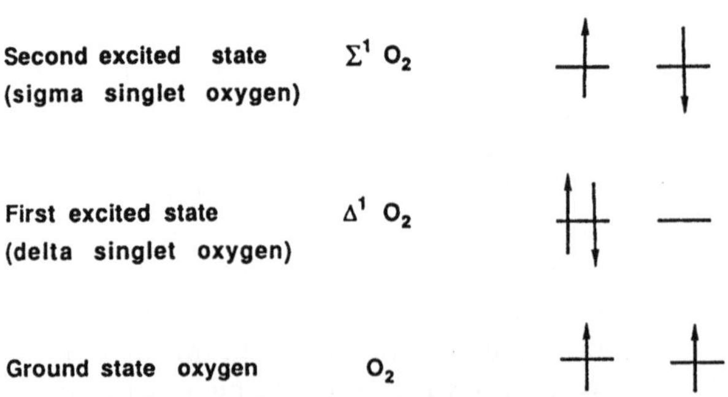

Fig. 2.4. Energy states of the oxygen molecule

$$^{1}O_2 \longrightarrow O_2 + h\nu \text{ (1268 nm)}$$

monomol emission

$$2\ ^{1}O_2 \longrightarrow 2\ O_2 + h\nu \text{ (634 nm)}$$

dimol emission

Fig. 2.5. Monomol and dimol emission of light

oxygen has a longer lifetime than superoxide anion radical (Kasha 1970). The lifetime of singlet oxygen varies considerably with the polarity of the environment. The half-life in solution of water is 7×10^{-6} s, in comparison with 5×10^{-2} s in perfluorohexane (Corey 1987). It is thought to be capable of diffusing appreciable distances in biomembranes and causing damage at sites quite far from its point of origin (Ito 1978). Besides reaction with lipids, inactivation of several enzymes by means of amino acid modification and damage of DNA constituents have been reported. Singlet oxygen is detected by means of chemiluminescence measurements, and monomol emission in the infrared is frequently used for detection in biological samples. Dimol emission produces light at 634 nm (and 703 nm) (Fig. 2.5).

V. Transition Metals

The reaction of oxygen with most organic molecules is kinetically forbidden and requires catalysis. Activation of molecular oxygen in biological systems can be achieved by its interaction with a paramagnetic transition metal ion and exchange coupling, which overcomes the spin restriction. Transition metals have an unpaired electron in an inner electron orbital. Enzymatic activation of oxygen by oxidases or oxygenases depends in general upon the transition metal ions associated with the enzymes. Transition metals were suggested to play a central role in the manifestation of oxygen toxicity (Halliwell and Gutteridge 1984).

In particular, iron (Halliwell and Gutteridge 1986a, b) and copper may be involved in oxidative stress by: (a) accelerating autoxidation of various small molecular substances such as phenols and thiols, (b) participating as a catalyst for the Haber-Weiss and the Fenton reactions, and (c) decomposing lipid hydroperoxides to peroxyl radicals (Davies and Slater 1987). Copper (I) salts react with hydrogen peroxide to yield hydroxyl radicals with a much greater rate constant than do iron (II) salts.

In biological systems, the actual concentrations of redox active transition metals capable of catalyzing the formation of reactive oxygen species is usually relatively low. However, under certain pathological conditions concentrations of redox active metal ions can be increased (Aust et al. 1985). The role of transition metals in oxidative injury in skin has not been analyzed yet. Transition metals of biological importance are iron, copper, zinc, manganese, vanadium, cobalt, nickel, chromium, and molybdenum.

Iron and Copper. The quantitatively most important transition metal ion in humans is iron. An adult human contains about 4 g of iron, but only 80 mg of the transition metal copper. The distribution of iron in the human body is as follows: approximately 60% in hemoglobin, 25% in storage proteins (hemosiderin, ferritin), 10% in myoglobin, and 0.15% in hemoproteins and iron-sulfur clusters. Body iron is tightly bound to proteins (transferrin in plasma, lactoferrin in some body fluids, ferritin and hemosiderin in tissues). Under physiological conditions, transferrin is saturated to only 30% with iron, providing a larger "iron buffer capacity." This indicates that under normal conditions the actual free concentration of iron in the human blood is virtually zero. Iron can be released form transferrin in a pH-dependent manner (acidosis favoring release) and by ascorbate and glutathione. Superoxide itself can release iron from ferritin in vivo (Mazur et al 1958) and in vitro (Thomas et al. 1985); this mechanism is proposed to potentiate the formation of reactive oxygen species in inflammatory states via iron mobilization from ferritin by superoxide radicals derived from activated neutrophils (Biemond et al. 1984).

Although a large amount of information is available on the distribution of transition metal ions in human tissues, few comparative data are available involving skin. The copper content of human skin (epidermis and dermis) is roughly between 0.8 and 8 mg% (wet weight). The average copper content of the adult human dermis is about half the epidermal value. Total skin iron content is about 0.9–6 mg% (dry weight) (Stüttgen and Schaefer 1974). These data imply that skin contains significant amounts of redox active transition metal ions, and that the skin copper content is about equal to the skin iron content.

Dualistic Mode of Action. Transition metal complexes may protect against oxygen damage or sensitize the effects of oxygen. Metal chelates (Cu, Zn-superoxide dismutase, SOD) can either serve as catalysts of superoxide anion radical dismutation or as catalysts (ethylenediaminetetraacetic acid EDTA) of the reaction between hydrogen peroxide and superoxide anion to form hydroxyl radicals (Haber-Weiss reaction) (Czapski and Goldstein 1986). The reaction type depends on many factors, one of them being chelator composition. Chelators may alter the redox potential of the metal ion and thereby control the reaction thermodynamics.

VI. Radical Chelates

The mean lifetime of a free radical species is determined by its chemical reactivity and depends primarily on the delocalization of the unpaired electron and the physical properties of the environment. In the presence of transition metals the lifetime of very reactive radicals such as $HO\cdot$, $HO_2\cdot$, $RO\cdot$, $RO_2\cdot$, $HO_3\cdot$, and $RO_3\cdot$ is altered.

Alkoxyl radicals can coordinate in an oxidative-addition reaction to transition metals, forming the highest oxidation state of the metal. An intermolecular d-p electron transfer of an unpaired electron from low spin transition chelates to the -O-O- bond of the alkoxyl radical occurs. The driving force of this reaction seems to be the tendency to achieve the stable 18 valence electron configuration (Tkac 1975, 1987).

The mean lifetime of free radicals increases by coordination to transition metals, including those in hemoproteins, e.g., hemoglobin and cytochromes. Radical chelates are stable in nonpolar media at physiological temperature. In polar media (e.g., water), or in the presence of strong, competitive ligating agents (e.g., bases of nucleic acids), or at slightly elevated temperatures ($>37\,^{\circ}C$), the stabilized oxygen-centered radicals are liberated from the complex, and the original high reactivity of the free radical is renewed.

In hydrophobic membrane domains, coordinated peroxyl radicals can survive as long as they do not contact effective H-atom donors such as HO, HS, or H_2N groups (Tkac 1988). It is assumed that in this way sterically unhindered free radicals with high reactivity could pass through hydrophobic cell membranes and may exert damage at sites distant from their point of origin (Tkac, personal communication). The pathophysiological role of radical chelates in skin has not been investigated yet.

VII. Hydroperoxides and Lipid Radicals

Polyunsaturated fatty acids are readily oxidized in vivo and in vitro. Lipid autoxidation occurs slowly in vivo and is a strictly controlled process. However, an uncontrolled and rapid peroxidation may occur if a lipid alkyl radical is formed, e.g., through hydrogen abstraction from the lipid by a reactive oxygen radical. The mechanism of lipid peroxidation involves three distinct reaction steps, initiation, propagation, and termination.

The *initiation step* (Fig. 2.6) occurs when an allylic proton is abstracted from a polyunsaturated lipid, e.g., by a hydroxyl radical, a thiyl radical, or a carbon-centered radical, and subsequently a lipid-alkyl radical is formed. The lipid-alkyl radical stabilizes by forming a conjugated diene radical which is subject to further reactions.

The *propagation step* (Fig. 2.7) follows, in which the lipid-alkyl radical reacts with oxygen to from a lipid-peroxyl radical. This can abstract a second allylic hydrogen atom from a polyunsaturated lipid, resulting in a free radical

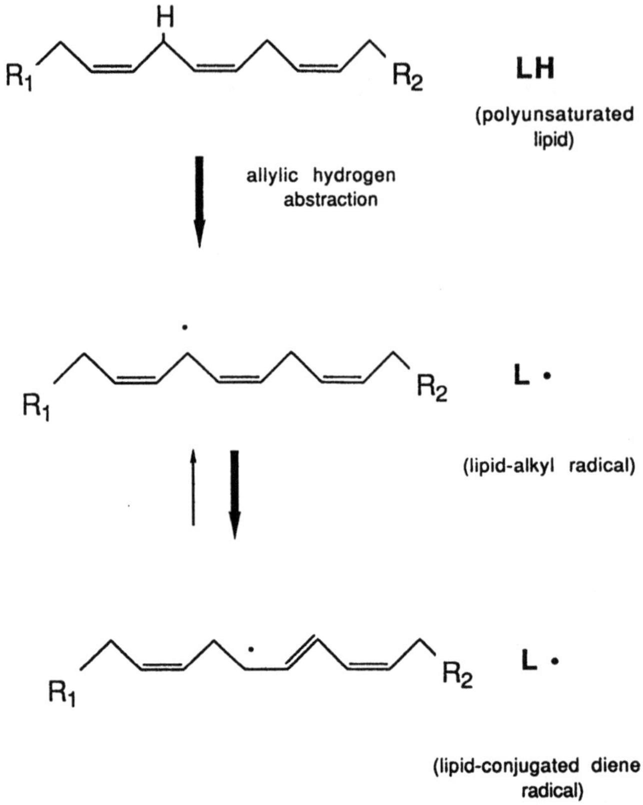

Fig. 2.6. Initiation of lipid peroxidation

chain reaction. Finally, lipid hydroperoxides are formed which can be detected by chromatographic or fluorimetric methods.

The final step is *termination* of the free radical chain reaction (Fig. 2.8), by bond rearrangement to form conjugated dienes or by degradation of peroxyl and alkoxyl radicals, yielding alkenals, hydroxyalkenals, and alkanes (ethane, penthane). Radical-radical addition reaction, resulting in the formation of di- or polymers, is a further possibility. Primary antioxidants which inhibit the radical chain reaction scavenge the propagating species, e.g., the peroxyl or alkyl radical.

Lipid Peroxidation Products. It is extremely difficult to detect the primary products of lipid peroxidation, lipid-alkyl and lipid-peroxyl radicals directly. However, stable degradation products of lipid peroxidation can be readily analyzed by different methods (Fig. 2.9). The volatile degradation products ethane and penthane are detected by gas chromatography. Ethane and penthane analysis in oxidatively stressed animals and humans is a noninvasive in vivo technique to measure lipid peroxidation. Alkenals, and hydroxyalkenals (malondialdehyde and 4-hydroxynonenal) are quantitated by chromato-

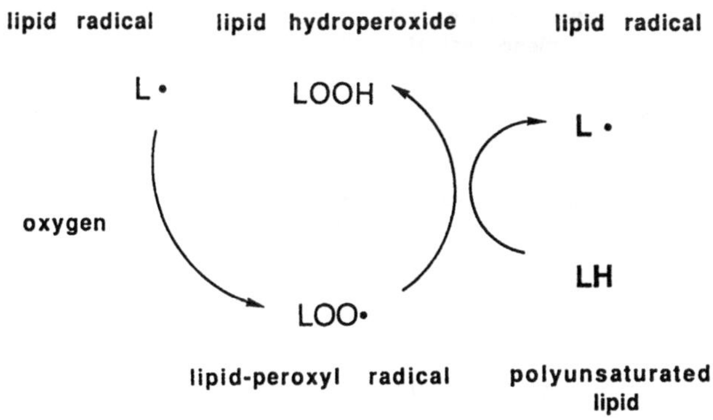

Fig. 2.7. Propagation of lipid peroxidation

graphy. Conjugated dienes are measured by their characteristic spectral absorbance at 230–235 nm.

Toxicity of Lipid Peroxidation Products. Hydroperoxide metabolism and toxicity has been well characterized in mammalian cells (Chance et al. 1979; Sies et al. 1982); it is closely linked to the cell thiol-disulfide equilibrium. Highly reactive products of lipid peroxidation, such as lipid-alkyl radicals (R·), alkylperoxyl radicals (R-O-O·), and alkoxyl radicals (R-O·) may damage all basic biomolecules in their ultimate vicinity. Lipid hydroperoxides R-O-O-H and their degradation products, e.g., 2-alkenals and 4-hydroxyalkenals, can damage cells and tissues at more distant sites not directly exposed to lipid peroxidation. 4-Hydroxyalkenals react rapidly with protein SH groups to form cyclic products (e.g., α-hydroxyfuran) (Fig. 2.10)

Bifunctional alkenals, e.g., malonic dialdehyde, can react with protein thiol and amino groups to yield stable covalent adducts. This may contribute to protein inactivation and crosslinking of peptides (Fig. 2.11).

2 LOO• ⟶ LOOL + oxygen
lipid-peroxyl radical — lipid peroxide

LOO• + XH ⟶ LOOH + X•
lipid-peroxyl radical — lipid hydroperoxide

2 L• ⟶ L-L
lipid-alkyl radical — lipid dimer

L• + XH ⟶ LH + X•
lipid-alkyl radical — regenerated lipid

Fig. 2.8. Termination of lipid peroxidation (*XH,* electron donor, e.g., antioxidant

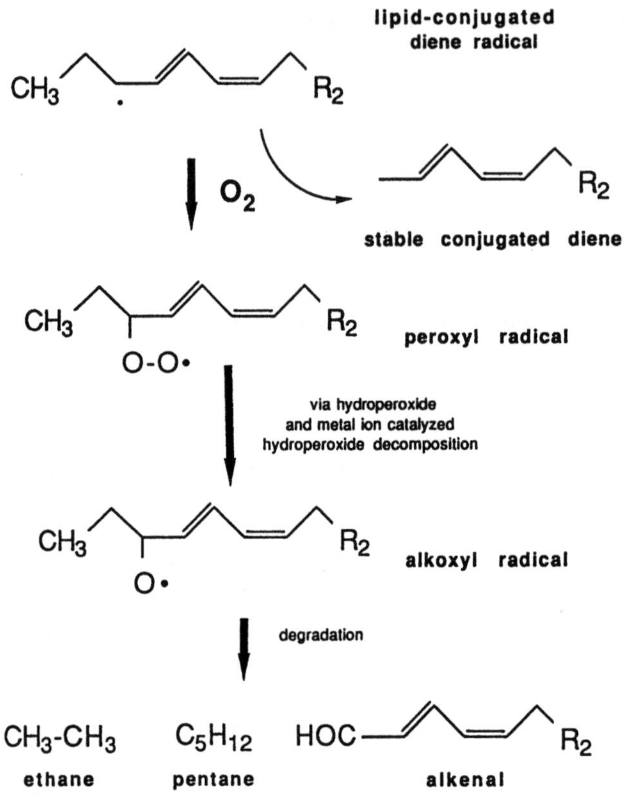

Fig. 2.9. Degradation products of lipid peroxidation

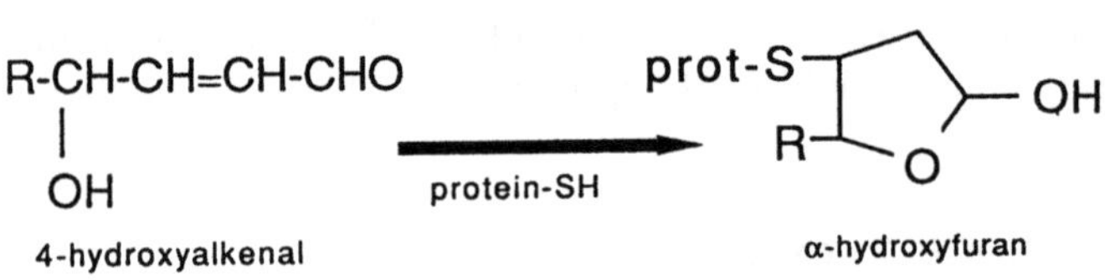

Fig. 2.10. Reaction of 4-hydroxyalkenals with protein SH groups

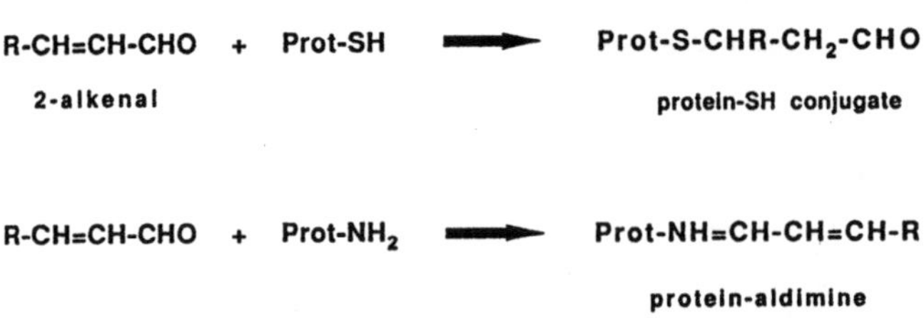

Fig. 2.11. Reaction of bifunctional alkenals with protein thiol *(Prot-SH)* and amino groups *(Prot-NH_2)*

$$2\ LOOH \xrightarrow{\text{dismutation}} RO\cdot + ROO\cdot + H_2O$$

$$LOOH \xrightarrow[\text{thermolysis}]{\text{photolysis}} \begin{matrix} LOO\cdot + H\cdot \\ LO\cdot + HO\cdot \end{matrix}$$

$$LOOH + Me^{2+} \longrightarrow LO\cdot + OH^- + Me^{3+}$$

$$LOOH + O_2^{-\cdot} \longrightarrow LO\cdot + OH^- + O_2$$

$$LOOH + X\cdot \longrightarrow LOO\cdot + XH$$

Fig. 2.12. Decomposition of lipid hydroperoxides into alkoxyl and peroxyl radicals

The more stable hydroperoxides (LOOH) can be transformed into highly reactive radical species (LO·, LOO·, HO·, H·) by transition metal ions (Me^{n+}), superoxide anion radical $O_2^{-\cdot}$), other free radicals (X·), thermolysis, photolysis, and dismutation (Fig. 2.12).

VIII. Thiyl Radicals

Thiyl radicals (RS·) are formed nonenzymatically by reaction of thiols (RSH) with free radicals such as allyl-type radicals (e.g., lipid radicals), superoxide anion radical, and transition metal ions (Fig. 2.13). In the reaction of a nonprotein thiol with a lipid radical, the thiyl radical formed is able to abstract allylic hydrogen from polyunsaturated fatty acids relatively quickly to yield lipid radicals again (Schöneich et al. 1989). Therefore, thiyl free radicals must be considered as a potential source of lipid peroxidation in certain cases. Because of the abundance of sulfhydryl groups in skin, their involvement in complex biochemical reactions, and their high reactivity, thiyl radicals may act significantly in free radical reactions in skin.

Thiyl radicals can also be formed enzymatically by xanthine oxidase, hydrogen peroxide being the major determinant of thiol oxidation (Ross et al. 1985), prostaglandin synthase (Eling et al. 1986), horseradish peroxidase (Harman et al. 1984; Medeiros et al. 1987), and xenobiotic metabolism (Ross et al. 1984, Ross and Moldeus 1985).

$$RSH + Fe^{3+} \longrightarrow RS\cdot + Fe^{2+} + H^+$$

$$2\ RSH + O_2^{-\cdot} \longrightarrow 2\ RS\cdot + H_2O_2$$

$$RSH + lipid\cdot \longrightarrow RS\cdot + lipid\text{-}H$$

Fig. 2.13. Production of thiyl radicals

$$RS\cdot + O_2 \longrightarrow RSOO\cdot \quad \text{sulfurperoxyl radical}$$

$$RS\cdot + RSH \longrightarrow RSSR^{-\cdot} + H^+ \quad \text{disulfide anion radical}$$

$$RSSR^{-\cdot} + O_2 \longrightarrow RSSR + O_2^{-\cdot}$$

$$RSSR^{-\cdot} + H_2O_2 \longrightarrow RSSR + OH^- + OH\cdot$$

$$\underset{\text{sulfur hydroperoxide}}{RSOOH} + H_2O \longrightarrow \underset{\text{sulfenic acid}}{RSOH} + H_2O_2$$

Fig. 2.14. Reactions of thiyl radicals

Thiyl radicals react rapidly with other thiols, e.g., protein thiols, to form disulfides via disulfide anion radicals. Disulfide formation may represent a further major mechanism whereby superoxide anion radicals exert their toxic effects in vivo. Disulfide anion radicals are a further source of superoxide anion radicals; by reacting with molecular oxygen, sulfurperoxyl radicals and subsequently sulfur hydroperoxides are formed. Thiyl radicals rapidly oxidize ferrocytochrome *c* (Forni and Willson 1986a), ascorbate, and NADH (Forni et al. 1986b); in the last instance the NAD$\cdot$ radical is formed. The cellular reducing potential can be decreased by these reactions, and this may contribute to the prooxidant effects of thiols (Fig. 2.14).

Nonprotein thiols such as glutathione react with superoxide anion radical with a reactivity of $5 \times 10^4\ M^{-1}\ s^{-1}$. At a thiol concentration 1000–10000 times the concentration of superoxide dismutase (SOD), an effective scavenging effect of thiols can be expected (Asada and Kanomatsu 1976). This reaction may contribute to the antioxidant effects of thiols.

C. Production Sites of Reactive Oxidants in Skin

Cells of the epidermis, dermis, and subcutis can be both the source and the target of biological oxidants during normal or altered metabolic activity. The epidermis and dermis are a complex system of distinct cell populations, and within a single cell population various differentiation states must be distinguished (Fig. 2.15). In the epidermis, a cornified and a noncornified layer can be distinguished. The stratum corneum (cornified layer) is entirely composed of corneocytes, the final cell differentiation product of keratinocytes. In the noncornified layer, various developmental stages of keratinocytes are found. Keratinocytes undergo dramatic metabolic changes during differentiation from a basal cell to a corneocyte. The epidermis has a high cell volume (approximately >90%); the cell volume is composed of about 85% keratinocytes, 10% melanocytes, and 5% Langerhans cells. The cell volume of the

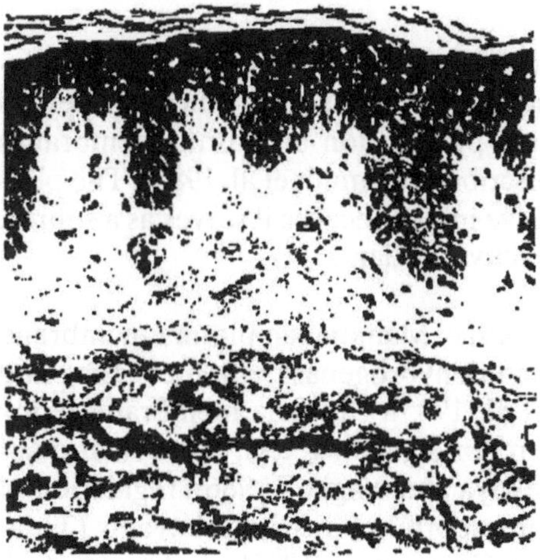

Fig. 2.15. The layers of the skin

dermis is considerably smaller and estimated to be about 20%; it is mainly composed of fibroblasts (80%), mast cells (20%), some myocytes, and a few white blood cells which physiologically reside in the dermis. The underlying subcutis is mainly built up of fat cells. Dermis and subcutis also contain some endothelial cells (blood and lymph vessels). During inflammatory reactions, skin can be significantly infiltrated with leukocytes.

Reactive oxidant mediated diseases can be categorized by the major site of pathologically increased generation of reactive species. Biological sources of reactive oxidants in skin can be endogenous or exogenous. Exogenous sources comprise redox cycling xenobiotics and electromagnetic irradiation. Endogenous free radical production may derive from metabolic processes which normally generate reactive oxygen as a byproduct. Intracellular production of free radical species can also be increased by modification of the electron transport in mitochondria, microsomes, nuclear envelope, and plasma membrane as well as by modifying cell metabolism in peroxisomes and cytosol. Increased intracellular flux of free radical species is observed in hyper- and hypooxygenation syndromes, intoxication by various xenobiotics, such as carbon tetrachloride, paraquat, and dibenzo-*p*-dioxins, and deficiency in lipophilic antioxidants such as tocopherol, ubiquinols, and vitamin A. Increased extracellular free radical generation is predominantly observed in inflammatory conditions involving oxidative burst of granulocytes or macrophages. Exposure to gamma, ultraviolet, visible, and infrared irradiation may result in increased intra- and extracellular flux of reactive oxygen species.

I. Plasma Membrane

The plasma membrane of the keratinocyte is morphologically well characterized (Brody 1968), although its histochemical and biochemical properties are only poorly understood. High yield purification of plasma membranes from human keratinocytes has been described (Schmidt et al. 1985). The plasma membrane is a critical site in oxidative injury because it serves as a source, a gating mechanism, and a target of reactive oxidants.

Plasma Membrane as a Free Radical Source. Mammalian plasma membrane contains a variety of redox systems and dehydrogenase enzymes (Löw and Crane 1978; Sun et al. 1984; Crane et al. 1985 a). Plasma membrane redox systems participate in transmembrane redox signalling, e.g., in modulation of cytosolic protein kinase C activity (Malviya and Anglard 1986), regulation of cell growth (Sun et al. 1985), and other important biochemical events (Tarakhovsky et al. 1985; Crane et al. 1982; Crane et al. 1985 b; Laliberte et al. 1987). Plasma membrane redox systems have been characterized as endo-, trans-, and ectodehydrogenases (Crane et al. 1985 a). *Endodehydrogenases* comprise NAD (P) H semidehydroascorbate reductase, xanthine oxidase, and NADH cytochrome b_5 reductase. *Transdehydrogenase* are NADH dehydrogenase and NADPH oxidase. *Ectodehydrogenase* consist of NADH dehydrogenase, glutathione (GSH) oxidase, D-amino acid oxidase.

Plasma membranes also contain various small molecular redox active compounds like cytochromes, flavins, iron and copper proteins, protein thiols, ubiquinones, and lipid hydroperoxides. They can be effective sources of free radicals and reactive oxygen species. Plasma-membrane-located cyclooxygenase and lipoxygenase are involved in the production of free radical intermediates of arachidonic acid metabolism. A plasma membrane thiol oxidase catalyzing the reaction:

$$2\,RSH + O_2 = R\text{-}S\text{-}S\text{-}R + H_2O_2$$

and thereby producing hydrogen peroxide was identified in the kidney (Lash and Jones 1986). In skin homogenate a sulfhydryl oxidase was identified three decades ago (Rony et al. 1958). Skin sulfhydryl oxidase (Takamori et al. 1980) shows its highest activity in the stratum granulosum, uses molecular oxygen (Yamada et al. 1987), catalyzes oxidation of thiols to disulfides, and reduces oxygen to hydrogen peroxide (Yamada et al. 1989), similar to the kidney enzyme.

Plasma Membrane as a Free Radical Gate. Highly reactive free radicals may pass the plasma membrane as radical chelates and travel in the cytoskeleton or endoplasmic reticulum to the nuclear membrane (Tkac, personal communication). Singlet oxygen is able to diffuse into phospholipid bilayers, and hydrogen peroxide passes the cell membrane nearly as freely as water.

Plasma Membrane as a Free Radical Target. Reactive oxygen species can cause damage to epithelial plasma membranes (Scott et al. 1989) and the associated cytoskeleton (Hansson et al. 1987). It was pointed out that the plasma membrane potential is altered very early in the course of epithelial cell injury induced by reactive oxygen species and that membrane potential changes may serve as a key link in the expression of oxygen-radical-mediated effects in the cell (Scott et al. 1987, Scott and Rabito 1988).

II. Mitochondria

Mitochondria as a Free Radical Source. Mitochondria are morphologically and biochemically well characterized organelles and their role in epidermal metabolism has been extensively studied. In 1923, it was discovered that skin consumes oxygen (Gans 1923), and this finding was subsequently confirmed by other investigators (Wohlgemut and Klopstock 1926; Glasenapp and Leonhardi 1953). Mitochondrial electron transport is mainly responsible for epidermal oxygen consumption, and oxidative phosphorylation generates the majority of energy for the basal cell layer of the epidermis (Decker and Andersen 1965). Under physiological conditions, the main biological source of superoxide anion radicals is the mitochondrion (Chance and Sies 1979) (Fig. 2.16). Epidermal mitochondria may be the main generation site of physi-

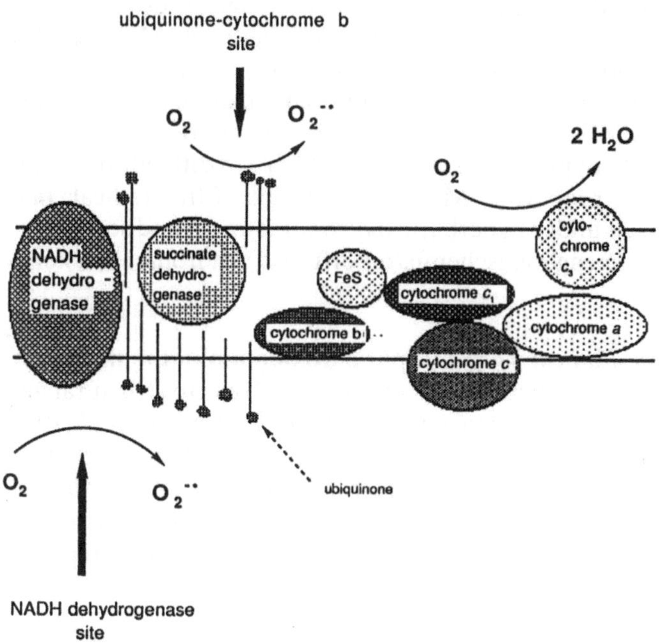

Fig. 2.16. Production of superoxide anion radicals at the inner mitochondrial membrane

ologically produced reactive oxidants in skin. Oxygen is converted by the mitochondrial cytochrome *c* oxidase via tetravalent reduction into water. It is still questioned whether or not cytochrome *c* oxidase produces superoxide anion radicals; it was suggested that cytochrome *c* oxidase itself is a scavenger of superoxide (Naqui et al. 1986). Presumably, cytochrome *c* oxidase does not release any oxygen radicals into its immediate vicinity. Several complexes of the electron transfer chain of mitochondria are sources of superoxide anion radical and subsequently hydrogen peroxide (Cadenas et al. 1977, Nohl et al. 1986). The main leakage site seems to be the NADH dehydrogenase complex.

Electron shuttle out of sequence from the respiratory chain may result in the formation of superoxide anion radicals. Under conditions of loose coupling, mitochondrial production of reactive oxygen species may therefore be significantly increased. Uncoupling of oxidative phosphorylation results in increased mitochondrial respiration and is associated with overproportional formation of reactive oxygen species due to a less rigid control of electron flux through components of the respiratory chain (Nohl 1990). Morphologically intact mitochondria have been isolated from the rat epidermis, but differed functionally from those of the liver (Rosett et al. 1967), exhibiting partially uncoupled oxidative phosphorylation. Mitochondria isolated from neonatal rat epidermis also have a low efficiency of oxidative phosphorylation (Bagatell et al. 1967). These preparations, however, were contaminated with cellular debris. Morphologically intact mitochondria isolated from human epidermis were also reported to have impaired coupling of oxidative phosphorylation (Gray and Yardley 1975 a, b).

Later, however it was demonstrated that skin mitochondria do have intact coupling of oxidative phosphorylation to electron transport. Mitochondria isolated from human fibroblasts (Millis and Pious 1973) and rat oral mucosa (Fine et al. 1978) have an intact electron transport chain, three coupling sites, theoretically correct ADP/O ratios, and a high degree of coupling. Human epidermis contains well-coupled mitochondria, which are difficult to isolate due to the specific tissue properties. Mitochondrial flux of free radicals may be significantly increased under hyperbaric oxygen treatment and several pathological conditions, e.g., aging, ischemia/reperfusion syndrome, and xenobiotic intoxication.

Mitochondria as a Free Radical Target. Lipid peroxidation of the mitochondrial membrane is under tight antioxidant control and leads to structural and functional damage if the subtle balance is disturbed (Vladimirov et al. 1980; Bindoli 1988). Mitochondria become more susceptible to oxidant attack if the tight control of oxygen reduction via the electron transport chain is affected. Membrane pertubation by membranotropic agents, uncoupling of oxidative phosphorylation, and inhibition of intramitochondrial antioxidant systems are mechanisms which may increase the oxidative susceptibility of the inner mitochondrial membrane.

Oxidant injury of the mitochondrial ATPase/synthetase complex by reactive oxygen species (Hyslop et al. 1988) results in the inhibition of oxidative phosphorylation and consequently in depletion of cellular ATP levels. Mitochondria accumulate photosensitizers such as rhodamine (Summerhayes et al. 1982) and porphyrins (Salet et al. 1987). The inner mitochondrial membrane is highly susceptible to singlet oxygen produced by photoactivation of sequestered porphyrins (Murant et al. 1987). The in vivo relevance of mitochondria as a damage site in porphyrin photochemotherapy of tumors is, however, unclear (Dougherty 1987).

Mitochondrial Epidermal Gradient. Mitochondria are heterogeneously distributed in distinct layers of the epidermis and dermis. In human epidermis, the stratum oxybioticum (basal cell layers) is differentiated from the stratum anoxybioticum (stratum granulosum and above). Anaerobic glycolytic energy production predominates in the stratum anoxybioticum; no mitochondrial activity is found in the stratum corneum (Braun-Falco 1961). Mitochondrial density decreases from the stratum basale via the stratum spinosum to the stratum granulosum (Rupec et al. 1969). However, the respiratory rates of basal and differentiated epidermal cells apparently do not vary (Erum 1969). Cellular phospholipids constitute the major substrate for oxidative phosphorylation in the epidermis (Decker 1971; Herndon and McGuire 1967; Cruickshank et al. 1962) (Fig. 2.17).

Respiration and glycolysis rates are higher in the epidermis than in the dermis (Stüttgen 1955); the respiratory activity of the epidermis is estimated to be about 10-fold greater than that of the associated dermis (Leibsohn et al. 1958). Oxygen consumption in the upper dermis is lower than in the deeper dermis. This is compatible with the low number of nucleated cells in the upper dermis. The higher oxygen consumption in the lower dermis is probably due to the presence of striated muscle fibers (Erum 1969).

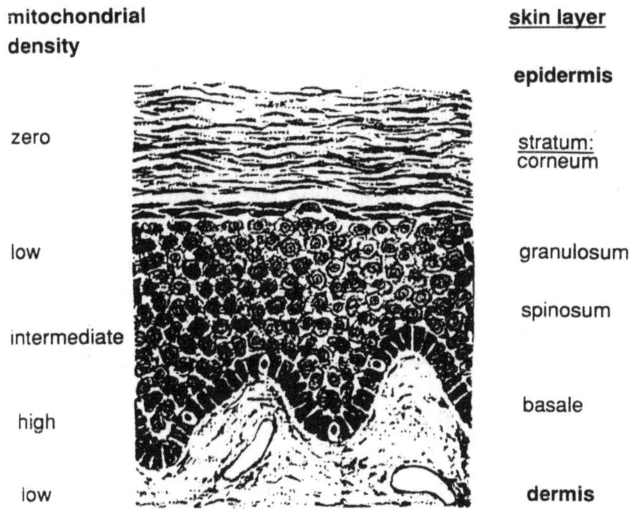

Fig. 2.17. Mitochondrial density in different skin layers

The level of skin oxygen consumption correlates to diurnal and nocturnal periods of cellular activity (Franz et al. 1978), indicating a central role of oxidative phosphorylation in mitotic activity. The distribution pattern of epidermal and dermal mitochondria may be linked to a tissue gradient of reactive oxidants produced under physiological conditions.

III. Microsomes

Numerous studies have investigated the involvement of skin microsomes in drug metabolism. Skin has many of the same drug metabolizing enzymes as the liver; for example, mouse skin contains multiple forms of cytochrome P-450 (Rettie et al. 1986; Thompson and Slaga 1976). The activities of several cutaneous enzymes have been compared with their activities in the liver and were found to be about 2%–6% of the hepatic values (Pohl et al. 1976). These data may indicate that skin metabolism is low. However, cutaneous enzyme activities reported were measured in total skin homogenates (epidermis plus dermis). Since the epidermis makes up only about 3% of the total skin homogenate and metabolic activity (e.g., oxygen consumption) is much higher in the epidermis than in the dermis, real enzyme activities in the epidermis may be considerably higher. Assuming that almost all of the microsomal enzymes are constrained to the epidermal layer, specific activities for mixed function oxidase in the epidermis is calculated to be in the range of the hepatic levels (Noonan and Wester 1983). However, it was pointed out that the unit reference point is critical in the measurement of enzyme activities in skin, because skin contains large amounts of metabolically inert material. If the contribution of inert protein to total protein is minimized (Wiebel et al. 1975), monooxygenase enzyme activity per milligram protein is highest in the dermis and considerably less than that in the liver (Finnen et al. 1985). In conclusion, the quantitative turnover of systemically administered drugs in skin is low, unless the enzymes are induced. Microsomal drug oxidation may play a significant role in the metabolism of topically applied drugs. Reactive oxidants produced in this scenario may participate in drug-mediated cutaneous toxicity.

Microsomes as a Free Radical Source. In protecting the organism from hazardous xenobiotics, microsomal cytochrome P-450 also potentially endangers the organism by forming reactive oxygen species, such as superoxide anion radical and hydrogen peroxide (Bast 1986). Microsomal hydroxylation of xenobiotics involves hydroxylation of a carbon, nitrogen, or sulfur atom using molecular oxygen and the formation of water as a by-product (mixed function oxidase or monooxygenase).

Microsomal NAD(P)H oxidizing flavoproteins are in general one-electron transfer enzymes and as such give rise to free radicals. Induction of epidermal monooxygenase activity, e.g., by dibenzo-*p*-dioxines (Pohl et al.

1976), may result in the production of prooxidant semiquinone radicals. These semiquinone radicals are easily autoxidizable and are a potent source of microsomal superoxide anion radicals (Nohl et al. 1986) and subsequently hydrogen peroxide. Cytochrome P-450 isoenzymes in the epidermis of rats are markedly induced by nitropyrenes (Asokan et al. 1985) and polycyclic aromatic hydrocarbons (Thompson and Slaga 1976), both important carcinogenic environmental pollutants. It must be pointed out, however, that isolated microsomes are artifacts, and due to the isolation procedure components of the electron transport chain may by dearranged, resulting in an increased release of free radicals. The true free radical formation in the endoplasmic reticulum in vivo may be significantly lower than in vitro.

Microsomes as Free Radical Targets. Microsome preparations are frequently used as biological model systems to study mechanisms of lipid peroxidation. The endoplasmic reticulum contains a high content of polyunsaturated phospholipids which renders the membrane highly susceptible to lipid peroxidation. Reactive-oxidant-mediated deterioration of microsomal structure and function severely impairs the metabolizing apparatus of the cell.

IV. Peroxisomes

The presence of peroxisomes has been documented in the liver, kidney, and bone. In the morphological dermatology literature, the presence of mitochondria, endoplasmic reticulum, and Golgi apparatus in keratinizing epithelial cells is amply illustrated; the knowledge about skin peroxisomes, however, is limited (Matoltsy and Parakkal 1967; Matoltsy 1986). Recently, it was shown that skin contains considerable catalase activity (Fuchs and Milbradt 1989b). Since catalase is mainly compartmentalized in peroxisomes in mammalian cells, this would indicate that peroxisomes are present in skin. Peroxisomes were found in lipid-secreting cells in mammalian skin, particularly in the epithelia of sebaceous glands (Gorgas 1987). The presence of peroxisomes in keratinocytes was recently documented by electron microscopy study in animal (Gorgas, personal communication) and human keratinocytes (Menon et al. 1989). Fatty-acid-oxidizing enzymes are present in peroxisomes, and hydrogen peroxide is a product of normal fatty acid metabolism. Keratinocyte peroxisomes can be induced by the inhibitor of hydroxy methyl glutaryl (HMG)-reductase, lovastatin. Lovastatin treatment could serve as a model system for increasing hydrogen peroxide production in keratinocytes.

Peroxisomes as a Free Radical Source. Peroxisomes (De Duve and Baudhuin 1966; Goldfischer and Reddy 1984) are potent sources of hydrogen peroxide since they contain a variety of H_2O_2-generating oxidases such as *D*-amino acid oxidase, urate oxidase, fatty-acyl coenzyme-A oxidase, and α-hydroxy-acid oxidase (Kramer 1986).

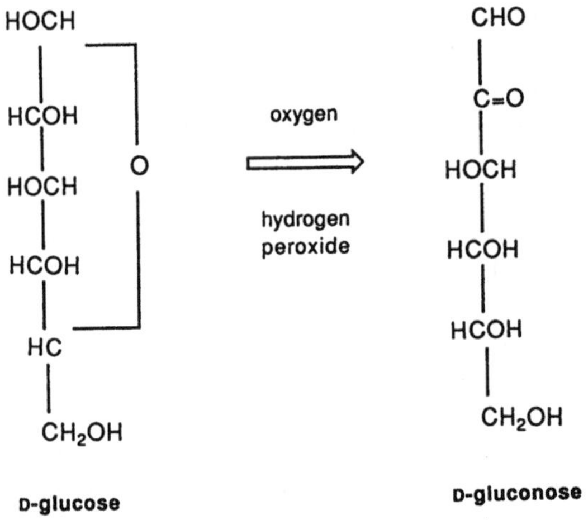

Fig. 2.18. Production of D-gluconase from D-glucose

V. Cytosol

Cytosolic autoxidation of other small molecular compounds of cell metabolism such as monosaccharides, hydroquinones, catecholamines, thiols, flavins, and tetrahydropterins may yield superoxide anion, subsequently hydrogen peroxide, and probably the hydroxyl radical. Various cytosolic skin oxygenases like xanthine oxidase, dihydroorotate dehydrogenase, and flavoprotein dehydrogenase may also generate superoxide anion radical in the cytosol.

In the cytosol *monosaccharide autoxidation* catalyzed by transition metal ions can produce reactive oxidants such as hydrogen peroxide, superoxide anion, and hydroxyl radical, as well as reactive carbonyls (Wolf et al. 1991). In an enolization reaction the α-hydroxyaldehyde is in equilibrium with its endiol anion, which readily reacts with transition metal ions via a radical intermediate to yield dicarbonyl products. The reactive dicarbonyls may induce protein crosslinking under physiological conditions. Autoxidation reactions or monosaccharides are probably chronic oxidative processes occurring in tissues which significantly utilize simple monosaccharides, e.g., in glycolysis and gluconeogenesis (Thornalley 1985). D-Gluconose (an α-oxoaldehyde = dicarbonyl) is the autoxidation product of D-Glucose (Fig. 2.18).

D. Targets of Reactive Oxidants in Skin

The structural and functional environment along with the reactivity of the activated oxygen species or the free radical determines the type of cellular damage sustained. Different cell types in the epidermis and dermis may have varying susceptibilities to reactive oxidants. Hydrogen peroxide is the oxygen species that causes the death of normal human fibroblasts exposed to an external source of oxygen radicals. In contrast, singlet oxygen, superoxide anion radical, and hydroxyl radical have no direct effect (Simon et al. 1981). The relative susceptibility of Langerhans cells, keratinocytes, melanocytes, and epidermal lymphocytes in comparison with fibroblasts towards different reactive oxidants is not known.

Membranes as the outer protective layer of mammalian cells and their subcellular organelles represent the first line of defense against any external oxidative injury. Impairment of membrane barrier function is the key event in the development of many diseases (Trump and Arstila 1975, 1980). Lipid peroxidation and activation of endogenous membrane phospholipases are mechanisms of major significance by which the integrity of the biomembranes is destroyed. The extracellular molecular targets of reactive oxidants are space components such as collagen and glycoproteins. Intracellular targets of reactive oxidants comprise unsaturated membrane lipids of microsomes and mitochondria, as well as other macromolecules such as proteins, polysaccharides, and nucleic acids, and a variety of small molecular compounds.

It is generally considered that the toxicity of reactive oxygen species may be due to their nonspecific attack upon cellular components such as lipids, nucleic acids, and proteins. Relatively low concentrations of reactive oxygen species, however, may cause biological phenomena by inducing specific physiologic reactions, e.g., modulation of cell growth and intracellular signalling. Reactive oxygen species, e.g., the superoxide anion radical, may act in vivo as an inter- and intracellular chemical messenger (Saran and Bors 1989). The stimulated release of low amounts of superoxide anion radicals from fibroblasts is thought rather to have a regulatory function, e.g., by modulating arachidonic acid metabolism (Meier et al. 1988), than to cause cell damage. Nitrous oxide (NO·), which is identical to endothelial relaxing factor (EDRF), plays an important part in control of vascular function (Bin 1991; Vanhoutee 1989).

I. Lipids

Lipids are one of the chemically best characterized macromolecular constituents of skin, and many reports concerning oxidative injury of this molecular component in skin exist. They have important biochemical and physiological functions. From a functional aspect, epidermal lipids are essential to the epidermal water barrier and modulate desquamation and cohesion of the stratum corneum (Elias 1981). They are also a major energy source for the viable epidermal cell, represent a precursor for prostaglandin and leukotriene synthesis, contribute to the epidermal antimicrobial activity, and participate in the metabolism of xenobiotics. Abnormalities in epidermal lipids are associated with an increased transepidermal water loss (essential fatty acid deficiency), disturbances of cornification with the clinical picture of ichthyosis, and alterations in cell proliferation.

1. Skin Lipid Composition

Skin lipids can artificially be divided into those of the epidermis, epidermal appendages (hair, nail), sebaceous glands, and subcutaneous tissue. Human skin surface lipids emanate from two sources, the keratinizing epidermis and the sebaceous gland (Downing et al. 1987).

Epidermal lipids account for about 8% of the cell dry weight (Gray and Yardley 1975b). The lipids composition is quite similar between various species but markedly different in the successive layers of the epidermal cells (Yardley 1983). As epidermal cells progress from the basal cell layer and keratinize to form the stratum corneum, they undergo major changes in their lipid composition. Epidermal phospholipids represent a major part of the basal cell lipids (45%), but phospholipase-mediated degradation results in a low phospholipid content in the upper cell layers. In contrast, neutral lipids increase in successive layers. During the process of keratinization, elongation and desaturation of fatty acids occurs, sterols are esterified and sulfated, and sphingolipids become acylated.

Striking amounts of sphingolipids are present as the cells mature (Lampe et al. 1983). These lipids consist predominantly of sphingomyelins, glucosylceramides, and ceramides. Glucosylceramide A accounts for about 50% of the total mammalian epidermal glycolipids. Some 50% of the glucosylceramides are acylglycosylceramides, and the most prominent fatty acid esterified to this carbohydrate is the essential fatty acid linoleic acid (C18:2) (Wertz and Downing 1984). This unsaturated fatty acid, essential to normal epidermal barrier function, is highly susceptible to oxidative injury. All other ceramides are highly saturated; therefore they seem to be well suited to resist oxidative damage on exposure to the atmosphere.

Quantitatively, *lipids produced by sebaceous glands* (Sebum) are much more significant than those derived from the keratinizing epidermis, although on a weight basis, epidermal cells are more active in lipid biosynthesis than dermal cells. The amount of surface lipids recovered is directly proportional to the abundance of sebaceous glands in the skin, which vary widely in size and density of distribution on the surface of the body. Surface lipids collected from areas rich in sebaceous glands represent the approximate composition of the sebum: triacylglycerols and free fatty acids 58%, wax esters 26%, squalene 12%, cholesterol esters 3%, and cholesterol 1.5%. Pure sebum has no free fatty acids. Fatty acids results primarily from triacylglycerol hydrolysis by propionibacteria in the sebaceous glands and on the skin surface (Shalita 1974).

It cannot be overemphasized that sebum composition is species-specific (Wheatley and James 1957; Nicolaides et al. 1968). Differences also exist among the subfamilies of nonhuman primates, e.g., Ponginae (gorilla, chimpanzee, orangutan) and Cercapiteanae (mandrillas, papio, macaca, cercopithecus). Genera belonging to the same subfamily have a similar sebum composition (De Luca et al. 1989). Ponginae, the largest and the most manlike of nonhuman primates, show a composition of skin surface lipids that are intermediate between those of man and other monkeys. In humans, sebum fatty acid composition may even differ greatly between individuals (Green et al. 1984).

The fatty acids of human sebaceous glands consist of saturated and unsaturated, branched and unbranched structures with chain lengths ranging from 6 to 28 carbons (Nicolaides and Apon 1977). However, most of these fatty acids are composed of C12 or C14 forms. The concentration of triacylglycerols and fatty acids varies, depending upon the activity of the bacteria. Fatty acids may comprise up to 20% of the sebum lipid mixture in humans, and about 50% of the sebum fatty acids are unsaturated, providing a substantial substrate for the lipid peroxidation processes.

In humans, surface wax esters and squalenes originate uniquely from the sebaceous glands (Greene et al. 1970). In contrast to most mammalian skin, large amounts of squalene, a sterol compound with six unsaturated double bonds, accumulates in human skin. Squalene represents about 10%–15% of the total skin surface lipids.

Since 25% of the total skin surface lipids is unsaturated, the physiological presence of lipid peroxidation products at the epidermal surface should be suspected. Indeed, in healthy human skin, they are detected (Sugiura et al. 1985a, 1986a, b). It should be pointed out that species differences do exist in the skin content of lipid peroxidation products, e.g., in contrast to the rat, only small amounts are found in mice, guinea pigs, and hamsters. The exceedingly high lipid peroxide content of the rat is explained by the fact that it is the only common rodent that excretes large amounts of linoleic acid onto the skin surface (Glavind and Christensen 1967; Glavind 1969).

2. Lipid Peroxidation in Skin

Lipid peroxidation may be induced by physiological and pathophysiological reactions. Nonenzymatic peroxidation of lipids can be triggered by heat, UV and ionizing irradiation, redox active metal ions, redox cycling drugs, and free radical initiators. Enzymatic peroxidation may involve the neutrophil plasma-membrane-associated NADPH dehydrogenase, microsomal NADPH dependent cytochrome P-450 reductase, and cyclooxygenase/lipoxygenase.

Lipid peroxidation has been suggested to play a significant role in basic pathological processes in humans such as aging, carcinogenesis, inflammation, and xenobiotic toxicity (Yagi 1987). Participation of lipid peroxidation reactions in the pathophysiology of gastrointestinal, myocardial, vascular, neural, and pulmonar diseases is suspected. In skin, lipid peroxidation is considered to take a major part in the pathogenesis of skin aging, burns, ultraviolet and ionizing irradiation damage, and inflammatory skin disorders (Miyachi et al. 1988; Fuchs and Schürer 1991).

Although lipid peroxidation is one of the consequences of oxidative stress in skin, numerous molecular changes occurring during oxidative injury are not induced by lipid peroxidation products. Most of the experimental results obtained so far on lipid peroxidation in skin is based on indirect evidence. Qualitative and quantitative analysis of lipid peroxidation in biological material is difficult. Primary products of lipid peroxidation, phospholipid and triglyceride hydroperoxides, are rarely measured.

Numerous data in the literature on lipid peroxidation in skin derive from the measurement of thiobarbituric-acid-reactive substances. Although this assay is simple and sensitive, it is not specific for lipid peroxides as it also measures ketoacids and some nucleic acid bases, and cannot be used to measure peroxidation kinetics. The possible pitfalls of this method have been discussed by Gutteridge (1986). Measurement of conjugated dienes, analysis of the alkanes *n*-pentane and ethane, and estimation of chemiluminescence also have considerable drawbacks in practical applicability (Smith and Anderson 1987; Iwaoka et al. 1987).

Due to the problems in methodology, it is likely that some data on cutaneous lipid peroxidation, especially using the thiobarbituric acid test, have not shown peroxidation of skin lipids per se but rather some artifacts (Kappus 1986a). On the other hand, lack of detection of its primary products does not mean that lipid peroxidation does not occur. Absence of evidence is not necessarily evidence of absence.

Lipid peroxidation is not only the cause of cell and tissue injury but may also be the consequence of it, caused by other factors. The interrelationship between tissue injury and lipid peroxidation poses an interesting "chicken and egg" puzzle. Thus, it is difficult to analyze whether tissue injury due to lipid peroxidation products is the primary cause or whether it is secondary or resembles an epiphenomenon. Similar objections can be raised with regard to oxidative damage to proteins, nucleic acids, and polysaccharides.

The products of lipid peroxidation are physiologically present in normal, healthy human skin. They may have important physiological functions, such as in the regulation of prostanoid synthesis, chemoattraction of leukocytes, and antimicrobial activity. Their content is significantly influenced by nutritional factors and the metabolic activity level of skin.

Lipid peroxidation products in rat skin were found to increase whith a tocopherol-deficient diet and decrease when the diet is deficient in essential fatty acids by Glavind and Christensen (1967). These results were confirmed by other authors. In tocopherol-deficient rats, the lipid peroxide content in skin and subcutaneous tissue rises (Igarashi et al. 1989), and topical application of α-tocopherol reduces the content of lipid peroxidation products in human skin (Hayakawa 1971).

The epidermal content of lipid peroxidation products in influenced by the cell turnover rate: A high mitotic activity is correlated with low levels of lipid peroxidation (Logani et al. 1982).

Antimicrobial Activity. The physiological presence of lipid peroxidation products in the outermost skin layers (stratum corneum) may be a regulatory device in the control of antimicrobial defense. The bactericidal activity of UV-irradiated fats was discovered about 6 decades ago (Harris et al. 1932a), and it was suspected that lipid peroxidation products are involved (Harris et al. 1932b). Germicidal activity against bacterial spores was also reported (Roth and Halvorson 1952). Recently, evidence was provided that epidermis-derived human skin lipids contribute to cutaneous antimicrobial resistance (Miller et al. 1988). The molecular basis for this is, however, unknown. The antimicrobial potency of epidermal lipids varies with chain length, *cis/trans* configuration, and degree of saturation (Miller et al. 1988).

Lipid peroxidation products may also contribute to the bactericidal activity of reactive oxygen species during phagocytosis. Human monocytes, possibly because of a greater membrane content of polyunsaturated fatty acids, peroxidize a greater amount of endogenous lipids during endocytosis of heat-killed *Staphylococcus epidermidis* than human granulocytes (Stossel et al. 1974). The capacity of human granulocytes to peroxidize ingested lipids is strongly inhibited in patients with chronic granulomatous skin disease, a pathological condition in which granulocyte metabolism is impaired. Malondialdehyde, purified from extracts of phagocytizing rabbit alveolar macrophages, has bactericidal activities against *Escherichia coli* and *S. epidermidis* (Stossel et al. 1974).

II. Proteins

Oxidized proteins may account for up to 20%–25% of the soluble protein in cells of old animals. Under conditions of oxidative stress, proteins are modified. Ionizing irradiation and metal ion-mediated oxidation cause modification of some amino acid residues, intra- and interchain cross-linking of tyrosine and cysteine residues, and cleavage of peptide bonds. Mixed function oxidase induces formation of carbonyl derivatives of proline and arginine. Because introduction of carbonyl groups into amino acid residues of proteins is a hallmark of oxidative modification, it can be used to detect oxidatively modified protein (Stadtman 1990).

A variety of amino acids such as cysteine, methionine, proline, histidine, and tryptophan are susceptible to oxidative attack. The reversible oxidation-reduction of thiol groups in cysteine and glutathione is an important regulating mechanism in the cellular redox status. Interaction of metal-binding sites of enzymes may generate hydroxyl radicals, which subsequently oxidize amino acids, in particular lysine, histidine, arginine, proline, and serine. Modification of the enzyme could be followed by inactivation or lead to the formation of autoantibodies in response to the modified protein. Histidine endoperoxide, tryptophan hydroperoxide, and methionine sulfoxide are well-characterized amino acid oxidation products (Fig. 2.19).

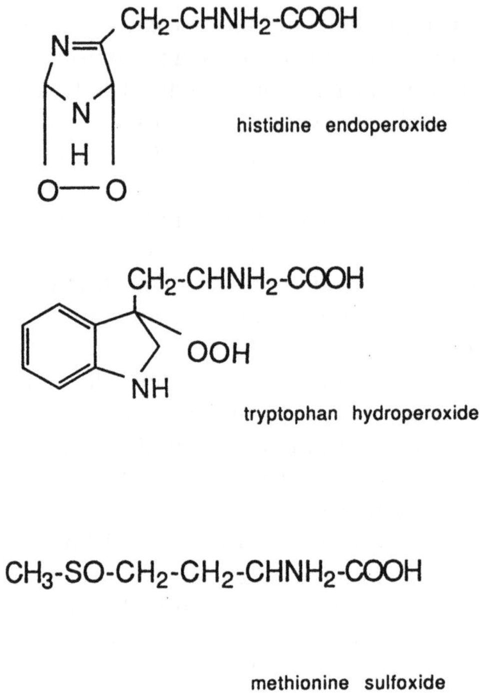

Fig. 2.19. Amino acid oxidation products

There are several examples of oxidative protein inactivation. The existence of a natural, antiinflammatory protein has been proposed, and it is thought to participate in the regulation of endothelial cell contraction. It is readily inactivated by reactive oxygen species. The oxidative damage of this protein may increase the vascular permeability and thus contribute to the inflammatory tissue response (Oyanagui 1981).

1. Collagen

Collagens are the main structural components of connective tissue and account for about 90% of the protein in the human dermis. It is very resistant to proteolysis and can be degraded under physiological conditions only by specific collgenases. There are at least five different types of collagen: Types I (about 80%) and III (about 20%) are found in dermal collagen, and type IV is found in basement membranes.

The collagen type I fiber, which is the one best investigated, is composed of penta-microfibrils arranged in a spiraling, five-stranded, ropelike structure. The microfibril is composed of a triple helix of three polypeptide chains. Three different chains (α_{1-3}) are known, and each polypeptide is determined by a single structural gene. At both ends, short nonhelical sequences are involved in cross-linking between the molecules.

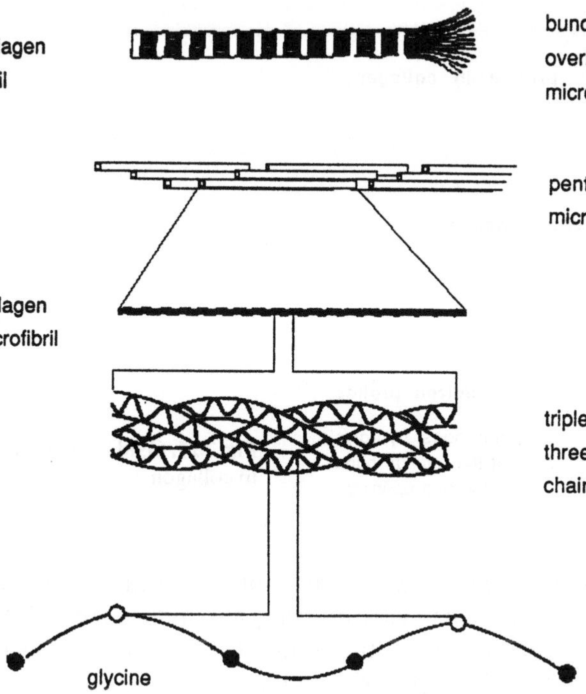

Fig. 2.20. Collagen

The fibroblast synthesizes a collagen precursor form that has to undergo several posttranslational modifications. Procollagen molecules are synthesized at the rough endoplasmic reticulum, where the subsequent enzymatic modification, e. g., prolyl and lysyl hydroxylation and glycosylation of certain hydroxylysyl residues, takes place.

After posttranslational modification, including disulfide bond and triple-helix formation, the completed procollagen molecules are secreted via microtubules into the extracellular space. The procollagen is then converted to collagen by cleavage of the propeptides, fibrillogenesis, and cross-linking. The enzyme lysyl oxidase is involved in the extracellular cross-linking and requires molecular oxygen to form reactive aldehydes. Copper ions are also necessary; chelation of copper depresses covalent cross-linking (Fig. 2.20). The structure and molecular biology of collagen were recently reviewed (Krieg and Aumailley 1990).

Collagen Metabolism and Reactive Oxidants. Regulation of collagen biosynthesis by reactive oxygen species may occur at different cellular levels, e. g., at the gene level or in the posttranslational phase. The effects of reactive oxygen species on mesenchymal metabolism are quite complex. Superoxide anion radicals may degrade collagen by attacking critical amino acid residues or induce increased collagen production.

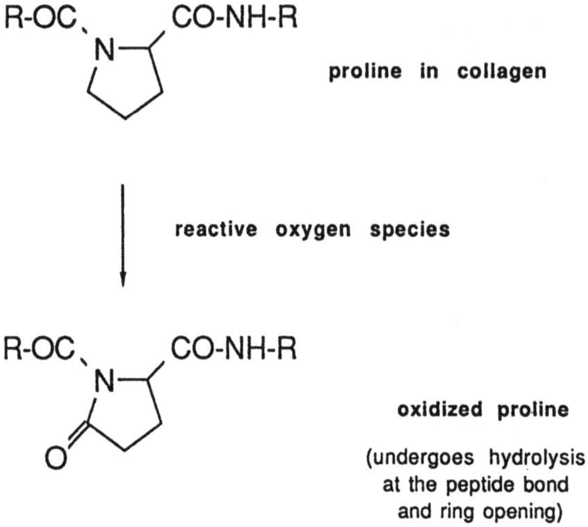

Fig. 2.21. Oxidation of proline in collagen

Modulation of the Collagen Molecule. Superoxide anion radicals can depolymerize collagen and hyaluronic acid in vitro and may also induce cross-linking of collagen molecules, thereby changing the solubility. The triple-helical region of collagen can be cleaved by superoxide anion radicals, and the susceptibility of collagen to oxidant attack is probably due to radical-induced hydrolysis of the proline peptide bond (Monboisse et al. 1983, 1984) (Fig. 2.21).

Stimulation of Collagen Production. Superoxide is a stimulant for collagen synthesis in vitro, probably by mediating ascorbate-dependent enzymic proline hydroxylation. The superoxide anion radical activates procollagen proline hydroxylase, which catalyzes the hydroxylation of proline, suggesting the involement of superoxide in collagen synthesis (Bhatnagar et Liu 1972). Prolyl hydroxylation is considered the rate-limiting step of collagen formation in vivo, but prolyl hydroxylase activity and collagen production may be regulated by different mechanisms (Murad et al. 1981 a, b).

Hydroxylation of proline and lysine requires the same cofactors, namely molecular oxygen, metal (II) ion, α-ketoglutarate, and ascorbate. Ascorbate acts as a reducing agent but can be replaced by other reductants (Mata et al. 1981). Prolyl hydroxylation is inhibited by hypoxia (Juva et al. 1966), ascorbate depletion (Murad et al. 1981 a), metal ion chelation (Jimenez et al. 1973), and Zinc at physiological concentrations (Anttinen et al. 1980).

Ascorbate stimulates total collagen synthesis in human skin fibroblasts without affecting noncollagen protein synthesis via a mechanism unrelated to its cofactor function (Murad et al. 1981 a, b). Ascorbate may act as an inducer of the collagen pathway by modulating collagen transcription and translation (Harwood et al. 1974; Schwarz et al. 1987). Ascorbate stimulates collagen synthesis in cultured human dermal fibroblasts by increasing transcription of the collagen genes. At the same concentration, ascorbate stimulates lipid peroxidation, and antioxidants, such as retinoic acid and retinol, which inhibit lipid peroxidation also inhibit collagen synthesis (Geesin et al. 1989). Tocopherol deficiency promotes lipid peroxidation in skin and accelerates the cross-linking of collagen (Igarashi et al. 1989). This would indicate that, with the exception of ascorbate, antioxidants may have some antifibrotic activity. The experimental finding that lipid peroxidation stimulates collagen synthesis in human dermal fibroblasts suggests an explanation for the fibrosis occurring in oxidant-induced tissue injury (Geesin et al. 1990).

Profibrotic Potential of Superoxide. To explain the profibrotic potential of the superoxide anion radical and the clinical antifibrotic activity of superoxide dismutase (SOD), it seems feasible to suggest that the initiation and growth of fibrotic tissue results from the action of superoxide anion radicals to produce cross-linking of mesenchymal proteins and other macromolecules. In addition, a direct effect on the induction of collagen synthesis may contribute. If the continuing effect of polymerization by superoxide is stopped by SOD, physiological depolymerization processes (collagenase, elastase) may predominate and result in clearing of the fibrotic lesion (Baillet et al. 1986). Whenever there is an increase in fibrogenesis, collagenase activity rises also (Crystal et al. 1958; Okazaki and Moruyama 1974), providing a well-balanced homeostasis.

High steady state concentrations of reactive oxygen species cause cellular damage in cultured human fibroblasts. Cell density is reduced and thymidine incorporation inhibited. Low steady state concentrations stimulate thymidine incorporation and increase cell density. Reactive oxygen species may provide a specific and sensitive trigger for fibroblast proliferation; prolonged stimulation may result in fibrosis (Murrell et al. 1990a).

2. Proteases and Antiproteases

Proteases are acute phase proteins whose concentration increases in serum when the individual is stressed, e.g., by trauma or illness. Activation of proteases without proper control by antiproteases can render the skin vulnerable to uncontrolled tissue damage. A system of antiproteases is present in the circulation and tissue fluids which functions to inactivate proteases released from inflammatory cells.

Various plasma proteins have antiprotease activity, such as α_1-antitrypsin, antithrombin III, α_1-antichymotrypsin, α_2-macroglobulin, α_2-plasmin inhibitor, and others. α_1-Proteinase inhibitor is one of the most important components of this antiprotease system and is capable of inhibiting both neutrophil elastase and collagenase. Serum protease inhibitors also antagonize the respiratory burst in human neutrophils and rat macrophages (Goldstein et al. 1979).

Inactivation of Antiproteases by Reactive Oxidants. Antiproteases can be inactivated by reactive oxygen species. Leukocyte-derived oxidants, especially the

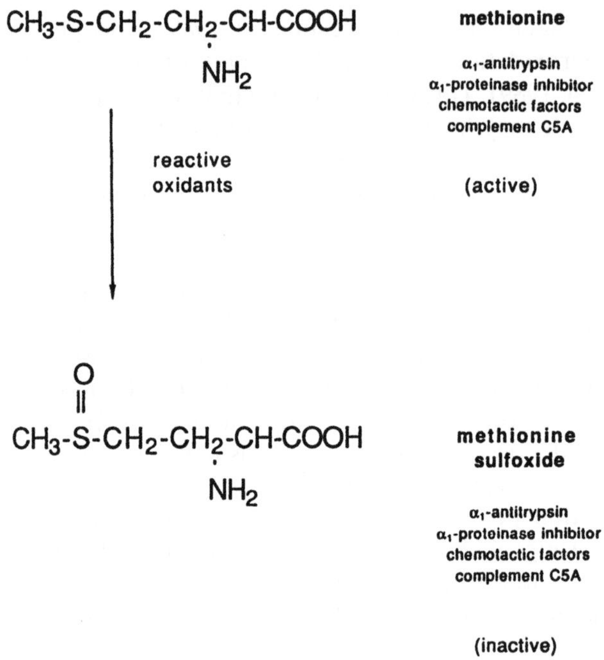

Fig. 2.22. Methionine oxidation in proteins and enzymes

myeloperoxidase system, may contribute to proteolytic tissue injury by oxidative inactivation of protective antiproteases (Clark et al. 1981). Human neutrophils phagocytizing opsonized antigen-antibody complexes produce activated oxygen species which are capable of partially suppressing the elastase-inhibiting capacity of whole human serum or purified human proteinase inhibitor. α_1-Proteinase inhibitor is inactivated through modification of the reactive site methionine of this protein by reactive oxygen species (Mathewson et al. 1981). Methionine oxidation is also the key biochemical event in the oxidative inactivation of α_1-antitrypsin, chemotactic factors, and complement C5A (Fig. 2.22).

This mechanism is considered to be involved in the pathogenesis of pulmonary emphysema. Oxidative inactivation in the microenvironment of accumulating neutrophils and sites of inflammation may allow proteases released from the cells to damage adjacent connective tissue structures more readily (Carp and Janoff 1979).

Skin Antiproteases. Protease inhibitors, including α_1-antitrypsin, have been identified in the dermis and play an important, although poorly defined, role in cutaneous homeostasis (Fräki et al. 1983 a, b; Hopsu-Havu and Fräki 1981). Proteolytic enzyme concentrations are elevated during burn injury and UV and X-radiation damage and may contribute to blister formation in skin.

Bullous Diseases. In immunologically mediated bullous skin disease, proteinases of the complement system cyn be activated. Furthermore, proteinases from skin cells and neutrophils can participate in blister formation. Pemphigus vulgaris immunoglobulin can induce the synthesis of proteinases in human keratinocytes (Singer et al. 1980) and may therefore contribute to acantholysis. Blister formation occurring at the basement membrane level in, e.g., bullous pemphigoid, dermatitis herpetiformis, and erythema multiforme could be a result of proteinase release from invading neutrophils or skin cells. In epidermolysis bullosa dystrophica, elevated proteinase activity is found in the blister fluid (Eisen 1969), and this serine proteinase may play an important role in the pathogenesis of recessive epidermolysis bullosa dystrophica (Takamori et al. 1985; Ikeda et al. 1985).

Panniculitis. Panniculitis associated with antiprotease (α_1-antitrypsin) deficiency is a chronic, relapsing, and disseminated disease with new lesions appearing as old lesions resolve. The dermis is also involved, showing invasion with neutrophils and macrophages and later dermal necrosis. The destructive and inflammatory changes extend to the fibrous septa of the adipose tissue (Hendrick et al. 1988).

Treatment Possibilities. Restoring the physiological balance between proteinases and antiproteinases by pharmacological intervention may be a promising therapeutic approach in bullous skin diseases and panniculitis associated with antiproteinase deficiency. This may be achieved by supplementation with antiproteases or antioxidants. However, the oxidative sensitivity of skin antiproteases is not known. A proteinase inhibitor was shown to be clinically effective in the treatment of patients with epidermolysis bullosa dystrophica (Ikeda et al. 1988). In addition, the antioxidant tocopherol is clinically efficient in the treatment of patients with epidermolysis bullosa dystrophica (Sehgral et al. 1972; Michelson et al. 1974; Degreef and Fluor 1974). Panniculitis associated with α_1-antitrypsin deficiency can be treated successfully by the administration of colchicine, dapsone, and proteinase inhibitors (Hendrick et al. 1988).

3. Amyloid

Amyloid is a proteinaceous tissue deposit, composed of fibrils with distinctive ultrastructural characteristics. A variety of proteins such as immunoglobulin light chains, polypeptide hormones, and other serum proteins are precursors of amyloid. A classification according to protein precursor can be made, and the chemically different types can be assigned to distinct clinical forms of amyloidosis. AL amyloid is derived from immunoglobulin light chains and mainly occurs in primary amyloidosis accompanying, e.g., multiple myeloma. AA amyloid is designated as secondary and is associated with prolonged inflammatory diseases. Amyloidosis of the skin can be associated with systemic amyloidosis (AL and AA amyloid) or may be localized. Localized skin amyloid may also derive from degradation products of keratin.

AA amyloid has been suggested to be a product of oxidized protein components; in vitro studies have shown that amyloid can be formed by the degradation of serum proteins through macrophages (Lavie et al. 1978). It was suggested that free radicals probably play a role in secondary amyloidosis, since amyloid accumulates during the aging process and in chronic inflammatory diseases. Antioxidants (ethoxyquin, butylhydroxyltoluene) prevent spontaneous amyloidosis in LAF1 mice (Harman et al. 1976). It was concluded that free radical reactions are involved in the pathogenesis of amyloidosis in LAF1 mice (Harmann 1980a, b).

In human serum the presence of an amyloid degrading and inhibiting factor was shown, and it was hypothesized that amyloid is a physiologically occurring protein with a high metabolic turnover. In the serum of amyloidotic patients the amyloid-degrading activity was markedly restored to normal by the metal ion chelator EDTA, citrate, and ascorbate (Kedar et al. 1982). This could indirectly suggest the involvement of reactive oxidants in amyloidogenesis in humans. Deposition of AA amyloid can be prevented in the mouse by colchicine (Shirahama and Cohen 1974) and dimethylsulfoxide (Kedar et al. 1977; Harmann et al. 1976). Colchicine is the drug of choice in the treatment of renal amyloidosis associated with familial Mediterranean fever, and dimethylsulfoxide is useful in the management of patients with secondary systemic amyloidosis (Breathnach 1988); however, treatment of macular amyloid and lichen amyloidosus is rather disappointing.

4. Amino Acid Racemization

Free radical damage of proteins can cause racemization of amino acids by attacking peptide backbones. The α-hydrogen of an amino acid is readily abstracted by, for example, the hydroxyl radical. The peptide free radical produced most likely is quickly repaired by nearby hydrogen donors, e.g., thiols. The repaired product would have a 50% chance of being racemized, a mechanisms which is similar to the formation of amino acid carbanion intermediates and proton readdition (Bada 1975). In the case of a free radical attack occuring with a protein, such a radical-induced racemization would probably inactivate the protein if the error significantly altered the quaternary structure. Amino acid racemization could be a major mechanism of free radical damage in proteins, and D-amino acids could be valuable as aging markers in skin (Mehlhorn, personal communication). Several proteins with a slow turnover, e.g., collagen, may contain D-amino acids, and these will accumulate as a direct function of time. D-Amino acid oxidase in peroxisomes may serve to recycle D-Amino acids generated through free radical processes (Fig. 2.23).

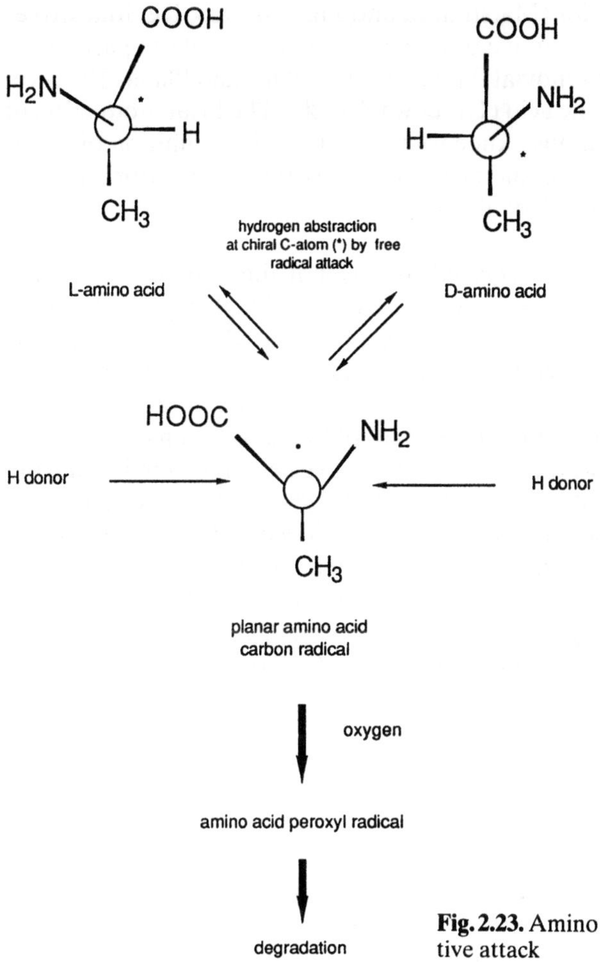

Fig. 2.23. Amino acid racemization by oxidative attack

III. Carbohydrates

Glycoproteins and glycolipids are constituents of the plasma membrane and desmosomes and may play an important role in many aspects of cell function such as differentiation, growth, cell-cell interaction, and receptor response. Large amounts of carbohydrates present in skin contribute to the matrix of connective tissue in the dermis. Glycosaminoglycans present in human skin are hyaluronic acid, dermatan, heparin, chondroitin, and their respective sulfates. Glycosaminoglycans can be linked to a protein core to form glycoproteins.

Carbohydrates (Harris et al. 1971 a, b), glycoproteins, and glycosaminoglycans are degraded by oxidative attack. Hyaluronic acid (a glycosaminoglycan), which is an essential component of the interstitial space, is depolymerized by reactive oxygen species (Matsumura and Pigman 1965; Del Maestro et al. 1980; Bragt et al. 1980). Reactive oxygen species are thought to participate in the depolymerization of synovial fluid (Puig-Parellada and Planas 1978) via the degradation of hyaluronic acid (Greenwald 1980). The local production of hydrogen peroxide in the mouse knee joint results in the inhibition of chondrocyte proteoglycan synthesis and is accompanied by inflammatory and degenerative changes (Schalkwijk et al. 1986).

Anionic Basement Membrane Zone. At the present time, there are few experimental data concerning oxidative injury of skin glycoproteins. One possible mechanism of tissue injury in leukocyte-mediated inflammation near the skin basement membrane zone is a loss of anionic sites resulting in the disturbance of the filtration barrier. Anionic membrane sites in rat skin are located on both the epidermal and dermal sides of the lamina densa in the dermoepidermal junction (Kazama et al. 1985) and have been suspected of playing a role in vascular permeability (Vehaskari et al. 1984) and deposition of immune complexes (Joselow and Mannik 1984). Glycoproteins in basement membranes may serve as a charge-selective permeability barrier to serum proteins as a result of possessing strongly anionic charges. Reactive oxygen species, particularly hydroxyl radicals, are suggested to be involved in the degradation of basal-lamina-associated anionic sites (Manabe and Ogawa 1987).

IV. Nucleic Acids

Oxidative damage to nucleic acids can be induced by energetic electromagnetic irradiation such as X-rays, UV, and visible light in the presence of a photosensitizer and by organic hydroperoxides, singlet oxygen, and hydroxyl radicals. Reactive oxygen species induce all forms of DNA damage, including base modifications, base free sites, strand breakes and cross links (Breimer 1991). Although nuclear DNA is well protected from oxidative injury by a multilevel defense system (compartmentalization, shielding by histones and polyamines, efficient enzymic repair system), it was calculated that in normal individuals the total number of all types of oxidative damage to DNA per cell per day may be much more than 10000 (Ames 1990). Hydroxyl radical DNA damage products are thymine glycol, 5-hydroxymethyluracil, and 8-hydroxyguanine (Fig. 2.24). These products can be detected in human urine and offer the potential to screen for oxidative damage in vivo (Cathcart et al. 1984). Mammals with the fastest metabolic rate and therefore the greatest rate of production of endogenous reactive oxidants also have the highest thymine and thymidine glycol levels in the urine. The endogenous background level of oxidant-induced DNA damage can be assayed in vivo by measuring 8-hydroxydeoxyguanosine, thymine glycol, and thymidine glycol in human urine (Ames 1989).

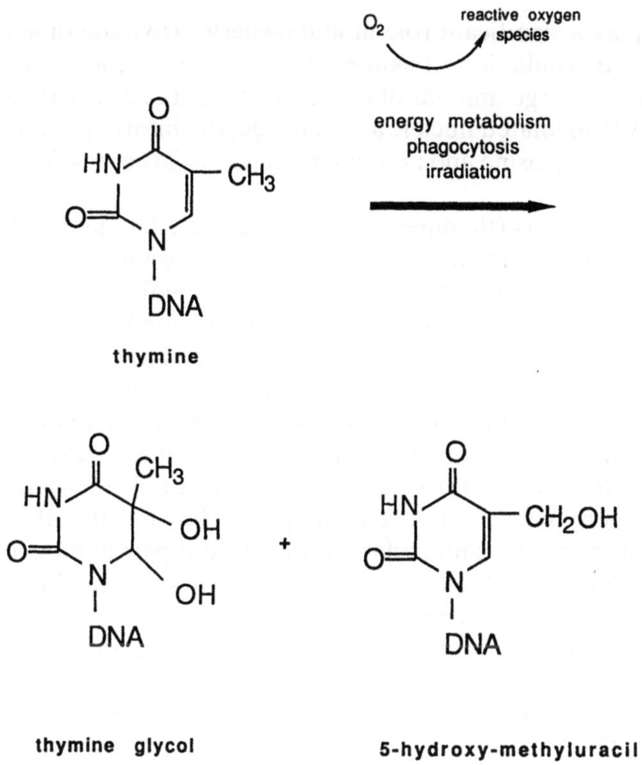

Fig. 2.24. Oxidant-induced DNA damage

Endonuclease III Sensitive Sites. Endonuclease III cleaves DNA at apurinic/apyrimidinic sites and is involved in the enzymic removal of thymine glycol residues from DNA. Exposure of plasmid DNA to a reactive oxygen species generating system results in the formation of endonuclease III sensitive sites, indicating participation of reactive oxidants in DNA base damage (Denq and Fridovich 1989).

Damage by Ultraviolet Light. DNA damage in skin has been extensively studied in vitro and in vivo using UV radiation as the damaging agent. Far (200–290 nm) and near (290–400 nm) UV irradiation can cause DNA damage. Exposure of keratinocytes to UV (248 nm) results in the formation of endonuclease sensitive sites in DNA and unscheduled DNA synthesis (Schothorst et al. 1985), which is considered as an indicator of DNA excision repair (Green et al. 1987).

At 260 nm the principal photoproducts in most types of DNA are cyclobutyl pyrimidine dimers. At longer wavelenghts it is likely that thymine dimers may not be the most biologically important lesions; pyrimidine-cytosine adducts formed by UVB irradiation (280–320 nm) or nondimer adducts formed by UVA (320–400 nm) (DNA protein cross-links, guanine and thymine oxidation products) may play a significant role in mutagenesis. Thymine dimer formation following UVB irradiation of skin cells and intact skin has been studied for several years. A large amount of experimental data has accumulated on this type of UVB-mediated nucleic acid damage; the information on UVB-mediated pyrimidine-cytosine and nondimer adducts is, in comparison, rather scarce.

Cyclobutyl pyrimidine dimers (thymine dimers) are detected in the DNA of human skin following in vivo irradiation with suberythemogenic doses of UVB light (Sutherland et al. 1980). Dimer formation rises with increasing UVB dose and correlates with the erythemal response. Augmented dimer formation is found in individuals with high UVB sensitivity (Freeman et al. 1986).

Photodimerization of pyrimidine residues in DNA has been causally related to photocarcinogenesis (Jung 1978; Doniger et al. 1981; Sutherland et al. 1980) (Fig. 2.25). The dimer photoproducts in human skin produced by simulated solar irradiation (Bruze et al. 1989) or UVB (Reusch et al. 1988) are rapidly removed, with approximately 30% remaining after 4 h and 10% after 24 h, indicating efficient secondary antioxidant nucleic acid repair in human skin. Little is known about the oxidative sensitivity of secondary antioxidant defense systems in skin, e.g., nucleic acid repair.

Cyclobutyl pyrimidine dimers are significantly induced in human skin by UVA irradiation at an erythemogenic dose (Freeman et al. 1987). Cyclobutyl pyrimidine dimer formation with UVA light involves an unknown photosensitizer, since in "naked" DNA such base lesions are not observed. UVA radiation can be lethal and mutagenic in mammalian cells (Wells and Han 1984;

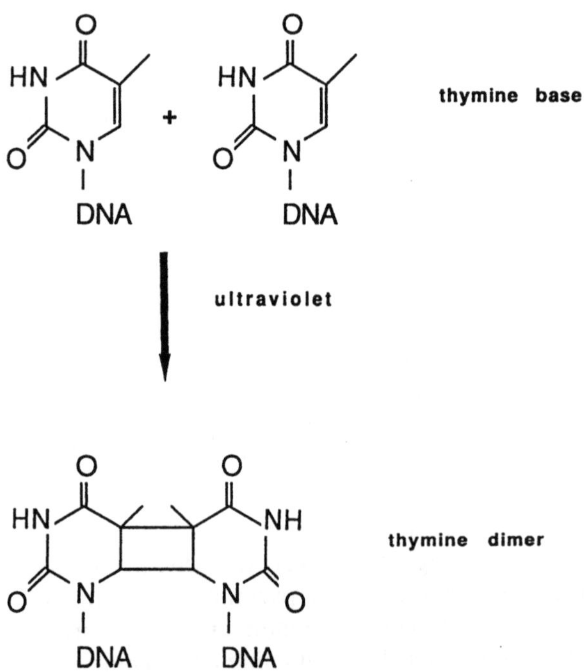

Fig. 2.25. Thymine dimer formation

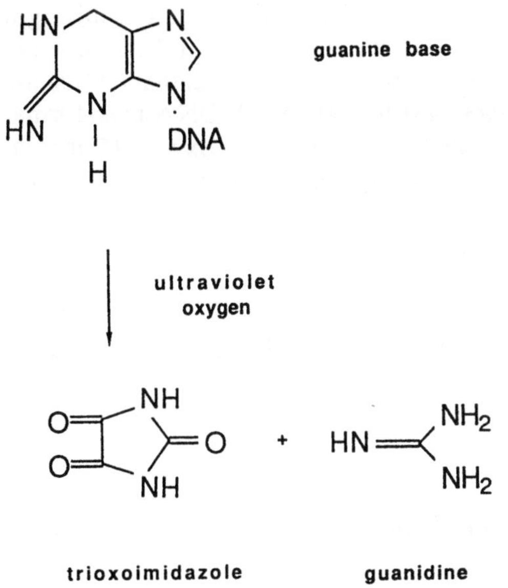

Fig. 2.26. Reaction products of guanine photooxidation

Tyrrell et al. 1984). UVA-induced photoproducts include cyclobutyl pyrimidine dimers, single-strand breaks, and DNA protein cross-links.

UVA-induced DNA strand breaks in *E. coli* and UVA-induced DNA protein cross-links in human cell lines are strongly oxygen dependent. UVA-induced single-strand breaks in fibroblasts are inhibited by catalase, but cell lethality is not decreased by the enzyme (Roza et al. 1985).

In contrast to several nondimer base lesions such as thymine and guanine oxidation, reactive oxygen species are probably not involved in cyclobutyl pyrimidine dimer formation. It was suggested that an endogenous triplet sensitizer may mediate cyclobutyl pyrimidine dimer formation in the near UV (Peak et al. 1984).

In some cases guanine oxidation products were detected in UVA-irradiated mammalian cells. Singlet oxygen reacts specifically with the guanine moiety in DNA; it is the DNA base most sensitive to photodynamic action. Guanine bases can be inactivated by UV in a photosensitized reaction involving singlet oxygen and forming guanidine and trioxoimidazole (Fig. 2.26).

Near-UV irradiation (290–400 nm) produces fewer mutagenic DNA alterations in comparison with *far-UV* (200–290 nm). Althoug the DNA molecule (maximal absorbance at 260 nm) has been identified as the chromophore for the mutagenic and lethal effects of far-UV irradiation, it absorbs at 320 nm only 0.1% of that which is absorbed at 260 nm. The absorption of UV by DNA falls by at least 3 orders of magnitude between 260 and 320 nm and by a further order of magnitude between 320 and 365 nm (Sutherland and Griffin 1981). Therefore, it was suggested that while DNA is not the critical chromophore for near-UV irradiation, it is the main target of it (Eisenstark 1987).

Molecular DNA damage upon exposure to near-UV and visible light probably operates through the intermediary of endogenous or exogenous cellular photosensitizers. A photodynamic effect of near-visible and visible light on *nucleic* acids was proposed (Cadet et al. 1986). Indeed, several studies have shown the participation of reactive oxygen species in DNA damage (Nishida et al. 1981; Breimer and Lindahl 1985; Sestili et al. 1986). DNA transformation and mutagenesis occur more readily in the presence of oxygen (Cabrera-Juarez 1981; Nishida et al. 1981), indicating the involvement of reactive oxygen species.

Near-UV also has some has some mutagenic and lethal effects even under anaerobic conditions (Peak et al. 1983). This points out that near-UV may also have a direct action on DNA. The exact biological role and the relative contribution of DNA photoreactions of types I and II remain to be characterized.

E. The Antioxidant System of the Skin

A full understanding of the causal factors of skin diseases and the role that oxidative stress and endogenous antioxidants play in this scenario requires knowledge of the similarities and differences in the defense mechanisms and oxidative sensitivity between species. There is only a limited amount of comparative information available on skin antioxidants and skin vulnerability to oxidative injury. Although data are gradually emerging from some species, e.g., the hairless mouse, the individual results only allow glimpses of the complex overall picture.

The biology and comparative microanatomy of nonvertebrate (Bereiter-Hahn et al. 1984) and vertebrate skin (Maderson 1972; Breiter-Hahn et al. 1986) have been reviewed extensively. However, biochemical and physiological data on vertebrate skin are scarce. Comparative analysis of the cutaneous antioxidant capacity in various vertebrate animals could provide new insights into the phylogentic development of the skin's antioxidant defense system and contribute to a better understanding of adaptational skin biology. Furthermore, a correlation of skin antioxidant potential with skin disease incidence in a given species could increase our biochemical knowledge of skin pathophysiology.

Over the past 2 decades, a great amount of results has been published on tissue antioxidant concentration and activities. The data are not always easily comparable because of different activity units, reference points, or experimental conditions. Enzymic antioxidants such as SOD, catalase, and glutathione peroxidase show great tissue and species heterogeneity (Matkovics et al. 1977; De Marchenna et al. 1974; Lawrence and Burk 1978; Westman and Marklund 1981; Tolmasoff et al. 1980; Hartz et al. 1973; Scholz et al. 1981 a, b). This may also hold true for small molecular nonenzymic antioxidants such as α-tocopherol and glutathione. The heterogeneous tissue distribution of antioxidant factors may possibly reflect the greater vulnerability of certain tissues to oxidative stress and therefore a greater need for antioxidant protection (Tappel 1980).

A detailed analysis of enzymic and nonenzymic antioxidant activities in the epidermis, dermis, and subcutis in humans is desirable. Unfortunately, comparative studies of enzymic and nonenzymic antioxidant activities in various organs frequently exclude skin. Nevertheless, attempts were made to characterize the antioxidant defense mechanism in skin (Fuchs et al. 1989a) and isolated skin cells (Yohn et al. 1991).

Each tissue has an "antioxygenic potential" which is determined by the balance between factors promoting autoxidation and those exerting an antioxidant action. The antioxidants of the adrenal gland are probably one of the best characterized tissue antioxidant defense systems investigated so far. Comparative data on adrenal gland antioxidants are available; however, they do not include skin antioxidants. The adrenal cortex is remarkably well protected against lipid peroxidation and oxygen toxicity. The high levels of enzymic and nonenzymic antioxidants indicate a need for such protection since the adrenal gland has a very high oxygenase content, necessary for steroid metabolism (Hornsby and Crivello 1983).

Mechanisms of Antioxidant Defense. Multiple lines of antioxidant defense have evolved and serve to protect human skin from oxidative stress, including prevention, interception, and repair.

Primary defense mechanisms *prevent* oxidative damage by scavenging reactive species directly. Secondary defense mechanisms *combat* processes elicited by reactive oxygen species, such as lipid peroxidation. The enzymes catalase and SOD scavenge hydrogen peroxide and superoxide anion directly. Tocopherol, β-carotene, and bilirubin are chain-breaking antioxidants that react with peroxyl and alkoxyl radicals in phospholipid membranes. However, they also quench singlet oxygen directly.

Oxidative damage in lipids, proteins, and nucleic acids can be *repaired* by a system of enzymic and nonenzymic processes. Membrane-bound lipid hydroperoxides are released by phospholipases, and the free hydroperoxide is then reduced to an alcohol by glutathione peroxidase. Thioredoxin reductase provides reduced thioredoxin that serves as a hydrogen donor for methionine-sulfoxide reductase, thereby reversing oxidative damage to the amino acid methionine in proteins. Proteolytic enzymes cleave oxidatively damaged amino acids or peptide chains. A variety of DNA repair mechanism exist to repair damage on the genome mediated by reactive oxygen species.

Ancillary Enzymes. Of additional crucial importance are those enzymes that prevent the formation and/or metabolism of prooxidants. Examples include NADPH-ubiquinone reductase and glutathione-S transferases. Other enzymes are involved in regenerative reactions, e.g., glutathione reductase in reduction of glutathione disulfide to glutathione and glucose-6-phosphate dehydrogenase in providing reducing equivalents. Removal of glutathione disulfide via a transport system is essential to prevent mixed disulfide formation with other enzymes, which leads to enzyme inactivation.

Cell Renewal. Antioxidants are effective in protecting biological tissues below a critical threshold of reactive oxygen species. If steady-state free radical concentrations exceed this threshold, then autocatalytic cell injury will result. Cell renewal constitutes the last line of defense, when other mechanisms fail to prevent oxidative injury caused by free radicals and reactive oxygen species.

Antioxidants may exert *paradoxical prooxidative effects.* The antioxidant radical which is formed upon reaction of the antioxidant with the radical to be scavenged may be sufficiently reactive to continue the chain reaction of lipid peroxidation when present in high concentrations. Biotransformation of the antioxidant may yield metabolites capable of producing reactive oxygen species (Kahl and Weinke 1988). Reduction of redox active transition metal ions by the antioxidant can lead to the formation of reactive oxidants. The ultimate effect exerted by an antioxidant in vivo depends on numerous factors, such as its microenvironment concentration, tissue oxygen tension, presence of free transition metal ions, and physicochemical properties of the reactant.

I. Superoxide Dismutase

SOD is ubiquitous in mammalian tissues. It is a metal-protein complex and catalyzes the dismutation of the superoxide anion radical into hydrogen peroxide and oxygen. The rate constant of this reaction is about $1.8 \times 10^9\ M^{-1}\,s^{-1}$. Some other copper-containing proteins also catalytically dismutate superoxide anion radical; however, the rate constant of cytochrome c oxidase and ceruloplasmin is $2 \times 10^7\ M^{-1}\,s^{-1}$ and $6 \times 10^5\ M^{-1}\,s^{-1}$, respectively. In mammals, two forms of the enzyme are reported: Cu, Zn-SOD is mostly localized within the cytosol, whereas Mn-SOD is predominantly present within mitochondria but is also found in the cytosol of higher primates.

In blood, most of the SOD is located in blood cells, and only a minor fraction is found in the serum, indicating that extracellular SOD probably plays no significant role as a superoxide radical scavenger. A considerable part of the human serum antioxidant potential is attributed to the acute phase protein ceruloplasmin (Vidlakova et al. 1972; Barber 1977; Stocks et al. 1974), which has SOD-like activity. Extracellular ceruloplasmin can act through dismutation of superoxide anion, similar to SOD (Goldstein et al. 1979, Plonka and Metodiewa 1979); however, a separate mechanism has been discussed (Bannister et al. 1980).

Cu, Zn-SOD is detected in the epidermis of guinea pigs, rats, mice, and humans. Mn-SOD is absent in albino CD1 mouse epidermis but present in human and guinea pig epidermis (Carraro and Pathak 1988). It is known that both types of SOD have a different alteration pattern under various pathological conditions. Therefore, determination of both enzyme activities is necessary for evaluating the physiologic and pathologic significance of SOD in keratinocyte function (Ohkuma et al. 1987). Epidermal SOD activity,

Table 2.2. Superoxide dismutase activity in the skin of various species

Antioxidant	Species	Tissue	Activity	Reference
Total superoxide dismutase (SOD)	Human	Total skin	12.6 U/mg protein	Niwa 1987
		Epidermis	12.0 U/mg protein	Kim and Lee 1987
		Dermis	10.5 U/mg protein	Kim and Lee 1987
		Epidermis	1–3.5 µg/mg protein	Carraro and Pathak 1988
		Epidermis	44 ng/mg epidermis	Van Baar et al. 1987a
	Mouse	Epidermis	4–10 U/mg protein	Solanki et al. 1982
		Total skin	3.0–4.3 U/mg protein	Fuchs et al. 1989a Fuchs et al. 1989c
		Total skin	9.5 U/mg protein	Miyachi et al. 1987b
		Epidermis	0.6 U/mg protein	Carraro and Pathak 1988
	Pig	Epidermis	11.4 U/mg protein	Ohkuma et al. 1987
	Guinea pig	Epidermis	0.5 U/mg protein	Carraro and Pathak 1988
	Cow	Epidermis	4.6–36.6 U/mg protein	Ogura et al. 1990
Mn-SOD	Human	Epidermis	0.2–2.2 U/mg protein	Sugiura et al. 1985a
	Pig	Epidermis	2.8 U/mg protein	Ohkuma et al. 1987
Cu, Zn-SOD	Human	Epidermis	1.6–8.6 U/mg protein	Sugiura et al. 1985a
	Pig	Epidermis	8.6 U/mg protein	Ohkuma et al. 1987

measured by an EPR spin trapping technique, increases successively from the upper to the lower epidermal layer (Ogura et al. 1990). When comparing assays (ESR spin trapping versus xanthine oxidase/nitroblue tetrazolium), SOD activity in whole epidermis was seen to vary from 4.6 to 36.6 U/mg protein in one species (Ogura et al. 1990) (Table 2.2).

SOD in human epidermis is reported to be low when compared with that found in other tissues, e.g., liver (Borello et al. 1984). Another study indicates that SOD content in human, guinea pig, and mouse epidermis is 5–10 times lower than that in the liver, heart, kidney, skeletal muscle, lung, and other tissues (Pathak and Carraro 1987a; Carraro and Pathak 1988). The total SOD content of fatty tissue is about 80 times lower than in liver (Marklund 1980). These reports indicate that skin SOD activity is low with respect to other organs (see Table 2.3).

Enzymatic and Immunoassays. There is a dilemma in the analysis of SOD activity in biological material. Enzymatic SOD assays usually measure a total superoxide anion radical scavenging activity. The exact quantitative contribution of SOD to this total activity is frequently not known. Compounds with SOD-like activity (SOD mimetics) contribute to the activity to an undefined extent. SOD immunossays are highly specific for the SOD enzyme and

Table 2.3. Superoxide dismutase activity and lipid peroxide content in different human tissues and cells (with permission, from Niwa 1987b)

Tissue or cell	SOD activity (U/mg protein)	Lipid peroxides (nmol/mg protein)
Epidermis/dermis	8.4	0.72
Dermis/subcutis	6.5	0.61
Muscle	2.9	0.55
Liver	66.7	1.12
Kidney	60.2	1.04
Lung	37.5	0.79
Erythrocyte	89	0.08
Neutrophil	2.0	0.59
Macrophage	1.2	0.44

measure the actual amount of SOD protein. However, they do not allow analysis of the enzyme activity (Sugiura et al. 1985b). Recently differential staining (immunohistochemical analysis) for each type of SOD in human skin was reported (Kobayashi et al. 1991).

Prooxidant Activity. SOD also possesses prooxidant activity. SOD enhances autoxidation of the antiinflammatory quinone-type drug rifamycine SV in vitro (Michelson 1986a). In trisomy 21, Cu, Zn-SOD activity is significantly elevated in various tissues, and lipid peroxidation products are increased in these organs. This is considered indirect evidence for a prooxidant action of SOD in vivo, which may result in the increased formation of hydrogen peroxide.

II. Catalase

Catalase is one of the oldest known enzymes. It is an ubiquitous heme enzyme which functions in two ways. First, it catalytically decomposes two molecules of hydrogen peroxide to water and oxygen (catalytic activity). Second, it oxidizes a substrate (AH_2) using a hydroperoxide (ROOH) as a hydrogen donor (peroxidic activity) (Fig. 2.27). The biological significance of the peroxidase activity in vivo is questionable.

$$2\ H_2O_2 \xrightarrow[\text{[catalytic activity]}]{\text{catalase}} 2H_2O + O_2$$

$$ROOH + AH_2 \xrightarrow[\text{[peroxidic activity]}]{\text{catalase}} H_2O + ROH + A$$

Fig. 2.27. Catalytic and peroxidic catalase activity

Table 2.4. Catalase activity in the skin

Antioxidant	Species	Tissues	Activity	Reference
Catalase	Human	Skin fibroblasts	7–12 μmol O_2/min per mg protein	Wanders et al. 1984 Wanders et al. 1987
	Human	Subcutis	270–560 U/mg protein	Marklund et al. 1982
	Mouse	Epidermis	78–175 U/mg protein	Solanki et al. 1982
	Mouse	Total skin	7.6–10.4 μmol O_2/min per mg protein	Fuchs et al. 1989 a, c

The enzyme is largely located within the peroxisomes where it constitutes about 50% of the protein in this organelle. It has a high turnover and is probably synthesized by rough endoplasmatic reticulum, conveyed and concentrated in the Golgi apparatus, and finally released within small coated vesicles.

Catalase has a relatively low affinity for its substrate but a high capacity for the reaction. A genetic deficiency of catalase in humans does not usually result in the clinical manifestation of organ pathology (Aebi et al. 1964), indicating the presence of compensating mechanisms or still effective residual amounts of catalase. Catalase may have a role in the detoxification of hydrogen peroxide only under conditions of severe oxidative stress. During mild oxidative stress, its protective role is less important than that of other antioxidant factors (Sammartano et al. 1986; Carlsson and Carpenter 1980; Eisenstark and Perrot 1987).

The significance of catalase in cellular hydrogen peroxide metabolism must, however, not be underestimated. Although present at low levels of activity in the heart compared with the erythrocyte and liver, catalase is thought to be the major route of hydrogen peroxide detoxification in rat myocardium (Thayer 1986). The biological significance of catalase in skin antioxidant defense is not known (Table 2.4).

Catalase assays based on the decomposition of hydrogen peroxide measure an overall hydrogen peroxide decomposing activity in biological material. Various peroxidases with affinity for hydrogen peroxide may contribute to this activity; therefore, these assays are not specific for catalase activity.

III. Peroxidases

A range of peroxidases with affinity for hydrogen peroxide, such as cytochrome *c*, NADPH peroxidase, and hemoglobin, may be involved in detoxification. Other enzymes that might play a part in hydroperoxide metabolism are cytochrome P-450, cytochrome oxidase, and heme peroxidase. In human skin, endogenous peroxidase activity is demonstrated by histochemical methods in the perinuclear envelope and in the endoplasmic reticulum of non-keratinized keratinocytes, Langerhans cells (Dubertret et al. 1982; Walzer and Frank 1983, 1986), and cutaneous mast cells (Escribano et al. 1987). Peroxidase activity is also detected in dermal fibroblasts and dermal macrophages (Coulomb et al. 1983). A variety of pro- and antioxidant peroxidases may contribute to this overall activity; some peroxidases are known to activate xenobiotics to prooxidants. Although catalase may decompose hydrogen peroxide produced by, e.g., near UV irradiation, it may also act as a photosensitizer (Eisenstark and Perrot 1987; Peak et al. 1983; Krämer and Ames 1987).

IV. The Enzymic Glutathione System

The enzymic glutathione system comprises glutathione peroxidase, reductase, and transferase. Selenium-dependent glutathione peroxidases reduce both hydrogen peroxide and free organic hydroperoxides and are found in the cytosol and within mitochondria. Non-selenium-dependent glutathione peroxidases (glutathione-S transferases) reduce only free organic hydroperoxides and do not metabolize hydrogen peroxide (Fridovich 1976). These processes are dependent on an adequate supply of glutathione and NADPH and are therefore glutathione reductase dependent. Glutathione reductase also reduces various mixed disulfides.

Glutathione (γ-glutamyl-cysteinyl-glycine; GSH) is rapidly oxidized by free radicals, hydroperoxides (ROOH), and reactive oxygen species to glutathione disulfide (GSSG). Glutathione reductase catalyzes the reduction of GSSG using NADPH generated, for example, by glucose-6-phosphate dehydrogenase (Sies and Cadenas 1983) (Fig. 2.28).

In human erythrocytes the major pathway of hydrogen peroxide metabolism is glutathione-peroxidase-dependent elimination. At higher concentrations of hydrogen peroxide the effect of catalase becomes increasingly prominent (Cohen and Hochstein 1963; Fantone and Ward 1982). In contrast to catalase, glutathione peroxidase has a low capacity for its substrate but a high affinity. A comparative study of the enzymic antioxidant defense system in human fibroblasts cultured under hyperbaric oxygen tension reveals that glutathione peroxidase is much more effective than catalase and SOD in protect-

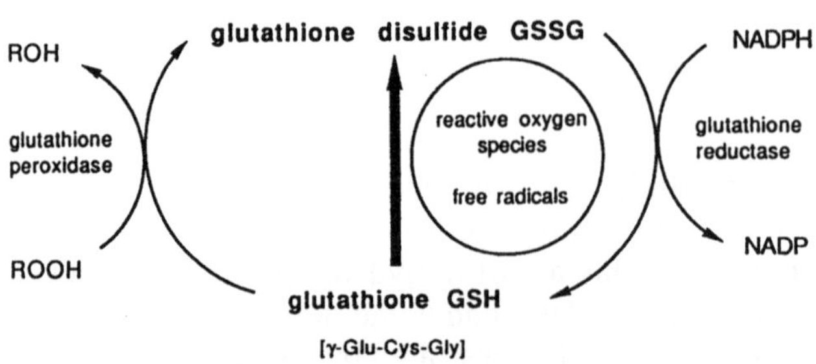

Fig. 2.28. Glutathione oxidation

ing the cells (Raes et al. 1987). This may indicate that glutathione peroxidase is an important fibroblast antioxidant enzyme.

The distribution of selenium-independent glutathione peroxidase is species and tissue dependent. In comparison with liver, kidney, and brain, skin glutathione peroxidase and glutathione S transferase activity are markedly lower (Borello et al. 1984; Singh and Clausen 1980). Glutathione-S transferase is present at all different stages of human keratinocyte differentiation. In contrast to other cell types, in keratinocytes only a single form of the enzyme is found that has a limited substrate specificity and lacks activity against organic hydroperoxides. Consequently, the epidermis may be especially susceptible to injury by electrophiles and organic hydroperoxides (Blacker et al. 1989; Blacker et al. 1991) (Table 2.5).

Table 2.5. Activity of the enzymes of the glutathione system in the skin

Antioxidant	Species	Tissue	Activity	Reference
Glutathione (GSH) peroxidase	Mouse	Epidermis	80.2	Wheeler et al. 1986
		Dermis	37.0	Wheeler et al. 1986
		Epidermis	80.2	Connor and Wheeler 1987
		Dermis	36.5	Connor and Wheeler 1987
		Total skin	35	Fuchs et al. 1989 a, c
GSH transferase	Mouse	Epidermis	53.3	Wheeler et al. 1986
		Dermis	64.8	Wheeler et al. 1986
		Epidermis	19.6	Connor and Wheeler 1987
		Dermis	33.8	Connor and Wheeler 1987
Glucose-6-phosphate dehydrogenase	Mouse	Epidermis	109.8	Connor and Wheeler 1987
GSH reductase	Mouse	Epidermis	22.5–31.6	Connor and Wheeler 1987
		Dermis	14.3–27.6	Connor and Wheeler 1987
		Total skin	20	Fuchs et al. 1989 a, c
GHS reductase[a]	Human (negro)	Total skin	0.21	Jocelyn 1972 a
	Human	Total skin	0.31	Jocelyn 1972 a

Enzyme activity units:
GSH peroxidase = nmol NADPH oxidized/mg protein · min; GSH transferase = nmol GSH/mg protein · min; glucose-6-phosphate dehydrogenase = nmol NADPH reduced/mg protein · min; GSH reductase = nmol GSH/mg protein · min; GSH reductase[a] = μmol GSSG reduced/min · g skin

V. Thioredoxin Reductase System

The mammalian thioredoxin system (thioredoxin, thioredoxin reductase, and NADPH) is an ubiquitous, thiol-dependent, electron transport system (Fig. 2.29). Mammalian thioredoxin is a small protein, which in its oxidized form contains a redox active disulfide bridge. The reduced form of thioredoxin contains a corresponding dithiol and is generated by the flavin-containing redox enzyme thioredoxin reductase and NADPH (Holmgren 1985).

Reduced thioredoxin is an important hydrogen donor for ribonucleotide reductase and methionine-sulfoxide reductase and is a protein disulfide reductase, e.g., it reduces insulin disulfide bonds (Holmgren 1979). The rate constant for disulfide reduction with nonprotein disulfides, e.g., cystine or glutathione disulfide, is about 1000-fold slower than with peptide disulfides, e.g., insulin (Holmgren 1985).

Tissue Distribution. Thioredoxin and thioredoxin reductase are widely distributed in mammalian tissues and organs and are generally present at a high activity level in epithelial cells, whereas mesenchymal cells exhibit a low activity (Rozell and Hansson 1985). Moderate to intense immunoreactivity for both proteins is observed in the stratum germinativum, hair follicles, nail beds, and sweat glands in rat skin. Fully cornified parts of the skin lack reactivity to either protein. Keratinizing cells are positive only for thioredoxin; Langerhans cells and melanocytes are positive for thioredoxin and thioredoxin reductase (Rozell and Hansson 1985). High immunoreactivity is found in the cytosol, at the plasma membrane, and in the subplasma membrane zone; however, no extracellular immunoreactivity is observed. Thioredoxin is probably membrane-associated and present in subcellular fractions such as the nucleus, microsomes, and mitochondria (Holmgren and Luthman 1978; Hansson et al. 1986).

Antioxidant Properties. Thioredoxin prevents hydrogen peroxide damage in lens epithelial cells (Spector et al. 1988a) and acts as a free radical quencher, based on studies with a stable free radical system, generated by ascorbate and 2,6-dimethoxy-*p*-benzoquinone. Other persistent free radicals such as nitroxides are reduced by thioredoxin reductase isolated from *E. coli* and from human keratinocytes (Schallreuter and Wood 1986; Schallreuter et al. 1986a). It was proposed that thioredoxin reductase may also reduce superoxide anion

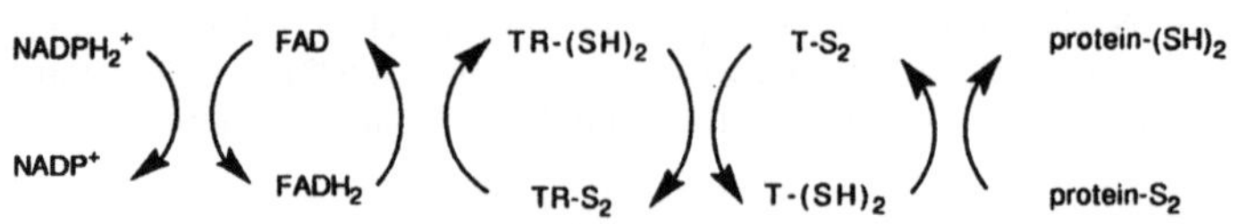

Fig. 2.29. The mammalian thioredoxin system. *TR-S2,* thioredoxin reductase (oxidized); *TR-(SH)2,* thioredoxin reductase (reduced); *T-S2,* thioredoxin (oxidized); *T-(SH)2,* thioredoxin (reduced); *protein-S2,* oxidized protein; *protein-(SH)2,* reduced protein; *FAD,* flavin adenine dinucleotide

radicals in the epidermis and constitutes a first line of defense against free radical damage at the epidermal surface (Schallreuter and Wood 1986; Schallreuter et al. 1986a, 1987). Its role in free radical reduction in skin was recently reviewed (Schallreuter and Wood 1989). Criticism was raised on the interpretation of results considering thioredoxin reductase a free radical scavenging enzyme in skin and on a putative specific bioassay.

It can be misleading to extrapolate from the nitroxide radical reducing activity of thioredoxin reductase to its scavenging properties with highly reactive oxygen species or other types of free radicals. Experiments revealed that in comparison with SOD, thioredoxin reductase has very little SOD-like activity in different assays (Freisleben, personal communication). The antioxidant properties of the thioredoxin reductase system in skin await further characterization.

A specific bioassay for thioredoxin reductase in skin cells and biopsies based on the reduction of a thioredoxin-reductase-binding cationic nitroxide was developed by Schallreuter and coworkers. Multiple biochemical and clinical studies have been performed based on this bioassay (Schallreuter et al. 1986b, 1987; Schallreuter and Pittelkow 1987; Schallreuter and Wood 1987; Schallreuter and Witkop 1988; Schallreuter and Wood 1990; Sundaram et al. 1990). Its selectivity and specificity, however, have been criticized (Fuchs 1988a, b, et al. 1989b), and caution should be exercised in interpreting reports that rely upon it (Fuchs et al. 1990c).

VI. Lipoamide System

The lipoamide system (lipoamide, lipoamide dehydrogenase, and NADH) is a physiological constituent of mitochondrial dehydrogenase complexes. Lipoamide dehydrogenase functions in the oxidation/reduction of lipoate covalently bound in amide linkage to a protein amino acid (lysine) and transfers electrons from dihydrolipoate (dihydrolipoamide) to NAD with a tightly bound flavin adenine dinucleotide (FAD) as a cofactor. The lipoamide system is involved in the reduction of oxidized thioredoxin fragments (Holmgren 1979; Spector et al. 1988b). Interaction with the thioredoxin system may participate in the cellular sulfhydryl redox control. Possibly 50% of the lipoamide in *E. coli* may have an unidentified electron acceptor (Frey et al. 1978), and similar conditions may prevail in mammals.

Antioxidant Effects of Dihydrolipoate/Lipoate. The redox couple dihydrolipoate/lipoate has antioxidant properties. Lipoate is a singlet oxygen scavenger (Stary et al. 1975) and inhibits photoperoxidation of unsaturated organic molecules (Stevens et al. 1974). Physically, the singlet oxygen quenching activity of lipoate is greater than that of glutathione; the chemical scavenging activity, however, is in the reverse order. Dihydrolipoate does not scavenge singlet oxygen (Devasgayam et al. 1989). Because of its standard

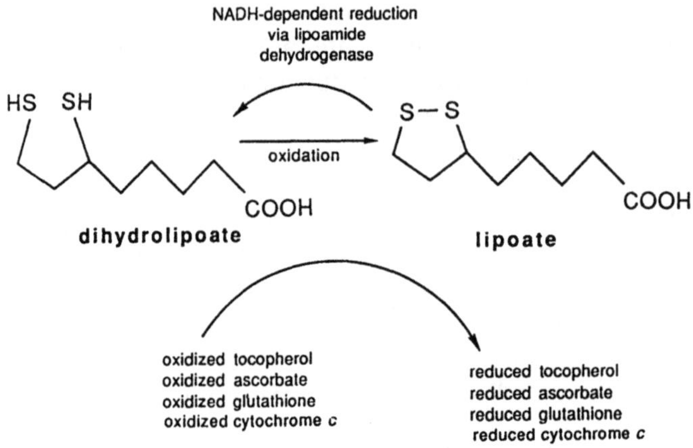

Fig. 2.30. Oxidation and reduction of dihydrolipoate and lipoate

redox potential of (–)0.32 V, dihydrolipoate is capable of reducing glutathione disulfide and dehydroascorbate (Haenen and Bast 1989). Dihydrolipoate also reduces organic peroxyacetyl nitrate (Leh and Mudd 1974) and cytochrome *c* at physiological pH (Matuda 1982). It was suggested that dihydrolipoate may regenerate tocopherol in biological membranes (Fuchs, unpublished results; Scholich et al. 1989). The antioxidant effects of dihydrolipoate in biological membranes may be partially due to its ability to regenerate vitamin E from the tocopheroxyl radical. Neither dihydrolipoate nor lipoate was able to recycle chromanoxyl radicals in rat liver microsomes unless NADPH or NADH were present. In their presence, dihydrolipoate acted synergistically to produce recycling of chromanoxyl radicals, whereas lipoate was ineffective (Kagan et al. 1990) (Fig. 2.30).

In guinea pigs deficient of either vitamin C or vitamin E, administration of lipoate alleviates the symptoms of ascorbate and tocopherol deficiencies (Rosenberg and Culik 1959), indicating regeneration of the vitamins and/or protection from oxidative vitamin destruction. Dihydrolipoate protects rat microsomes (Scholich et al. 1989) and rat skin lipids (Packer, personal communication) from peroxidation. Dihydrolipoate inhibits severe dermatitis in mice exposed to different reactive oxidant generating systems (Fuchs et al. 1990d), and lipoate was successfully used in the treatment of patients with dermatitis herpetiformis Duhring, an inflammatory skin disease (Mirzoev et al. 1988).

Prooxidant Effects of Dihydrolipoate/Lipoate. Thiols may undergo autoxidation to disulfides, while the latter are readily reduced to thiols. Thiyl radicals are involved in thiol/disulfide redox cycling, and reactive oxygen species can be formed. This mechanism is thought to be responsible for causing tissue injury provoked by thiols and disulfides (Munday 1989).

A prooxidant effect of dihydrolipoate causing microsomal lipid peroxidation (Bast and Haenen 1988), inactivation of glutamine synthetase (Kim et al. 1988), and single-strand breaks of DNA in the presence of Fe(III)ions was reported (Reed and Douglas 1989). Dihydrolipoate can release iron bound to ferritin (Bonomi and Pagani 1986; Bonomi et al. 1989), and autoxidation of dihydrolipoate in the presence of transition metal ions [Fe(III), Cu(II)] may generate reactive oxygen species. Generated Fe(II) or Cu(I) ions can then be used for the production of superoxide anion radicals by biological systems.

Lipoate is presumably converted into dihydrolipoate in vivo. In rats given lipoate supplement in their diet, no increased formation of lipid peroxidation products was detected in the brain, liver, heart, and skin tissue. On the contrary, the tissue lipids were highly resistant to nonenzymatically induced peroxidation (Packer, personal communication). Dihydrolipoate inhibits UV-induced skin inflammation in mice (Fuchs et al. 1990d) and humans (Fuchs and Milbradt 1989a). This clearly indicates an antioxidant effect of lipoate/dihydrolipoate in vivo.

A Free Radical Mechanism. Highly unstable sulfur radicals are formed by the reaction of lipoate/dihydrolipoate with oxidizing and reducing radicals. Lipoate rapidly reacts with oxidizing radicals to yield lipoate cation radical (Bonifacic and Asmus 1984). In reaction with reducing radicals, lipoate anion radical is formed (Chan et al. 1974; Faraggi et al. 1975), which is highly unstable and can transfer its electron to FAD. Dihydrolipoamide is readily oxidized by the hydroxyl radical to lipoamide anion radical (Wu et al. 1984). Lipoamide dehydrogenase is inactivated by oxidizing radicals (hydroxyl radicals) but is not measurably inhibited by reducing radicals (superoxide anion radical, free electron) (Elliot et al. 1980). It was concluded that a free radical mechanism may be involved in the lipoamide dehydrogenase reaction (Chan et al. 1974).

Prostanoid Metabolism. Lipoate stimulates the enzymatic oxygenation of arachidonic acid and increases reduction of prostaglandin intermediate PGG_2 (hydroperoxide-endoperoxide) to PGH_2 (alcohol-endoperoxide). Indirect stimulation of cyclooxygenase due to trapping of oxidants released during PGG_2 reduction is thought to be involved in the reaction mechanism. Lipoate mediates the nonenzymatic reduction of PGH_2 to $PGF_2\alpha$ (alcohol) (Marnett and Bienkowski 1977). It depresses phorbolmyristateacetate-caused mouse ear edema (Egan et al. 1978) and inhibits UVB erythema in mice (Fuchs et al. 1990d) and humans (Fuchs and Milbradt 1989a). Modulation of the arachidonic acid cascade presumably contributes to the lipoate/dihydrolipoate antiinflammatory effects. Direct scavenging of reactive oxidants may be a further mechanism involved in its antiinflammatory properties.

Lipoate Tissue Concentration. Cellular lipoate can be quantitated by chromatography (Stevenson et al. 1988; Hayakawa and Oizumi 1989). In mammalian tissues it is 5–25 nmol/g (Jocelyn 1972b) and about 50 μM in human serum (Loginov et al. 1989). Normal human urine may contain up to 200 nmol/ml lipoate in a conjugate form. No data are available on lipoate concentration in skin.

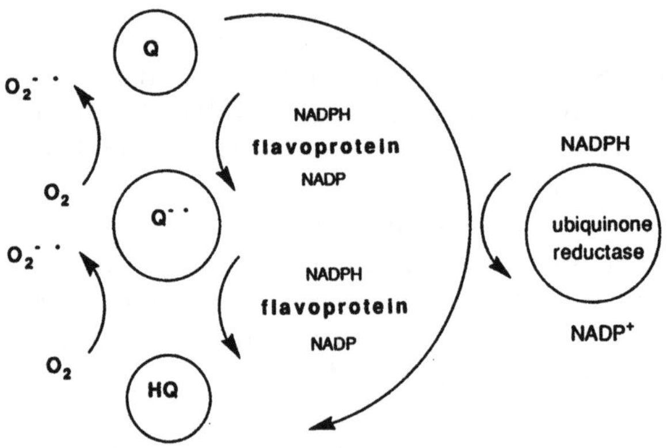

Fig. 2.31. One- and two-electron metabolism of quinones/hydroquinones

VII. NADPH Ubiquinone Reductase

Diminution of the steady state levels of compounds capable of generating reactive oxygen species is an important antioxidant mechanism. Quinones are readily reduced by various flavoproteins in a one-electron step to semiquinone radicals. Semiquinone radicals may then react with molecular oxygen to yield superoxide anion radicals and the quinone once again. By this redox cycling, reducing equivalents (NADPH) are used up, and the steady-state level of reactive oxidants is increased. NADPH ubiquinone reductase (DTdiaphorase) reduces quinones (Q) in a two-electron step to hydroquinones (HQ), omitting production of autooxidizing semiquinone radicals. Hydroquinones can then undergo conjugation reactions catalyzed by conjugation enzymes such as glutathione-S transferase, resulting in detoxified conjugation products. NADPH ubiquinone reductase is considered a cellular control device against semiquinone ($Q^{-\cdot}$) and superoxide radical formation (Lind et al. 1982) (Fig. 2.31). NADPH ubiquinone reductase has a higher activity in the epidermis than in the liver (Khan et al. 1987). The same enzymatic form is present in murine and human liver and skin, and the basal enzyme activity is similar in murine skin and liver. However, the induction of NADPH ubiquinone reductase with polycyclic hydrocarbons is stronger in the liver (Jugert and Merk 1990). Anticarcinogenic enzyme inducers elevate NADPH ubiquinone reductase. The enzyme level may therefore serve as a marker for the chemoprotected state.

VIII. Nonenzymic Protein Antioxidants

Metallothioneins are low molecular weight metal-(zinc, cadmium) and sulfhydryl-rich proteins which are widely distributed in mammalian tissues and are found in human keratinocytes and fibroblasts. They are present at 1.0–

1.1 pg/mg wet weight of total skin in mice (Frings et al. 1989; Mullin et al. 1987). It was suggested that they are involved in cellular adaptation and defense mechanisms (Hamer 1986). Proposed functions of metallothioneins include participation in metal metabolism and antioxidant functions. They are induced by various stresses and hormones which are elevated by inflammatory processes. Metallothionein gene expression is regulated by heavy metal ions, tumor promotors, glucocorticoid, and cytokines. The metallothionein content of rat liver is increased following exposure to high oxygen concentration or X-radiation (Shiraishi et al. 1983). This indicates that metallothionein is involved in the cellular response to environmental stress reactions such as oxidative stress. Metallothionein is inducible in mammalian skin and it has been suggested to be a photoprotective agent against UVB-induced oxidative skin damage (Hanada personal communication). In the skin of hairless mice metallothionein is induced after systemic administration of dexamethasone or zinc (Mullin et al. 1987). Metallothioneins are radioprotectors in animals and are scavengers of superoxide anion and hydroxyl radical. The rate constant of metallothionein with hydroxyl radicals (bimolecular rate constant $k = 10^{12} M^{-1} s^{-1}$) is much greater than that of other proteins (Thornalley and Vasak 1985). Besides its role as an antioxidant, skin metallothionein may play an important role in the process of keratinization.

The stratum corneum is the outermost layer of the epidermis and constitutes the main diffusion barrier for many xenobiotics and oxygen. It therefore requires efficient mechanisms to detoxify reactive oxygen species, other free radical compounds, and lipid peroxides. In newborn rat stratum corneum a membrane-coating protein has been identified that is an effective scavenger of lipid peroxides (Tezuka and Takahashi 1987) when compared with keratin and albumin (Yoshino et al. 1981). This cysteine-rich protein, which has many free sulfhydryl and disulfide bonds, is probably located in the cytoplasm of corneocytes. By scavenging lipid peroxide radicals, its sulfhydryl groups become oxidized. The characterization of a similar free radical scavenging protein isolated from yeast has recently been described (Kim et al. 1988).

Many intra- and extracellular proteins containing clustered tyrosine and cysteine residues are thought to provide antioxidant activity in biological material (Grant and Williamson 1989). The epidermis structural protein keratin, which is a sulfhydryl-rich protein, probably constitutes a skin antioxidant.

Extracellular space glycoproteins in skin may exert a protective function by scavenging reactive oxygen species, resulting in damage to noncritical cellular sites.

Since free transition metal ions such as Fe^{2+} and Cu^{+} catalyze Fenton-type reactions in biological tissue, metal-binding proteins such as ceruloplasmin, transferrin, and lactoferrin may contribute to the extracellular host defense in skin.

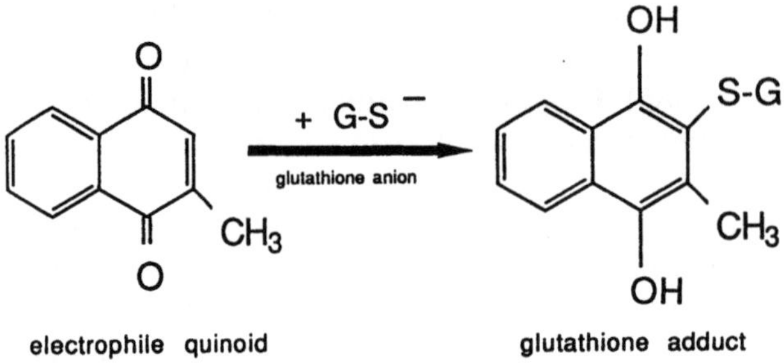

Fig. 2.32. Glutathione reacting as a nucleophile

IX. Hydrophilic Antioxidants

1. Thiols

Glutathione. Present at a millimolar concentration in the cytosol, glutathione (GSH) may directly react with species such as hydrogen peroxide, superoxide anion, hydroxyl radical, and alkoxyl radical by a radical transfer process to yield the thiyl radical GS· and subsequently glutathione disulfide (GSSG). It also quenches triplet (excited) carbonyls. Glutathione may react as a nucleophile (GS^-, thiolate anion of glutathione) with electrophiles (e.g., quinoid compound) to form covalent adducts catalyzed by glutathione transferase (Michael-like reaction) (Fig. 2.32).

The glutathione redox cycle has an important role in protecting the lens epithelium against hydrogen peroxide damage (Giblin et al. 1985) and intesti-

Table 2.6. Glutathione and glutathione disulfide activity in the skin

Antioxidant	Species	Tissue	Activity	Reference
Glutathione	Human	Epidermis	1.8 μmol/g tissue	Halprin and Ohkawara 1967
		Epidermis	1.2 μmol/g tissue	Connor and Wheeler 1987
		Epidermis	0.7 m*M*	Kingston et al. 1985
		Total skin	2–3 μmol/g wet weight	Jocelyn 1972a
	Mouse	Epidermis	0.75 μmol/g tissue	Wheeler et al. 1986
		Dermis	0.32 μmol/g tissue	Wheeler et al. 1986
		Total skin	3.9–6.3 nmol/mg protein	Fuchs et al. 1989a, c
	Guinea pig	Total skin	0.7–1.1 μmol/g tissue	Benedetto et al. 1981
Glutathione disulfide	Human	Epidermis	0.09 μmol/g tissue	Halprin and Ohkawara 1967
	Guinea pig	Total skin	1.4–1.5 μmol/g tissue	Benedetto et al. 1981
	Mouse	Total skin	1.0–1.5 nmol/mg protein	Fuchs et al. 1989a, c

nal epithelia from oxidative injury (Lash et al. 1986). Skin fibroblasts are strongly sensitized to the cytotoxic action of near visible (405 nm), UVA, and UVB radiation if depleted of glutathione. For UVB the magnitude of protection afforded by endogenous glutathione approaches that provided by excision repair (Tyrrell and Pidoux 1986). This indicates an important antioxidant function of the skin glutathione system.

In rats, the liver and the eye lens are the organs most abundantly supplied with glutathione. Its concentration is low in blood plasma and adipose tissue. One-third of the total cellular supply is contained in mixed disulfides, e.g., coenzyme A, other small molecular thiols, and proteins. Most glutathione is present as GSH rather than GSSG. Glutathione disulfide can inactivate enzymes by forming mixed disulfides with them. Therefore, its concentration is held low in cells either by effective reducing mechanisms or by excretion. A comparative analysis of the glutathione/glutathione disulfide distribution in different animal or human organs including skin is not available. The large variations of glutathione/glutathione disulfide concentration in skin in several studies (Halprin and Ohkawara 1967; Benedetto et al. 1981) may be due to species differences or differences in assay protocols (Table 2.6).

Skin Thiol/Disulfide Distribution. Skin is a sulfhydryl-rich organ; free sulfhydryl groups are abundant in epidermal membranes and in the cytosol. The thiol content of the epidermis varies among the distinct epidermal layers and may indicate a close relationship of the skin's thiol/disulfide status to epidermal differentiation (Fig. 2.33). The thiol/disulfide redox status is an important control mechanism in cellular homeostasis.

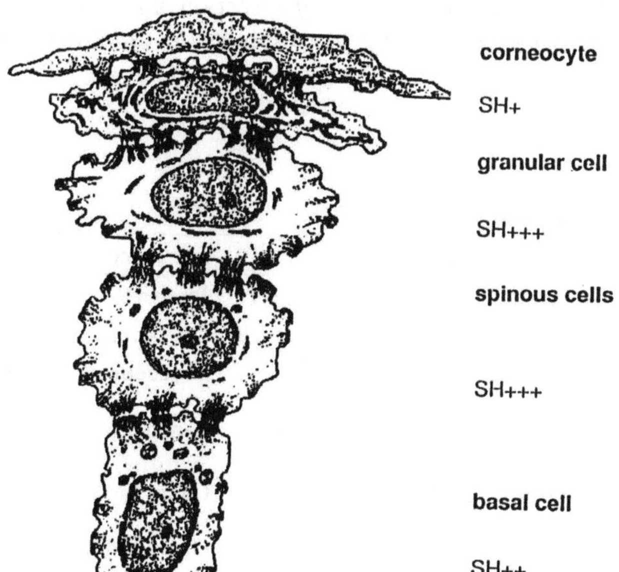

Fig. 2.33. Epidermal sulfhydryl distribution

Thiol groups comprise about 10% of the total sulfur content of stratum corneum (Stüttgen and Schoefer 1974); their presence in the stratum corneum is about 60% of that in the noncornified layers (Flesch and Satanoce 1955). A decrease in sulfhydryl groups in the stratum corneum compared with the deeper skin layers has also been noted by others (van Scott and Flesch 1954; Ogawa et al. 1979). The sulfhydryl concentration measured in keratinocytes isolated from the stratum basale is lower than in keratinocytes from the stratum spinosum and stratum granulosum (Nomura et al. 1985). The level of disulfides is very low in the cytoplasm of keratinized cells (Ogawa et al. 1979).

Protein SH and Nonprotein SH Groups. About 75% of total skin SH reactivity in the guinea pig is comprised of nonprotein SH groups, the majority of which is glutathione (Bendetto et al. 1981). In the epidermis of rats the protein thiol content is about halt that of nonprotein thiol, suggesting a potential involvement of protein thiols as antioxidants (Ogura and Knox 1967). A thorough analysis of the total sulfhydryl (SH) and disulfide (SS) content in the human epidermis, as measured by amperometric titration, revealed that (a) there are no significant differences in both parameters obtained from various anatomical regions (except palms and glabella), (b) Negro skin exhibited a lower epidermal SH content than Caucasian skin, and (c) infants have higher SH and lower SS concentrations (Ogura et al. 1962).

Ogura differentiated between immediate reactive epidermal SH groups and denaturated epidermal SH groups. The latter are reactive only after epidermal protein denaturation with urea. The average ratio of denaturated/immediate SH reactivity is about 1.7, indicating a large amount of masked protein SH groups. In human (Ogura et al. 1962) and rat (Ogura and Knox 1967) epidermis, the disulfide content is only a fraction of the total thiol concentration, indicating a highly reducing environment at the epidermis (Table 2.7).

Table 2.7. SH group activity in the skin

Antioxidant	Species	Tissue	Activity	Reference
Total SH groups	Human	Total skin	7–9 µmol/g wet weight	Jocelyn 1972a
		Epidermis	900 µ*M*/100 g dry weight	Ogura et al. 1962
	Guinea pig	Total skin	1.8–3.6 µmol/g tissue	Benedetto et al. 1981
Non-protein SH groups	Guinea pig	Total skin	1.3–3.2 µmol/g tissue	Benedetto et al. 1981
	Rat	Epidermis	258 µ*M*/100 g dry weight	Ogura and Knox 1967
Protein SH groups	Rat	Epidermis	413 µ*M*/100 g dry weight	Ogura and Knox 1967

Melanin Biosynthesis and Skin Thiol/Disulfide Redox Status. Rothman reported that aqueous extracts of human epidermis inhibited the tyrosinase system (Rothman et al. 1946), and it was demonstrated that nonprotein thiols present in the human epidermis inhibit melanin formation (Flesch and Rothman 1948). This indicates involvement of skin thiols in the pigmentation process.

Animal and human studies later supported this concept. In guinea pigs, the skin glutathione redox status is different in pigmented and nonpigmented skin. The tortoiseshell guinea pig is an animal model which allows comparative studies on eu- and pheomelanogenesis, owing to the presence of melanocytes producing both types of pigments in different regions of the same animal. In tortoiseshell guinea pigs, the glutathione/glutathione disulfide ratio is higher in red skin (pheomelanin) (0.9) than in black (eumelanin) (0.5) or white (0.5) skin. Since the content of total SH groups and total glutathione is not significantly different in black and red guinea pig skin, it was concluded that the glutathione reductase activity is augmented in pheomelanin-rich skin (Benedetto et al. 1981).

Negro skin contains less glutathione and more glutathione disulfide and has a lower glutathione reductase activity than Caucasian skin (Jocelyn 1972 a). Regulation of melanin synthesis by skin redox metabolism and modulation by thioredoxin reductase have been hypothesized (Schallreuter et al. 1989).

2. Ascorbate

Ascorbate (AH) was isolated in 1928–1932 by Szent-Györgyi, and later by Waugh and King. Ascorbate reacts directly with a variety of reactive oxidants such as peroxyl radical, superoxide anion radical, hypochloric acid, and thiyl radicals, with the concurrent formation of oxidized ascorbate (dehydroascorbate; A), via the ascorbyl anion radical ($A^{-\bullet}$) (Fig. 2.34). It also rapidly quenches excited carbonyls and singlet oxygen. The ascorbyl radical can be converted back to ascorbate by a glutathione/NADPH-dependent dehydroascorbate reductase or irreversibly decompose into diketogulonic acid, which is unstable and subject to further decomposition. NADH semidehydroascorbate reductase is an enzyme that reduces the ascorbyl anion radical back to ascorbate and concomitantly oxidizes NADH to NAD^{+}. Both ascorbate and dehydroascorbate have vitamin C activity in bioassays. The ascorbyl anion radical is not very reactive and mainly undergoes a disproportionation reation to yield ascorbate and dehydroascorbate.

Ascorbate prevents photooxidative injury in the eye lens (Varma et al. 1982). Although plasma sulfhydryl groups are considered the first line of antioxidant defense in human plasma (Wayner et al. 1986), ascorbate is considered the most effective aqueous antioxidant in human serum (Frei et al. 1988, 1989). The daily requirement of ascorbate is about 75 mg. High-dose ascorbate supplementation reduces the serum peroxide content in an elderly popu-

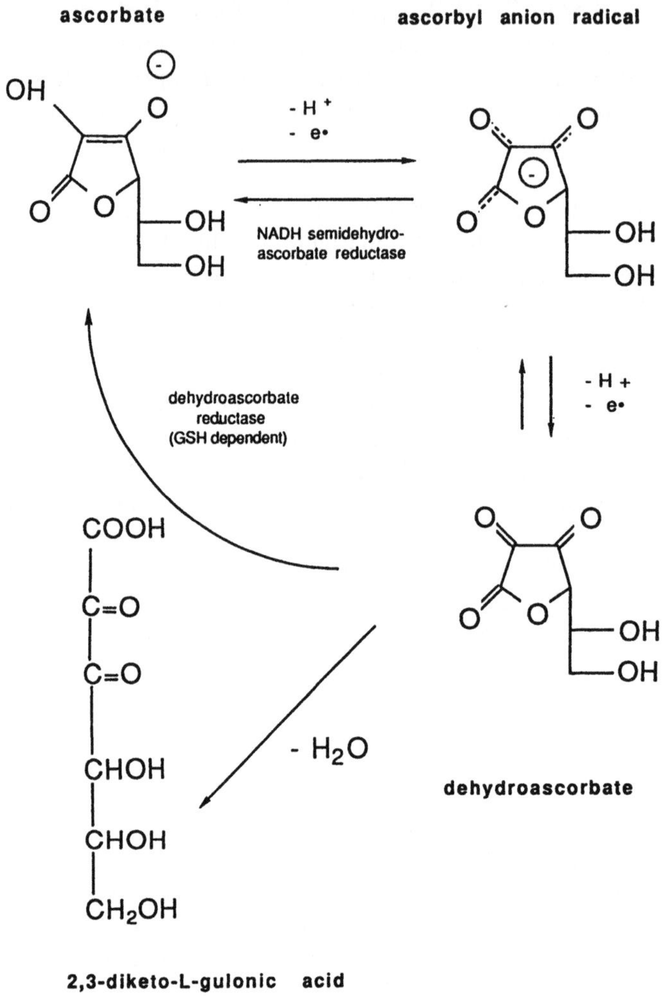

Fig. 2.34. Ascorbate oxidation

Table 2.8. Ascorbate activity in the skin

Antioxidant	Species	Tissue	Activity	Reference
Ascorbate	Human	Total skin	41 µg/g dry weight	Stüttgen and Schaefer 1974
		Stratum corneum	7 µg/g dry weight	Pascher and Röckl 1960
	Rat	Total skin	0.2 g/kg tissue weight	Salomon and Stubbs 1961
	Mouse	Total skin	6–7 nmol/mg protein	Fuchs et al. 1989 a, c

Table 2.9. Ascorbate content of various rat and human organs

Tissue	Rat	Human
Adrenal gland	331–576	30–40
Brain	3129	13–15
Liver	29–40	10–16
Lung	45	7
Kidney	7–19	5–15
Heart	3–15	5–15
Skeletal muscle	8	3–4
Skin	22	0.4–1
Plasma	4	1

lation (Wantanowicz et al. 1984) and reduces expired ethane (an indicator of in vivo lipid peroxidation) in guinea pigs exposed to carbon tetrachloride.

In skin, the antioxidant efficacy of ascorbate in relation to other water-soluble antioxidants is not known. The outer surface of human skin, the stratum corneum, contains low amounts of ascorbate (see Tables 2.8, 2.9). In human plasma, the ascorbate concentration is 4–15 μg/ml and dehydroascorbate 0–1 μg/ml (Clemetson and Andersen 1966), indicating a high reducing potential in blood. In a separate study, the ratio of dehydroascorbate to total ascorbate showed great individual variation ranging from 0.4 to 1.00 (Sasaki et al. 1983). The ratio of reduced to oxidized ascorbate is about 15–66 μg reduced/41 μg oxidized ascorbate per 100 g sweat (Pascher and Röckl 1960). The ratio of ascorbate to dehydroascorbate in rats varies considerably between 12–14 (adrenal, brain, liver) and 1–4 (heart, liver, kidney, plasma) (Behrens and Madere 1987).

The ascorbate content of different human (Levine and Hartzell 1987) and rat organs (Salomon and Stubbs 1961; Behrens and Madere 1987) has also been analyzed (see Table 2.9). Values are expressed in mg/100 g tissue wet weight.

Prooxidant Effects. Ascorbate is a water-soluble antioxidant whose antioxidant efficiency is strongly concentration dependent (Wayner et al. 1986). In the presence of metal ions, such as ferrous and copper ions, ascorbate is a prooxidant. The catalytic effect of these metals on the rate of lipid peroxidation is strongly enhanced by ascorbate. Ascorbate may act as a prooxidant by converting iron(III) into iron(II) or copper(II) into copper(I). Ascorbate releases iron(II) bound as iron(III) in ferritin and hemosiderin. In addition, the prooxidant effects of ascorbate could be related to its ability to promote the formation of a proposed Fe(II):Fe(III) complex.

Although ascorbate is a polar molecule and not membrane permeable at physiological pH, it scavenges propagating lipid peroxyl radicals in biological membranes by intercepting the reactive species at the membrane interface. At high ascorbate and consequently high ascorbyl radical concentrations, prooxidant effects may occur.

Ascorbate and the ascorbyl radical can autooxidize in the presence of

$$A^{\cdot -} + O_2 \xrightarrow{Me^{n+}} A + O_2^{\cdot -}$$

$$AH^- + O_2^{\cdot -} \xrightarrow{H^+} A^{\cdot -} + H_2O_2$$

$$AH^- + HO_2^{\cdot -} \longrightarrow A^{\cdot -} + H_2O_2$$

Fig. 2.35. Prooxidant effects of ascorbate. *AH*, ascorbate; *A*, dehydroascorbate; Me^{n+}, transition metal ion; $A^{-\cdot}$, ascorbyl radical

redox active transition metal ions to yield superoxide anion radicals and hydrogen peroxide. The direct formation of reactive oxygen species through the metal-catalyzed autoxidation of ascorbate or ascorbyl radical is, however, a reaction unlikely to contribute significantly to the prooxidant effects of ascorbate (Miller and Aust 1989) (Fig. 2.35).

The biological significance of the prooxidant ascorbate effects in vivo is not known. It was demonstrated that in guinea pigs deficient in ascorbate, lipid peroxidation is greatly increased and normalized after substitution with the vitamin. This indicates that in vivo ascorbate acts as an antioxidant at physiological concentrations, although its prooxidant properties have been clearly established in vitro.

3. Urate

Uric acid is a potent antioxidant in human blood (Wayner et al. 1987). Urate scavenges peroxyl and alkoxyl radicals in the aqueous phase more efficiently than ascorbate. It reacts with hypochlorous acid and ozone. Urate is much more easily oxidized than deoxynucleosides by singlet oxygen and reacts with hydroxyl radicals at a comparable rate in comparison with nucleosides (Ames et al. 1981). Free radical attack upon uric acid generates allantoin (Fig. 2.36), which is considered a potential indicator of free radical reactions in vivo (Grootveld and Halliwell 1987). Few data are available on skin urate concentration. In humans, it is 16.9 ng/mg tissue (Aoki and Iriyama 1986), which is about one-third of the serum concentration.

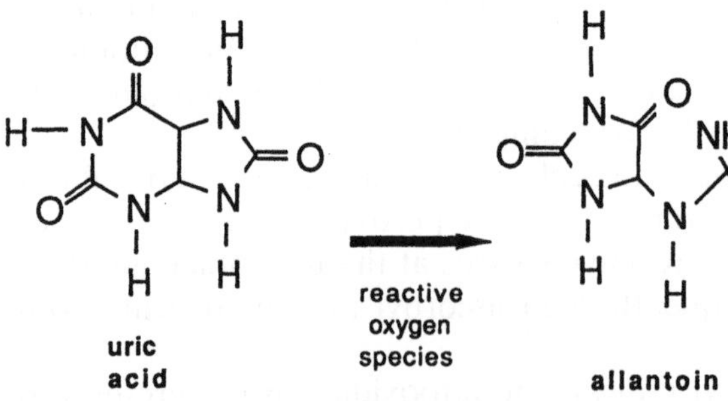

Fig. 2.36. Production of allantion from uric acid

X. Lipophilic Antioxidants

1. Tocopherol

Vitamin E is an essential nutrient that is receiving growing attention because of its antioxidant function in human biological systems. In 1922, an unknown substance was discovered in lettuce by Evans and Bishop, which prevented fetal death in animals fed a rancid diet. Subsequently, in 1924, Sure designated this unknown compound a fertility vitamin. The antioxidant properties of the new vitamin were first studied by Olcott and Emerson. In 1936, Evans isolated the fertility vitamin from wheat germ oil. Fernholtz revealed the chemical structure of vitamin E in 1938, and shortly thereafter it was synthesized by Karrer.

Vitamin E is a generic term that includes all entities that exhibit the biological activity of natural vitamin E, *d-α*-tocopherol. Biological activity is assessed by studying inhibition of reactive oxidant-mediated erythrocyte hemolysis in vitro, or by analyzing the pregnancy-protecting effect (prevention of fetal reabsorption in the rat) in animals deficient of vitamin E. Eight natural substances have vitamin E activity: *d-α*-, d-*β*-, d-*γ*-, and d-*δ*-tocopherol and *d-α*-, d-*β*-, d-*γ*-, and d-*δ*-tocotrienol. The formulas of four tocopherol homologues are shown in Fig. 2.37.

Synthetic tocopherols also have vitamin E activity, as do the succinate and acetate derivatives of tocopherols. In the rat resorption gestation test, the natural *d-α*-tocopherol (*R,R,R-α*-tocopherol) has the highest biopotency (1.49 IU/mg = 100%), while the *l*-enantiomer of *α*-tocopherol (*S,R,R-α*-tocopherol) has only 31% bioactivity (0.46 IU/mg) (Weiser and Vecchi 1982). Synthetic vitamin E (*d,l-α*-tocopherol), a mixture of eight stereoisomers, has a biopotency of 74%.

Antioxidant Effects. Vitamin E is among the oldest recognized biological antioxidants and its redox and free radical chemistry are well documented (Bur-

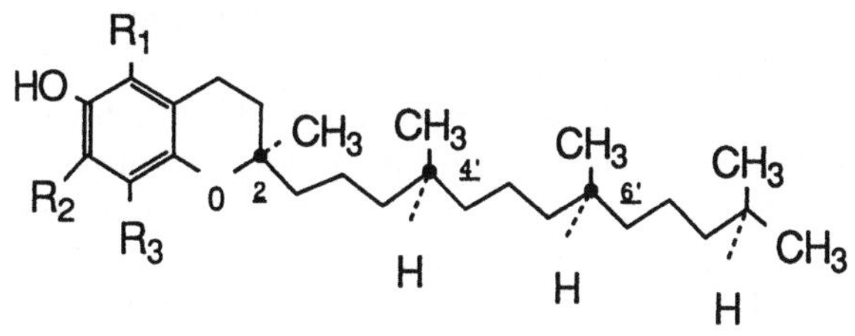

tocopherol	R_1	R_2	R_3
α	CH_3	CH_3	CH_3
β	CH_3	H	CH_3
γ	H	CH_3	CH_3
δ	H	H	CH_3

Fig. 2.37. Tocopherol homologues

ton et al. 1983; Burton and Ingold 1986). Direct and indirect evidence that α-tocopherol is a powerful antioxidant in subcellular structures, in isolated cells and organs, and in animals and in humans is convincing. Vitamin E is the major lipophilic antioxidant in tissues (Bieri and Anderson 1960). It is a primary and secondary antioxidant. It reacts directly with reactive oxygen species in biological membranes and inhibits already initiated lipid peroxidation (radical chain reaction) by capturing the propagating species (peroxyl radical). The superoxide anion radical readily oxidizes tocopherol (Yagi et al. 1978; Nishikimi et al. 1982). Tocopherol quenches triplet carbonyls and reacts with singlet oxygen, either in a reaction without formation of products or in an oxidation reaction (Foote et al. 1978; McCay et al. 1978). Highly toxic free radical species of xenobiotics such as trichloromethyl and trichloroperoxymethyl radical are also scavenged and detoxified rapidly by tocopherol (Simic 1981). Furthermore, tocopherols possess iron ion binding activity, thereby limiting the amount of biologically active transition metal ions in biomembranes (Stoyanovsky et al. 1989).

The initial oxidation product of tocopherol is the metastable tocopheroxyl radical, which can be further oxidized to tocopherolquinone. Other oxidation products include epoxides as well as oligomers. The tocopheroxyl radical is central to the antioxidant potency. It can be reduced under physiological conditions to tocopherol, thus regenerating the antioxidant. Since the physiological molar ratio of tocopherol to polyunsaturated phospholipid is less than about 1 : 1000 in most biological membranes, regeneration of tocopherol is essential for its high antioxidant efficiency in vivo. Small physiological molecules such as ascorbate, glutathione, dihydrolipoate, and ubiquinoles contribute to tocopherol regeneration. The synergistic effects of ascorbate in vitro (Golumbic and Mattill 1941; Packer et al. 1979; Scarpa et al. 1984; Niki et al. 1985; Doba et al. 1985) and in vivo (Kunert and Tappel 1983; McCay 1985) are well documented. It has also been suggested that tocopherol is regenerated in biological membranes by dihydrolipoate (Fuchs, unpublished results; Scholich et al. 1989). In guinea pigs, lipoate alleviates the symptoms of tocopherol deficiency (Rosenberg and Culik 1959). In vitro studies on the interaction of tocopherol with small molecular hydrophilic antioxidants, such as glutathione and uric acid, indicate that these antioxidants are apparently of no importance in non-enzymatic tocopherol regeneration (Ingold 1986). This conclusion is in contrast to previous hypotheses (Niki et al. 1982; Tsuchiya et al. 1985). Because of the long half-life of the tocopheroxyl radical (Rousseau-Richard et al. 1988), enzymatic regenerative processes seem to be plausible. Enzymatic regeneration may occur via microsomal and mitochondrial factors (Reddy et al. 1982; Haenen and Bast 1983; Hill and Burk 1984; McCay et al. 1986; Fuchs and Packer 1987; Packer et al. 1989; Maguire et al. 1989). The exact understanding of tocopherol recycling is of great interest (Fig. 2.38). Dysfunction of the regeneration process may cause tocopherol deficiency in cellular microcompartments and result in a prooxidative environment.

Under certain conditions in vitro, tocopherol may also exert prooxidative effects, and these have also been described for other antioxidants such

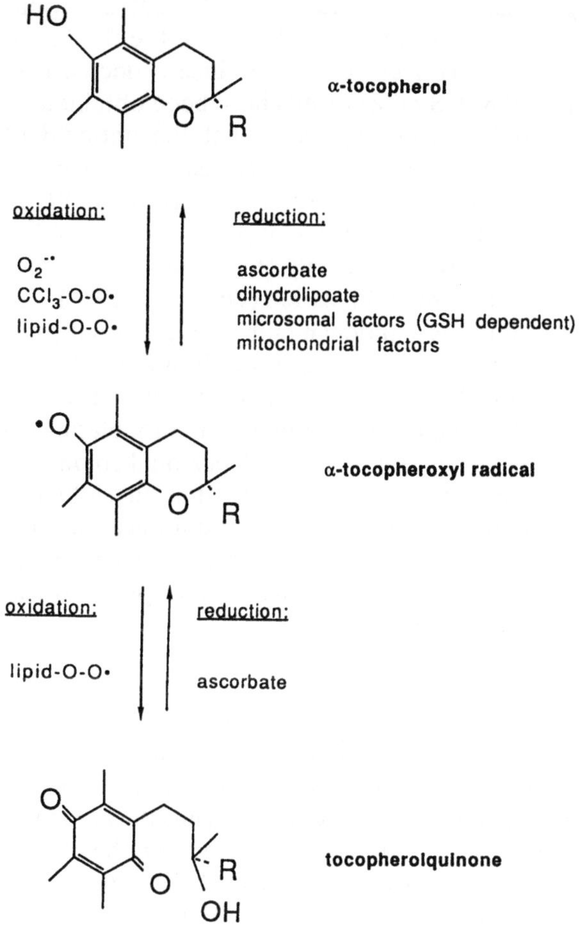

Fig. 2.38. Recycling of tocopherol

as ascorbate and dihydrolipoate. High concentrations of α-tocopherol ($\gg$0.005 mol tocopherol/mol unsaturated fatty acid) accelerate lipid autooxidation in vitro (Cillard et al. 1980; Husain et al. 1987a). Other authors have also reported prooxidant effects of α-tocopherol in vitro (Khrapova 1977; Hicks and Gebicki 1981). Although it seems reasonable to speculate that high tocopherol concentrations may also cause prooxidative effects in vivo, no animal experiments or clinical reports support this assumption.

Membrane Modulating Effects. The effect of α-tocopherol on membrane fluidity and permeability is quite complex and depends upon the composition of the phospholipid and the molar tocopherol/phospholipid ratio (Lai et al. 1985). α-Tocopherol decreases the membrane fluidity of phosphatidylcholine liposomes (Urano et al. 1988) and reduces the permeability of the membrane to uncharged molecules such as glucose (Stillwell and Bryant 1983). In high concentrations, however, tocopherol increases membrane fluidity of human erythrocytes and thrombocytes (Ernst and Matrai 1985). By contrast, it has been reported that low concentrations of tocopherol increase the fluidity of mitochondrial membranes and high concentrations produce a decrease in lipid movement (Kerimov et al. 1987). Similar results have been obtained by Zimmer and Thürich (1991). It has been suggested that the antioxidant properties of tocopherol in biomembranes depend on its free radical-capturing activity as well as on membrane stabilizing effects (Gutteridge 1978). However, the inhibition of microsomal lipid peroxidation by tocopherol has been suggested to result from the antioxidant activity of the compound rather than from an indirect effect via membrane stabilization (Ohki et al. 1984). Tocopherol-induced membrane alterations in model membranes are usually observed only at very high unphysiological concentrations. Physiological tocopherol concentrations are thought likely to influence membrane structures only marginally, so the biological relevance of the membrane-modulating effects are rather questionable (Massey et al. 1982; Ingold, personal communication). The recent studies of Zimmer, however, suggest that physiological concentrations of tocopherol significantly influence the biophysical properties of membranes (Zimmer and Thürich 1991). However, the mechanism by which one molecule of tocopherol per 1000 molecules of phospholipid can modulate membrane structure is not yet understood.

Distribution, Tissue Uptake, and Turnover. The total tocopherol tissue pool in an healthy adult varies between 4 and 8 g. Tissue concentration is highest in the adrenal gland and fat tissue. The adrenal gland is highly vulnerable to oxidative injury due to its high content of oxygenases, and fat tissue is readily susceptible to oxidative attack due to its high content of polyunsaturated fatty acids. In humans, mobilizitation of tocopherol occurs readily from the plasma and liver pools, to an intermediate degree from skeletal and heart muscle, and only little from fat tissue. In the rat, about 90% of the total body tocopherol is distributed in the liver (425 nmol/organ), skeletal muscle (622 nmol/organ), and fat tissue (319 nmol/organ) (Bjorneboe et al. 1986). Serum, skin, and skeletal muscle have the lowest tocopherol concentrations (Table 2.10).

Table 2.10. Concentrations of α-tocopherol in rat tissues (from Vatassery et al. 1988; Fuchs, unpublished observations)

Organ	α-Tocopherol (nmol(g tissue wet weight)
Suprarenal glands	169
Liver	35
Brain	33
Lung	32
Kidney	30
Heart	20
Fat	10
Serum	8
Total skin	7.5
Skeletal muscle	5

Oral administration of vitamin E (300 mg/day) for 10 days does not result in a increase in tocopherol in human skin lipids (Stüttgen et al. 1957). Feeding rats with extremely high doses of *d*-α-tocopherol acetate (10000 IU/kg diet; a 200-g rat consumes about 20 g food in 24 h = 2500 IU/kg body weight) increases the skin tocopherol content from a basal level of 5 nmol/g skin (wet weight), to 40 nmol/g after 10 days, 60 nmol/g after 20 days, and a plateau and maximum of 80 nmol/g after 30 days (Packer 1991). Oral administration of 300 or 900 mg *d*-α-tocopherol/day to humans increases heart tissue concentration after 2 weeks from 60 nmol/g (wet weight) to 120 nmol/g. No significant differences in tissue concentration were found between the groups receiving 300 or 900 mg tocopherol (Ingold 1991). Remarkably higher tissue levels (liver, kidney, heart, lung, fatty tissue) are achieved in sheep after intravenous administration of *d*-α-tocopherol than after intramuscular or oral administration (Hidiroglou 1986). Shiratori (1974) reported that intravenous administration of radioactive α-tocopherol in rats resulted in a 10% increase 2 h after injection, and 20 days after the injection, 40% of the dose administered had accumulated in the in skin. Preferential uptake of the radioactivity by the skin and subsequent transepidermal elimination was noted (Shiratori 1974).

Autoradiographic investigations in human skin revealed that α-tocopherol acetate permeates rapidly into epidermis after topical application. Two permeation pathways have been characterized: (1) via stratum corneum and (2) via hair follicles. The apocrine and eccrine glands probably do not play a significant role in permeation. A high affinity of tocopherol acetate for the small dermal vessels and the septa in the subcutis has been observed (Kamimura and Matsuzawa 1968). Quantitative penetration studies in rat skin showed that 7% of the topically applied dose (5% tocopherol solution in ethanol) was detected in skin layers below the stratum corneum, which is the penetration barrier, after 1 h and 39% after 6 h (Djerassi et al. 1986). The vehicle significantly influences skin penetration of *d*-α-tocopherol; penetration is better from an oil/water microemulsion than from a water/oil emulsion (Martini et al. 1984). Penetration of topical α-tocopherol into epidermis was also documented by Ohzawa et al. (1984). An increase in serum α-tocopherol following topical application was not observed (Stüttgen et al. 1957). However, Martini et al. (1984) reported an increase in tissue tocopherol concentration (liver, kidney, muscle, and fat tissue) starting 6 h after topical application of radioactively labeled tocopherol (Martini et al. 1984). There are only few data on tocopherol concentration in skin of humans and rodents (Tables 2.10, 2.11).

Table 2.11. Tocopherol activity in the skin

Species	Tissue	Activity	Reference
Human	Epidermis	180–675 ng/mg protein	De Simone (1987)
Mouse	Whole skin	200 pmol/mg protein	Fuchs et al. (1989 a, c)
Mouse	Whole skin	5 nmol/g (wet weight)	Packer (1991)

Natural *d*-*α*-tocopherol is better retained in the tissues than synthetic *dl*-*α*-tocopherol (Hung et al. 1982; Cheng et al. 1987; Ingold et al. 1987). Ingold investigated how quickly tocopherol gets into different tissues in the absence of any metabolic stress. The biological half-life of *α*-tocopherol was determined in different tissues in the rat after oral feeding with deuterated *d*-*α*-tocopherol acetate. The physiological tocopherol turnover is dependent on the type of tissue. The tocopherol half life in rat skin (23.4 days) is between the minimal value, (lung, 7.6 days), and the maximal value (spinal cord, 76.3 days). The half-life in plasma is 10.9 days and in erythrocytes 12.5 days (Ingold et al. 1987). Extrapolating to humans from these results, one may assume that changes in skin tocopherol content are to be expected only several weeks after oral tocopherol therapy (assuming a physiological tocopherol turnover). *d*-*α*-Tocopherol acetate has a similar tissue uptake rate in humans to *d*-*α*-tocopherol (Ingold 1991).

Metabolism. Tocopherol is subject to oxidative metabolism; however, it is uncertain whether certain tissue metabolites (tocopherol quinone, tocopherol oligiomers) represent isolation artifacts. In the skin of hairless mice, bioconversion of the tocopherol ester *dl*-*α*-tocopherol-*l*-ascorbate phosphodiester to *α*-tocopherol and ascorbate was reported (Tojo and Lee 1987). In human skin some nonspecific ali-esterase activity is found, which probably induces activation of the prodrug. The pharmacodynamic activity of *α*-tocopherol esters in the skin depends on this enzyme-induced bioactivation.

2. Vitamin A and Carotenoids

Carotene was isolated by Euler in 1928 and show to have vitamin A activity. Carotenoid pigments, including hydrocarbons such as *β*-carotene, lycopene, cryptoxanthine, and lutein, are important lipophilic antioxidants present in human tissues. *β*-Carotene is the precursor of vitamin A (retinol; Fig. 2.39). Although one molecule of *β*-carotene theoretically yields two molecules of vitamin A, in humans only about 10% of *β*-carotene is converted into vitamin A. The daily requirement of *β*-carotene is 6 mg, and of all carotenoids about 15 mg.

In comparison with *α*-tocopherol, the *β*-carotene membrane concentration is severalfold lower. However, *β*-carotene significantly accumulates in skin and may achieve concentrations far exceeding those of *α*-tocopherol in patients on a *β*-carotene-supplemented diet. *β*-Carotene prevents lipid peroxidation in model membranes (Caprini et al. 1986) and directly scavenges reactive oxygen species, e.g., lipid peroxyl radicals and singlet oxygen. The carotenoids lycopene and *β*-carotene are better singlet oxygen quenchers than *α*-tocopherol and lipoate, which have similar quenching capabilities. Since tocopherols occur at about a 10–100 fold higher plasma concentration in humans, their overall singlet oxygen quenching capacity is comparable to that of carotenoids (Sies and Murphy 1991). Through reaction with a peroxyl radical,

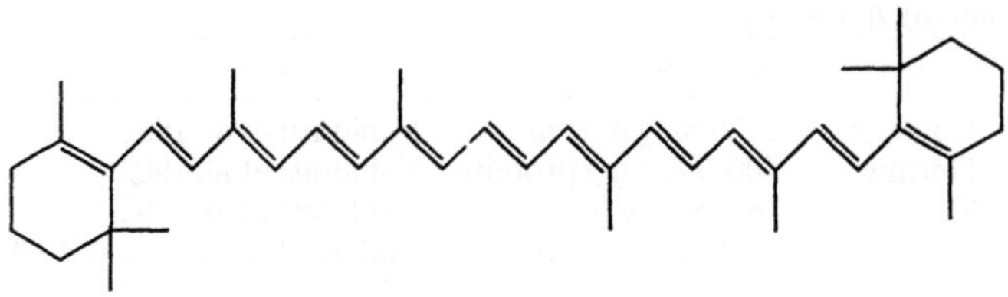

ß-carotene

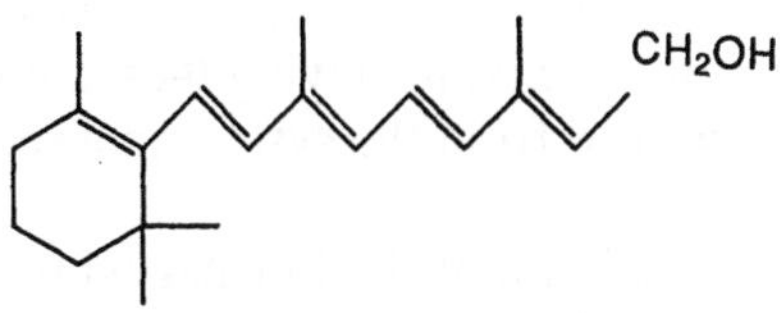

vitamin A

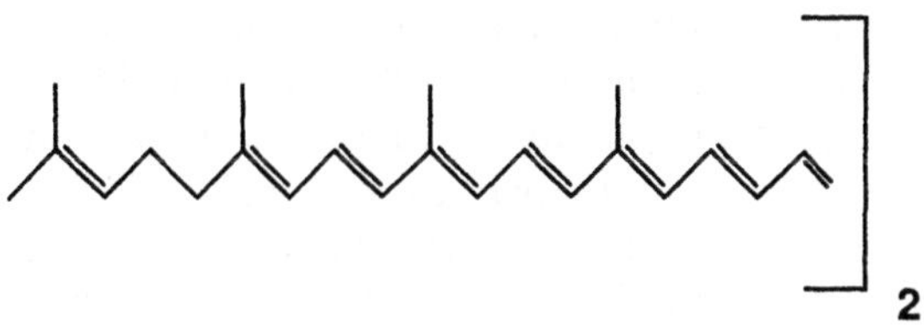

lycopene

Fig. 2.39. Carotenoids and vitamin A

ß-carotene + ROO• ⟶ ß-car• + ROOH

<u>carotene radical</u>

ß-car• + O_2 ⇌ ß-car-O-O •

<u>carotene peroxyl radical</u>

ß-car-OO• + lipid-H ⟶ ß-car-OOH + lipid•

<u>carotene hydroperoxide</u>

ß-car• + ROO• ⟶ nonreactive products

Fig. 2.40. Carotene radical

Table 2.12. *β*-Carotene activity in the skin

Antioxidant	Species	Tissue	Activity	Reference
β-Carotene	Human	Epidermis	2220 ng/g protein	Vahlquist et al. 1982
		Dermis	700–775 ng/g protein	Vahlquist et al. 1982
		Subcutis	18885 ng/g protein	Vahlquist et al. 1982
		Total skin	0.23 μg/g protein	Culling-Berglund et al. 1989

the resulting resonance-stabilized *β*-carotene radical can readily undergo autoxidation to yield the chain-carrying *β*-carotene-peroxyl radical. At low oxygen concentrations, however, formation of the chain-carrying *β*-carotene-peroxyl radical is not favored and can be removed from the system in a reaction with another peroxyl radical (Fig. 2.40).

β-Carotene is a good radical-trapping antioxidant at the low partial pressures of oxygen which are found in most tissues under physiological conditions. At higher oxygen pressures, it shows rather a prooxidant effect in vitro (Burton and Ingold 1984). At low partial pressures of oxygen, it has the potential to act as a chain-breaking antioxidant. It complements the role of tocopherol at high oxygen pressures. A prooxidant activity of *β*-carotene in vitro was described in the 1930s (Monaghan and Schmitt 1932; Olcovich and Mattill 1930). Several other reports have raised the possibility that carotenoids might enhance free radical reactions (Terao et al. 1980; Faria and Mukai 1983). To our knowledge, there are no reports of prooxidant in vivo effects of *β*-carotene.

A large pool of carotenoids is found in fat tissue and the supra-adrenal gland. In autopsy samples the total carotene content varies between 8–98 μg/g fat tissue (Dagadu 1967) and 5–6 μg/g fat tissue (Pierce 1954). *β*-carotene is also detected in human oral mucosal cells and is reported to increase several-fold in the mucosa cells of individuals whose diet is supplemented with *β*-carotene (Stich et al. 1986).

The concentration of *β*-carotene in different skin layers (epidermis, dermis, subcutis; Table 2.12) is always higher than that of vitamin A (retinol; Table 2.13). Skin surface lipids contain both carotene and retinol but not dehydroretinol, a retinol metabolite present in small quantities (about 1/10 of the retinol concentration) in the epidermis and dermis (Vahlquist et al. 1982).

Table 2.13. Vitamin A activity in the skin

Antioxidant	Species	Tissue	Activity	Reference
Vitamin A	Human	Epidermis	1.3 ng/mg protein	Tormä and Valquist 1984
		Total skin	0.9 ng/mg protein	Tormä and Valquist 1984
		Epidermis	294 ng/g protein	Vahlquist et al. 1982
		Dermis	186–393 ng/g protein	Vahlquist et al. 1982
		Subcutis	6350 ng/g protein	Vahlquist et al. 1982
	Mouse	Total skin	4 μg/100 g wet weight	Mathews-Roth et al. 1977

3. Ubiquinols/Ubiquinones

One of the major functions of ubiquinones and ubiquinols is to transmit protons and electrons across membranes during a redox cycle (Nohl et al. 1986). Ubiquinones/ols have a lipophilic sidechain and a redox active quinone/hydroquinone head group (Fig. 2.41). As electron carriers ubiquinones/-ols also have antioxidant properties. Although ubiquinone cannot prevent autocatalytic free radical reactions by donating a phenolic hydrogen atom (unlike ubiquinol and tocopherol), it scavenges singlet oxygen and inhibits lipid peroxidation in model membranes (Caprini et al. 1986). The reduced form of ubiquinone protects biological membranes against lipid peroxidation by directly scavenging the initiating species. Mitochondrial lipid peroxidation is prevented by electron-transport-linked reduction of ubiquinone to ubiquinol (Mellors and Tappel 1966; Takayanagi et al. 1980; Bindoli et al. 1982). Ubiquinone in the epidermis is bound to mitochondria as well as to nonmitochondrial cytoplasmic components (Brody 1968). Presumably, however, ubiquinones play a main metabolic role only in mitochondria. The ubiquinol/one content of a large number of tissues of different animals have been investigated and compiled by Ramasarma (1985), but no data are given on the skin content in any species (see Table 2.14 for known data). Ubiquinone 10 is usually undetectable in rat skin, while ubiquinone 9 is found in skin (Giovannini et al. 1988).

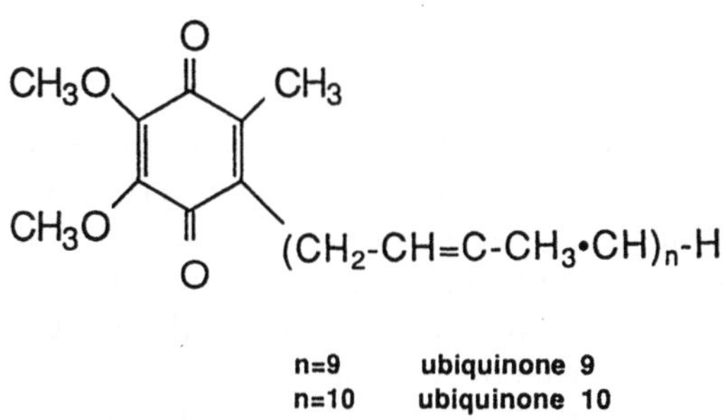

Fig. 2.41. Ubiquinones

Table 2.14. Ubiquinone/ol activity in the skin and brown adipose tissue

Antioxidant	Species	Tissue	Activity	Reference
Ubiquinol 9	Mouse	Total skin	20–48 pmol/mg protein	Fuchs et al. 1989 a, c
Ubiquinone 9	Mouse	Total skin	98–136 pmol/mg protein	Fuchs et al. 1989 a, c
	Hamster	Brown adipose tissue	0.54 nmol/mg mitochondrial protein	Battino et al. 1990
Ubiquinone 10	Human	Epidermis	212–637 ng/mg protein	De Simone 1987
	Human	Total skin	20–115 μg/mg protein	Oradei et al. 1990
	Hamster	Brown adipose tissue	0.21 nmol/mg mitochondrial protein	Battino et al. 1990

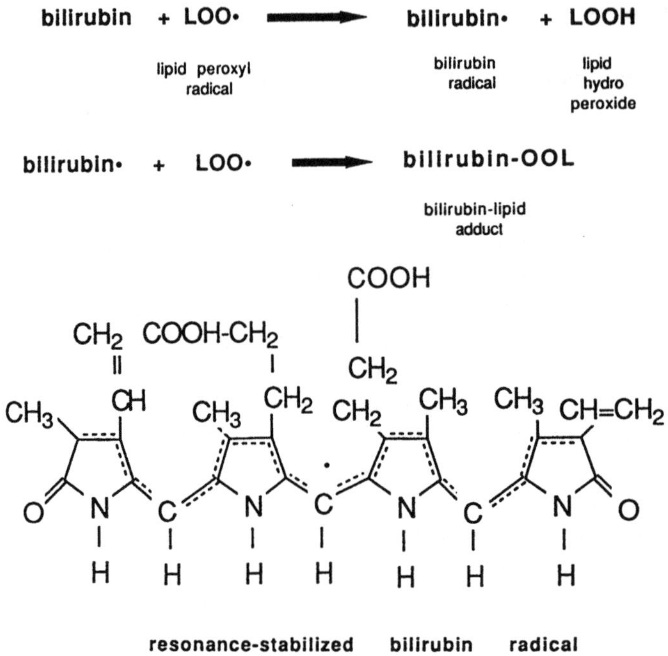

Fig. 2.42. Bilirubin radical

4. Bilirubin

The lipid-soluble bilirubin, the end product of heme metabolism in mammals, has antioxidant activity in vitro. It scavenges lipid peroxyl radicals (LOO·), whereby the resonance-stabilized bilirubin radical is formed. Bilirubin radicals and lipid-peroxyl radicals can also form adducts by radical-radical addition reactions (Fig. 2.42). Its antioxidant potential increases as the oxygen concentration is decreased from 20% (partial pressure in air) to 2% (partial pressure in tissues). At low oxygen concentrations, bilirubin is a more potent antioxidant than α-tocopherol (Stocker et al. 1987). Conjugated bilirubin inhibits the stimulated chemiluminescence of human neutrophils; it does not react readily with either superoxide anion radical or hydrogen peroxide but scavenges efficiently hypochloric acid, HOCl (Stocker and Peterhans 1989). It is not known whether bilirubin contributes to skin antioxidant capacity significantly under physiological conditions.

XI. Antioxidant Capacity of Skin

In conclusion, skin is equipped with prevention, interception, and repair mechanisms. Although it appears to be endowed with a variety of enzymic and nonenzymic protective systems, the effectiveness of skin antioxidant systems relative to other tissues has not been established. Several reports indicate a limited availability of enzymic and nonenzymic defense mechanisms in skin against oxygen toxicity. However, a strong antioxidant potential of skin was noted in comparison with liver. This conclusion was obtained from experiments studying the spontaneous autoxidation of skin and liver homgenates (Meffert and Lohrisch 1971 b).

The antioxidant potential of a tissue was determined by incubating its homogenate in air and measuring the spontaneous (Bieri and Anderson 1960) or hydrogen-peroxide-induced (McMurray and Dormandy 1974) lipid peroxidation. Autoxidation mechanisms in tissue homogenates are very complex (Wills 1966), e.g., anatomically and functionally distinct muscle groups have different susceptibilities to autoxidation, and no correlation between the total lipid content, α-tocopherol concentration, and malonic dialdehyde formation was found (McMurray and Dormandy 1974). Multiple factors can be significant determinants in skin homogenate autoxidation, and further experiments are warranted to evaluate the antioxidant capacity of skin in comparison with other organs.

It must be pointed out that no analytical assay exists for measuring the "total antioxidant capacity" of serum or organ homogenate. Different assays measure different "antioxidant activities." Thus, measurements of the superoxide radical-trapping activity or the singlet quenching capacity do not provide information about the peroxyl trapping potential. Similarly measuring the total peroxyl radical trapping activity of serum (Wayner et al. 1987) does not show the scavenging activity for singlet oxygen, superoxide, or hydrogen peroxide. In order to estimate accurately the "total antioxidant capacity," the activity of all the small molecular lipophilic (and hydrophilic) and enzymic antioxidants must be analyzed individually.

1. Regulation of the Skin Antioxidant Potential

Very little is known about the acute and chronic responses of enzymic and nonenzymic antioxidants in skin to various oxidant stress factors and the overall regulation of the skin antioxidant potential. Regulation of the antioxidant potential of human blood was suggested to be like an "acute phase reaction" after challenge with chemical, biological, physical, emotional, and immunological stimuli (Popov et al. 1988 a).

Environmental stress factors such as hot or cold environment, hyper- and hypooxygenation, UV, microbial invasion, strenuous exercise, and xenobiotic toxicity can induce antioxidant systems in pro- and eukaryotes. SOD (Loven 1988) and glutathione peroxidase (Kapoor and Sveenivasan 1988) can be induced by heat shock and may also be regulated by other oxidative stress factors. It was suggested that heat shock protein genes are regulated by free-radical-mediated reactions and that lipid peroxidation could be a crucial mechanism in heat shock protein gene activation (Burdon et al. 1987).

Distinct antioxidant systems respond differently to separate environmental stress factors. The zinc-sulfhydryl protein metallothionein is induced in mammals only by some environmental stress factors (Oh et al. 1978). The induction effect on liver metallothionein is much greater than on the kidney enzyme, indicating significant tissue differences.

The capacity for SOD induction following oxygen stress is different from species to species; the adaptive changes in mice or guinea pigs is weaker than in bacteria and humans (Kawai et al. 1988). Factor analysis (Harman 1976) revealed that regulation of Cu,Zn-SOD, catalase, and glutathione peroxidase activity in normal human tissues is influenced by a single factor. The activity of Mn-SOD is influenced by a separate factor. In neoplastic human cell lines two different factors appear to influence Cu,Zn-SOD, Mn-SOD, catalase, and glutathione peroxidase activities (Bannister and Bannister 1987). This indicates that antioxidant induction may be regulated differently in physiological and pathological states.

The nature of the regulating mechanism(s) is not known exactly. In general, exposure to increased oxygen concentration, metal ion perturbation, and exposure to certain environmental prooxidants influence the rate of SOD synthesis (Hassan 1988). Regulation of the SOD genes is considered to play an important role in antioxidant adaptation. Some studies indicate that skin SOD activity is initially reduced after acute oxidant injury but may be elevated after chronic exposure to oxidants. Antioxidant modulation on the gene level as well as posttranslational control may participate in the overall regulation of the antioxidant potential.

Heme oxygenase, a heme catabolic enzyme, is an inducible pathway of skin antioxidant defense. UV radiation, reactive oxidants, and agents that modify the cellular glutathione level stimulate heme oxygenase gene expression in cultured human skin cells. It has been suggested that heme oygenase plays a protective role by reducing cellular heme pools and thereby suppressing the Fenton reaction (Tyrrell et al. 1990).

Skin Heat Shock. In clinical dermatology, hyperthermia has been used in the treatment of certain hyperproliferative states (neoplasia, warts, psoriasis) and infections (Dover et al. 1989). The heat stress induces transcriptional induction of specific protein synthesis (heat-shock proteins), while normal protein synthesis is blocked. Different molecular weight classes of heat-shock proteins can be distinguished. They may play a general role in cellular protection not only from thermal injury but also from various other types of cellular injury, including oxidative injury. Heat-shock proteins are induced also by UV, reactive oxidants, ischemia reperfusion damage, inflammation, and wounding (Polla 1990).

The heat-shock response is observed in tissues and cells after exposure to an elevated temperature (2°–10°C above normal). The existence of a functional thermotolerance response in human keratinocytes was recently reported at temperatures that are readily found in viable human skin (Maytin et al. 1990). Heat shock can induce antioxidant systems in various cell lines

(Loven 1988; Kapoor and Sveenivasan 1988). However, hyperthermia treatment of mouse hind legs (43 °C for 45 min) resulted in a remarkably strong decrease in skin SOD activity 1 day after treatment. The maximum of depression was reached after 4 days (inhibition to about 15 % of the control activity); the enzyme activity remained severely depressed over several days and slowly recovered, still being significantly decreased 14 days after hyperthermia exposure (Miyachi et al. 1988).

Skin Aging. Conflicting results have been reported regarding the activity of antioxidants in the aging process. The content of SOD and glutathione peroxidase decreases with aging in a large variety of tissues (Fridovich 1978; Imre et al. 1984; Hazelton and Lang 1985; Cand and Verdetti 1989). However, few reports indicate that the activity of SOD in aged animals and humans remains normal. Different organs may be affected in various ways. During aging of mice, catalase activity increases in the heart but decreases in the skeletal muscle (Salminen et al. 1988), liver, kidney, and brain (Cand and Verdetti 1989).

Skin SOD activity in elderly animals and humans is generally comparable with levels observed in younger ones. However, the capacity for skin antioxidant induction following oxidative stress is limited in aged animals and humans. SOD activities in skin lesions (grade II) of burn patients are significantly different in aged and younger patients. Total skin SOD activity between younger healthy individuals and aged healthy individuals is not significantly different. Younger burn victims have total SOD activities of 22.9 U/mg protein, whereas aged ones show activities comparable with aged or younger healthy controls (11 U/mg protein). Similar results were obtained with guinea pigs. It was suggested that although the activity of SOD in elderly animals and humans is generally comparable with levels observed in younger ones, in the former the capacity for further induction is limited (Niwa et al. 1988).

In human lymphocytes, an age-dependent induction of SOD activity by paraquat, a superoxide anion radical generating compound, was shown. In older subjects (65–79 years) there was only a very small increase (8 %) in lymphocyte SOD activity, while that of young subjects (20–40 years) showed an increase of 85 %. In a follow-up study of the older patients it appeared that preservation of SOD inducibility is correlated with longevity and may be of value in predicting resistance to malignancies (Niwa et al. 1990).

Skin Inflammation. SOD activity is impaired during acute toxic contact dermatitis in guinea pigs. It is significantly depressed 1 day after irritation and normalizes after about 8 days (Miyachi et al. 1986 a). In humans, the severity of skin inflammation may in part be determined by the degree to which SOD activity is up-regulated in response to the generation of reactive oxidants (Niwa et al. 1987 a; Miyachi et al. 1985).

F. Biological Models for Studying Oxygen Toxicity

In skin, the role of oxygen and reactive oxygen species in cutaneous pathology is only poorly understood at the moment. Adequate model systems are needed to study oxygen toxicity in skin cells, epidermis, dermis, and intact skin. Model systems commonly employed to study in vivo pathology induced by reactive oxidants comprise exhaustive exercise training and hyperbaric oxygen (HBO) treatment. Exercise training is not a suitable method for studying oxidative injury in skin, because skin is not a target organ of this particular kind of oxidative stress. Model systems more relevant for studying oxidative injury in skin are HBO treatment and inflammatory and photooxidative stress. In these cases, however, besides reactive oxidants, other significant factors, e. g., leukocyte-derived proteases and photochemical reactions not involving oxygen, participate in the pathophysiology. This causes a dilemma and presently there is no in vivo model available which allows selective analysis of reactive oxidants in skin pathophysiology.

I. Exercise Training

A common in vivo model for studying oxidative injury in animals and humans is exercise training. Severe physical exercise which increases total body oxygen uptake by more than ten fold can be potentially harmful to tissues through the increased formation of reactive oxygen species and changes in the antioxidant status. Knowledge concerning the effects of exercise and endurance training on the production of reactive oxidants and modulation of tissue antioxidants is quite limited and the results sometimes rather conflicting. However, this experimental approach allows studies on humans and has a direct relationship to a physiological situation.

Animal Studies. Free radicals can be detected in normal skeletal muscle and are assigned to lipid peroxidation products as well as semiquinone type radicals of mitochondrial origin (Blokha et al. 1972; Koren et al. 1980). Free radical generation in skeletal muscle is probably a physiological event that is influenced by stimulation frequency, temperature, and mitochondrial activity (Solic 1979, 1977; Koren et al. 1983). Biochemical studies indicate that endurance exercise training in rats increases free radical production in skeletal muscle (Davies et al. 1982; Gohil et al. 1987).

Although endurance training increases oxygen metabolism in the heart muscle, no evidence for increased free radical formation was found in this tissue. In rat myocardium, endurance training by swimming causes a significant decrease of the activities of catalase, Cu,Zn-SOD, thioredoxin reductase, glutathione reductase, and tocopherol. However, the level of glutathione and glucose-6-phosphate dehydrogenase activity increased and of thiobarbituric acid reactive products decreased (Kihlström and Salminen 1988).

The antioxidant enzymes in skeletal muscle are capable of adapting to exercise to minimize oxidative injury caused by free radicals. Cytosolic and mitochondrial SOD activities in skeletal muscle were investigated in trained and untrained rats after an acute exercise bout. Training itself has no significant influence on SOD and catalase activities in these animals. However, selenium-dependent glutathione peroxidase activity in muscle mitochondria is doubled in trained animals, and selenium-independent mitochondrial glutathione peroxidase is also increased (Ji et al. 1988).

Human Studies. During prolonged submaximal exercise in humans, the blood glutathione level decreases, and that of glutathione disulfide increases, suggesting increased formation of reactive oxygen species. Tocopherol serum levels, however, do not change. In the recovery phase after submaximal physical exercise in humans, the glutathione concentration increases from exercise levels and overshoots preexercise levels, reaching maximal values after 3 days (Gohil et al. 1988).

II. Hyperbaric Oxygen Treatment

Hyperbaric oxygen treatment of skin cells and intact skin is a model system that produces an increased steady state flux of reactive oxygen species without contamination of other reactants which may be difficult to evaluate. Although (similar to exercise training) skin is not a target organ of oxygen toxicity, this approach allows a detailed analysis of events relevant for the understanding of oxidative skin injury in a well defined and characterized in vitro system.

In the eukaryotic cell, most oxygen reduction is carried out by mitochondrial cytochrome *c* oxidase, which reduces oxygen tetravalently to water and minimizes the release of univalent oygen intermediates. During hyperbaric oxygen treatment, univalent reduction of oxygen may occur in significant amounts and cause acute or chronic tissue damage. It is believed that hyperbaric oxygen toxicity is due to the formation of reactive oxygen species at a rate in excess of the capacity of the cellular antioxidant defense system to detoxify these reactants (Jamieson et al. 1986).

Plots of the logarithm of survival time against the logarithm of oxygen pressure show an inverse, approximately linear relationship for mammals. The lifespan of human diploid fibroblasts in culture is extended by growing them under low oxygen concentration (10%) (Packer and Fuehr 1977). Because oxygen toxicity was first reported in the lung and has significant health implications, hyperbaric oxygen treatment of animals serves as a common model system for studying pulmonary oxidative injury. Therefore, prooxidant and antioxidant mechanisms have been extensively analyzed in this organ.

If mammalian cell suspensions or subcellular organelles are exposed to 20% oxygen in vitro, their environment is hyperoxic compared with their physiological norm. The exception to this is lung tissue, in which cells are normally exposed to oxygen tensions of about 100 torr (13%). Other organs have oxygen tensions around 30 torr in normoxia (Jamieson and van den Brenk 1965). In hyperoxia, the intracellular oxygen metabolism is altered, and total oxygen consumption of lung cells in elevated (Freeman and Crapo 1981), resulting in an increased formation of reactive oxygen species.

Different stages in lethal pulmonary oxygen toxicity can be distinguished (Crapo 1986). The initiation phase is characterized by an increased production of reactive oxygen species and is succeeded by an inflammatory and a destructive phase. The earliest morphological alterations occur at the lung capillary endothelium, followed by platelet accumulation and subsequently by the rapid invasion of neutrophils. Neutrophils start the final stage of lethal pulmonary oxygen toxicity by releasing further reactive oxygen species and inflammatory mediators (Crapo 1986). The production rate of reactive oxygen species in lung microsomes and mitochondria (Turrens et al. 1982a, b) as well as nuclear membranes (Yusa et al. 1984) increases in hyperoxia. Heart, liver, and lung mitochondria show a linear increase in both superoxide anion radical and hydrogen peroxide production with increasing oxygen concentration. Similar studies using skin organelles have not yet been performed.

Adaptive Response to Tissue Hyperoxia. Tissue tolerance to oxidative stress and induction of antioxidants has been extensively investigated in the rodent lung. Adaptive pulmonary responses to hyperbaric oxygen, ozone, nitrogen dioxide, and paraquat were studied.

Continuous exposure of mice to high levels of ozone is more injurious than intermittent exposures to the same levels (Stokinger et al. 1956; Stokinger 1965). Rodents made tolerant to ozone develop a cross-tolerance to other oxidant gases. This is manifested biochemically by an increase in the activities of protective antioxidant enzymes in lung tissues (Chow 1982). Nonenzymic antioxidants in lung tissue, such as glutathione, are also induced by ozone (Forman et al. 1983).

Adaptation to oxidant stressors such as ozone has the obvious short-term advantage of survival. However, in animals made tolerant to ozone, premature development of degenerative diseases was found. It was suggested that the limit in the maintenance of tolerance by one organ system to oxidants is ultimately related to the total pool of antioxidants available in the body (Levine and Kidd 1985b). If the oxidative stress persists, the endogenous reserves may be exhausted, and organ pathology is the consequence.

Tolerance to hyperoxia in mammals can be achieved by pretreatment with sublethal hyperoxic oxygen concentrations, and this correlates with increased SOD (Kimball et al. 1976; Crapo and Tiernly 1974, Stevens and Autor 1980) and catalase activity (Crapo et al. 1978; Stevens and Autor 1980) in the lung. Elevated SOD activities were found under hyperoxic conditions in neutrophils (Rister and Baehner 1976), alveolar macrophages (Roos et al. 1980), and pneumocytes (Forman and Fischer 1981). Catalase activity was increased under these conditions in alveolar macrophages and pneumocytes (Roos et al. 1980; Forman and Fischer 1981). In bacteria, SOD can be induced by exposure to oxygen (Gregory and Fridovich 1973, Gregory et al 1973). Inhibition of SOD by disulfiram greatly increases the susceptibility to hyperoxic-mediated lung injury in rats (Forman et al. 1980).

Hyperoxygenation and Skin Metabolism. The stratum corneum has a major role as a diffusion barrier for oxygen in murine (Evans et al. 1981) and human epidermis (Baumgardner et al. 1985), so that an oxygen tension of 3% in the stratum basale and 7% in the upper layers is maintained (Silver 1972; Evans et al. 1981). These conditions seem to be optimal for keratinocyte function. The physiological oxygen gradient in skin, which may be accompanied by a counterbalancing antioxidant gradient, has important regulatory functions in the epidermis. Hyperoxygenation of keratinocyte cultures has significant effects on epidermal metabolism such as DNA and prostanoid synthesis and cell respiration.

Keratinocyte Proliferation. Best plating efficiency of keratinocytes is observed at 2% oxygen tension. Exposure of keratinocytes to oxygen tensions >7% results in keratinocyte proliferation (Horikoshi et al. 1986; Taylor and Camalier 1982), with maximal growth occurring at 18% oxygen. Oxygen tension above 20% inhibits the growth of keratinocytes in culture and diminishes the rate of DNA synthesis in mouse keratinocyte cultures by an undefined mechanism (Horikoshi et al. 1986). Low oxygen concentration is also favorable for both melanocyte proliferation and tyrosinase activity (Horikoshi et al. 1991). In keratinocyte cultures, the rate of DNA synthesis increases when the gas phase partial pressure of oxygen is changed from 7.5% to 21%. Increasing the carbon dioxide partial pressure from 5% to 10% stimulates prostaglandin synthesis; this effect is greatest at high oxygen tension (Pentland et al. 1986).

Keratinocyte Respiration. Oxygen uptake in the epidermis increases in several species with a moderate increase in the blood oxygen tension. Elevating oxygen tension increases the respiratory rate in guinea pig skin, maximal respiration occurring at 75% oxygen. A pronounced fall in epidermal respiration occurs after exposure to 1.5 atm of oxygen for 24 h. After this time, histological examination shows extensive abnormality (Cruickshank and Trotter 1956).

Inhibition of cutaneous respiration in vivo is observed in humans when exposed for 10–20 min to pure oxygen (Brinkman effect; Brinkman and Lamberts 1958). This is also observed with traces of ozone and UVA irradiation. Air may contain ozone concentrations ranging from 0.1 to 1.0 ppm. A concentration of about 1.0 ppm ozone respired ober 10 min is sufficient to decrease human epidermal oxygen consumption. The antioxidant cysteamine counteracts the inhibitory effects of ozone in skin (Brinkman and Lamberts 1958). The Brinkman effect of oxygen could not be reproduced by another group (Findlay 1964).

[illegible] under these conditions an aerobic macrophage and microbe [illegible] 1986; Forthum and Fischer 1991). [illegible] can be [illegible]

[illegible]

human epidermal oxygen consumption. The antioxidant cysteamine counteracts the inhibitory effects of ozone in situ (Brinkman and Lambers 1988). The Brinkman effect of oxygen could not be reproduced by another group (Findlay 1984).

Chapter 3

Reactive Oxidants and Antioxidants in Skin Pathophysiology

Generation of reactive oxidants in skin is a consequence of various pathophysiological events. In general, skin oxidative injury may occur under identical conditions which are of significant pathophysiological importance for other organ systems. Examples of such conditions are ischemia reperfusion injury of the brain, heart, or gastrointestinal tract, oxidative drug toxicity in the liver, hyperoxygenation damage in the eye, lung, and brain, and photooxidative injury of the eye. Due to its organ characteristics, skin oxidative pathophysiology may be subject to organ-specific reaction patterns.

A. Electromagnetic Radiation

I. Ionizing Radiation

1. Formation of Reactive Species

Biological radiation damage can be induced directly, e.g., by ionization of nucleic acids and other macromolecules and by an indirect action involving chemically highly reactive molecules such as electrophiles or free radicals (Rajewsky et al. 1959). Ionizing radiation can induce cell pathology by the ejection of electrons from molecules in tissues. The ejected electrons interact with other molecules and cause ionization and excitation. Much of the absorbed radiation energy produces excited molecules rather than ionization. In biological systems, the effect of ionizing radiation differs qualitatively and quantitatively with the radiation dose and radiation quality. At small energy doses indirect effects predominate; at large doses both indirect and direct effects are observed. Ionizing radiation produces cation radicals and electrons as primary species.

Irradiation of water with X-rays, gamma-rays, or alpha- or beta-particles results in the formation of the cation radical H_2O^+ and one electron as primary species (Fig. 3.1). These species then react with water and each other to produce various reactive intermediates, such as solvated electrons, hydrogen peroxide, hydroxyl radicals, and hydrogen atoms. Since water is the most abundant chemical substance in biological tissue, it is an important endogen-

$$H_2O \longrightarrow H_2O^+ + e^-$$

$$H_2O^+ \longrightarrow H^+ + OH^{\cdot}$$

$$e^- + H_2O \longrightarrow H^{\cdot} + OH^-$$

in presence of oxygen

$$O_2 + H\cdot \longrightarrow HO_2^{\cdot}$$

$$HO_2^{\cdot} + H\cdot \longrightarrow H_2O_2$$

Fig. 3.1. Actions of ionizing radiation on water

ous source of radiation-induced free radical formation by radiolysis of the hydrogen-oxygen bond. In the presence of dissolved oxygen, hydroperoxyl radicals and hydrogen peroxide are formed in biological tissues (Fig. 3.1).

Hydrogen atoms and electrons are reducing species. Hydrogen atoms readily combine with oxygen to form hydroperoxyl radicals. The oxidizing species hydroxyl radical, hydroperoxyl radical, and hydrogen peroxide may subsequently react with biological target molecules. Hydroxyl radicals may be primarily responsible for radiation-induced DNA single-strand breaks. For bacteria, it has been demonstrated that hydroxyl radicals are involved in radiation killing (Johansen and Howard-Flanders 1965). Other possible highly reactive free radical intermediates formed by ionizing radiation in vivo are thiyl radicals (Guzman-Barron 1954).

2. Skin Damage

The character and magnitude of the cutaneous response to ionizing irradiation depend on the skin area irradiated, anatomical site and its vasculature, age of the subject, radiation dose, mode of application, and radiation quality. The early skin reaction is characterized by erythema, epilation, dry desquamation with or without hyperpigmentation, moist desquamation, erosions, and healing. The late response involves typically atrophy, teleangiectasia, fibrosis, hypo- and hyperpigmentation, ulcer development, keratosis, and perhaps skin cancer (Malkinson and Keane 1981). Acute skin damage induced by ionizing radiation results primarily in injury to the epidermis and the appendages while the later chronic changes result from dermal insults, particularly to the vasculature (Cipolaro and Crossland 1967; Rubin and Casarett 1968). Release of histamine, pH changes, alterations of membrane permeability, and antigen-antibody reactions have been discussed as factors of radiation-induced early reaction in skin (Rajewsky et al. 1959). In addition, reactive oxygen species and other free radicals are important mediators of acute and chronic ionizing radiation damage in the skin.

3. Ionizing Radiation and Lipid Peroxidation

In their short lifetime of nano- and milliseconds, free radicals may produce a wide variety of chemical reactions in critical and less critical molecules. It is well-known that a low dose of ionizing radiation results in peroxidation of membrane lipids (Wills and Rotblat 1965; Petkau 1980). Few lipid peroxidation products are found in subcellular fractions immediately after exposure, while in the postirradiation period up to several hours their concentration increases dramatically (Wills and Rotblat 1964; Konings and Oosterloo 1980) and is accompanied by a decrease in the antioxidant capacity.

A great similarity exists in the time course and extent of lipid peroxidation caused by ionizing irradiation, UVB, and ozone exposure. In the liver, antioxidant capacity decreases immediately after total body irradiation with X-rays and continues to decline for several hours after irradiation (Konings and Oosterloo 1980). Levels of lipid peroxides increase in the skin of rats irradiated with cobalt 60 gamma-radiation, relatively more so at low than at high dose levels. The rise in lipid peroxidation products with a low dose of gamma-irradiation is highest in adipose tissue compared with skin (Glavind and Faber 1967).

However, ionizing irradiation can also induce the antioxidant enzyme SOD, and it was suggested that, like many enzymes, SOD activity is induced by the increased availability of its substrate. A dose-response study revealed an all-or-none response of the enzyme: The increase in total SOD activity in rat intestinal smooth muscle was significant 20 h after ionizing radiation exposure (Summers et al. 1989).

4. Oxygen as a Radiation Sensitizer

Certain chemicals and pharmacological and physical agents can modify the biological effects of irradiation. Oxygen (Holthusen 1921), electron affinity agents, and hyperthermia (Malkinson 1981) are radiosensitizers. The radiation-enhancing effect of oxygen is observed from very low to partial tissue pressures of 30 mm Hg. Above 30 mm Hg, no further enhancing effect is observed. The radiosensitizing effects are probably mediated by free radicals such as superoxide anion and hydroxyl radical. (Baxendale 1964; Cerutti 1974). Oxygen also reacts directly with organic free radicals produced by irradiation, forming peroxyl radicals.

The oxygen enhancement ratio (OER) is defined as the ratio between the doses required to induce a given degree of biological damage in the presence or absence of oxygen. The OER depends on the type of ionizing radiation. For ionizing electromagnetic radiation, e.g., gamma-rays, deposition of energy occurs in discrete packets distributed at a low density, up to a few hundred nanometers between events. Particulate radiation such as alpha-particles causes energy deposition at a much higher density than nonparticulate radiation, down to a fraction of 1 nm. For highly ionizing radiation (alpha-particles), there is no significant radiation effect of oxygen (OER = 1.0). For radiation of intermediate ionizing density (neutrons) the OER is about 1.6, and for radiation of low ionizing density (X-ray, gamma-ray) the ratio is in the range of 2.5–3.0 (Malkinson and Keane 1983).

It was discovered by the dermatologist Gans that there is a converse relationship between tissue oxygen consumption and radiosensitivity of skin. Highly active respiring embryonal skin tissue is more radiosensitive than the more slowly respiring adult skin (Gans 1923). Studies in patients breathing air or hyperbaric oxygen and receiving strontium 90 beta-irradiation revealed increased skin damage at all dose levels for hyperbaric oxygen. In this study, oxygen enhancement ratios of 1.3 or greater were found and increased with higher radiation dosages (Malkinson and Keane 1983). The occurrence of acute radiodermatitis in rat and human skin is significantly decreased by a hypoxic gas mixture (Strelkoy et al. 1983).

5. Skin Radioprotection by Antioxidants

The redox state of the cell represents the net balance between reducing and oxidizing equivalents in dynamic equilibrium. The cellular redox state determines and is determined by the viability, adaptability, and metabolic activity of the cell (Biaglow et al. 1969). Any shift in the redox state of a cell will alter its sensitivity to free radicals and reactive oxygen species and the extent of damage associated with oxidative stress arising from pathophysiological processes (Duchesne 1977).

Many radioprotectors are redox active compounds, for example, cysteine and cysteamine (Fig. 3.2). Both SH-compounds protect animal skin from the effects of sparsely ionizing radiation such as gamma- or X-rays when applied topically or intracutaneously before irradiation (Forssberg 1950; Fogh 1960; Lowy and Baker 1972). Topical cysteamine also protects animal skin from the effects of alpha-particles (Baker and Leith 1975). Radioprotection in human skin was also demonstrated using cysteamine (Bianchi and Gasparini 1955); however, no significant effect on erythema and pigmentation was found in a human study by other authors (Peruzzi and Corsi 1957). Even an enhancement of the radiation-induced reaction was observed when treating human skin with cysteamine (Farris and Oliva 1957), which was probably due to irritation from the high concentration (10%) in the vesicle used.

Topical skin treatment with *N*-acetylcysteine (Fig. 3.2) of patients undergoing megavoltage gamma-irradiation for the radiotherapy of cancer offers significant radioprotection. It reduces the severity of moist desquamation and results in a more rapid reepithelialization of radiation-induced skin erosions (Kim et al. 1983).

Most of the radioprotectors characterized by a free SH-group and a basic amine or guanidine group act largely, but not exclusively, as free radical scavengers. Other mechanisms, e. g., scavenging electrophiles, may contribute to their radioprotective properties.

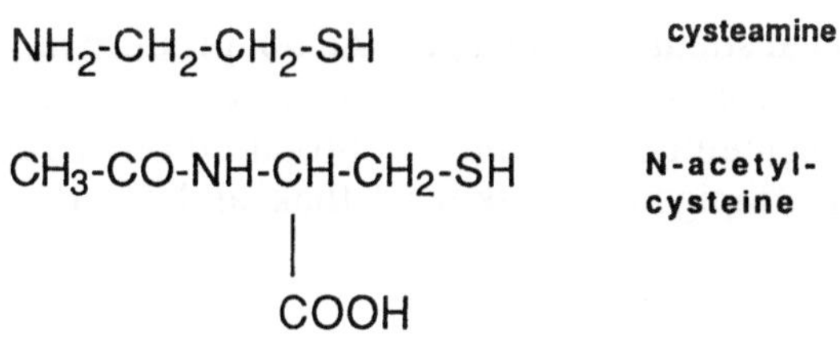

Fig. 3.2. Cysteamine and *N*-acetylcysteine

The data indicate that reactive species, particularly reactive oxygen species, and presumably lipid peroxidation participate in the pathophysiology of acute radiodermatitis and its sequelae. However, many radiation effects lessen but do not disappear in the absence of oxygen. Overemphasizing the role of oxygen as a radiosensitizer neglects the understanding that ionizing radiation has an effect of its own, namely a direct radiation effect, that does not require the presence of a mediator molecule (Dale 1954). Antioxidants are effective in animal and some clinical studies in the prevention of acute radiodermatitis. They may provide a treatment regimen of help prevent skin damage in radiation therapy.

II. Nonionizing Radiation

1. Formation of Reactive Oxidants by Ultraviolet Light

Several years ago, it was pointed out that there is no direct evidence relating free radical or reactive oxygen species formation to acute or chronic UV effects in skin (Epstein 1977a). Recently, however, a considerable body of circumstantial evidence has accumulated that strongly infers that reactive oxygen species or reactions initiated by them are responsible for some of the deleterious effects of UV upon skin (Black 1987a). Reactive oxygen species are formed by UVC through water photolysis and by UVB and UVA through photodynamic action and, e.g., photodissociation of hydrogen peroxide. Direct and indirect evidence for UV-mediated free radical generation in skin has been obtained by electron paramagnetic resonance (EPR) spectroscopy.

Low-Temperature EPR Studies. Direct evidence for UV-induced free radical formation in skin has been found by low-temperature EPR. UV irradiation (254 nm) of human skin at liquid nitrogen temperature results in the formation of a broad EPR resonance signal, which could not be assigned to a defined chemical species (Norins 1962). UV (240–585 nm) irradiation of lyophilized guinea pig epidermis results in the formation of lipid-derived free radicals which decay at temperature above $-20\,^{\circ}C$ (Nishi et al. 1991b; Ogura et al. 1991). Irradiation of human skin at wavelengths greater than 320 nm enhanced the endogenous paramagnetic melanin signal in pigmented skin samples. Nonpigmented human skin samples gave off no endogenous melanin free radical signal, and irradiation at similar wavelengths did not produce any detectable free radicals. UV radiation at wavelengths shorter than 320 nm enhanced the melanin signal in pigmented skin and also produced other radical species in pigmented and nonpigmented skin (Pathak and Stratton 1968).

Ambient-Temperature EPR Studies. UVA irradiation of mouse skin results in the formation of the endogenous ascorbyl radical as seen with EPR spectroscopy at ambient temperature (Buettner et al. 1987). It lyophilized epidermis a free radical can be detected at room temperature without UVB exposure. Following UVB exposure the radical concentration increased in an aerobic environment but was inhibited in an anaerobic environment (Ogura et al. 1987). Since the freeze-drying technique can produce paramagnetic artifacts in biological tissues (Lohmann and Neubacher 1985; Heckly 1980) these results must be interpreted with caution. Irradiation of rhinoceros-horn-derived keratin with UV (325–400 nm) and visible light results in the formation of several unidentified free radicals as photochemical intermediates. These species are of a more stable nature than those formed by UVC irradiation of keratin. The latter type of keratin radicals is associated with cysteine residues and decay in the presence of oxygen (Dunlop and Nicholls 1965).

Spin Trapping. Only few data are available on the spin trapping of reactive oxidants in UV-irradiated skin. Although singlet oxygen can be detected in photosensitized, isolated epithelial cells using a singlet oxygen trap (anthracenyl nitroxide) (Nye et al. 1987), no information is available on the detection of singlet oxygen in UV-irradiated intact skin or skin cells. By this method, lipid radicals were detected in UV-exposed (240–585 nm) epidermis (Nishi et al. 1991 b, Ogura et al. 1991) and unidentified thiyl radicals after photosensitization in intact skin (Li, personal communication).

2. Ultraviolet-Light-Induced Skin Damage

The reaction pattern and the time course of acute UV photodamage, e. g., skin inflammation, is wavelength dependent. This may be due to differences in the penetration of UV into skin layers (UVA > UVB > UVC) and from specific reaction cascades initiated by a specific wavelength band. The inflammatory response to UVA, UVB and UVC irradiation may occur along separate molecular pathways differing with respect to inflammatory mediators and oxygen dependence.

UVA causes pigmentation and erythema, but the dose required to produce erythema is about 2000 times greater than that for UVB-induced erythema. UVA-induced erythema appears during irradiation, disappears subsequently for several hours, and then slowly increases to a maximum which may persist for more than 48 h. Mediators of this erythema are histamine and prostaglandins.

UVB causes pigmentation and erythema. In humans, erythema appears 2–4 h after irradiation, reaches its maximum by 24–48 h, and then gradually disappears. Prostaglandins of the E series appear to be important mediators of UVB-induced skin inflammation in the early phase (< 24 h) Eaglstein et al. 1979). Lysosomal enzyme release and the synergistic action of prostaglandins with other vasoactive mediators, e.g., kinins, may also contribute to the erythema pathogenesis.

UVC does not cause pigmentation, and erythema is less intense than with similar doses of UVB. Erythema appears 2 h after irradiation, is maximal at 12–24 h, and then gradually fades. Prostaglandins contribute to this response.

Oxygen Dependence of Visible Light UVA/UVB Response. Early tissue culture studies have demonstrated the deleterious effect of visible light of mammalian cells, particularly in the presence of a photosensitizer (Earle 1928). Hydrogen peroxide was identified as the toxic photoproduct responsible for cell death in human tissue cultures irradiated with daylight (Wang et al. 1978). It was shown that riboflavin, tryptophan, and tyrosine in the medium were responsible for the hydrogen peroxide generation (Stoien and Wang 1974, Wang 1975, 1976). Riboflavin-mediated photooxidation of tryptophan and tyrosin via activated oxygen results in the formation of hydrogen peroxide.

The phototoxicity of visible light and UVA is oxygen dependent, indicating the involvement of reactive oxygen species. UVA caused inactivation of mammalian cells is strongly oxygen dependent, and singlet oxygen is suggested to be involved in the damage to cultured human fibroblasts by UVA irradiation (Tyrrell and Pidoux 1989).

In contrast to UVA irradiation (Tegner et al. 1983, Pathak et al. 1962), UVB-induced erythema and pigment responses are not oxygen dependent (Auletta et al. 1986). This suggests that reactive oxygen species are not directly involved. However, several other studies do not support this conclusion (see below). Glutathione plays a major role in the protection of cultured human fibroblasts against both UVB and UVA irradiation (Tyrrell and Pidoux 1986, 1988), providing additional evidence that reactive oxidants are involved.

Sunburn Cells. Major skin cell types, such as keratinocytes, Langerhans cells, and fibroblasts, are damaged by UVB light exposure, and reactive oxygen species are thought to be responsible for some of the UVB-induced biological damage in skin. The micromorphology and biochemistry of the human sunburn reaction have been analyzed. Major histologic alterations in the epidermis are the formation of sunburn cells and the disappearance of Langerhans cells. In the dermis, endothelial cell camage and degranulation of mast cells are apparent (Gilchrest et al. 1981).

Free radicals and reactive oxygen species generated by UV radiation are suggested to participate in sunburn cell formation but are not apparently involved in UV-induced skin edema (Miyachi et al. 1983; Danno and Horio 1987). The sunburn cell is a histological characteristic of UV-induced epidermal injury (Danno et al. 1984) and is a type of individual keratinocyte cell death appearing in the epidermis principally after exposure to middle wavelength UV irradiation. They are formed if reactive oxygen species production overcomes the capacity of the antioxidants normally present in skin. Sunburn cell formation is increased in mouse skin depleted of glutathione (Hanada et al. 1991) and inhibited by hypoxia (Youn et al. 1988). Oxygen intermediates such as superoxide anion or the subsequent species generated by UVB irradiation also damage epidermal Langerhans cells (Horio and Okamoto 1987).

Lysosomal Damage. Ultrastructural studies of UVB-exposed mouse skin reveals specific damage to epidermal lysosomes 4 h after irradiation (Johnson and Farrington 1969). Lysosomes are more susceptible to UV-induced lipid peroxidation than mitochondria or the plasma membrane. Lysosomes of different tissues have widely differing susceptibilities to photodamage, epidermal lysosomes being more sensitive than liver lysosomes, for example (Slater and Riley 1966). Damage to the lysosomal membrane will release digestive enzymes leading to cell autolysis (Packer et al. 1966) and indeed, lysosomal enzymes are released during UV irradiation of skin (Johnson 1968).

The participation of lysosomal disruption in the sunburn reaction has been considered (Johnson and Farrington 1969). Exposure of lysosomes to reactive oxygen species in vitro rapidly results in lysosome membrane damage (Brunmark and Brunk 1990). Antioxidant protection of lysosomes from UV-induced lipid peroxidation may be a significant mechanism in photoprotection.

Phospholipase A_2 Activation. Membrane phospholipids were suggested to play some part in the primary response of skin to UV (Johnson and Mier 1962). UVB increases the release of arachidonic acid from membrane phospholipids in guinea pig skin (Ziboh et al. 1978), murine fibroblasts (De Leo et al. 1985), and human keratinocytes (De Leo et al. 1984). UVA stimulates phospholipase A_2 activity in vitro severalfold and increases phospholipase activity in human keratinocytes (Hanson et al. 1989, 1990) and fibroblasts (Hanson and De Leo 1989). The keratinocyte response was inhibited in a hypoxic environment (Hanson and De). In isolated brain capillaries, the addition of a free radical generating system stimulates phospholipase A_2 activity and degrades membrane phospholipids (Au et al. 1985). In cultured human keratinocytes, UVA and/or PUVA therapy stimulates liberation of arachidonic acid from membranes (Punnonen et al. 1991 a). Data indicate that reactive oxidants are probably involved in the stimulation of phospholipase A_2 activity by UV. Photoactivation of phospholipase may play a central role in the pathogenesis of UV-caused inflammation via formation of prostaglandins and leukotrienes.

Lipid Peroxidation. Experimental results indicate that UV irradiation of intact skin, skin homogenate, isolated skin cells, and extracted skin surface lipids causes lipid peroxidation.

UVA irradiation of guinea pig skin enhances lipid peroxidation (Akimov 1987). Exposure of human skin to an artificial UV source (mercury lamp) results in a significant increase in lipid peroxidation products (Meffert and Reich 1969 a), which are also elevated in chronically sun-exposed human skin (Meffert et al. 1976 a).

Irradiation of human epidermis, dermis, and subcutis homogenates with UVB leads to the formation of cytotoxic malondialdehyde (Meffert and Lohrisch 1971 b; Meffert et al. 1972).

In cultured human keratinocytes, UVA and/or PUVA therapy increases the formation of thiobarbiturate reactive material (Punnonen et al. 1991 a), although conjugated dienes were unaffected. In UVB-exposed keratinocytes, both conjugated dienes and thiobarbituric acid reactive material accumulate (Punnonen et al. 1991 b). These differences may be explained by wavelength-dependent reaction pathways of keratinocyte lipid peroxidation.

Solar irradiation of extracted human skin surface lipids results in a strong decrease in unsaturated fatty acids. Exposure of human skin surface lipids to UVC (260 nm) and UVA (360 nm) leads to only a modest decrease. No changes were observed with infrared, beta- or gamma-rays of comparable energy. The total amount of free fatty acid was unchanged (Horacek and Cernikova-Brünn 1961).

Near-UV light is capable of peroxidizing human skin surface lipids (Nazzaro-Porro et al. 1986). Lipid hydroperoxide levels, as determined by gas chromatography, rose in human skin surface lipids after UVB irradiation, and a direct correlation of hydroperoxide concentration with the minimal erythemogenic dose was found. The lipid degradation product malondialhdehyde did not increase in skin. It was concluded that lipid peroxidation of skin surface lipids is not involved in the acute erythema reaction to UVB, because lipid hydroperoxides failed to degrade into cytotoxic products (malondialdehyde and other alkenals) (Meffert and Reichmann 1972 b). However, hydroperoxides themselves are cytotoxic compounds, and malondialdehyde can be metabolized or react with epidermal amino groups and may thereby circumvent detection.

Time Course of Lipid Peroxidation. Analysis of the time course of UVB-induced lipid peroxidation in skin of hairless mice reveals a significant elevation of lipid peroxidation products 3 h after irradiation. Immediately after exposure, the content of lipid peroxidation products in the epidermis is normal (Ogura et al. 1987). Similar results were obtained in animal (Maisuradze et al. 1987; Fuchs et al. 1989 a) and in vitro studies. Photooxidation of squalene, one of the major skin surface lipid components, results in a significant formation of peroxidized products only 2–4 h after UVB irradiation (Ohsawa et al. 1984).

Thiobarbituric acid reactive substances are detected in the epidermis of hairless mice after UVB irradiation, with a maximum at 9 h after irradiation, returning to preirradiation values after about 20 h. There is a nonlinear increase in thiobarbiturate reactive products with radiation dose, the maximal response occurring at 200 mJ/cm^2; at higher energy doses, the lipid peroxide content decreases (Pugliese and Lampley 1985). However, detection of thiobarbituric acid reactive stubstances before and after skin irradiation is a rather obsolete method for analyzing the time course of lipid peroxidation. The assay is not specific and, besides other products, only degradation products of lipid hydroperoxides are measured. A much better method is detection of the initial persistent peroxidation product, lipid hydroperoxides. A very sensitive HPLC assay for lipid hydroperoxides in skin following UV exposure was used. This method is based on the technique described by Yamamoto et al. (1990). Before skin irradiation there are absolutely no organic hydroperoxides detectable in skin, whereas after UV exposure (solar simulator) they increase enormously. Furthermore, the HPLC profiles are diagnostic, indicating that many subspecies arise, and this gives us the opportunity to identify which may be the critical and most important ones (Packer 1991; Witt et al. 1991).

UVB Inflammation and Lipid Peroxidation Products. Hydroperoxides are important regulators of cyclooxygenase and lipoxygenase activity. Therefore, modulation of the lipid peroxide baseline level will influence prostaglandin and leukotriene metabolism. Cyclooxygenase activation and an increased availability of substrate via stimulation of phospholipase A_2 activity could contribute to elevated concentrations of proinflammatory prostaglandins. The cytotoxic and proinflammatory potency of lipid peroxidation products is suggested to play an important role in UV-induced skin inflammation (Ohsawa et al. 1984; Tankov 1989).

3. Photoprotection by Antioxidants

Antioxidants inhibit UV-induced lipid peroxidation in vitro and in vivo, possess photoprotective properties in animals and humans, and reduce chronic cutaneous actinic damage (Birset et al. 1990). Orally administered and topically applied antioxidants provide protection from UV-induced erythema and cytotoxicity. An important role in the development of UVB erythema is due to photochemical reactions after irradiation, most probably the development of peroxidation processes (Potapenko et al. 1980). The observed decrease in erythema response in skin brought about by antioxidants is related to the inhibition of lipid peroxidation processes initiated by UV radiation (Roschchupkin et al. 1979).

Several studies indicate that *tocopherol* protects keratinocytes and intact skin from damage caused by solar radiation, UVA-C, and PUVA treatment. Liposomal tocopherol preparations protect cultured human keratinocytes exposed to UV derived from a solar simulator (Werninghaus et al. 1990), and systemic application of α-tocopherol reduces UV-induced skin damage in humans (Namigata 1960).

Actinic reticuloid is a rare disorder characterized by an extreme skin sensitivity to UV and VIS radiation. The presence of a yet unidentified photosensitizer and/or a possible defect in the antioxidant defense system have been discussed. The water-soluble tocopherol analogue trolox protects actinic reticuloid cells from UVA-induced lethality, which indicates free radical involvement in the etiology of the disease (Kralli and Moss 1987). Tocopherol also inhibits UVA-induced photohemolysis of erythrocytes in patients with erythropoietic protoporphyria (Goldstein and Harber 1972).

The relative photoprotective potency of the natural *d*-tocopherols in UVB-induced edema of the hairless mouse (topical application of tocopherol prior to irradiation) is in the same order as their reactivity to peroxyl radicals: *d*-α-tocopherol (100%) > *d*-γ-tocopherol (72%) > *d*-β-tocopherol (42%) > *d*-δ-tocopherol (40%) (Potokar et al. 1988). Topical application results in a significant reduction of lipid peroxide formation after UVA exposure of hairless mice (Pugliese and Lampley 1985). Skin ornithine decarboxylase activity increases in response to UV irradiation. Application of the prodrug α-tocopherol acetate, which is presumably hydrolyzed in the epidermis to the active α-tocopherol (Tojo and Lee 1987), significantly reduces UV-induced ornithine decarboxylase activity in skin (Pugliese et al. 1983). Topical application of tocopherol reduces free radical formation in UVB-irradiation skin of hairless mice and reduces increased polyamine levels (Khettab et al. 1988). In vitro, α-tocopherol inhibits UVB-induced peroxidation of squalene, one of the major skin surface lipid components (Ohsawa et al. 1984). In contrast to inhibitors of cyclooxygenase such as indomethacin (Snyder 1975; Kaidbey and Kurban 1976), the local application of tocopherol results in a significant reduction of sunburn cell formation from UVB in human skin when administered before irradiation (Msika et al. 1988). This indicates that tocopherol inhibits both cellular damage and the inflammatory reaction. Cyclooxygenase inhibitors do not prevent UVB-mediated keratinocyte damage.

Electron microscopic and light microscopic studies indicate that the inflammatory response in mice irradiated with UVC (254 nm) was significantly ameliorated by topical treatment with vitamin E after irradiation (Watabiki and Ogawa 1975).

α-Tocopherol protects the skin against PUVA erythema (Pliquett et al. 1984). PUVA erythema in rabbits is inhibited by tocopherol at low-dose UVA irradiation (3 minimal erythema dose, MED), if applied before irradiation. If it is applied shortly after irradiation, its inhibitory effect is greatly reduced. α-Tocopherol application prior to PUVA treatment with a large UV influence (10 MED) has virtually no inhibitory action (Potapenko et al. 1980). α-Tocopherol has a weak UV filter effect; however, it inhibits UVB-induced erythema irrespective of whether it is applied before or immediately after irradiation. At a high tocopherol concentration on the skin surface, the protective action is decreased, which is explained by prooxidant action at higher concentrations (Potapenko et al. 1984). A nonspecific, local, irritating effect at high concentrations is another explanation of this phenomenon.

β-Carotene and *canthaxanthin* have protected in animal models against UV-induced skin cancer (Epstein 1977b) as well as against some chemically induced tumors. β-Carotene prevented the lethal photosensitization induced in mice by hematoporphyrin (Mathews 1964), and topical application signifi-

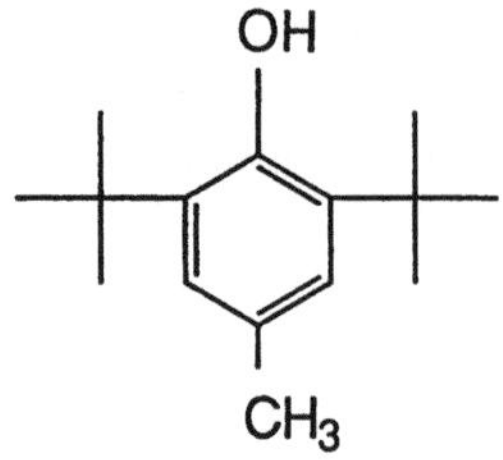

Fig. 3.3. Tertiary butylhydroxytoluene (BHT)

cantly reduced UVB-induced lipid peroxidation in hairless mice (Pugliese and Lampley 1985). A photoprotective mechanism of β-carotene in skin which involves scavenging of singlet oxygen (Foote and Denny 1968; Pollitt 1975) and free radicals (Krinsky and Deneke 1982) was suggested.

Dihydrolipoate provides local photoprotection in human skin when applied before or immediately after UVB irradiation; no photoprotection is observed when the drug is applied 4 h after irradiation (Fuchs and Milbradt 1989 a). Photoprotective properties of dihydrolipoate are also reported in the hairless mouse (Fuchs et al. 1990 d).

Pretreatment of the epidermal surface of guinea pigs before exposure to UVB with liposomal *SOD* prevents the increase in lipid peroxide formation (Ogura et al. 1987).

Systemic administration of *tertiary butylhydroxytoluene* (BHT; Fig. 3.3) caused significant photoprotection (against mercury arc lamp) in rats. In contrast, glutathione, ascorbate, and tocopherol provided negligible protection (De Rios et al. 1978). Oral administration of BHT to hairles mice resulted in decreased transmission in the stratum corneum of treated animals. Photoprotection is suggested to be related to changes in the physicochemical properties of the stratum corneum rather than to changes in physical parameters (Koone and Black 1986). Structure-activity evaluations of chemically related phenols reveal that the antioxidant properties of BHT are not significant determinants of photoprotective activity (Black and Tigges 1986). A large component of the photoprotective effect of BHT is thought to be diminishing levels of UV irradiation reaching epidermal target sites by altering the physicochemical properties of the stratum corneum (Black et al. 1984).

A combination of BHT and tocopherol applied on skin before and within 2 min after irradiation greatly inhibited the erythemal response to UV (280–365 nm). The antierythema efficiency of BHT and tocopherol applied 2 min postirradiation was not weaker than that observed in the case of application before irradiation. This means that the antioxidants retard the dark stages of erythema. Antioxidant treatment of skin 5 h postirradiation or later has almost no influence on the erythema response (Potapenko et al. 1980; Roschchupkin et al. 1979).

4. Ultraviolet Light Effects on Skin Antioxidants

Photoactivation and photodeactivation of enzymes may play a considerably role in the photobiological responses of biological tissue (Hug 1978, 1981). One example of enzyme photoactivation is UV-induced stimulation of phospholipase A_2 activity. Several other enzymes including those involved in deactivation of reactive oxygen species can be inhibited by UV. Photoinactivation of catalase is a well-investigated example. Antioxidants can be depleted by a high flux of reactive oxidants, generated, e.g., by UV irradiation. Several enzymic and nonenzymic skin antioxidants are susceptible to UV deactivation (Fuchs and Packer 1990). With the exception of a few studies (Miyachi et a. 1987a; Punnonen et al. 1991b), most reports allow only a glimpse of the overall picture, because antioxidants were determined only at one single time point after irradiation. An investigation of the dynamic responses of various skin antioxidants after acute and chronic UV irradiation is desirable to obtain a more complete picture of the short- and long-term effects on the skin's antioxidants capacity.

Inactivation of Catalase, SOD, and Glutathione Peroxidase. Antioxidant enzymes can be inactivated during oxidative exposure; photoinactivation of catalase in rat hepatocytes (Cheng and Packer 1979) and in fibroblasts (Vuillaume et al. 1986) has been described. Catalase is photoinactivated by near UV (Freierabend and Engel 1986) as well as by visible light (Mitchell and Anderson 1965). Photooxidation of one of the porphyrin rings in the heme enzyme results in complete inhibition of enzyme activity (Aronoff 1965). It was also reported that superoxide anion radicals (Kono and Fridovich 1982; Sun and Oberley 1989) and ozone (Whiteside and Hassan 1987) inactivate catalase. Reactive oxygen species were suggested to be involved in the mechanism of catalase inactivation (Whiteside and Hassan 1988). SOD is readily inactivated by hydrogen peroxide (Bray and Cockle 1974) and ozone (Whiteside and Hassan 1987). Glutathione peroxidase is quite resistant to oxidative damage by a variety of organic peroxides and hydrogen peroxide (Condell and Tappel 1983); however, it is also considered vulnerable to oxidative injury (Flohe 1982; Ganther and Kraus 1981). Glutathione peroxidase can be inactivated by superoxide anion radicals (Blum and Fridovich 1985).

Skin Enzymic Antioxidants. In cultured human keratinocytes, UVB (Punnonen et al. 1991a, UVA, or PUVA therapy (Punnonen et al. 1991b) decreases the activity of the antioxidant enzymes catalase and SOD. In UVA- or PUVA-treated keratinocytes, antioxidant enzyme activities have already decreased 30 min after irradiation and were partially restored after 24 h.

Immediately after UV irradiation of excised mouse skin (> 280 nm, Fuchs et al. 1989a; > 320 nm, Fuchs et al. 1989c), a small but significant inhibition of glutathione reductase and catalase activity was seen; glutathione peroxidase and SOD are not affected. Similar results were obtained in hairless mice irradiated with a solar simulator (Packer et al. 1990). In UVB-exposed rat skin, the activity of SOD, glutathione peroxidase, and glutathione reductase is inhibited (Maisuradze et al. 1987). Significant decreases in skin SOD and catalase activity were observed 12 h after a single exposure to UVB radiation of hairless mouse skin, and they remained depressed for up to 72 h (Pence and Naylo 1990). A single exposure of mouse skin to UVB radiation resulted in a significant decline of SOD activity 24–48 h after exposure, returning to the normal level 72 h after irradiation. At 3 h postirradiation, the SOD activity was unaltered in skin (Miyachi et al. 1987 a). Accordingly, in guinea pig skin, no change in SOD activity is observed immediately after UVB exposure (Ogura et al. 1987). Cu, Zn-SOD and Mn-SOD activities are not significantly different in sun-exposed and nonexposed skin of healthy individuals (Sugiura et al. 1985 a).

Hydrophilic Antioxidants. UV irradiation (> 280 nm) of excised mouse skin decreases the glutathione and conversely increases the glutathione disulfide concentration. A slight depletion of the total glutathione content was noted, while the total ascorbate concentration remained unchanged (Fuchs et al. 1989 a). UV (> 320 nm) did not cause a change in the cutaneous glutathione/glutathione disulfide ratio (Fuchs et al. 1989 c). Treatment of hairless mice with 8-methoxypsoralen and skin exposure to 5 J/cm^2 UVA results in a depletion of cutaneous glutathione; the minimum is reached after 24–48 h (Wheeler et al. 1986). Irradiation of skin of hairless mice with 90 mJ/cm^2 UVB leads to a decrease in epidermal glutathione concentration to 60 % of its control level within 10 min and an increase in glutathione disulfide level. This returns to control levels within 30 min (Connor and Wheeler 1987).

A single exposure of rat skin to UV derived from a broad-spectrum quartz lamp causes a strong decline in epidermal nonprotein SH compounds 1 h after irradiation which does not normalize within 4 weeks after exposure. Levels of rat epidermal disulfide groups are also diminished but normalize within several days (Ogura and Knox 1967). UVC (254 nm) induces photolysis of disulfide bonds in rat and human epidermis; this is repaired starting 6 h after irradiation and reaches normal values within 24 h. There is a subsequent increase in epidermal disulfide groups which is normalized 10 days after irradiation (Jung and Bohnert 1979).

Lipophilic Antioxidants. UV irradiation (> 280 nm) of excised mouse skin results in an immediate decrease in α-tocopherol, ubiquinol-9, and ubiquinone-9 (Fuchs et al. 1989 a). These antioxidants were not affected by near UV (> 320 nm) (Fuchs et al. 1989 c). However, the tocopherol and ubiquinone-10 content in healthy, unexposed, human epidermis was severalfold lower than in sun-exposed epidermis of the same individual (De Simone et al. 1987). Although the ubiquinone-10 concentration in human skin is highly variable (9–101 μg/mg protein in sun-exposed skin and 20–115 μg/mg protein in unexposed skin), for the same subject it is statistically significantly higher in exposed than in unexposed skin (Oradei et al. 1990). It was demonstrated that upon topical application, ubiquinone-10 penetrates the skin of rats, and this was suggested to be of pharmacological interst in dermatology (Giovannini et al. 1988). The data indicate that chronic UV exposure may result in the induction of compensating mechanisms, increasing lipophilic antioxidants in the epidermis. Topical antioxidant therapy may help to prevent the depletion of lipophilic skin antioxidants.

One-Electron Reducing Activity at the Epidermal Surface. Erythemogenic UVB irradiation of hairless mouse skin inhibits the free radical scavenging activity at the epidermal surface, as analyzed by scavenging of stable nitroxide radicals (Fuchs et al. 1989 a, b). In guinea pig and human skin, UVA and UVB (below the minimal erythemogenic dose) also inhibit free radical scavenging activity at the epidermis, using a similar assay. UV-induced deactivation of thioredoxin reductase was suggested as the cause of this inhibition (Schallreuter et al. 1989). However, nitroxide radical reduction by thioredoxin reductase is stimulated in vitro by UV (> 280 nm; Fuchs, unpublished results). Depletion of epidermal ascorbate by photooxidation is a more likely explanation for UV-caused inhibition of free radical scavenging activity at the epidermal surface (Darr 1991).

Photohemotherapy. Reinfusion of UV-irradiated blood was employed in the treatment of a number of diseases by Knott (1948). Various studies indicate that UV caused the formation of reactive oxidants and peroxidation products in blood, and it is speculated that biological effects caused by photohemotherapy are mediated by these species.

UV irradiation of blood leads to lipid peroxidation of the erythrocyte membrane and a decrease in erythrocyte aggregation (Murina and Roschupkin 1983, 1984). In irradiated, thrombocyte-rich plasma, the ADP-induced thrombocyte aggregation is inhibited (Roschupkin et al. 1983). Low-dose UV irradiation increases the ability of T lymphocytes to form rosettes with sheep lymphocytes (Krylenkov and Malygin 1982). This effect was attributed to UV-induced lipid peroxidation products and was eliminated by adding antioxidants to UV-irradiated lymphocytes (Krylenkov et al. 1984). UVB irradiation of murine and human blood in vitro results in a continuous decrease of the total antioxidant blood capacity. UV-mediated oxidation of hemoglobin was suggested to be the source of reactive oxidants which consume blood antioxidants and cause lipid peroxidation (Lewin and Popov 1989):

$$\text{Hb-Fe}^{2+} + O_2 + h\nu = \text{Hb-Fe}^{3+} + O_2^{-\cdot}$$

Extracorporeal UVB irradiation of blood from patients with psoriasis and reinfusion of irradiated autologous blood results in a small but significant increase in the blood antioxidant capacity and is accompanied by an initial improvement of skin lesions. It was suggested that the patients' antioxidant system is induced by UVB-generated lipid peroxidation products. In most patients with psoriasis, however, the increase of the blood antioxidant capacity and improvement of skin lesions were only transient (Popov et al. 1988 b).

5. Infrared Radiation

Infrared radiation can be divided into near (800–1400 nm), middle (3–30 μm), and far (30–1000 μm). Most of the infrared radiation that reaches the earth's surface penetrates skin readily and can cause infrared erythema and elevation of arachidonic acid and prostaglandins (Juhlin et al. 1983). It increases the incidence of UVB-induced skin cancer in mice (Bain et al. 1943; Freeman and Nox 1964) and can cause skin photoaging. Infrared-generated skin cancers in humans have been known for a long time: Kang cancer (North China; from sleeping on hot bricks), Kangri cancer (Kashmir, from burning coal close to the body), Kario cancer (Japan, from petrol burner beneath the clothing), turf or peat cancer (Ireland, from sitting in front of an open fire) (Kligman and Kligman 1984).

Infrared radiation typically causes molecular vibrations at wavelengths down to 800 nm in the very near infrared; however, photochemical reactions may occur. Near-infrared light can induce the excitation of electrons in various synthetic organic molecules (Fabian and Zahradnik 1989) and under special conditions may presumably lead to the formation of singlet oxygen (Kasha and Khan 1970). Photochemical reactions may contribute to near-infrared photodamage in skin. However, participation of reactive oxidants in infrared-caused tissue injury is only hypothetical at the present time. Thermal cell damage may also be caused by inactivation of essential enzymes, some of which are involved in nucleic acid repair.

6. Ultrasound

Although ultrasound is, of course, not an electromagnetic wave but sound waves, it is discussed here as this seems the most appropriate point in the book. Ultrasound diagnostic thechniques have been introduced in dermatology for noninvasively measuring skin thickness (Alexander and Miller 1979; Tan et al. 1982; Miyauchi and Miki 1983) and structural pathological changes such as cutaneous tumors (Rukavina and Mohat 1979; Weinraub et al. 1982). The applications of ultrasound in dermatology have been reviewed recently (Payne 1985). Under certain physicochemical conditions (cavitation), ultrasound is known to generate free radicals in aqueous solutions (Riesz et al. 1985). Sonolysis of aqueous solutions can result in the formation of hydroxyl radicals and hydrogen atoms (Makino et al. 1982, 1983a, b); in nonaqueous solutions organic radicals can be formed (Riesz et al. 1985). There are no confirmed reports of ultrasound damage in mammals, including humans. It is not known whether sonolysis results in the generation of reactive species in skin; if so, this presumably has no clinical significance.

III. Photosensitization

Photosensitization is a cutaneous reaction to electromagnetic irradiation. The reaction is distinguished from normal photodermatitis by an amplification of the physiological response due to endogenous or exogenous chemical substances. Drug photosensitivity, for example, is the adverse cutaneous reaction due to the combined action of light and an applied chemical substance. Either agent alone evokes no abnormal response, but the two in combination result in an additive or overadditive skin reaction. The mechanisms of photosensitivity reactions are quite complex. Drug-induced phototoxicity must be differentiated from drug-induced photoallergy. Epstein was a pioneering photobiologist who first differentiated between phototoxic and photoallergic skin reactions in humans (Epstein 1939). Although two distinct mechanisms can be postulated, they share common manifestations. An inflammatory response and cell damage result in both cases, and free radicals are frequently involved in both processes.

The basic *photochemical mechanism* in relation to photochemotherapy is excitation of organic molecules by nonionizing electromagnetic irradiation to electronically excited states, the singlet or triplet states, from which photochemistry may result. The exited states are short-lived (singlet, 10^{-9} s; triplet, 10^{-3} s) and can give rise to free radical formation. The electronically excited species react readily with a substrate molecule, and the nonreactive precursor molecule is defined as a photosensitizer. The reaction pathways of photosensitizers usually proceed via the triplet state and can be characterized into type I and type II. In a type I reaction, the activated photosensitizer reacts directly with a substrate molecule via electron or hydrogen atom transfer and may give rise to free radical formation. In a type II reaction (photodynamic), the electronically excited photosensitizer reacts with oxygen, thereby forming reactive oxygen species as reactive intermediates.

Synthetic dyes (Kearns et al. 1967) oxidize substrates by the pathway involving singlet oxygen. If the concentrations and reaction rates of reductants with triplet state dyes are sufficiently high, the direct reduction pathway will predominate over that producing singlet oxygen. Most photosensitized reactions are oxygen dependent and may cause molecular damage mediated by reactive oxygen species at various cellular sites.

Participation of reactive species in photosensitivity reactions due to drugs has been described (Miyachi et al. 1986a; Torinuki et al. 1980; Pathak et al. 1961). Endogenous photosensitizers may play an important role in the etiopathology of some photosensitivity skin diseases. Activated oxygen species can be considered important mediators in some photodermatose or light-aggravated skin diseases.

1. Endogenous Photosensitizers

Although photolysis of water by UVC leads to the direct formation of superoxide anion radical (McCord and Fridovich 1973), the direct generation of reactive oxygen species in water is not observed during irradiation with UVB, UVA, or visible light. Several endogenous metabolites like flavins, NADH, NADPH, and nucleosides can sensitize the transmission of photon energy from solar radiation to oxygen, resulting in the formation of reactive oxygen species (Cunningham et al. 1985; Ballou et al. 1969; Massey et al. 1969).

Endogenous photosensitizers may play a role in photodermatoses. Photodermatoses are pathological skin conditions triggered or aggravated by light and can be classified into genetic, metabolic, idiopathic, chemical, degenerative, and neoplastic disorders. The most commonly known photosensitizers comprise porphyrins, flavins, melanins, photosensitizing molecules present in natural substances (Ames 1983; Straight and Spikes 1985), and tryptophan metabolites. Tryptophan and *N*-formylkynurenin (McCormick et al. 1976) are endogenous sources of hydrogen peroxide upon irradiation with near UV light.

Porphyrins. It was shown in 1908 that hematoporphyrin is a photosensitizer in rat blood cells and that this reaction requires oxygen (Hausmann 1908). In 1909, it was suggested that this property may be responsible for the cutaneous manifestations of vesicobullous lesions of the skin related to sun exposure (Ehrmann 1909). This hypothesis was tested by a self-experiment of the medical student Meyer-Betz (1913). He developed erythema and edema in sun-exposed skin area within minutes after intravenous injection of hematoporphyrin prepared from hemoglobin treated with sulfuric acid. The photosensitivity remained for more than 2 months.

Porphyrias result from acquired or inherited disturbances in heme biosynthesis with the consequence of metabolite accumulation in the tissues. Toxic effects of the heme intermediates are then responsible for the clinical expression of the disease. Different types of human porphyrias are characterized by the specific enzyme defect and by the tissue origin of this specific defect (hepatic, erythropoetic, or both). Classifications of porphyria are based on the pathological porphyrin pattern found in the urine, feces, and erythrocytes, although in some cases the biochemical parameters can overlap.

Photosensitivity is a common clinical feature in porphyria cutanea tarda, erythropoetic protophorphyria, porphyria variegata, and hepatoerythropoietic porphyria. Acute intermittent porphyria is not associated with photosensitivity reactions because the accumulating porphyrin precursors (δ-aminolevulinic acid and porphobilinogen) are not photosensitizers. However, δ-aminolevulinic acid easily autooxidizes in the presence of oxyhemoglobin, forming methemoglobin and reactive oxygen species (Monteiro et al. 1989). This reaction could be involved in the neurological-visceral symptom complex of the disease.

It was demonstrated that the lipid solubility of porphyrins controls their phototoxicity in model membranes (Emiliani and Delmelle 1983). Differences in lipophilicity may account for the contrasting signs of cutaneous photosensitivity in patients wiht porphyria cutanea tarda (predominance of hydrophilic urophorphyrin) and erythropoietic porphyria (predominance of lipophilic protoporphyrin). In the latter, photosensitivity is characterized by succulent plaques, erythema, and purpura (acute reactions), diffuse skin thickening and scars (chronic reactions), whereas in the former skin fragility is increased, vesicular, bullous, and ulcerative lesions occur, and finally hyperpigmentation, scleroderma-like plaques, and hypertrichosis appear.

Fluoroscopic investigation of skin biopsies of patients with cutaneous porphyria revealed the presence of porphyrins in the epidermis and dermis. The mechanisms of cutaneous photosensitization by porphyrins may be mediated by type I and type II photoreactions. Porphyrins absorb in the 400 nm range, and the molecule is photoexcited to a metastable triplet state. In type I reactions, direct interaction of photoexcited porphyrin (triplet state) may occur with a biological substrate. In type II, reactive oxygen species may be formed by the interaction of molecular oxygen with porphyrin in the triplet state (Cauzzo et al. 1977; Kessel and Rossi 1982). Porphyrin-induced skin photosensitization reactions involve the production of singlet oxygen but not superoxide anion radical (Pathak and Carraro 1987b) (Fig. 3.4).

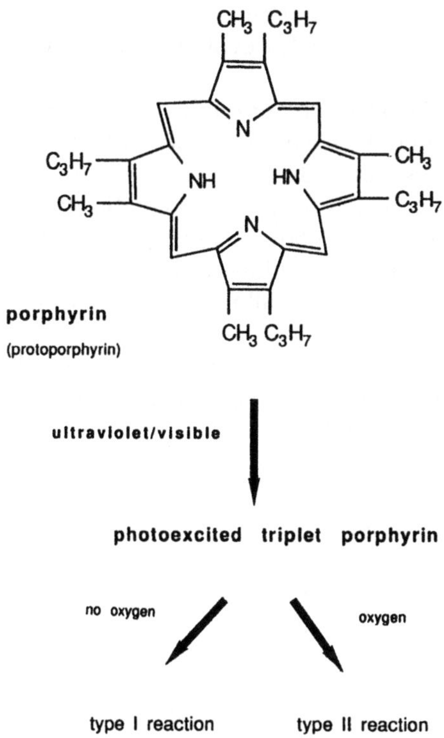

Fig. 3.4. Porphyrin photochemistry

Evidence for the production of oxidative injury stems from experiments demonstrating porphyrin-caused photoperoxidation of red cell (Dubbelman et al. 1977, 1978; Girotti 1979), lysosomal, mitochondrial, and microsomal membranes (Allison et al. 1966; Sandberg 1981, Dixit et al. 1983 b) and oxidation of fibroblast membrane thiol groups (Schothorst et al. 1980). An additional mechanism may be activation of the complement system by porphyrin and light (Gigli et al. 1980; Lim and Gigli 1981; Lim et al. 1981). The biological significance of oxidative injury as one of the key pathophysiological mechanisms in porphyrin photosensitivity is underlined by findings that the antioxidant β-carotene and carotenoid pigments show photoprotective effects in patients with erythropoietic protophorphyria (Mathews 1964; Mathews-Roth et al. 1970, 1974; Mathews-Roth 1987). Furthermore, tocopherol prevents photo lipid peroxidation of red blood cells from patients with erythropoietic protophorphyria (Goldstein and Harber 1972).

Flavins. Because riboflavin is present free and as a conjugate in skin, it was hypothesized to be an endogenous photosensitizer (Tuveson and Sammartano 1986). Flavins are among potential candidates which mediate UVA-induced DNA cross-linking via generation of reactive oxygen species. The formation of DNA-protein cross-links by UVA irradiation in human fibroblasts was suggested to be an indirect reaction involving endogenous photosensitizers and reactive oxygen species (Peak et al. 1985). Flavins were identified as photosensitizers in human fibroblasts (Pereira et al. 1976). Although the involvement of flavins in any photodermatosis has not been established, it was suggested that flavins are potential chromophores for evoking UV radiation effects in human skin (Joshi 1989).

Flavin is readily converted to the radical form by enzymic or nonenzymic reactions. The resonance-stabilized flavin radical can then react with molecular oxygen, resulting in the formation of superoxide anion radicals (Fig. 3.5). Nonenzymic activation of flavin can be induced by electromagnetic irradiation. Flavins have an absorption maximum in the 450–460 nm region and sensitize the transmission of solar energy to oxygen, leading to the formation of superoxide anion radicals (Cunningham et al. 1985; Ballou et al. 1969; Massey et al. 1969).

Oxidative damage may play a key role in flavin photoinactivation of cells and organelles. Human bone marrow progenitor cells (Petkau et al. 1987) and human diploid lung fibroblasts (Pereira et al. 1976) are inactivated in vitro by photosensitizing the cells with flavins in a reaction involving superoxide anion.

Mitochondrial electron transport is inhibited and oxidative phosphorylation partially uncoupled in isolated mitochondria exposed to visible light in the presence of riboflavin (Aggarwal et al. 1976). Flavin-photosensitized mitochondrial damage is oxygen dependent, indicating involvement of reactive oxygen species (Aggarwal et al. 1978).

Photoinactivation of microsomal NADPH cytochrome P-450 reductase by visible light is oxygen dependent and correlates with lipid peroxidation.

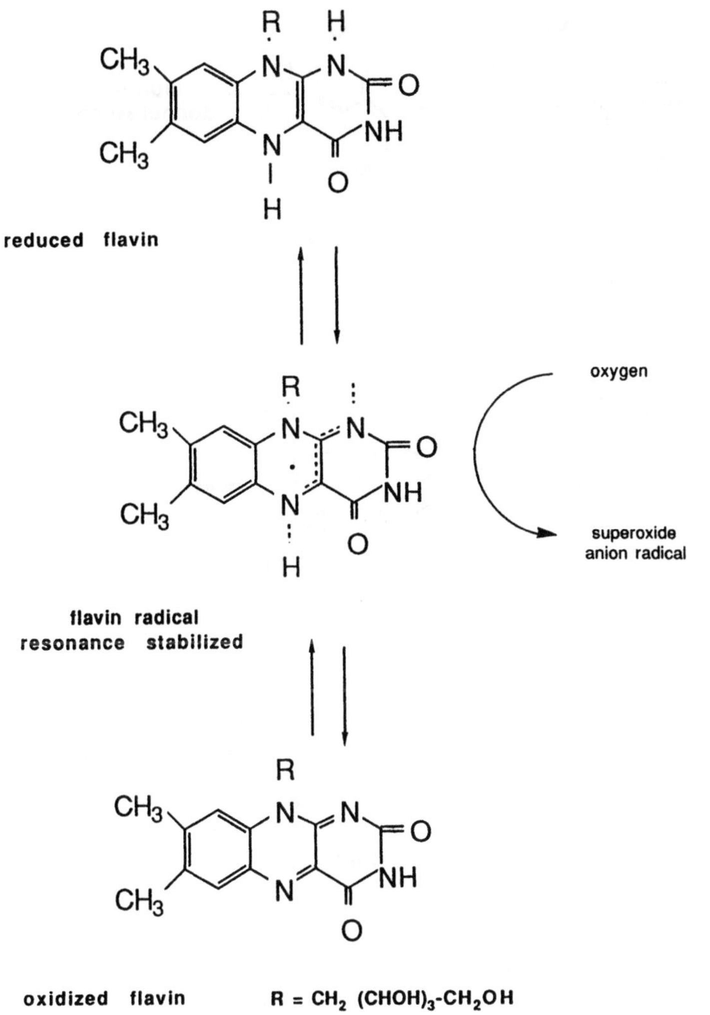

Fig. 3.5. Flavin oxidation

Flavin-mediated photosensitization is suggested to cause inactivation of the enzyme (Augusto and Packer 1981).

Melanins. Melanins are redox polymers containing high concentrations of *o*-quinones (oxidizing groups) and *o*-hydroquinones (reducing groups). According to the presumed chemical precursors, melanins can be classified into eumelanins (tyrosine/dopa), pheomelanins (cysteine/dopa), and allomelanins (catechol or other polyhydroxy-aromatics). Tyrosinase is the rate-limiting enzyme of melanin biosynthesis which hydroxylates tyrosine and oxidizes the catechol dopa to dopaquinone (Fig. 3.6). Dopaquinone cyclizes to leucodopachrome, which is oxidized to dopachrome, decarboxylated to 5,6-dihydroxyindole, and oxidized to indole-5,6-quinone. Indol-5,6-quinone polymerizes

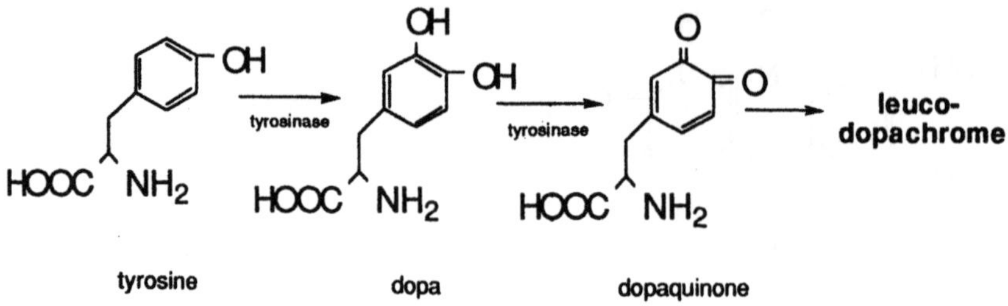

Fig. 3.6. Formation of dopaquinone

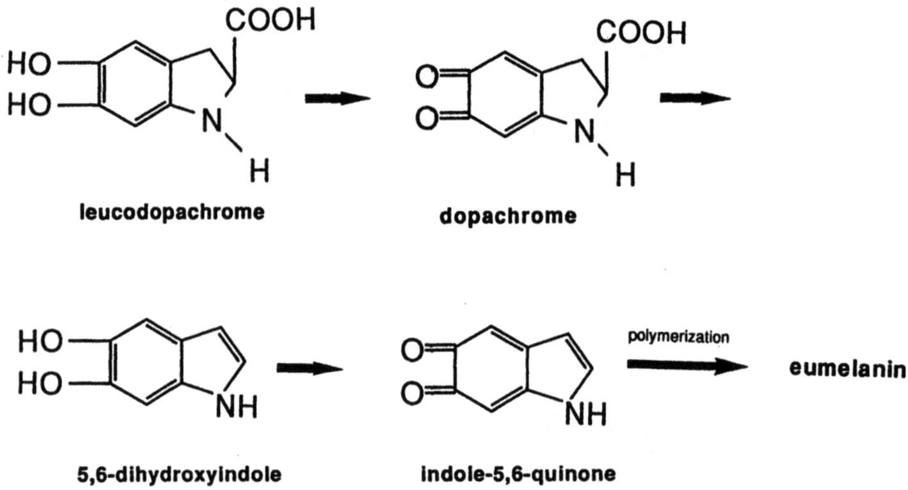

Fig. 3.7. Formation of eumelanin

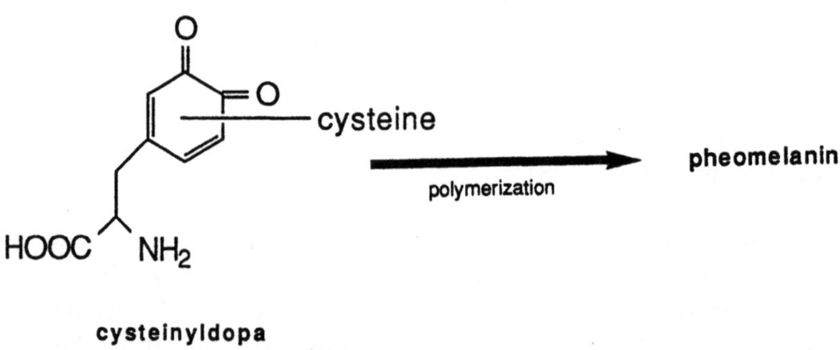

Fig. 3.8. Formation of pheomelanin

to eumelanin (Fig. 3.7). Pheomelanin is the polymerization product of the cysteinyl conjugate of dopa (Fig. 3.8.).

The free radical properties of melanin in biological tissue was reported in 1954 (Commoner et al. 1954) and later confirmed by others (Blois et al. 1964; Mason et al. 1960). A heterogeneous population of intrinsic free radicals can be found in melanins. Additional free radicals may be induced by UV (Stratton and Pathak 1968). The intrinsic ESR signal of melanin can either be increaseed or decreased by metal ions, depending on the magnetic nature of the nuclei (Sealy 1984). Melanins contain functional groups capable of metal binding and may fill biological roles as a metal ion exchanger. A variety of drugs bind to melanin, probably forming charge transfer complexes in which melanin serves as an electron acceptor (Sealy et al. 1980).

Low-temperature and ambient-temperature EPR studies have contributed significantly to our present understanding of melanin photochemistry (Sealy et al. 1980). A large number of EPR studies on melanin have been conducted recently and are compiled in the yearly Specialist Periodical Reports by the Royal Society of Chemistry (e.g., Symons 1990).

Melanins can undergo reversible redox processes and are able to scavenge persistent as well as highly reactive free radical species. However, during autooxidation of melanin pigments, redox active metal ions like iron and copper can catalyze the formation of hydroxyl radicals and hydrogen peroxide (Hintz et al. 1986). Hydrogen peroxide is considered the main reaction product of photoinduced reduction of oxygen by eumelanin (Sarna and Sealy 1984 a).

Although the main function of melanin is considered to be protection against UV irradiation (Lukiewicz 1972), some results show that melanins may also act as photosensitizers. Photosensitization involves the formation of metastable triplet melanin via singlet melanin, and the subsequent reaction of the triplet melanin with a substrate, e.g., oxygen. The reaction with molecular oxygen yields superoxide anion radicals and ground state melanin with an increased content of semiquinone radicals. This melanin can directly dismutate or react in a second step with oxygen. Formation of singlet oxygen does not involve production of increased semiquinone radicals (Fig. 3.9).

Melanin may act as a photoprotective agent at long UV and visible wavelengths, while it may act as a photosensitizer in the short UV range. The chromophore responsible for photoinduced oxygen consumption in melanins is not identical with the major melanin chromophore responsible for absorption of visible light. Quantum yields for oxygen consumption in eumelanin are low with visible light but increase sharply with light of shorter wavelength (Sarna and Sealy 1984 a, b). This indicates increased generation of reactive oxygen species in the short UV. It was also discussed that a lower doses of UV irradiation melanin may have a protective effect on the pigmented cell, and at higher doses it could become cytotoxic (Menon and Habermann 1977).

Radical species like superoxide anion, hydrated electrons, and hydrogen atoms appear to be involved in melanin photochemistry (Sealy 1984; Kalyanaraman et al. 1984). Reactive oxygen species are formed during irradiation of

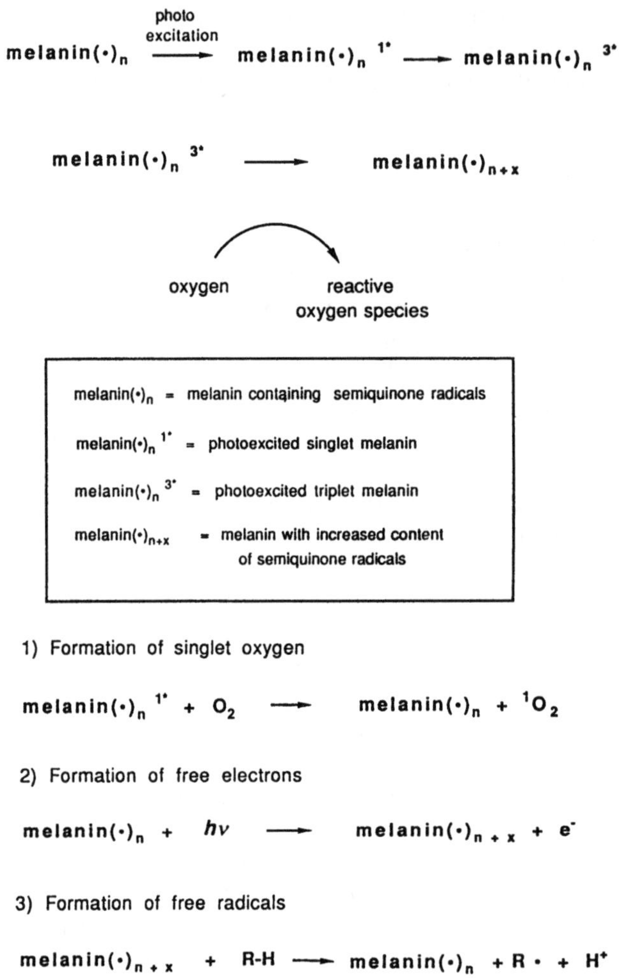

Fig. 3.9. Melanin radical

melanogenic precursors as well as during radiation of melanins (Tomita et al. 1984). Singlet oxygen and superoxide anion radical participate in the photopolymerization of melanogenic precursors in the form of immediate and delayed tanning reactions stimulated by UVA (Joshi et al. 1987). Reactive oxidants derived from melanin (Ranadive et al. 1986) and melanogenic precursors (Koch and Chedekel 1987; Chedekel and Zeise 1988) might play a significant role in the pathogenesis of acute solar responses, such as inflammation, as well as in chronic sequelae, such as premature aging and carcinogenesis.

Irradiation of mast cells in the presence of *pheomelanin* releases large amounts of histamine, while with *eumelanin* it does not (Ranadive et al. 1986). It is suggested that the sunburn reaction more frequently observed in celtic skin may be related to the type of melanin present in skin. Pheomelanins have been suggested to be more efficient oxygen radical producers than eumelanins. Irradiation of pheomelanin can generate large amounts of superoxide radicals (Chedekel et al. 1978). However, in another study pheomelanins were not found to be more effective than eumelanin in promoting oxygen consumption or to be more susceptible to photooxidation (Sarna et al. 1984c).

For a variety of different melanins, free radical production and oxygen consumption can be enhanced in the presence of other photosensitizers (Sealy et al. 1984). Whereas oxygen consumption and free radical production induced by direct irradiation appear to be closely related for eumelanins and pheomelanins, sensitized reactions reveal a greater reactivity of pheomelanins (Sarna et al. 1985a). Thus, pheomelanins are likely to be more potent oxygen radical generators than eumelanins in the presence of exogenous or endogenous sensitizers. This might be important in view of the putative link between pheomelanin in the skin of red-haired individuals and the incidence of skin cancer (Sarna et al. 1985a).

Reaction analysis of melanin with radicals generated in aqueous media by pulse radiolysis reveals that melanin reacts rapidly with either oxidizing (hydroxyl radical) or reducing radicals (hydrated electrons and superoxide anion). The reaction of melanin with hydroxyl radical has the highest rate constant, hydrated electrons having a slightly lower reactivity. The superoxide radical seems to be less reactive with melanins. Since superoxide can act both as an oxidizing and as a reducing agent, it is possible that it reduces the quinone moiety of melanin, liberating molecular oxygen and melanin radical, or oxidizes the catechol moiety, forming hydrogen peroxide and the melanin radical (Fig. 3.10).

The ability of any agent to scavenge short-lived radicals depends on both the effective rate constants of the specific reaction and the local concentration of the scavenger. Melanin's insolubility at physiological pH and its structural organization in the form of melanosomes and pigment granules are mainly responsible for its heterogeneous distribution in pigmented cells. These factors have to be considered carefully when speculating about the possible radiation-protective role of melanin in pigmented cells (Sarna et al. 1986).

Melanization protects human cells from solar UVB by a mechanism that is associated with guarding from reactive oxygen species rather than direct shielding of the DNA (Musk and Parsons 1987). Scavenging of superoxide is a possible factor in the photoprotection afforded by the melanin pigments.

$$O_2^{\cdot\,\cdot} + \text{melanin}(\cdot)_n \xrightarrow{\text{oxidation}} H_2O_2 + \text{melanin}\ (\cdot)_n + 1$$

$$O_2^{\cdot\,\cdot} + \text{melanin}(\cdot)_n \xrightarrow{\text{reduction}} O_2 + \text{melanin}\ (\cdot)_{n-1}$$

Fig. 3.10. Melanin redox chemistry. *Melanin* $(\cdot)_{n+1}$, *melanin* $(\cdot)_{n-1}$, oxidized and reduced melanin

Melanin can compete with SOD for scavenging of the superoxide radical (Korytowski et al. 1985, 1986). It is proposed that, in melanized cells, melanin can act as a radioprotector in a similar fashion to the observed SOD protection afforded to other mammalian cells (Goodchild et al. 1981). The scavenging properties of melanin may be related to the relatively high resistance to ionizing radiation of melanin-containing cells. The scavenging of superoxide by melanin may also be involved in processes other than photoprotection. The melanin concentration in the pigmented livers of several mammalian species is inversely proportional to the SOD content. It is speculated that in tissue with a low SOD content, melanin protects cells against oxygen toxicity by scavenging the superoxide anion (Korytowski et al. 1986).

Furthermore, it is also suggested that part of the function of melanin may be to remove transient free radicals. Melanins can scavenge persistent free radicals such as diphenylpicrylhydrazyl (DPPH) and nitroxides. Nitroxide free radicals are reduced by melanins (Sarna et al. 1985b). During this reaction, the hydroquinone groups of melanin are oxidized to quinone groups via transient semiquinone radicals. The oxidized polymer can be recycled using reductants such as ascorbate or NADH (Sealy et al. 1980).

2. Exogenous Photosensitizers

Toward the end of the nineteenth century it was discovered by Raab (1900) that the microbial toxicity of acridine dyes depended on the intensity of the sunlight in the laboratory. Raab also reported that eosin (Fig. 3.11) can cause a photosensitizing effect and noted that oxygen was involved in the photosensitivity reaction. Eosin was probably the first photosensitizing drug administered orally in humans. A french neurologist noted in 1900 that patients with epilepsy treated with eosin responded with a dermatitis in light-exposed body areas. Eosin derivatives are suggested to cause photosensitization via the formation of singlet oxygen (Pooler 1989). The mechanism of eosin-mediated generation of reactive oxygen species may involve triplet eosin and/or resonance-stabilized eosin radicals.

A variety of drugs may cause photosensitivity reactions in skin, and different mechanisms are involved in the pathogenesis of photosensitivity reactions. Drug-induced photosensitivity involves phototoxic and photoallergic reactions. Phototoxic reactions occur most frequently and are nonimmuno-

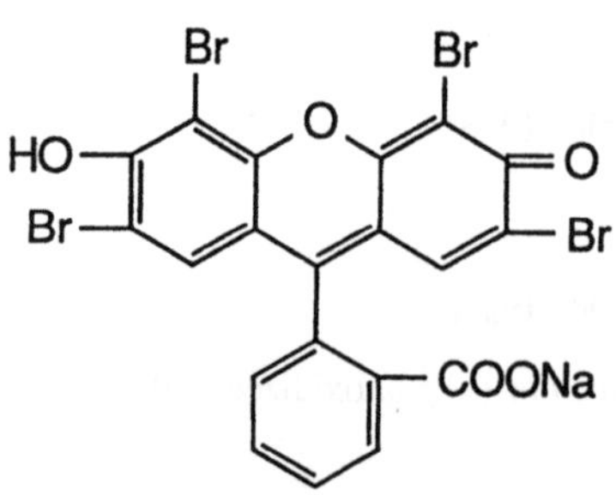

Fig. 3.11. Eosin

logical. Photoallergic reactions are defined as an acquired, delayed hypersensitivity dermatitis and are clinically rather uncommon. They involve photochemical modification of the haptene molecule to a reactive species that readily reacts with a macromolecule to yield an antigen. Phototoxic reactions, which are frequently seen by clinicians, usually occur with the first exposure to a photosensitizing substance and cause an acute, sunburnlike dermatitis. Photoallergy develops after an incubation period of about 10 days and is characterized by an urticarial, papular, or eczematoid-like morphology. Formation of free radicals and reactive oxygen species may be involved in both phototoxicity and photoallergy.

Basic Photochemistry. The ground-state photosensitizer absorbs energy, and the metastable triplet state is formed, via the short-lived singlet state. The triplet photosensitizer then reacts with a substrate (e.g., DNA base) via a hydrogen atom or electron transfer. Thereby, free radicals may be generated (type I photoreaction). The triplet photosensitizer may also react with oxygen, leading to reactive oxygen species (type II photoreaction). Both types of photoreaction include the formation of free radicals, activation of reactive oxygen species, photoactivation resulting in a toxic, nonradical product, and photoaddition of the drug to biological target molecules (Epstein 1977a, Epstein and Wintroub 1985). Types I and II reactions are known to be involved in photooxidation of amino acids, DNA base alterations, and lipid peroxidation.

Thiazide diuretics are cutaneous photosensitizers. Photosensitization studies of five structurally different thiazides revealed that only penflutizide (Fig. 3.12), the most potent phototoxic one, photooxidizes squalene in vitro to yield lipid peroxidation products (Matsuo et al. 1986). The mechanism of phototoxicity involves singlet oxygen, and membrane lipids may be one of the targets (Hayakawa et al. 1987).

Phototoxic and photoallergic reactions in skin are induced by *sulfanilamide,* probably via a free radical mechanism. Indirect evidence for the production of free radicals during irradiation of sulfonamides has been provided by using the method of spin trapping. A variety of reactive free radicals are formed upon irradiation of sulfanilamide (Chignell et al. 1980, 1981; Motten and Chignell 1983) (Fig. 3.13).

Aromatic sulfonyl radicals readily join to unsaturated hydrocarbons, and the amine radical easily binds to thiol groups. Photoinduced covalent binding of haptene sulfonamide radical to protein may result in the formation of a complete antigen responsible for the sulfonamide photoallergxy. Presently, it

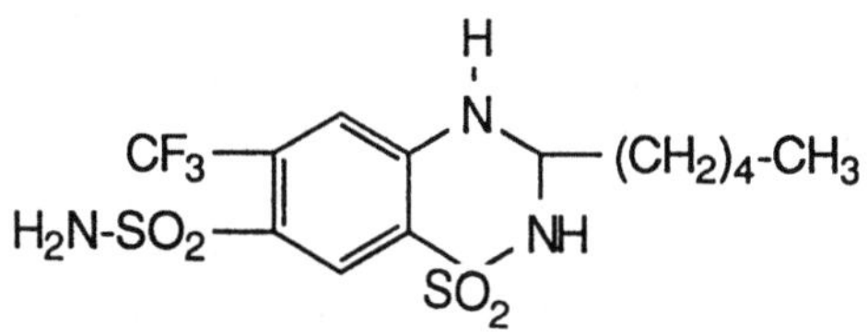

Fig. 3.12. Penflutizide

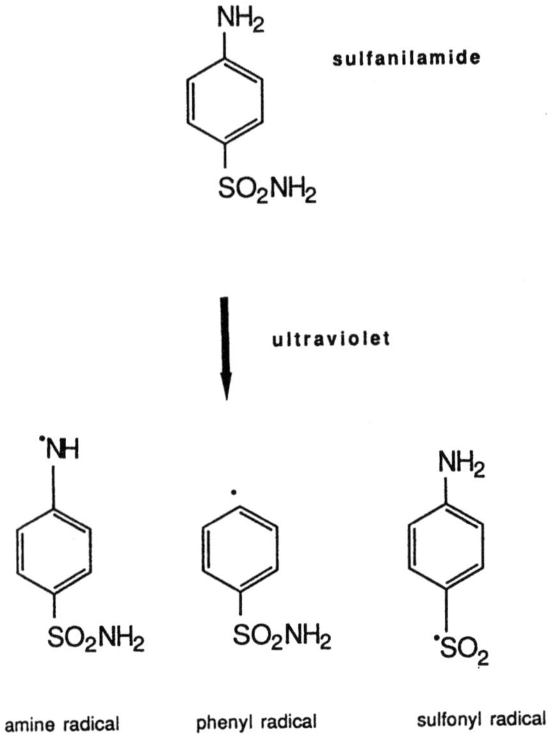

Fig. 3.13. Sulfanilamide-derived free radicals

is not known which of the sulfonamide radical photoproducts is ultimately responsible for the photosensitizing effect.

Phenothiazine drugs are sensitive to oxidation by photochemical, chemical, and metabolic mechanisms. Chlorpromazine and related phenothiazines cause cutaneous and ocular, phototoxic and photoallergic reactions in patients treated with these drugs. The wavelength range required to cause a photosensitive reaction is a topic of controversy. Chlorpromazine has an absorption maximum at 305 nm in aqueous solution. The action spectrum for phototoxicity in humans was reported to be below and above 320 nm. For phototoxicity in mice, the action spectrum was reported to be 320–340 nm (Hunter et al. 1970). Upon irradiation with UV, chlorpromazine yields a variety of free radicals. A cationic one is formed by UVC; UVB induces the formation of the neutral promazinyl radical and a sulfur-centered peroxyl radical.

The neutral promazinyl radical is suggested to be the important intermediate in chlorpromazine photosensitization (Chignell et al. 1985; Motten et al. 1985) and is held responsible for the photobinding to eye and skin (Schoonderwoerd et al. 1989). Metabolites of phenothiazines, e.g., the sulfoxides, have also been identified as photodynamic agents responsible for photosensitivity (Clare et al. 1947). Although oxygen-derived free radicals and singlet oxygen do not appear to be important (Chignell et al. 1985), hydroxyl radicals

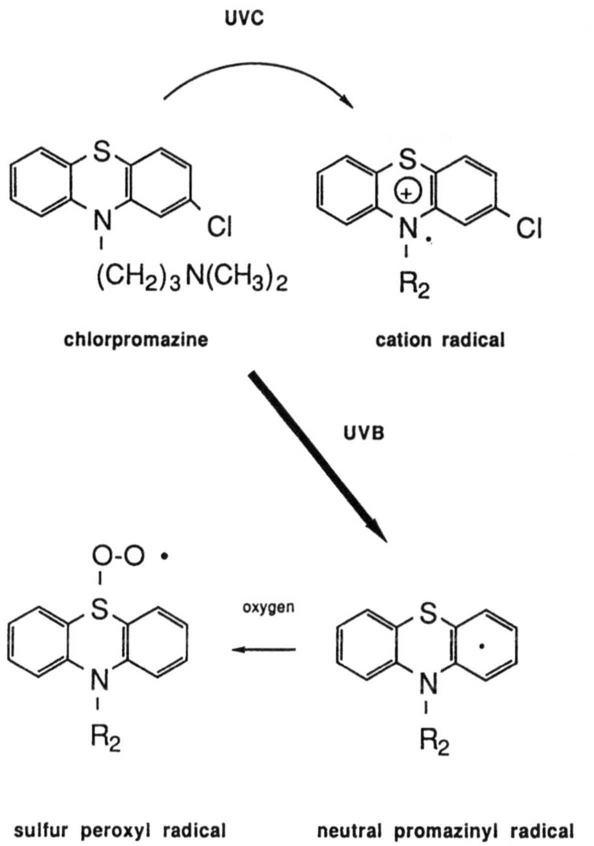

Fig. 3.14. Chlorpromazine photooxidation

are implicated in the phototoxicity of chlorpromazinesulfoxide (Buettner et al. 1986). In chlorpromazine photoallergy, the neutral promazinyl radical may be the reactive haptene, covalently conjugating to macromolecules and thereby forming a complete antigen (Fig. 3.14).

Piroxicam ist a nonsteroidal antiphlogistic agent that causes cutaneous photosensitivity via a phototoxic and a photoallergic mechanism. The low incidence and production of eczematous cutaneous reactions in patients rather indicate a photoallergic mechanism. However, piroxicam photosensitivity begins within a few days after starting application, which is inconsistent with a photoallergic reaction. A metabolite is probably responsible for the photosensitivity induced, involving UVA-mediated formation of singlet oxygen (Kochevar et al. 1986; Western et al. 1987) (Fig. 3.15).

Cutaneous photooxidation (UVA) to *tetracyclines* is mainly associated with doxycycline, demeclocycline, and chlortetracycline, and infrequently with oxytetracycline and tetracycline; minocycline is not phototoxic. The order of reasing potency in generating singlet oxygen is: demeclocycline > tetracycline > chlortetracycline > doxycycline > minocycline. Superoxide anion radicals (Li et al. 1987a, b) and singlet oxygen (Glette and Sandberg

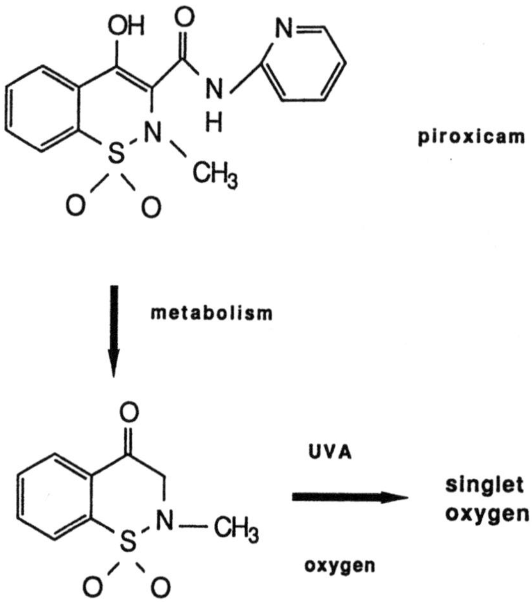

Fig. 3.15. Piroxicam photooxidation

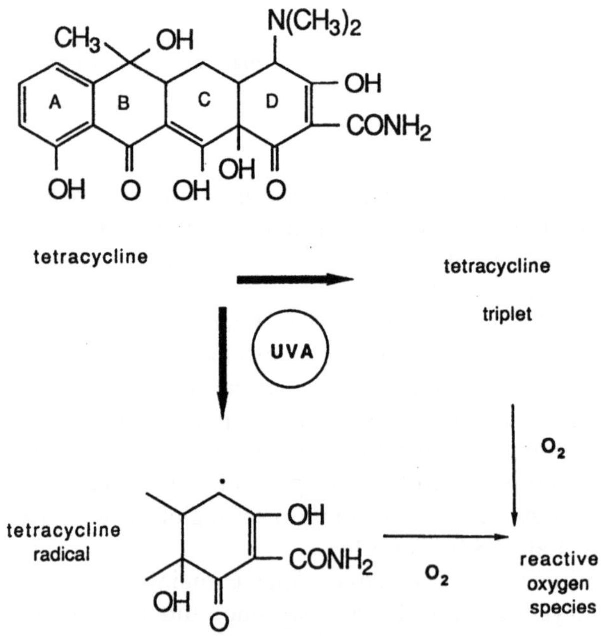

Fig. 3.16. Tetracycline photooxidation

1986) are thought to be involved in tetracycline-induced phototoxicity. There is also evidence for type I photoreactions in tetracycline photosensitivity. Chlortetracycline, but no demeclocycline, generates an aryl radical during UVA irradiation via dechlorination at the aromatic ring system A (Li et al. 1987 a, b) (Fig. 3.16).

The subcellular target site of doxycycline phototoxicity in human epithelial cells was identified as the mitochondrion (Shea et al. 1986). Inhibition of mitochondrial oxidative phosphorylation has been suggested as a key factor in the cutaneous phototoxicity of demeclocycline (demethylchlortetracycline) (Takagi et al. 1989). The high susceptibility of the mitochondrial ATPase complex to reactive oxidants (Hyslop et al. 1988) may contribute to the mechanism of mitochondrial phototoxicity of tetracyclines. In addition of protein and lipid alterations, tetracycline also photosensitizes DNA damage via the formation of reactive oxygen species (Piette et al. 1986).

Tar preparations, in particular coal tar, have been used in dermatological therapy for decades, and it is widely accepted that coal tar causes phototoxic reactions. The photosensitizing components of coal tar in humans are anthracene, pyrene, benzo[a]anthracene, fluoranthrene, and acridine (Fig. 3.17). The metastable triplet state of the polyaromatic hydrocarbons is probably involved in the sensitizing reaction. It was suggested that singlet oxygen partici-

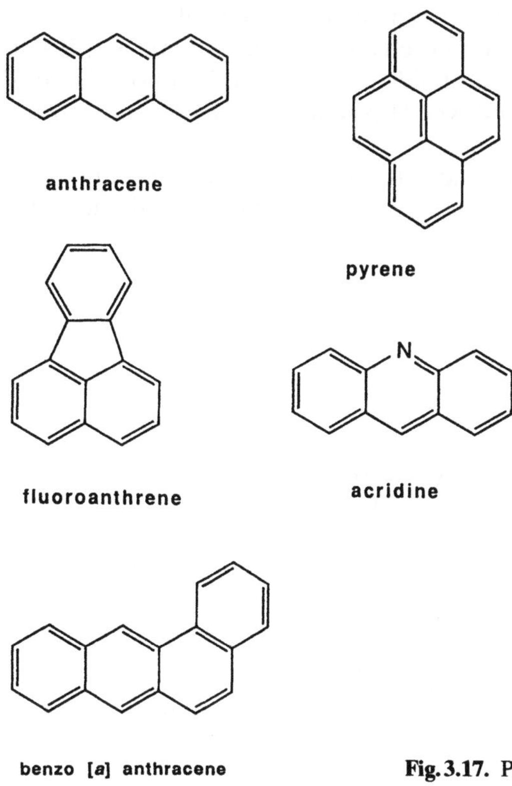

Fig. 3.17. Photosensitizing components of coal tar

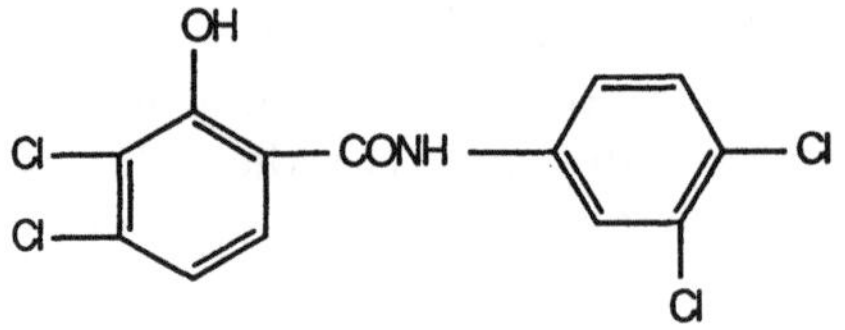

Fig. 3.18. Tetrachlorosalicylanilide

pates significantly in the phototoxicity of tar products (Joshi and Pathak 1983; Kochevar et al. 1982).

Halogenated salicylanilides, used topically as antibacterial agents, are important contact photoallergens (Epstein et al. 1968). Photodermatitis due to soap containing germicides like tetrachlorosalicylanilides is probably induced by a free radical mechanism (Jenkins et al. 1964) and is believed to constitute a photoallergic reaction. UV irradiation of 3,3′,4′,5-tetrachlorosalicylanilide (Fig. 3.18) results in the formation of aryl radicals (Chignell 1989), which can form covalent products with antigenic properties.

Dermal lipofuscinosis, indicating drug-induced lipid peroxidation in skin (Miller and McDonald 1984), is reported in cases of *amiodarone*-induced phototoxic skin reactions.

The binding of the photoallergen *fenticlor* to proteins was found to be partly oxygen dependent, demonstrating a role for semiquinone-type radicals in the photoallergy mechanism (Barratt; personal communication).

Griseofulvin is known to cause photosensitivity with systemic use. It sensitizes photooxidation of unsaturated lipids in vitro, and the photochemical mechanisms involve both type I and type II (singlet oxygen) reactions (Matsuo, personal communication).

Superoxide anion and hydroxyl radicals are the molecular agents of the phototoxicity mediated by certain *acridine dyes* (acriflavin, ethacridinlactate)

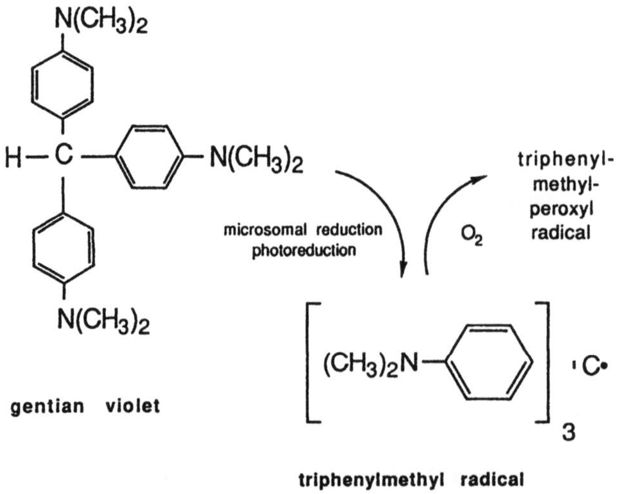

Fig. 3.19. Formation of radicals from gentian violet

(Martin and Logsdon 1987) and by the food and drug additive *tartrazine* (Merville et al. 1984).

Triphenylmethane dyes such as gentian violet, fuchsin, and brilliant green may cause photosensitivity and phototoxic reactions via the formation of triphenylmethyl and triphenylmethylperoxyl radicals or the triplet state (Fig. 3.19).

IV. Photoaging

It is established that intrinsic (chronological) aging and photoaging result in distinct pathomorphological changes in skin (Lavker 1979; Gilchrest et al. 1983; Braverman and Fonlerko 1982a,b; Kligman and Kligman 1985). As a corollary, the biochemical events responsible for the distinct morphological alterations are presumably different.

In intrinsic aged skin, epidermal thinning is characteristic, all cell types decrease in number in the dermis, and the horizontal vascular plexus remains intact. The number of elastin fibers is only slightly increased, with some morphological deterioration. The glycosaminoglycan content is slightly reduced, and no elastosis is found. Collagen fibers are not fewer but become more stable in intrinsic aged skin.

In photoaged skin, actinic damage results in epidermal thickening (acanthosis), enlargement of sebaceous glands, thickening, and inelasticity of the dermis because of massive elastosis. Fibroblasts are numerous (some are hyperplastic), histiocytes and other mononuclear cells increase in number, and mast cells, partially degranulated, are abundant. Fibroblasts become more active, resulting in augmented dermal collagen, glycosaminoglycan, and reticulin production. Later on, damage to preexisting mature collagen is also noted. Collagen fibers decrease, become solubilized, and are replenished by elastotic material and glycosaminoglycans. The elastotic material found in photodamaged dermis was suggested to represent altered elastin; however, its profusion supports the hypothesis that collagen is involved in its formation. Indeed, it was recently discovered that surface alterations of collagen fibers may play a central role in the development of elastotic material in photodamaged skin (Schwartz et al. 1989). However, biochemical analysis of several parameters in UVB-exposed mouse skin revealed that there is insufficient collagen present to support the increasing mass of the tissue (Chatterjee et al. 1990). In photoaged skin, the horizontal vascular plexus is obliterated, and there is a marked invasion of inflammatory cells, mainly lymphocytes, into the dermis. Skin atrophy is the end-stage of photodamage.

Reactive Oxidants in Skin Photoaging. The fact that free radicals generated in biological systems cause damage similar to aging is well recognized. Photochemically generated reactive oxygen species can induce protein denaturation and cross-links in collagen and are thought to contribute to the accelerated process of photoaging of mammalian skin (Dalle-Carbonare and Pathak 1989; Pathak et al. 1990). Histological and histochemical markers for UV-induced photoaging have been characterized. However, the biochemical changes associated with the photoaging process are only poorly understood, and the theory that accumulation of reactive-oxidant-induced cellular injury contributes significantly to the photoaging process in skin is open to debate.

Skin Autofluorescence. In vivo measurements of human skin revealed that sun exposure alters the spectral pattern of skin fluorescence. The differences in skin autofluorescence appear to be related to photoexposure and not to the intrinsic skin aging process. The solar-induced changes are located in the dermis and can probably be assigned to changes in elastin and/or collagen (Leffell and Stetz 1988). The specificity of the particular method used in this study has, however, been criticized (Anderson 1989a).

UVB is thought to be largely responsible for photoaging of skin (Pfau et al. 1986). However, it is entirely possible for humans to accumulate UVA dosages over a few years' time that accelerate the skin's aging process. UVB is implicated in elastosis, wrinkle production, and tumor formation, while UVA causes acceleration of skin changes similar to those occurring with intrinsic aging in the mouse (Bissett et al. 1987). Many of the UVB-induced alterations regress after UVB irradiation is stopped. However, no regression in UVA-induced skin alterations is observed when UVA irradiation ends (Bissett et al. 1987). In the mouse, the chronic effects of UVA and UVB on skin are quite distinguishable (Pfau et al. 1986). Topical treatment of mice with antioxidants such as tocopherol reduces UVB but not UVA induced skin photoaging (Bissett et al. 1990).

UVA and Infrared Radiation. As with erythema, the dose required to produce these changes with the less energetic UVA is 3 orders of magnitude greater than with UVB. In contrast to UVB, there is no significant infiltration by inflammatory cells, and collagen is not altered (Kligman and Kligman 1985). The combined irradiation of mouse skin with UVA and UVB results in a potentiation of most of the effects produced by UVB alone (Kligman and Kligman 1985). Infrared radiation may also contribute to photoaging. It has been shown that infrared light can substantially potentiate UVB-induced dermal damage in the hairless mouse (Kligman 1982). Application of a total dose of 3000 J/cm^2 infrared light (35 J/cm^2 delivered thrice weekly) produces changes in the dermis of hairless mice (elastosis) that are qualitatively similar to UVB-induced changes (Kligman and Kligman 1985).

V. Photocarcinogenesis

In 1884, Unna suggested a relationship of malignant skin diseases with solar exposure. Recently, a large number of dermatopathological conditions including skin neoplasms, such as basal cell carcinoma and squamous cell carcinoma, and their dependence on overexposure to sunlight were reviewed (Engel et al. 1988). The potential risks of therapeutic skin irradiation (UV and ionizing) have been discussed extensively (Burns 1989). The environmental hazard due to the depletion of stratospheric ozone is controversial (Frederick et al. 1989); however, it is believed that the consequent increase in solar irradiation will significantly elevate the incidence of skin cancer in the future (Urbach 1989).

Melanoma. Although the involvement of sunlight in the induction of squamous cell carcinoma and basaloma is established, its participation in the pathogenesis of melanoma is disputed, with the exception of lentigo maligna melanoma. Cumulative, lifelong, UV exposure is responsible for the development of lentigo maligna melanoma. The more common types of melanoma (superficial spreading and nodular) are typically not located in sun-exposed skin areas, are not associated with solar skin damage around the tumor, and do not occur more frequently among outdoor workers (or outdoor workers carry only a very small excess risk). However, the incidence of melanoma among white-skinned people increases as their residence approaches the equator, and it is very rare among dark-skinned people. The risk rises in lightly pigmented subjects who burn rapidly in the sun and do not tan well. People with an increased number of nevi are also at higher risk (Lee 1989). The incidence of malignant melanoma is growing worldwide, and this has been related to the increased exposure to solar radiation. High-dose UV exposure in early life is suggested to participate in the pathogenesis of superficial spreading and nodular melanoma, as well as in the activation of melanoma in dysplastic nevus syndrome (Jung 1990). However, analytical epidemiology does not support the hypothesis that malignant melanoma is UV dependent (Cascinelli and Marchesini 1989).

Since the theory of UV carcinogenesis of cutaneous melanoma exhibits some inconsistencies, it was hypothesized that an unknown xenobiotic is involved in induction (Rampen and Fleuren 1987). This unknown xenobiotic could work as a photosensitizer interacting with pheomelanin. An extremely high content of pheomelanin is reported in dysplastic melanocytic nevi, and it has been postulated that UV irradiation, through the generation of UV photoproducts upon exposure of pheomelanin, results in the transformation of dysplastic melanocytes (Yamada et al. 1989).

UVB and UVA. The most effective wavelengths for UV carcinogenesis are in the UVB. UVC is probably much less carcinogenic than UVB in vivo (Lill 1983; De Gruijl et al. 1983), due to the strong absorption of UVC in the superficial skin layers. Skin tumors can be induced in hairless mice following multiple small exposures to UVB as well as after a large single dose (0.36 J/cm^2) of UVB (Mathews-Roth 1983). In hairless mice, multiple small doses of UVB radiation are more effective in producing skin cancer than larger, less frequent ones. Exposure of human skin to UVA was considered to be completely harmless (Raab 1979). It is now evident that exposure to high-intensity UVA sources or to solar UVA for prolonged time periods can contribute significantly to skin carcinogenesis.

The rate of skin cancer induction in hairless mice by UVB is enhanced by subsequent UVA irradiation (Staberg et al. 1983a, b). However, irradiation with multiple doses of combined UVB and UVA (total dose UVB 20 J/cm^2, total dose UVA 60 J/cm^2) brings about a slightly lower tumor incidence compared with UVB alone (Staberg et al. 1983a). In another study (Willis et al. 1981), simultaneous UVB/UVA irradiation (total dose 270 J/cm^2; 16% UVB, 84% UVA) of hairless mice resulted in a significant increase in cancer induction. These discrepant results may be due to a time-dose relationship. It must be pointed out that different hairless mouse strains have different susceptibilities to UV carcinogenesis. This may partially explain the dose-response differences in various studies.

The first tumors induced by UVA light alone were reported by Forbes and Cole (1982). Irradiation of hairless mouse skin with multiple doses of UVA from which UVB had been rigorously excluded did induce tumors in mice (Van Weelden et al. 1983; Van Weelden and van der Leun 1985; Strickland 1986). The total UVA dose applied in this study was very large (2.5 J/cm^2 single exposure, 1170 J/cm^2 total exposure). Daily irradiation of hairless mice with a minimal erythemal dose of 22 J/cm^2 UVA ($>$340 nm) resulted in the formation of squamous cell carcinoma with a mean induction time of 265 days, corresponding to a total UVA dose of 5830 J/cm^2 (Sterenborg and van der Leun 1990). Exposure of hairless mice to multiple small doses of UVA with a total dose of 71 J/cm^2 (Staberg et al. 1983a, b), or 40.5 J/cm^2 (Bech-Thomsen et al. 1988) did not result in raised skin tumor incidence.

Generally, the UVA dose required to induce skin tumors is orders of magnitude larger than the dose of UVB required. The maximum UVA fluence recorded in 1 day at an elevation of 3576 m above sea level is 170 J/cm^2 (Blumenthaler et al. 1985), which is about tenfold the minimal eryhtema dose in skin type II patients. Large fluences of UVA light are also emitted by artificial suntanning devices. Since tanning requires much higher doses of UVA than UVB, the carcinogenic risks of tanning by UVA and of tanning by UVB are reported to be of the same order of magnitude (Engel et al. 1988). A compilation of the literature on the dose-response relationship of UVA photocarcinogenesis was recently presented (Sterenborg and van der Leun 1990). Treatment of hairless mice with 42 J/cm^2 UVA preceding exposure to broadband UV irradiation (UVB + UVA) significantly delays skin tumor development (Bech-Thomsen et al. 1988). Pretreatment with UVA did not alter the epidermal thickness or melanin content. It was suggested that UVA, in addition to its carcinogenic properties (photoaugmentation of UVB-induced skin tumors), may also act as an antipromotor.

1. Photocarcinogenesis and Lipid Peroxidation

It has been proposed that carcinogenesis in the skin may involve free-radical-mediated lipid peroxidation reactions (Black et al. 1985; Black 1987a). UVA and UVB are capable of peroxidizing human skin lipids, and lipid peroxides can be linked to carcinogenesis. Some lipid peroxidation products can react with DNA (Ames et al. 1982), and malonic dialdehyde is a mutagen. The dietary lipid intake and the degree of lipid saturation have pronounced effects on photoinduced skin cancer, with increasing unsaturated fat intake enhancing cancer expression. The level of intake of these lipids is also manifested in the level of epidermal lipid peroxidation. A role of dietary unsaturated lipids in the activation of carcinogens, with particular reference to polycyclic aromatic compounds, has been discussed (Gower 1988). Conversely, dietary antioxidants inhibit both lipid peroxidation and photocarcinogenesis in skin.

2. Photocarcinogenesis and Antioxidants

Antioxidants may play a role in both the inhibition of the primary events involved in UV-mediated tumor induction and the subsequent development of precancerous lesions into tumors (Black and Chan 1975). Dietary administration of BHT inhibits UVB skin carcinogenesis, erythema, and ornithine decarboxylase activity (Koone and Black 1986). Cholesterol oxide, which is a carcinogen, is formed in skin of hairless mice upon UV irradiation and might be involved in the etiology of UV-induced carcinogenesis (Black and Lo 1971;

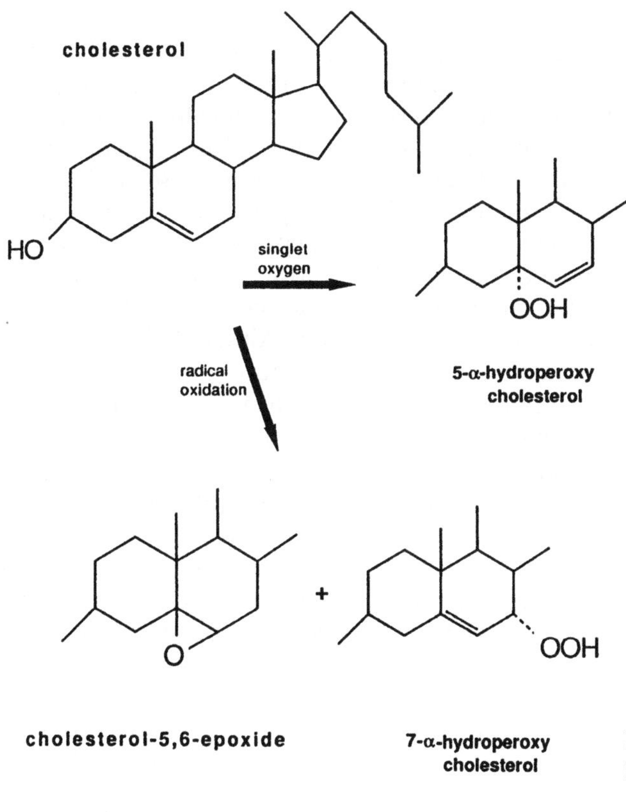

Fig. 3.20. Cholesterol photooxidation

Black and Douglas 1972, 1973). Cholesterol-5,6-epoxide and 5-α-hydroperoxycholesterol are reaction products of cholesterol with reactive oxygen species. Singlet oxygen reacts with cholesterol and the 5-α-hydroperoxide is formed, while reaction of cholesterol with oxidizing radicals yields 7-α-hydroperoxide and other compounds (Santus and Reyftmann 1986) (Fig. 3.20). Systemic application of tocopherol inhibits the photochemical conversion of skin cholesterol to cholesterol oxide. This suggests a possible prophylactic effect of systemic antioxidants on the formation of this carcinogen and the subsequent pathological conditions that may result (Lo and Black 1973).

It is believed that agents capable of enhancing natural defense mechanisms, e.g., melanization and antiradical activity, hold the greatest promise for future investigations on systemic photoprotective agents. Photoprotectants should possess the following characteristics: (a) absorbing or acting otherwise as a barrier to UV, (b) competing with target molecules for the damaging products, (c) providing restorative protection, and (d) suppressing the inflammatory response (Black 1987b, c). An antioxidant with antiinflammatory potency that absorbs in the UVB and UVA ranges and regenerates other endogenous antioxidants may fulfill these criteria.

VI. Photoimmunology

In addition to its direct carcinogenic properties via modification of nucleic acids and lipid peroxidation, UV irradiation causes (with a few exceptions) a systemic suppression of the immune response. The development of skin cancer is thought to be closely associated with a failure of the immune system to recognize and destroy tumor antigens (Cruz and Bergstresser 1989). UV-caused impairment of the immune system may significantly contribute to photocarcinogenesis in humans. UV-induced suppression of immunity is highly selective in experimental animals. Low-dose UVB irradiation produces suppression of contact dermatitis when a specific hapten is subsequently applied to the skin to induce sensitization. Although this selectivity is not yet established in humans, UV-caused immunosuppression is distinct from viral- or drug-induced suppression, which are unspecific and affect most functions of the immune system. The effects of UVB, UVA, and PUVA on immune function are qualitatively different (Morison 1981). Much is known about UVB-induced suppression of contact dermatitis, while data on the immunomodulatory effects of UVA and PUVA are gradually emerging. PUVA treatment of the recipients of murine tail skin grafts leads to prolongation of transplant survival time (Gruner et al. 1990). The effects of UV radiation on the human immune system were recently reviewed by Morison (1989).

Exposure to UV radiation can alter the function and viability of Langerhans cells, neutrophils, and peripheral blood lymphocytes in humans. UVB-peroxidated skin surface lipid fractions (squalene, cholesterol) and synthetic lipid peroxides are able to interfere with the natural killer activity of human peripheral blood lymphocytes in vitro. In vivo studies in mice have revealed that topical application of peroxidized skin surface lipid fractions induces immunosuppression (Picardo et al. 1989). Exposure of murine epidermal cells to UV radiation in vitro causes the release of immunoregulatory factors that mimic some of the immunosuppressive effects of in vivo UV irradiation (Kim et al. 1990). Neutrophils are more susceptible to damage by UV than lymphocytes (Gast et al. 1988). UVB irradiation decreases the natural killer activity of human peripheral blood mononuclear cells, and this suppression is improved by the addition of SOD, indicating involvement of the superoxide anion radical (Toda et al. 1986).

VII. Skin Diseases with Abnormal Reactions to Light

UV and visible light can cause quantiatively and qualitatively abnormal reactions in susceptible patients. Skin diseases with abnormal reactions to light can be categorized into disorders primarily caused by UV (photodermatoses) and those precipitated and/or exacerbated by light. Photodermatoses can be classified into idiopathic (unknown etiology), metabolic and genetic, chemical, neoplastic, and degenerative ones (Table 3.1). Skin diseases which are precipitated or exacerbated by UV are very heterogeneous, and individual susceptibility varies greatly.

Involvement of reactive oxidants in skin diseases with abnormal reactions to light is speculative in all cases. Indirect evidence mainly stems from in vitro demonstrations which show that reactive oxidants and free radical species are involved in important mechanisms relevant to the particular disease. Reaction products of reactive oxidants can be detected in diseased skin or systemically, skin antioxidant activity is found to be impaired, and clinical symptoms can be ameliorated by antioxidants. Involvement of reactive oxidants in photooxida-

Table 3.1. Groups of photodermatoses

Type of photodermatosis	Disease
Idiopathic	Polymorphous light eruption Solar urticaria Actinic reticuloid Hydroa vacciniforme
Chemical	Phototoxicity Photoallergy Persistent light reaction
Metabolic and genetic	Porphyrias Disorders of tryptophan metabolism Melanin deficiency syndromes Skin diseases with increased cellular susceptibility[a]

[a] Not all of these diseases have an abnormal reaction to UV.

tive stress was studied in patients with lupus erythematosus and with a group of diseases with increased cellular sensitivity of DNA to UV.

1. Lupus Erythematosus

Lupus erythematosus is an autoimmune disease of unknown origin, characterized by major alterations of both humoral and cellular immunity. The disease can involve multiple internal organ systems (systemic lupus erythematosus, SLE) or the skin only (chronic discoid lupus erythematosus, CDLE). Photosensitivity is one of the major clinical symptoms of the disease; skin lesions are frequently associated with solar exposure and are also aggravated by UV. UVB and sunlight can precipitate or exacerbate the disease. High doses of UVA and visible light may also be effective in inducing skin lesions in patients (Van Weelden et al. 1989). The mechanism of photosensitivity in lupus erythematosus patients is not well understood, but formation of antigenic material in skin by UV is involved.

Autoantibodies to keratinocyte histone nucleoprotein antigens (SS-A/Ro, SS-B/La, and RNP) are strongly associated with photosensitive cutaneous lupus erythematosus. UV-induced damage to histone nucleoproteins may result in a transcellular elimination of the chemically altered macromolecules. By this mechanism, UVB could induce expression of these nuclear antigens on the keratinocyte cell surface. It is speculated that this results in cell susceptibility to the effects of specific antibodies and subsequently leads to immunological keratinocyte damage (Furukawa et al. 1990). Several studies indicate involvement of reactive oxidants in the pathogenesis of lupus erythematosus, in particular with reference to vascular injury and photosensitivity.

Clastogenic Factors. SLE patients have increased numbers of chromosome breaks (Tuschl et al. 1984). This correlates with a low molecular weight chromosome-damaging agent being released from lymphocytes into the serum. The clastogenic factor was detected in the plasma of SLE but not of CDLE patients (Emerit, personal communication). The presence of clastogenic products in plasma is not specific for lupus erythematosus. They were first described in plasma from patients given ionizing radiation and are also found in patients with spontaneous chromosomal instability, as well as chronic inflammatory and autoimmune diseases such as progressive systemic sclerosis, dermatomyositis, periarteriitis nodosa, rheumatoid arthritis, and Crohn's disease.

Lymphocytes from SLE patients have an abnormal sensitivity to UVA, and UV-induced formation of clastogenic products was suggested to be involved in the photosensitivity (Emerit and Michelson 1981). Generation of reactive oxidants in the medium of lymphocyte cultures either by photoreduction of flavins or by the xanthine-xanthine oxidase system leads to the production of chromosome breakage factors (clastogenic material) which induces chromosome breaks and sister chromatid exchanges (Emerit 1982;

Emerit and Cerutti 1982). This clastogenic material may be composed of lipid peroxidation products; clastogenic factor isolated from lymphocyte cultures contains thiobarbituric acid reactive substances and conjugated dienes (Emerit et al. 1985; Khan and Emerit 1985). Chromosome breaks are produced in normal lymphocytes when they are incubated in the presence of (1) SLE patients' serum, (b) SLE patients' lymphocytes, or (c) purified clastogenic factor (Emerit and Michelson 1981). Chromosome breakage rates in lymphocytes are influenced by exogenous SOD in vitro and were also reduced by in vivo injection of SOD in lupus erythematosus patients and New Zealand Black (NZB) mice, an animal model for SLE (Emerit and Michelson 1980). Tocopherol provides significant systemic photoprotection in patients with cutaneous upus erythematosus (Jordan and Wulf 1950), although it did not significantly improve their existant skin lesions.

Reactive oxygen species have a short half-life, and their hazardous potential is limited to the ultimate vicinity in which they arise. Secondary, longer-lived clastogenic material, which is produced in reactions with short-lived, highly reactive oxidants, can exert their clastogenic effect via free radical intermediates at a distant site from their formation. The fact that patients with progressive systemic sclerosis, Crohn's disease, or rheumatoid arthritis have no significantly increased photosensitivity argues against a common mechanism of clastogenic-product-mediated photosensitivity.

2. Diseases with Increased Cellular Susceptibility

A group of heritable diseases can be linked with an increased cellular sensitivity of DNA to several physicochemical agents. The clinical picture is distinct for each condition, but the diseases often overlap with respect to some of the following clinical abnormalities: high rate of malignancy, paremature aging, skin photosensitivity, hyperpigmentation, telangiectasia, atrophy, and neurological dysfunctions. The underlying mechanism of increased cellular sensitivity to UV and gamma-irradiation and augmented rates of spontaneous chromosome breaks is not known in most cases, and participation of free radicals and reactive oxygen species are suspected in some forms.

The clinically heterogeneous diseases comprise, among others, xeroderma pigmentosum, Bloom's syndrome, Cockayne's syndrome, Fanconi's anemia, ataxia telangiectasia, and dyskeratosis congenita. UV-induced cellular hypermutability is observed in cultured cells from patients with dysplastic nevus syndrome (Jung et al. 1986) and nevoid basal cell carcinoma (Bohnert and Jung 1989). Although it is not known how all of the clinical features of these diseases are explained by the increased cellular sensitivity of DNA, it is evident from the literature that this dysfunction is of significant importance.

Xeroderma pigmentosum, an autosomal recessive disease, is characterized by increased sun sensitivity, freckles, and squamous cell carcinoma in sun-exposed areas. The photobiological wear and tear of suberythemal doses of solar radiation is evident from studies of these patients, who lack efficient excision repair mechanisms. They have a 2000-fold higher incidence of squamous cell carcinomas compared with the general population (Kraemer et al. 1984), and degenerative skin changes occur very early in life. Cellular hypersensitivity to UV radiation and to certain chemicals (4-nitroquinoline-1-oxide, PUVA therapy, nitrogen mustard) exists in association with abnormal DNA repair. Genetic heterogeneity has been demonstrated, and different molecular defects in DNA repair have been identified. Seven complementation groups of excision repair defective cells have been characterized. The cells differ in their degree of UV sensitivity and their relative rates of pyrimidine dimer excision. A variant was described with normal DNA excision repair but with defective postreplication repair. Usually, patients with xeroderma pigmentosum have an impaired excision repair mechanism. Cultured fibroblasts show markedly low catalase (Vuillaume et al. 1986) and SOD activities (Nishigori et al. 1989). Catalase, SOD, and thioredoxin reductase activity were also found to be decreased in the epidermis of xeroderma pigmentosum patients with complementation groups C and D, while glutathione reductase activity was not affected (Schallreuter et al. 1990b). These decreases in enzyme activity could form an additive effect for inducing the carcinogenic process, particularly for those patients with xeroderma pigmentosum and normal excision repair (Vuillaume et al. 1986).

Bloom's syndrome is an autosomal recessive disease characterized clinically by growth retardation, immunodeficiency, solar sensitivity, telangiectatic lesions in sun-exposed skin areas, and proneness to the development of cancer. Cells from these patients are not unusually sensitive to UV. Clastogenic activity was detected in the concentrated ultrafiltrates of culture media from fibroblasts of patients with Bloom's syndrome. These clastogenic components may be DNA-damaging agents, or they may inflict chromosomal damage indirectly by inducing the formation of a burst of reactive oxygen species (Emerit and Cerrutti 1981b). The clastogenic activity of the ultrafiltrate was strongly suppressed by SOD. Fibroblasts from patients with Bloom's syndrome have significantly low catalase activity (Vuillaume et al. 1986). The working hypothesis proposed was that these cells may be deficient in the detoxification of reactive oxygen species, and it is these species that lead to the formation of clastogenic components. Another hypothesis suggests that these cells suffer from an elevated endogenous generation of reactive oxygen species (Poot et al. 1988; Poot and Hoehn 1989).

Cockayne's syndrome is an autosomal recessive degenerative disorder and is not associated with an increased incidence of neoplasia; a DNA repair defect has been suggested as the cause. It is clinically characterized by increased skin photosensitivity, premature aging, pigmentary retinal degeneration, deafness, and dwarfism. Fibroblasts from these patients are hypersensitive to UV-induced growth inhibition. Although they are hypersensitive to UV, they have a similar susceptibility to DNA-damaging agents as do control cells.

Fanconi's anemia is an autosomal recessive disease characterized by pancytopenia, diffuse hyperpigmentation, café-au-lait macules, achromic patches, a high incidence of neoplasia, and multiple systemic developmental defects. It is associated with a high frequency of spontaneous chromosomal aberrations. Chromosomal aberrations are easily induced in Fanconi's anemia cells by ionizing radiation, PUVA therapy, nitrogen mustard, and DNA cross-linking agents. However, the cells are of similar sensitivity to UV compared with control cells. There is evidence for genetic heterogeneity in this disease, as described in xeroderma pigmentosum.

Ataxia telangiectasia is autosomal recessive trait and is among the most common of the heritable disorders with increased cellular sensitivity. Genetic heterogeneity has also been suggested for this disease. It is characterized by cerebellar ataxia, cutaneous and ocular telangiectasias, immune dysfunction, and a high incidence of neoplasia. Telangiectatic lesions develop preferably on skin exposed to sunlight and friction; however, photosensitivity is not increased. Patients have an especially increased sensitivity to ionizing radiation, and cytotoxicity of ataxia telangiectasia cells is further increased in the presence of oxygen. Susceptibility to various DNA-damaging chemicals is generally abnormal. Spontaneous chromosome breakage is found in cultured skin fibroblasts, and in the serum of these patients a clastogenic factor has been described (Shaham et al. 1980). Fibroblasts from patients with ataxia telangiectasia also have significantly low catalase activity (Vuillaume et al. 1986). In cultured human ataxia teleangiectasia cells glutathione biosynthesis is impaired, which is only evident after depletion of the thiol (Meredith and Dodson 1987).

B. Mechanical and Thermal Skin Trauma

I. Wound Healing

Wound healing is a complex biological process, and pathological situations exist with wound healing deficiency. Partial-thickness wounds (a portion of the dermis remains intact) heal by epithelialization; full-thickness wounds can heal by three mechanisms: primary intention, delayed primary closure, or secondary intention (Hunt 1980; Cohen and McCoy 1983). Epithelialization and wound contraction are of minor significance for primary wound healing of full-thickness wounds, e.g., healing of juxtaposed, surgically created wound margins without foreign bodies or infection. Synthesis, deposition, and cross-linking of collagen fibers are important biochemical events in primary or delayed primary wound healing. In the latter case, the delay is caused by infection or foreign bodies. Macrophages and lymphocytes are vital factors in the wound healing process. Epithelialization plays an important role in the healing of partial-thickness wounds such as second-degree burns and mechanical abrasions. Wound healing by secondary intention involves granulation tissue formation, wound contraction, collagen deposition, and reepithelialization. The wound contraction process is the most significant mechanism in the spontaneous closure of large open wounds and involves modification of the cytoskeletal and contractile elements in the fibroblasts. These cells acquire the morphological characteristics of smooth muscle cells and are called myofibroblasts, containing microfilament bundels rich in actin filaments.

The cytoskeletal apparatus of fibroblasts/myofibroblasts and keratinocytes is of great importance in the wound healing process. The cytoskeleton of keratinocytes is an essential factor in epithelial migration (Bereiter-Hahn 1986). Keratinocyte migration plays a large role in the reepithelialization of second-degree burns and mechanical abrasions. The cytoskeleton is maintained by cellular ATP. Susceptibility of the cytoskeleton to oxidative injury can be an important pathophysiological mechanism in wound healing disturbances. Reactive oxidants are generated by neutrophils during the early period of wound healing (Douglas et al. 1989).

The cytoskeleton of rabbit kidney cells is readily damaged during exposure to systems generating free radicals (Hansson et al. 1987). Reactive oxygen species generated during the metabolism of redox-cycling quinone drugs can cause disruption of the normal cytoskeleton organization, presumably by cross-linking actin protein SH groups (Bellomo et al. 1988). Cytoskeletal damage, oxidation of cytoskeletal polypeptides, is also observed in human platelets treated with redox-cycling drugs (Mirabelli et al. 1989). Disturbances in epithelial migration under zero gravity, increased radiation exposure to extremely penetrating charged particles, and T-lymphocyte dysfunction can presumably contribute to wound healing disturbances during space flight (Toback and Kohn 1989).

Antioxidants and Wound Healing. Agents wich reduce neutrophil influx have been suggested to prevent lipid peroxidation at the wound site and improve wound healing (Douglas et al. 1989). Wound healing of partial-thickness wounds in rat skin exposed to ionizing irradiation was improved by tocopherol application (Taren et al. 1987), and reepithelialization of burn wounds was accelerated by tocopherol (Melkumian et al. 1978). However, tocopherol also causes delay in full-thickness wound healing and inhibition of collagen biosynthesis (Rave et al. 1971; Ehrlich et al. 1972).

Hyperbaric Oxygen Treatment and Wound Healing. Hyperbaric oxygen is used in the treatment of gangrenous wounds, ulcers caused by ionizing radiation, and burns. Hyperbaric oxygen therapy of cryobiological wounds in rats significantly reduces the healing time (Delfino et al. 1990). Although it results in increased steady state concentrations of reactive oxygen species in skin, improved production of ATP and stimulated angiogenesis may contribute to its wound healing properties.

Clinical Significance of Reactive Oxidants in Wound Healing Disturbances. Lipid peroxidation products in wounds are chemotactic factors and promote tissue infiltration with white blood cells. This will subsequently result in digestion of tissue debris and destruction of microorganisms. If this process is uncontrolled, it may lead to autooxidative tissue injury and wound healing disturbances. The cytoskeleton of ketatinocytes/myofibroblasts is a target organelle of oxidative damage. Few reports exist on antioxidant treatment to improve wound healing, and hyperbaric oxygen is a clinically accepted regimen only in the treatment of severe wound healing disturbances. In conclusion, although reactive oxidants participate in the pathophysiology of wound healing and its disturbances, the clinical significance remains unknown.

II. Skin Burns

Experimental and clinical evidence is accumulating that reactive oxygen species play a significant role in the pathophysiology of thermal skin trauma and related pathology. Thermally injured rats show increased serum xanthine oxidase activity, which can induce the formation of reactive oxygen species and subsequently skin inflammation and edema (Till et al. 1988). The uric acid concentration in burned human skin is significantly elevated in comparison with unburned skin (Aoki and Iriyama 1986). This indicates the increased formation of reactive oxygen species in burned skin during xanthine-oxidase-mediated degradation of the purine intermediate hypoxanthine to xanthine

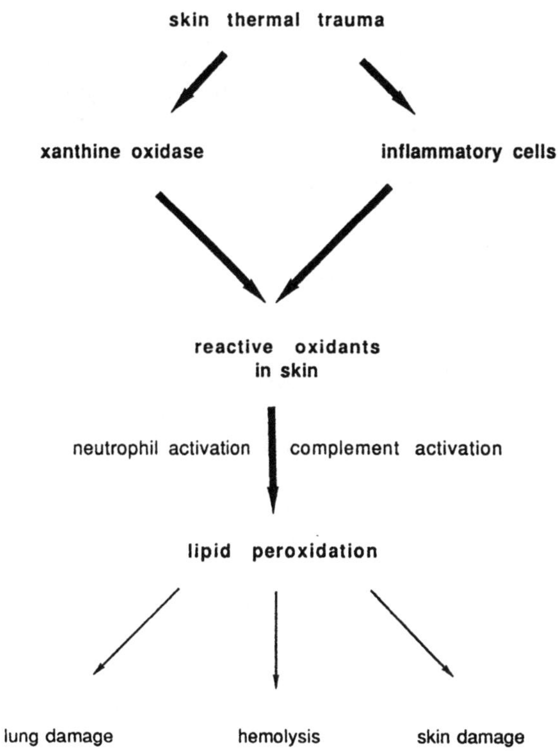

Fig. 3.21. Thermal trauma and oxidative tissue damage

and then to uric acid. Indeed, intravenous administration of hypoxanthine significantly enhances skin edema in burned rat skin (Till et al. 1988).

Skin Damage and Lung Injury. Thermal trauma to skin has been demonstrated to be associated with hemolysis and complement- and neutrophil-dependent acute lung injury. It can cause oxygen radical formation by complement-activated blood neutrophils with resulting microvascular injury in the lung interstitium (Annesely et al. 1985) (Fig. 3.21).

Protection from acute lung injury following remote thermal injury involving skin can be achieved by depleting animals of complement or blood neutrophils or by systemic treatment with a combination of catalase and SOD (Till et al. 1985). Intravenous administration of catalase and SOD reduces plasma volume loss in thermally injured dogs (Hilton 1980). Lipid peroxidation products generated in skin by thermal trauma can leak into the bloodstream, as has been observed in rats (Nishigaki et al. 1980), mice (Kawai et al. 1988), and humans (Hiramatsu et al. 1984). These products may cause hemolysis, complement activation, and microvascular lung damage.

The experimental data indiate that reactive oxidants are involved in the pathophysiology of thermal skin and lung injury. Reactive oxygen species and lipid peroxidation products can activate complement, which will subsequently result in neutrophil stimulation, depletion of antioxidants, and tissue damage. This scenario can be influenced in animals by antioxidant treatment. The promising results warrant clinical studies in burn patients.

Lipid Peroxidation – Animal Studies. Lipid peroxides increase in skin lesions after burns are inflicted on the skin of mice. SOD activity decreased, and glutathione peroxidase activity is almost undetectable in mouse skin after burning. In the serum of mice, lipid peroxide content and SOD activity rise after burning, and in the liver lipid peroxide levels, SOD activity, and glutathione peroxidase activity are also elevated. The increase in serum and liver lipid peroxides is explained by a release of lipid peroxides from skin lesions into the serum and subsequent trapping in the liver (Kawai et al. 1988).

Lipid Peroxidation – Human Studies. In the plasma of burn patients the level of lipid peroxidation products rises, while plasma tocopherol concentrations and glutathione peroxidase activity decrease (Yuan et al. 1988). In human burn victims, lipid peroxide concentrations are elevated in skin lesions. Lipid peroxides in burn lesions of aged individuals with impaired antioxidant compensating mechanisms are significantly higher than in younger burn patients (Niwa et al. 1988). In burned human skin, a lipid epoxide which has uncoupling activity is formed. It has been proposed that this uncoupler of oxidative phosphorylation may exert toxic effects in burned skin and could be transferred into the circulation and thus other tissues (Yokoo et al. 1986). Uncouplers of oxidative phosphorylation (Heytler 1979; Terada 1981; Hanstein 1976) stimulate mitochondrial respiration and may thereby increase the endogenous production of reactive oxygen species.

C. Skin Ischemia

Ischemia/reperfusion-induced formation of reactive oxidants has been extensively studied in muscle tissue, particularly in myocardium. Direct evidence has been obtained by EPR spectroscopy of freeze-clamped rabbit myocardium that oxygen radicals are generated in ischemia and that a burst of oxygen radicals occurs during initial reperfusion (Zweier et al. 1987). Although these results should be interpreted with caution because freezing is known to produce paramagnetic tissue species, this study and other reports (Das et al. 1986) support the free radical hypothesis of the ischemia/reperfusion syndrome. In a variety of model systems, reactive oxygen species have been implicated as mediators of postischemic tissue damage (McCord 1985). Disturbed function of the mitochondrial electron transport chain during reperfusion, ischemic activation of prooxidant enzymes, and inhibition of antioxidants may contribute to an elevated steady-state concentration of reactive oxygen species. Since a pathophysiological link between ischemia and inflammation is clearly established, leukocyte-derived reactive oxidants may also participate in this scenario (Weiss 1986). Thiol antioxidants can provide cardioprotection to ischemic and reperfused rat hearts (Fuchs et al. 1985a, b, 1987; Fuchs and Zimmer 1986a; Zimmer et al. 1985) and dog hearts (Beyersdorf et al. 1989).

Biochemical sequelae of reactive oxidant generation in ischemic myocardium include the formation of peroxidation products and inhibition of antioxidants, e.g., diminution of glutathione and inhibition of SOD activity (Röth et al. 1985). The antioxidant response may be quite heterogeneous. An increase of cytoplasmic and mitochondrial selenium-dependent glutathione peroxidase is observed during rat heart ischemia. In reperfusion, only the mitochondrial enzyme shows a further increase. Glutathione reductase is also found increased in ischemic hearts but returns to values observed in normoxic hearts during reperfusion. Glutathione and mitochondrial SOD activity decrease in ischemic as well as in reperfused hearts, although cytosolic SOD activity remains unmodified. Catalase and glutathione-S-transferase activities are unchanged in ischemia and reperfusion (Arduini et al. 1988).

Several studies indicate the involvement of reactive oxygen species in ischemic skin flap necrosis (Im and Hoopes 1984; Manson et al. 1983; Angel et al. 1986b; Miyachi et al. 1989). Oxidative stress is of pathophysiological significance for island skin flap and free flap reconstructive surgery, limb reimplantation (McCord 1985), and organ storage and subsequent reperfusion (Fuller and Green 1986). Transplanted vascularized skin flaps are always ischemic to some degree (Cohen et al. 1983c). Ischemic conditions in skin may also be prevalent in a variety of other pathophysiological conditions, such as skin burns, venous occlusion, arterial obliteration, and cutaneous vasculopathies. In experimental island skin flaps, tissue necrosis can be a consequence of ischemia and reperfusion. Pig skin, for example, may survive if the circulation is restored after total occlusion of up to 8 h (Milton 1972). However, even when the blood flow is resumed after several hours of ischemia, skin flaps can still necrose.

I. Acute Skin Response to Ischemia

Skin responds to oxygen deprivation and flow reduction by a rapid change in nucleoside and high energy phosphate metabolism. The endogenous level of skin cAMP increases rapidly and transiently after removal of skin from the body in a β-adrenergic adenylate-cyclase-dependent process (Ilizuka and Ohkaware 1986). Maximum accumulation occurs 2 min after incubation, and levels revert to near control values after 7 min of incubation. Pretreatment of skin with the tumor promotor 12-O-tetradecanoylphorbol-13-acetate greatly reduces this ischemic rise in epidermal cAMP (Solanki and Murray 1982b; Murray et al. 1977). It is proposed that modification of the redox status of the cells due to accumulation of NADH following ischemia may result in the accumulation of cAMP (Solanki and Murray 1982a). During ischemia of pig skin flaps, the ATP content decreases, and the concentration of ADP and AMP increases. Upon reperfusion, these changes are reversed, and normal epidermal concentrations of adenine nucleotides are restored within 30 min and remain stable therafter (Harmon et al. 1986).

Xanthine oxidase has been implicated as a factor in the pathogenesis of ischemic disorders since there is growing evidence that reactive oxygen species generated by the xanthine oxidase system are involved in cellular damage associated with the ischemia/reperfusion syndrome (Fig. 3.22). Xanthine oxidase is physiologically present in tissues as a dehydrogenase enzyme. It uses NAD^+ as an oxidant to convert hypoxanthine to xanthine, and then xanthine to uric acid. Xanthine dehydrogenase can be converted into the oxidase enzyme by limited proteolysis or oxidation of specific enzyme thiol groups. Xanthine oxidase uses molecular oxygen to oxidize xanthine or hypoxanthine, which results in the formation of the superoxide anion radical and hydrogen peroxide. Depletion of ATP in ischemic tissue causes accumulation of hypoxanthine, which serves as a substrate for xanthine oxidase. It has been proposed that the conversion of xanthine dehydrogenase to xanthine oxidase and the accumulation of hypoxanthine are involved in the pathophysiology of ischemic tissue (Granger et al. 1986; Jarasch et al. 1986; Weiss 1986).

Xanthine oxidase is located in the capillary endothelial cells of many tissues, but the amount present in other epithelial, endothelial, and mesenchymal cells is minor (Jarasch et al. 1986). Although endothelial cells constitute only a small fraction of the total tissue mass, self-destruction of capillary endothelial cells during ischemia/reperfusion injury due to conversion of xanthine dehydrogenase to xanthine oxidase and subsequent production of reactive oxidants may endanger the entire tissue.

Tissue disruption causes a conversion of the dehydrogenase to the oxidase form; therefore, in tissue homogenates the oxidase activity is usually measured.

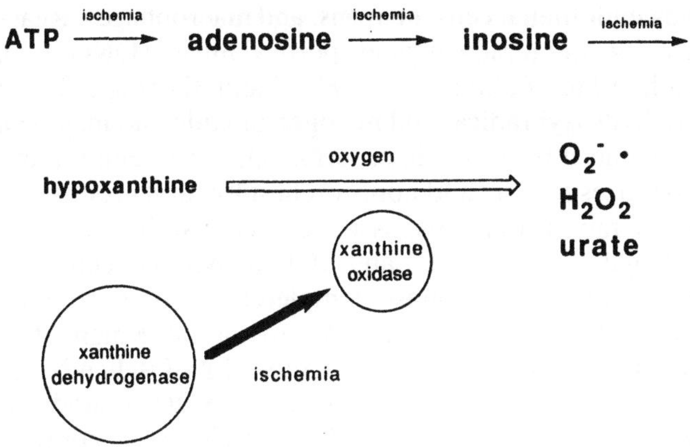

Fig. 3.22. Ischemia/reperfusion and production of reactive oxidants

In human skin xanthine oxidase activity (0.5 mU/g wet weight) is about 100- to 300-fold lower (Al-Khalidi and Chaglassian 1965a) than in liver or intestine (Parks and Granger 1986). However, it was reported that epidermal xanthine dehydrogenase is converted to xanthine oxidase during the later stages of epidermal differentiation (Reiners and Rupp 1989). This may indicate that xanthine oxidase is present in normal skin in significant quantities. Similar to the results of reports on xanthine-oxidase-mediated ischemic damage in the intestine and myocardium, the xanthine oxidase system was implicated as a factor in the pathogenesis of skin ischemia (Angel et al. 1986b; Im et al. 1984).

It has been documented that in venous-occluded neurocutaneous groin skin flaps of rats the xanthine oxidase activity in skin increases during ischemia and subsequent reperfusion (Im et al. 1984). In ischemic rat skin an increase in xanthine oxidase activity parallels elevation of lipid peroxidation products (Angel et al. 1986a). Improvement of survival rates of venous-occluded skin flaps is achieved with allopurinol, an inhibitor of xanthine oxidase. This indicates that xanthine oxidase may be one of the sources of superoxide anion radicals and hydrogen peroxide in skin and plays an important role in the development of necrosis following ischemia and reperfusion (Im et al. 1984).

Superoxide anion radical may be a potent source of the hydroxyl radical by reaction with endothelium – derived relaxing factor (EDRF), which is identical to nitrous oxide (Fig. 3.23). NO·

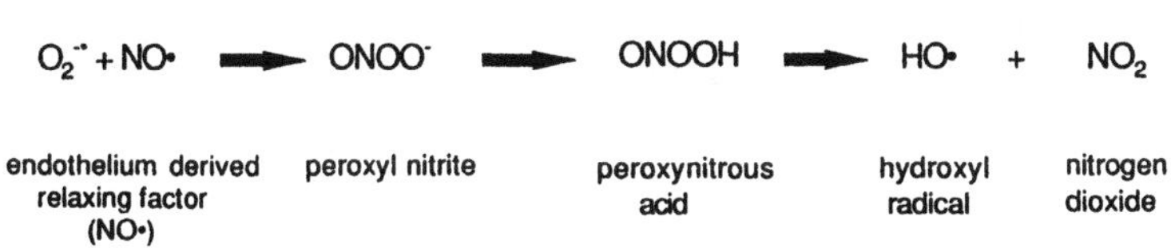

Fig. 3.23. Reaction of nitrous oxide with superoxide anion radical

(NO·) is produced by endothelial cells, neurons, and macrophages. Reaction of NO· with superoxide anion radical yields peroxyl nitrite (ONOO−), a potent oxidant with a half life of about 1 s at pH 7.4 and 37° (Fig. 3.23). It readily decomposes into hydroxyl radical and nitrogen dioxide and may be a major cytotoxic agent produced by ischemia/reperfusion, sepsis, and inflammation. Peroxylnitrite oxidizes protein and nonprotein sulfhydryl groups.

In reperfused, ischemic rabbit skin flaps, tissue levels of xanthine oxidase activity and malondialdehyde are increased and SOD activity is decreased. SOD treatment during the reperfusion phase completely restores xanthine oxidase, malondialdehyde, and SOD levels in the skin to normal (Knight et al. 1990). The total SOD content of skin is unchanged after 21 h of ischemia but drops to about 50% of its original value during reperfusion. SOD-treated animals have supranormal skin SOD activities (140% of preischemic controls and 290% of reperfused controls) in the reperfusion phase (Manson et al. 1986). This indicates that although the serum half-life of SOD is only some minutes, intravenous SOD administration is sufficient to achieve a protective effect in skin for some period longer than this. Derivatized SOD with an extended plasma half-life increases survival of ischemic skin flaps more efficiently than native SOD alone (Huang et al. 1987). Other reports confirm SOD-mediated protection of ischemic skin. SOD treatment of rat skin flaps subjected to venous occlusion leads to a significant reduction of macro- and micromorphological skin damage (Manson et al. 1983). Survival of free skin flap transfer by microvascular anastomosis in rats is significantly improved by intravenous administration of SOD (Manson et al. 1986). Skin flap survival in rats was also improved by intravenous or oral administration of natural antioxidants such as glutathione, vitamins C and E, and β-carotene (Hayden et al. 1987).

Desferrioxamine, a metal ion chelating agent, was also shown to increase skin flap survival in rats (Huang et al. 1987; Angel et al. 1986a). It is thought that tissue iron delocalization occurring during ischemia could be counteracted by this iron chelator. A role for released iron in hematoma-induced skin flap necrosis has been suspected (Angel et al. 1986b). Free hemoglobin is a Fenton reagent and generates hydroxyl radicals in the presence of hydrogen peroxide in vitro (Sadrazadeh et al. 1984). Hemoglobin has long been known to stimulate lipid peroxidation and was recently shown to generate reactive oxygen species in the presence of hydrogen peroxide (Puppo and Halliwell 1988). Formation of superoxide anion by stimulated neutrophils releases free iron by reduction (Thomas et al. 1985) from iron binding proteins, such as ferritin, and it is proposed that this may potentiate the formation of hydroxyl radicals in a Fenton type reaction at inflammatory sites (Biemond et al. 1984). Intramembranous formation of hydroxyl radical and initiatiation of lipid peroxidation via Fenton reactions in the lipid phases of membranes in vivo appear feasible (Schaich and Borg 1988).

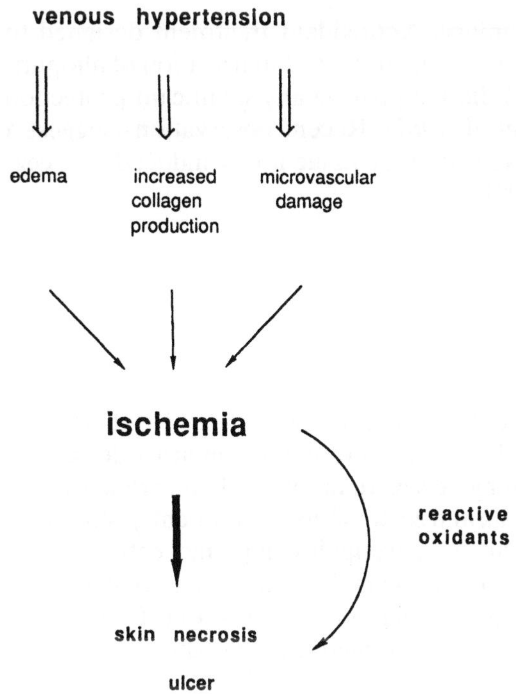

Fig. 3.24. Ischemia and venous hypertension

II. Hematoma and Venous Ulcers

Lipid peroxidation product concentrations are higher in ischemic skin flaps with hematoma than in those without hematoma under the flap (Angel et al. 1986b). Similarly, local iron accumulation in skin around venous ulcers may actively perpetuate tissue damage (Ackermann et al. 1988) and cause fibrosis. Participation of reactive oxygen species in the pathogenesis of skin ulcers associated with venous insufficiency is suspected. A hypothetical scheme is presented in Fig. 3.24 which shows the szenario of ischemia, production of reactive oxidants, microvascular damage, increased collagen production, and tissue edema caused by venous hypertension (Angel et al. 1987).

III. Skin Ischemia After Burn/Frostbite

The periphery of tissue coagulated by burn is a zone of heat-injured tissue within which the microvascular blood flow progressively diminishes to become static by the 4th day after insult. This zone of stasis is quite vulnerable to further injury, e.g., reperfusion. The hypothesis that ongoing skin damage occurring after burning may be in part due to oxygen-derived free radicals during the 16-h to 4-day postburn period of reperfusion was investigated. Wound healing was studied macro- and microscopically in burned skin of guinea pigs 1 and 3 weeks after injury. Antioxidant treatment designed to shield the zones of stasis from reperfusion effects (administration of allopurinol, SOD, and dimethylsulfoxide) did not provide any significant protection in this model system (Melikian et al. 1987). Recent observations suggest a potential role of reactive oxidants in ischemic tissue injury induced by freezing and thawing (Manson et al. 1991).

D. Microbial Skin Diseases

Skin manifestation is a prominent feature of many viral diseases. Erythematous rashes (rubella, measles, mononucleosis) indicate an inflammatory dermal reaction. Cutaneous inflammation may be severe in several hemorrhagic exanthema, e.g., those caused by varicella and zoster viruses. Frequently, skin is the primary target of bacterial infection, e.g., erysipelas, impetigo, ecthyma, and folliculitis. Reactive oxdiants may play a role in the pathophysiology of infectious skin diseases. Supportive arguments for this hypothesis come from studies on oxidative injury in infectious diseases of organs other than skin.

I. Autotoxicity

The concept of autotoxicity as a significant factor in the clinical pathology of *microbial* diseases, in particular viral ones, was introduced by Peterhans et al. (1988). Autotoxicity is defined as "self-inflicted damage as a result of activated and uncontrolled host effector mechanisms." Numerous literature reports indicate that fungal, bacterial, viral, and protozoal infections may directly or indirectly increase the production of neutrophil-generated reactive oxidants, which may be significantly involved in the pathological sequelae. There is convincing evidence that autotoxicity mediated by reactive oxidants is an important factor in the pathogenesis of bacterial infections and in certain parasitic diseases (Clark et al. 1986). It was noted in animal and human studies that bacterial infections generally elevate host cellular chemiluminescence, indicating an increased production of reactive oxidants. *Salmonella typhi* was found to depress cellular chemiluminescence; however, this is an exception rather than the rule. Endotoxin stimulates in vivo a massive oxygen burst by complement activation (Flohe and Giertz 1987), and generation of reactive oxygen species in neutrophils from patients with bacterial disease rises (Solberg et al. 1982). In candidiasis and sporotrichosis, tissue infiltration with neutrophils occurs at an early stage, and reactive oxygen species produced by the inflammatory cells are thought to play an important role in the pathology (Yoshioka et al. 1987 a).

The mechanism of *virus-induced autotoxicity* resembles the action of endotoxin (Peterhans et al. 1988) and involves the formation of reactive oxygen species by different mechanisms.

Formation of reactive oxidants by *phagocytes* can be triggered by the virus itself, virus-antibody immune complex, and certain mediators, e.g., complement 5a, TNF, and platelet activating factor. Different virus species, such as paramyxo-, influenza (Peterhans et al. 1987; Henricks et al. 1985; Mills et al. 1981), respiratory syncytial, and herpes simplex viruses (Bingham et al. 1985) stimulate reactive oxidant production in neutrophils. Activation of cyclooxygenase and lipoxygenase by neutrophil-derived reactive oxygen species can then lead to the increased generation of lipid peroxidation products, which autocatalytically stimulate prostanoid enzyme activity and sustain the inflammatory process.

Formation of *clastogenic factors* may participate in autooxidative tissue injury in viral diseases. Chromosomal breakage can be caused by tumor- and nontumor-associated viruses by means of clastogenic factors, which are presumably composed of lipid peroxidation products (Emerit et al. 1985; Khan and Emerit 1985). In the serum of patients with acute hepatitis B, a clastogenic factor was detected (El-Alfi 1965).

Induction of prooxidant enzymes in various tissues is a further mechanism involved in virus-induced autooxidative tissue injury. Interferon and interferon-inducing agents (Newcastle disease virus and chemicals) induce the prooxidant enzyme xanthine oxidase in various mouse organs (Ghezzi et al. 1984; Deloria et al. 1985). The antioxidant BHT protects chickens from the lethal Newcastle disease virus at a serum concentration sufficient to have antioxidant activity (Brugh 1977). Induction by interferon of xanthine oxidase in the liver is associated with decreased activity of cytochrome P-450. It was suggested that this reduction of liver enzyme activity is a consequence of the increased formation of reactive oxidants (Ghezzi et al. 1984, 1986; Deloria et al. 1985). Influenza-virus-infected mice have elevated xanthine oxidase activity in serum and lung, and lethal influenza virus infection in mice is prevented by systemic administration of SOD (Oda et al. 1989). Pulmonary indolamine-2,3-dioxygenase, is markedly induced in mice by viral infections or bacterial endotoxin. Interferon produced during the inflammatory reaction is suggested to mediate indolamine-2,3-dioxygenase induction in lung tissue (Hayaishi 1987). Influenza virus causes a 95-fold rise in mouse lung indolamine-2,3-dioxygenase, and a 50% fall in the glutathione content of the lung. Although indolamine-2,3-dioxygenase may generate reactive oxidants, it also uses superoxide anion radical as cosubstrate, thereby having antioxidant properties.

Viruses can also cause *inhibition of tissue antioxidants.* Viral infections, such as lethal parainfluenza infection of mice, cause a decrease of liver antioxidants, an organ which is primarily not affected by the virus (Hennet et al. 1988). Simian virus (SV)40-transformed human embryonic lung fibroblasts have reduced Mn-SOD enzyme activity as well as reduced Mn-SOD immunoreactivity (Oberley et al. 1989). In SV40-transformed human fibroblasts the amount of Mn-SOD is significantly depleted due to a fall in the level of translatable mRNA for Mn-SOD (Marlhens et al. 1985). SV40-transformed human fibroblasts also exhibit a decrease in catalase activity of about 80% compared with controls (Vuillaume et al. 1986), and transformed mouse embryo cells have 2–5-fold lower glutathione peroxidase, catalase, and SOD activities than control cells (Omar et al. 1987). In patients with bacterial infectious diseases (meningococcal and tuberculoid meningitis, tetanus, lobar pneumonia, typhoid fever), there is a large increase in blood dehydroascorbate and a remerkably diminished level of blood ascorbate (Chakrabarti and Banerjee 1955), indicating oxidative stress.

Infectious agents can also induce an *increase in antioxidant activity.* In rats, endotoxin treatment induces lung SOD, catalase, and glutathione peroxidase activity and protects the animals against hyperoxic toxicity (Frank et al. 1978; Frank and Massaro 1980). Endotoxin, however, does not induce antioxidant enzymes in the lungs of mice (Frank et al. 1980), indicating that SOD is more easily induced in rats than in other animals. SOD activity in peripheral human lymphocytes is elevated 2–5-fold during upper respiratory viral infections (McCue 1978).

Formation of reactive oxidants in neutrophils from patients with viral disease is found to be decreased (Solberg et al. 1982). T lymphocytes from patients with *infectious mononucleosis* suppress oxygen radical generation in neutrophils; monocytes, however, exhibit a normal level of reactive oxidant production (Niwa et al. 1984a). Measles patients with secondary infections show a significant drop in the generation of oxygen species. Normal neutrophils incubated with T lymphocytes from these patients generate significantly lower levels of reactive oxygen species than those incubated with T cells from a healthy control group (Niew et al. 1985c). T-cell-mediated suppression of reactive oxygen generation requires both OKT4 cells from patients with infectious mononucleosis and OKT8 cells from either patients or normal controls. In measles patients without secondary infection, no abnormal findings of neutrophil respiratory burst are observed. It is suggested that the secondary bacterial infections sometimes seen in measles patietns may derive from a decrease in oxygen radical generation presumably induced by suppressor T cells. Suppression of neutrophil production of reactive oxygen species in patients with infectious mononucleosis and measles may be due at least in part to interacting subsets of T lymphocytes (Niwa et al. 1984a, 1985c).

Mucocutaneous lymph node syndrome (Kawasaki disease) is a disease of children characterized by fever, polymorphous exanthema and enanthema, palmoplantar erythema, lymph node enlargement, conjunctivitis, and visceral manifestations, in particular coronaritis (Bligard 1987). A lymphotropic infectious agent such as a retrovirus is suspected (Burns et al. 1986; Shulman and Rowley 1986); however, one has not yet been isolated from patients with mucocutaneous lymph node syndrome (Rowley and Shulman 1987). Within 5 days of onset, the generation of reactive oxygen species by neutrophils of these patients is markedly potentiated in comparison with normal control subjects and with patients with bacterial disease, who also have elevated levels of reactive oxygen production. At about 7 days after onset, it becomes similar to that in healthy controls. It is suggested that the formation of immune complexes and increased generation of reactive oxygen species by activated neutrophils during the very early phase of the mucocutaneous lymph node syndrome may induce coronary vascular tissue damage and provide a site for platelet aggregation, resulting in thromboarteritis and coronary occlusion about a week after onset of the syndrome (Niwa and Sohmiya 1984).

The viral etiology of the *acquired immune deficiency syndrome (AIDS)* is clearly established; however, the mechanisms which contribute to the immunodeficiency state are not known. Oxidative factors were thought to be involved (Papadopulous-Eluopolos 1988). In particular, it has been proposed that oxidative imbalance in HIV-infected patients may be a consequence of the original infection, the subsequent opportunistic infections and chronic wasting (Fuchs et al. 1991a). Several reports implicate reactive oxidants as being involved in the pathogenesis of HIV infection. The presently available epidemiological and experimental evidence may justify clinical testing of physiological antioxidants in HIV-positive patients. However, it has been suggested that non-physiological scavengers of reactive oxidants should not be used clinically until more evidence is available (Halliwell and Cross 1991).

Studies with HIV-infected patients clearly indicate alterations in their *antioxidant/prooxidant balance.* About 30% of HIV-positive patients with lymphadenopathy have a subnormal plasma tocopherol level (Javier et al. 1990) and decreased plasma selenium (Dworkin et al. 1986). HIV-infected cells are reported to have low Mn-SOD activity (Wong et al. 1990). AIDS patients have a deficiency of the blood antioxidant ubiquinone 10, and the deficiency grows with increasing severity of the disease (Folkers 1988). Serum methionine and cysteine levels are decreased, and the glutathione concentration in neutrophils and monocytes in patients with LAS (lymphadenopathy syndrome) and AIDS are significantly diminished (Dröge et al. 1988; Eck et al. 1989; Eck and Dröge 1989). The glutathione and total glutathione concentrations in the plasma of AIDS patients are about 30% of those in the normal individual (Buhl et al. 1989). The thiol/disulfide redox status is an important control mechanism of cellular functions (Ziegler 1985). Lipoxygenase activity in human neutrophils is regulated by the glutathione status (Hatzelmann and Ullrich 1987), and in immunosuppressed mice the antibody response is augmented by lipoate, a glutathione-regenerating compound (Ohmori et al. 1986a, b). Glutathione can directly regulate proliferation of T cells in vitro, and this reaction is suggested to be of biological relevance in immunomodulation in HIV-infected patients (Suthanthiran et al. 1990). A disturbed thiol/disulfide status in HIV-infected individuals may have vital biomedical implications for lymphocyte dysfunction.

Serum ferritin, whose level is also increased in various chronic diseases, is elevated in AIDS patients, and a significant role for it in the suppression of the immune function has been discussed (Wigginton 1989). Elevated serum ferritin may contribute substantially to the formation of reactive oxidants by providing increased serum concentrations of the redox-active transition metal iron. Iron ions are released from ferritin by superoxide anion radicals in vitro (Thomas et al. 1985) and in vivo (Mazur et al. 1958). Therefore, superoxide anion radicals derived from activated neutrophils may potentiate the formation of reactive oxygen species via iron release from ferritin (Biemond et al. 1984).

Xanthine oxidase is an intracellular enzyme found mainly in liver, small intestine, heart, and lung. Although xanthine oxidase is normally undetectable in human plasma (Al-Khalidi and Chaglassian 1965), it is detected in patients with hepatitis (Ramboer et al. 1972; Shamma's et al. 1973) and adult respiratory distress syndrome (Grum et al. 1987). Recently, it was suggested that circulating xanthine oxidase could contribute to oxidative injury in AIDS patients (Flavin-König, personal communication). The HIV virus and secondary viruses associated with HIV-infected patients may cause an elevation of plasma xanthine oxidase activity, which is usually not present in plasma. The reason for this elevation of xanthine oxidase is not clear, but it may be released from infected cells or induced by interferon. Interferon and several viruses cause elevation of plasma xanthine oxidase in mice; the viruses tested include coxsackie B_1, herpes simplex, Nyamanini, Dengue III, and vaccinia viruses.

Malondialdehyde is elevated in HIV patients at different stages of the disease, indicating lipid peroxidation (Sönneborg et al. 1988). 4-Hydroxynonenal, a lipid peroxidation product, is also elevated in the blood of HIV-positive patients (Fuchs et al. 1991 b).

Reactive oxidant production in *neutrophils* is impaired in HIV-infected patients. Neutrophils from HIV-infected chimpanzees exhibit a suppressed secretory function similar to that observed in other nonprimate viral and retroviral infections (Lafrado et al. 1989). AIDS patients have a markedly reduced chemiluminescence activity of neutrophils stimulated by opsonized zymosan. Only some of the patients suffering from LAS show a reduced chemiluminescence activity of zymosan-stimulated neutrophils (Stöhr et al. 1984).

Monocyte-derived *tumor necrosis factor (TNF)* is a glycoprotein cytotoxic to tumor cells and virus-infected cells. It has been suggested that some of its biological effects (e.g., immunological killing) are mediated by reactive oxygen species (Jones 1987). Blood monocytes from HIV-infected patients spontaneously produce high livels of TNF-α (Roux-Lombard et al. 1989) and are hypersensitive to endotoxin stimulation, which results in enhanced synthesis of TNF (Lau and Livesey 1989). TNF and phorbol ester can upregulate HIV production and induce an oxygen burst in neutrophils. Intracellular thiol concentration influences HIF production in vitro. *N*-Acetylcysteine, which scavenges reactive oxidants directly and increases intracellular glutathione, inhibits TNF- and phorbol ester-mediated stimulation of HIV replication (Herzenberg et al. 1989; Roederer et al. 1990).

In vitro, the *production of HIV* in cells containing latent virus is stimulated by various cytokines, heterologous viruses, and physicochemical agents. The common denominator shared by all these factors is the potential to induce a cellular stress response.

Recently, it was demonstrated that oxidative stress can cause activation of HIV-1 in cells (Legrand-Poels et al. 1990), and this suggests that reactive oxidants can promote the expression of dormant viral genes in a host. UV light and sunlight significantly activate HIV gene expression (Valerie et al. 1988).

Hydrogen peroxide causes amplification of endogenous and viral gene sequences in mammalian cells (Weitzman et al. 1990), and it induces expression of viral antigen in various cell lines that harbor the Epstein-Barr virus. The mechanism of action is indirect and mediated by singlet oxygen and hydroxyl radicals (Oya et al. 1987). Hydrogen peroxide also induces cellular factors which reactivate HIV-1 and induce HIV expression (Piette et al. 1990). Chemically induced expression of human T-cell lymphoma virus (HTLV-1) genes is inhibited by the antioxidant vitamins E and C and by retinoids (Blakasbe et al. 1985). Vitamin A supplementation during retrovirus infection leading to murine AIDS enhances survival of infected mice and increases the number of activated macrophages (Watson 1988).

E. Skin Aging

Several theories have been developed to explain the phenomenon of aging. The hypothesis of programmed aging suggests the involvement of a few specific genes in determining the rate of aging and the species-specific maximal lifespan potential. The error catastrophe theory proposes a failure of protein transcription and translation. Harman put forth the free radical theory several years ago (Harman 1956), which postulates free radical reactions as the single basic course of aging, modified by genetic and environmental factors (Harman 1984). Involvement of free radicals in the aging process, however, remains obscure (Mehlhorn and Cole 1985).

Some human genetic diseases shorten the lifespan, such as progeria, Down's syndrome, Cockayne's syndrome, and Werner's syndrome, and are also associated with premature skin aging. The physiology of the intrinsic aging process of skin has recently been reviewed (Balin and Pratt 1989).

In *Down's syndrome*, increased production of reactive oxidants may play a role in the aging process. Cu,Zn-SOD is coded for by human chromosome 21. Down's syndrome (trisomy 21) is a human disease model for increased SOD activity. A significant role of oxidative stress in the pathophysiology of trisomy 21 is discussed based on the following biochemical findings. In fetal human trisomy 21 brains (Brooksbank and Balazs 1984), in human trisomy 21 fibroblasts (Anneren and Epstein 1987), and in the blood of trisomy 21 patients (Kedziora et al. 1986), elevated levels of lipid peroxidation products are found. It is suggested that the increased Cu,Zn-SOD activity (50% increase) results in an augmented generation of hydrogen peroxide and as a consequence in greater lipid peroxidation and other molecular damage. Glutathione peroxidase and glutathione-6-phosphate dehydrogenase activities are also elevated in the red and white blood cells of trisomy 21 patients (Sinet 1982).

It seems plausible that glutathione peroxidase and glutathione-6-phosphate dehydrogenase activities are stimulated adaptively, possibly from the greater availability of hydrogen peroxide generated by the raised SOD activity manifested by trisomy 21 cells (Levine and Kidd 1985a). At a later point, however, glutathione peroxidase activity may be depressed due to ongoing oxidative stress. Vulnerability to oxidative stress is an intrinsic feature of glutathione peroxidase (Flohe 1982; Ganther and Kraus 1981). The development of an animal model for trisomy 21, transgenic mice that carry the human Cu,Zn-SOD gene, will provide a unique system for studying the consequences of increased Cu,Zn-SOD activity in Down's syndrome and in a variety of other pathological diseases (Epstein et al. 1987).

I. Collagen

Type I collagen comprises about 75% of the dry weight of adult skin, and although it has a greater turnover than other collagen types, it is biochemically a relatively inert fiber. During intrinsic (chronological) skin aging, progressive collagen cross-linking occurs, the chemical nature of which is unknown. Age-related cross-link formation is probably chemically different from physiologically occurring cross-links by Schiff's base formation and aldol condensation. In aged skin, collagen solubility decreases, and a diminution in aldimine cross-links is observed (Robins et al. 1973). The aldimine double bond is susceptible to oxidation to a peptide-like linkage (Bailey et al. 1977). Calf skin collagen is

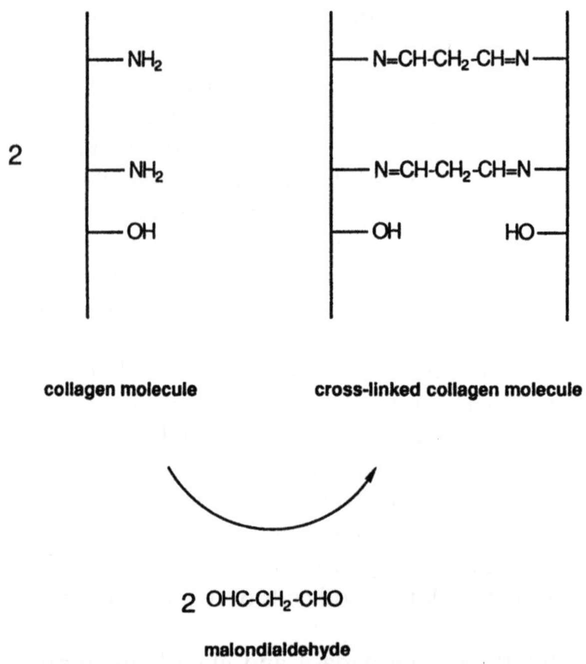

Fig. 3.25. Malondialdehyde and collagen cross-linking

degraded in vitro when exposed to a system generating reactive oxygen species (Monboisse et al. 1984).

It was suggested that the lipid peroxidation breakdown product malondialdehyde, which is a bifunctional reactive aldehyde, can induce increased collagen cross-linking. Aldehydes can react spontaneously with hydroxylysyl residues in the helical region of the collagen molecule. In contrast to most proteins, cross-links in collagen are derived from lysyl and hydroxylysyl residues and not from cysteine residues (Fig. 3.25).

II. Elastin

Skin elasticity is based on an intact elastic fiber network, and the tensile strength of skin primarily arises from collagen. During intrinsic aging there is a loss and fragmentation of elastic fibres, clinically resulting in wrinkling, thinning, and laxity. The relative oxidative susceptibility of elastin in comparison with collagen fiber is not known. Elastin has a slow metabolic turnover but is degraded by elastases. Serine elastases are under the strict control of serum antiproteases such as α_1-proteinase inhibitor or α_2-macroglobulin. Since antiproteases can be inhibited by reactive oxygen species, oxidative damage may result in the increased degradation of elastic fibers. Oxidative damage of elastin fibers and stimulated proteolysis may contribute to skin wrinkling in the physiological aging process (Fukuzumi 1986).

III. Glycosaminoglycans

During skin aging, striking changes take place in the soluble pool of dermal glycosaminoglycans. Their strictly age-dependent decrease can be correlated with a reduced mitotic potential of dermal fibroblasts. The lifespan of cultured fibroblasts is inversely proportional to the donor's age, and human fibroblasts exhibit prolonged growth and extended life-span at decreased oxygen concentrations. It is thought that the decrease of extracellular moisture and alteration of mucopolysaccharides occurring with age could be related to the formation of lipid peroxidation in skin (Fukuzumi 1986). Lipid peroxidation products are cytotoxic for fibroblasts (Meffert and Lohrisch 1971 a), and could inhibit glycosaminoglycan biosynthesis.

IV. Lipid Peroxidation

In vivo measurements in the epidermis of animals and humans using reflective near infrared spectroscopy (Lippman 1985) have clearly shown an age-related increase of lipid peroxidation products. Similar conclusions have been obtained by measurements in human skin (Meffert et al. 1976a). In human subjects, there is an age-dependent pattern of skin malondialdehyde concentration, maxima appearing in the age groups of 0–9 years and 50–59 years and a minimum at 20–30 years (Sugiura et al. 1985a). In aged patients, impaired compensating mechanisms for the clearance of skin lipid peroxidation products were described (Niwa et al. 1988). Although in aged rats lipid peroxidation of liver microsomes and mitochondria increases, in mitochondria isolated from human skin no age-related rise in lipid peroxidation products is noted. It was suggested that melanin may exert protection against mitochondrial lipid peroxidation in skin (Serri et al. 1979). Tocopherol does not affect the pattern of skin aging in rats as analyzed by biochemical and biophysical methods. It was, therefore, concluded that free radicals do not significantly participate in the aging process of connective tissue (Blackett and Hall 1980). Other reports, however, are in disagreement with this.

In the senescence-accelerated mouse, a murine animal model of intrinsic aging, aging is marked by early onset and irreversible advancement of senescence after a normal process of development (Takeda et al. 1981). During aging, the animals show loss of hair and skin glossiness, augmented skin coarseness, and increased frequency of skin ulceration. In senescence-accelerated mice, the serum and liver lipid peroxide levels rise significantly prior to the appearance of age-related clinical signs (Yagi et al. 1988); lipid peroxide levels in skin increase several weeks prior to the onset of cutaneous pathology (Komura et al. 1988). It was concluded that elevated serum lipid peroxide concentrations are transferred through the vascular wall into the skin tissue and provoke the skin symptoms involved in aging (Komura et al. 1988).

V. Fluorescent Pigments

Fluorescent pigments accumulate in aging tissues and are suspected to be reaction products of peroxidized lipids with proteins. The chemical nature of fluorescent products arising in peroxidized membranes is not clear, and the mechanisms of formation are quite complex. Lipofuscin and ceroid are probably two of the major components of the fluorescent pigments accumulating in tissues with age. They are chemically not identical. Fluorescent pigments contain conjugated Schiff's bases, dihydropyridine derivatives, and polymerized aldehydes.

Lipofuscin is a lipoproteinous substance of as yet poorly defined chemical composition that accumulates progressively with age in certain types of cells. It was recently shown for retinal pigment epithelium that the autofluorescent products arising after lipid peroxidation are probably not lipofuscin (Eldred and Katz 1989).

A variety of evidence suggests that lipofuscin formation may result from autoxidation of subcellular components. Lipofuscin is a product of autophagocytized membrane-rich material that is undergoing peroxidation, fragmentation, and polymerization within the lysosomal compartment. The mechanism of lipofuscinogenesis is presently unknown: One hypothesis considers oxidative stress as a causal factor. Lipofuscinogenesis may be regulated by a complex chain of interactions between pro- and antioxidation factors in the cell and indicates that oxidative stress is a major determinant. A substantial accumulation of lipofuscin in aged skin is not reported in the literature.

The fluorescent pigment ceroid increases in animals fed a diet rich in polyunsaturated fatty acids or deficient in tocopherol. Reactive oxidants are suggested to be involved in the pathogenesis of ceroid accumulation (Armstrong et al. 1982; Armstrong 1984).

VI. Amyloid

Amyloid is a proteinaceous tissue deposit which may vary widely in protein composition. It accumulates in several tissues during various disease processes. Amyloid deposits in various internal organs and the skin of the mice have been demonstrated to be related to the aging process (Shimizu et al. 1981; Higuchi et al. 1983). The appearance of amyloid fibril protein in the skin of senescence accelerated mice might be correlated with augmented lipid peroxidation (Komura et al. 1988).

VII. Antioxidants

Conflicting results have been published regarding the activity of antioxidants in aged individuals. It is reported that the SOD concentration remains normal in the elderly, provided they are not under oxidative stress. Others have claimed that its activity declines during the aging process.

Cytoplasmic SOD activities are considerably reduced in livers of aging rats and mice, a very small reduction in activity is observed in the heart, and no reduction is found in the brain. However, the activities show a considerable decline in catalytic activity per antigenic unit in all three organs in both aging rats and mice. The cells in old animals synthesize more enzyme molecules in order to compensate for the partial loss of activity and thus maintain a constant level of SOD activity (Reiss and Gershon 1976). The SOD activity of erythrocytes is reported to decrease with the age of the cell and animal (Glass and Gershon 1981). In rat kidneys and lungs, the age-dependent activity changes of some NADPH-consuming redox enzymes were studied (Maris and Machado 1986, 1987). Glutathione peroxidase and glutathione reductase activities rise during the aging process, while NADPH-cytochrome *c* reductase and thioredoxin reductase decline in activity.

With increasing age the SOD activity in rat skin decreases, implying an impairment of the defense mechanism against oxygen toxicity during the skin aging process (Im and Hoopes 1984a). Younger mice (7 weeks) in the resting phase of the hair cycle exhibit an average skin SOD activity of 6.5 U/mg protein, whereas older mice (11 weeks), which have begun their growth phase of the hair cycle, show an average of 3.8 U/mg protein (Solanki et al. 1982). Epidermal catalase activity in mouse skin is reported to vary from 78 to 175 U/mg protein (Solanki et al. 1982), younger mice having higher epidermal catalase activities than older ones. There was, however, no significant difference in skin SOD activity in healthy young and old guinea pigs (Yoshioka et al. 1987b). SOD activity in the epidermis of humans and mammals (Carraro and Pathak 1988) and in skin homogenates of humans (Niwa et al. 1988) displayed no age-dependent changes.

Skin SOD Induction. Although the activity of SOD in elderly animals and humans is generally comparable with levels observed in younger ones, in old age the capacity for further induction during oxidative stress is limited (Niwa et al. 1988). Intradermal injection of xanthine oxidase and hypoxanthine in guinea pigs of different age groups results in the production of erythema and causes a decline of SOD activity. The increased severity of skin erythema with age correlates with the extent of xanthine-oxidase-mediated inhibition of SOD activities in aged animals (Yoshioka et al. 1987b). The SOD activity of the skin appears to decline strongly in aged animals under oxidative stress that have little capacity for inducing SOD activity. The experimental evidence shows that the capacity to augment antioxidant activity in stressed skin appears to be subject to age-related changes. The imbalance between prooxidant factors and endogenous antioxidant capacity during the skin aging process may significantly contribute to pathophysiological events (Miyachi et al. 1988).

F. Skin Immunology

Reactive Oxidants as Immunomodulators. At picomolar concentrations, hydrogen peroxide is sufficient to convert soluble immune response suppressor (SRIS) to macrophage-derived suppressive factor (MPhi-SF). SRIS is a glycoprotein product of activated T cells that suppresses immune responses through its action on macrophages. SRIS–triggered macrophages release a second factor (MPhi-SF) that is directly responsible for the immune suppressive activity of SRIS. MPhi-SF suppresses primary and secondary IgM and IgG antibody responses in vitro. SRIS can be converted to MPhi-SF by macrophages or by very low concentrations of hydrogen peroxide in the absence of macrophages (Aune and Pierce 1981).

Histamine Release by Reactive Oxygen Species. Increased buffer oxygen tension (Damas et al. 1981) induces the release of histamine from isolated peritoneal mast cells. An enhanced basal release of histamine from isolated perfused rat lungs exposed to 100% oxygen is suggested to be the result of free-radical-induced peroxidation of the lung mast cell membranes (Januszkiewicz and Faiman 1984; Januszkiewicz et al. 1986). Chemically generated reactive oxygen species, presumably hydrogen peroxide, induce the release of histamine from isolated peritoneal mast cells (Ohmori et al. 1979). It has been postulated that the reaction of hydrogen peroxide with some cellular components may produce various changes leading to histamine release; however, it remains to be elucidated whether hydrogen peroxide directly functions as a releasing stimulus. Hydrogen-peroxide-induced release is not due to unspecific lysis of the cell. It requires the presence of calcium ions and is enhanced by deuterium oxide (Ohmori et al. 1980). Irradiation of mast cells in the presence of pheomelanin (photochemical generation of reactive oxidants) also releases large amounts of histamine (Ranadive and Menon 1986).

Microtubular functions were suggested to be involved in histamine release, because deuterium oxide augmented the hydrogen-peroxide-induced liberation of histamine. Several studies have shown that deuterium oxide enhances histamine secretion from stimulated cells, and similar effects were observed with agents that affect microtubular function (colchicine, vinblastine, griseofulvin), (Gillespie and Lichtenstein 1972; Gillespie et al. 1968; Kazimirczak et al. 1984). An alternative explanation would be the involvement of singlet oxygen in the release of histamine. The lifetime of singlet oxygen is extended severalfold in deuterium oxide (Ito 1978).

Histamine Release by Lipid Peroxidation Products. A role of lipid peroxidation products in histamine-induced allergic reactions in skin was suggested. Histamine release of mast cells is triggered by oxidation products of linoleic (Mannaioni and Masini 1988) and arachidonic acid. In the bronchial secretion of young bronchial asthma patients (5–14 years old), increased levels of hydroperoxides and malondialdehyde can be detected in the acute phase of the disease, whereas products of lipid peroxidation are decreased in the remission phase (Krylov et al. 1984). In contrast to oxidation products of arachidonic acid, histamine release induced by linoleic acid oxidation products is cytotoxic in nature.

Reactive Oxidants and Drug Intolerance Syndrome. Platelets of patients suffering from allergic asthma possess a greatly enhanced number of IgE receptors (Capron et al. 1985). Platelets of asthmatics with drug intolerance syndrome react in vitro to the causative allergen with the production of chemiluminescence, indicative of a respiratory burst with production of reactive oxygen species. Glutathione peroxidase activity is significantly lower in comparison with healthy individuals (Malmgren et al. 1986). This indicates that peroxide metabolism may be impaired in these patients. It seems feasible to suggest that reactive oxidants could be mediators in nonimmunological triggering (pseudoallergic reaction) of histamine release from mast cells and basophils. A role of free radicals in the pathogenesis in intolerance reactions

(Soloshenko 1989), e.g., to aspirin has been discussed (Moneret-Vautrin 1985; Moneret-Vautrin et al. 1986), and it was pointed out that high oral doses of the antioxidant ascorbate possess antiallergic activity in humans (Cathcart 1986). Indeed, the antioxidant vitamins E and C ameliorate pseudoallergic reactions, e.g., aspirin-induced asthma and Quincke's edema in some patients (Moneret-Vautrin 1985; Moneret-Vautrin et al. 1986).

G. Skin Inflammation

The molecular basis of inflammatory processes involves the interaction of multiple factors which act as mediators, regulators, and potentiators. Inflammation comprises a complex scenario of biochemical events with the ultimate aim of eliminating the eliciting inductor. Under normal circumstances the inflammatory process is host defensive in nature and occurs at a subclinical level. When the inflammatory response is abnormal, inflammatory diseases can ensue, and autooxidative tissue damage may result. Prooxidant factors are only one component in the pattern of cutaneous inflammation. Chemotactic factors for T lymphocytes, granulocytes, and macrophages and activation of epidermal cytokines may interact in a complicated way in skin (Kupper 1989).

Exogenous factors which may cause skin inflammation are physical (sunburn, radiodermatitis), chemical (toxic dermatitis), immunologigal (allergic dermatitis), and biological (erysipela, zoster). Endogenous noxious stimuli include mainly autoimmunological factors (lupus erythematosus) (Goerz and Merk 1983). Prooxidant factors involved in the pathogenesis of skin inflammation comprise macrophages, neutrophil and eosinophil granulocytes, proteases and antiproteases, immune complexes, products of lipid peroxidation, and prostanoids.

I. Phagocytes

Phagocytosis and intracellular killing of microorganisms are important functions in the antimicrobial defense. The accumulation of phagocytes at sites of inflammation is a mechansim by which the immune system localizes and destroys antigenic material. When inflammatory cells (macrophages, neutrophil and eosinophil granulocytes) are exposed to appropriate stimuli, a complex cascade of events is triggered (Bellavite 1988). Reactive oxygen species are generated during the respiratory burst, degranulation of lysosomes releases proteolytic enzymes, and finally degradation of the phagocyte plasma membrane may result in the formation of antimicrobial lipid peroxides. Besides macrophages, neutrophils, and eosinophils, B lymphocytes possess a superoxide anion radical generating system (Maly 1990).

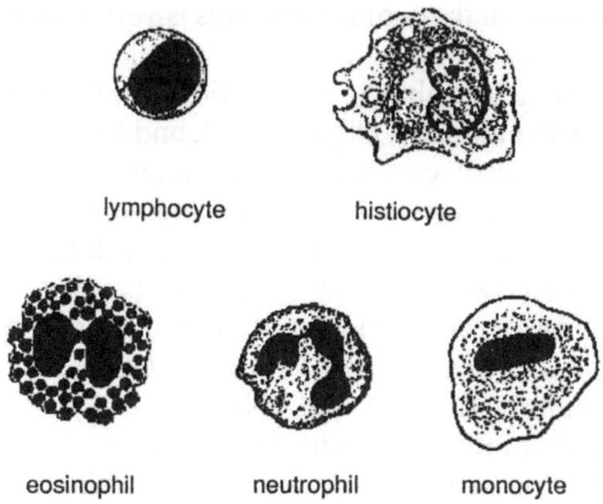

Fig. 3.26. Phagocytes

The important role of reactive oxygen species in leukocyte-dependent inflammatory reactions has been recently reviewed (Fantone and Ward 1982). Cellular chemiluminescence of phagocytic cells is an indicator of reactive oxidant production by these cells and can be associated with different disease status (Van Dyke and van Dyke 1986). The function of inflammatory cells, such as neutrophils, eosinophils, and mononuclear phagocytes, can be modified by various pharmacological agents. Autoaggressive tissue injury can therefore be inhibited by different mechanisms: (a) alteration of the production of chemoattractants, (b) inactivation of chemoattractants, (c) alteration of chemotactic factor receptor function and transduction, (d) alteration of the cytoskeletal motile apparatus, and (e) alteration of the leukocyte's metabolism (Snyderman 1983).

The short-lived superoxide anion radical, which is produced in large quantities by phagocytes, is the precursor of other reactive oxidants. The action of oxidants released from phagocytes (neutrophils, eosinophils, monocytes; Fig. 3.26) is not restricted to their ultimate environment but includes the tissues in the neighborhood. Therefore, inactivation of reactive oxidants released from inflammatory cells is an additional approach to combating autoaggressive/autooxidative tissue injury.

1. Neutrophil Granulocytes

Reactive oxygen species produced by activated neutrophils can exert autooxidative tissue damage at the site of inflammation (Bannister and Banister 1985) or mediate microbial killing (Babior 1978a, b). Both types of effect are mediated in a similar way. Neutrophil-mediated tissue injury is also thought to be a consequence of lysosomal degranulation. However, cytotoxic reactions due to the formation of reactive oxygen species can occur independently of the release of lysosomal enzymes. Luminol-enhanced chemiluminescence of activated neutrophils directly correlates with their cytotoxic activity.

The plasmalemmal NADPH oxidase activity of neutrophils is very low in resting cells but is stimulated by various humoral as well as cellular factors (Forman 1986). The responsiveness of neutrophils to stimulation is enhanced by priming agents. Priming results in an enhanced oxygen burst, and is caused by various agents, some of them are also respiratory burst inducers (see Fig. 3.27). Membrane stimulants may be surface active agents, e.g., digitonin and lectins (concanavalin). Neutrophils can also be activated by interleukins (Maly et al. 1987), human γ-interferon (Das et al. 1986), immune complexes, complement breakdown products, viruses (either via direct interaction of the viral antigen with the neutrophil plasma membrane or via an indirect, antibody-mediated mechanism) (Peterhans et al. 1987), and TNF.

Activation results in increased chemotactic activity and secretory functions of the neutrophil. Adherence of neutrophils to vascular endothelium is enhanced, and subsequently they migrate into the epidermis. Leukotriene LTB4 is a strong chemoattractant for human neutrophils but a weak stimulant of transient superoxide anion production (Serhan et al. 1982; Sumimoto et al. 1984). It is questionable whether LTB4 contributes to the production of reactive oxygen species in neutrophils at all (Ozaki et al. 1987).

Neutrophil activation may result in a bimodal chemiluminescence pattern. The first peak is of extracellular origin, the second peak of intracellular origin (Bender and van Epps 1983). The chmiluminescence reaction is dependent on the production of reactive oxygen species by a plasma-membrane-located NADPH oxidase and their reaction with an intracellular myeloperoxidase system. Hydrogen peroxide, which is generated during plasmalemmal NADPH oxidation via dismutation of the superoxide anion radical, is utilized by the myeloperoxidase, a heme-containing enzyme found within primary (azurophilic) granules of the neutrophil. Primary granules also contain the protease elastase; secondary granules house lactoferrin, an iron binding protein. Several lysosomal enzymes are sequestered in tertiary granules.

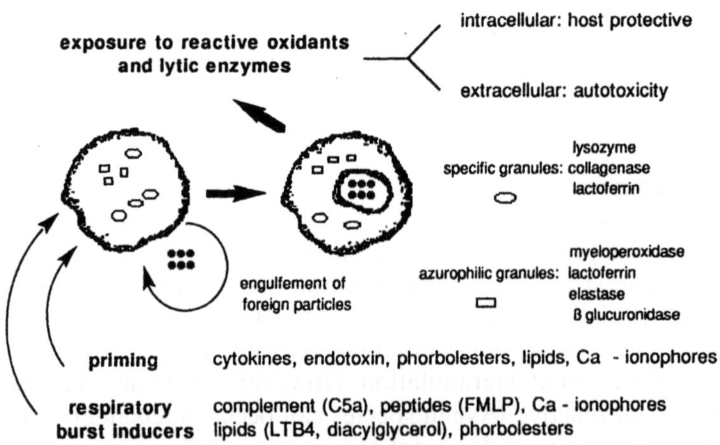

Fig. 3.27. Particle phagocytosis and neutrophil mediators

It was proposed that lysosomal fusion, bringing the NADPH oxidase and the myeloperoxidase systems together, is a prerequisite for the chemiluminescence reaction (Dahlgren 1989). When the foreign particle touches the neutrophil plasma membrane, phagocytosis begins: The particle becomes engulfed and exposed to a high flux of reactive oxidants inside the phagocytic vesicle.

About 90% of the oxygen consumed after initiation of the respiratory burst can be accounted for by secretion of superoxide anion radical (Root and Metcalf 1977; Roos et al. 1976). In vitro studies have shown that superoxide anion and its dismutation product, hydrogen peroxide, have only weak bactericidal activity (Mandell 1975; Gregory et al. 1973; Tauber and Babior 1978). Neutrophils produce two types of reactive oxidants, oxidizing radicals and oxidizing halogens (Fig. 3.28). Hydroxyl radicals may be formed in a metal-ion-catalyzed reaction, and hypohalous acid is produced, catalyzed by the myeloperoxidase-mediated halogenation of hydrogen peroxide inside the azurophilic granules of the neutrophil (Test and Weiss 1986). In the presence of halide ions such as bromine or chlorine ions, intracellular myeloperoxidase produces HOBr or HOCl. Hypochlorous acid can then react with amines to give oxidizing chloramines (R-NHCl) (Fig. 3.28).

The superoxide anion radical optimizes the production of hypohalous acids by neutrophils and potentiates oxidant damage (Kettle and Winterbourn 1988). Myeloperoxidase of neutrophils and macrophages probably plays the major role in the bactericidal action of hydrogen peroxide and superoxide anion through the formation of hypohalous acids and singlet oxygen (Klebanoff 1967a, b, 1968). Human neutrophils also produce the free radical nitrous oxide (Wright et al. 1989), which is identical with EDRF (Palmer et al. 1987) and participates in regulation of the vascular tone.

The amount of activated oxygen species produced by neutrophils is species-specific and depends on the nature of the stimulus and on cell differentiation. Differentiation of human promyelocytes into functional intact

$$O_2 + NADPH \xrightarrow[\text{NADPH oxidase}]{\text{neutrophil}} O_2^{-\cdot} + NADP + H^+$$

$$O_2^{-\cdot} \longrightarrow {}^1O_2,\ H_2O_2,\ OH\cdot$$

$$H_2O_2 + Cl^- \xrightarrow[\text{myeloperoxidase}]{\text{neutrophil}} HOCl + OH^-$$

$$HOCl + R\text{-}NH_2 \longrightarrow R\text{-}NH\text{-}Cl + H_2O$$

Fig. 3.28. Neutrophil-generated reactive oxidants

neutrophils results in a down-regulation of the antioxidant defense activity (Mn-SOD, glutathione reductase) and in an up-regulation of the capacity to generate reactive oxygen species. The consequence is an increased susceptibility of the neutrophil to autoxidation and thereby autoinactivation.

Defects in Oxidative Microbial Killing. Studies with patients deficient in (a) intact respiratory burst of neutrophils, (b) neutrophil NADPH oxidase, (c) myeloperoxidase, (d) (impaired) glucose-6-phosphate dehydrogenase, and (e) glutathione reductase system emphasize the importance of the respiratory burst and its metabolites in leukocyte-derived microbicidal activity and tissue injury.

Granulocytes from patients with chronic granulomatous disease have a defect in the generation of reactive oxygen species during phagocytosis (Holmes et al. 1967; Briggs et al. 1977) which impairs killing of certain microorganisms. Severe glucose-6-phosphate dehydrogenase deficiency also results in elevated susceptibility of the host to bacterial infections, because reducing equivalents for the NADPH oxidase are significantly diminished. Although oxidizing halogens are suggested to be important endogenous antimicrobial agents, myeloperoxidase deficiency is usually not associated with an increased incidence of bacterial infections in patients. This may be due to a compensatory rise in superoxide anion radical production and the subsequent formation of other reactive oxygen species. In specific diseases, granulocyte function may be impaired, resulting in decreased blood bactericidal activity and more bacterial superinfections. Neutrophils of patients with the T-cell lymphoma mycosis fungoides (stages III and IV) have normal basal chemiluminescence levels but depressed stimulated levels; low stimulated levels correspond to an advanced disease stage (Larussa et al. 1986). In the tumor stage, these patients frequently succumb to bacterial infections.

2. Eosinophil Granulocytes

Eosinophils are attracted by several humoral and cellular factors such as products of the complement cascade and arachidonic acid metabolism as well as by factors derived from lymphocytes, mast cells, and neutrophils. They con-

$$O_2 + NADPH \xrightarrow{\text{eosinophil NADPH oxidase}} O_2^{\cdot -} + NADP + H^+$$

$$H_2O_2 + Br^- \xrightarrow{\text{eosinophil myeloperoxidase}} HOBr + OH^-$$

$$HOBr + R\text{-}NH_2 \longrightarrow R\text{-}NH\text{-}Br + H_2O$$

Fig. 3.29. Eosinophil-generated reactive oxidants

tain several enzymes which inactivate mediators of inflammation released from mast cells and basophils (Czarnetzki 1980); they can also mediate proinflammatory tissue reactions (Kroegel et al. 1988b) via the release of prostanoids, basic proteins, and reactive oxygen species (Kroegel et al. 1988a). Eosinophils are able to produce two to three times as much superoxide anion radical as neutrophils (Mayeno et al. 1989). They contain a unique peroxidase that uses bromide instead of chloride as substrate. Consequently, eosinophils produce hypobromic acid (HOBr) and bromamines (R-NHBr) as oxidizing agents. Singlet oxygen may be formed in eosinophils in a reaction involving hydrogen peroxide and hypobromic acid (Fig. 3.29).

3. Macrophages

Macrophages play an important part in inflammatory processes such as phagocytosis, cytotoxicity, and immunomodulation. Among leukocytes, they are considered the most important active inflammatory cells in the secretion of degrading enzymes and production of reactive oxygen species (Nathan et al. 1980; Gemsa et al. 1982). In response to activation by a stimulus, macrophages undergo a respiratory burst. A 2–20-fold increase in oxygen consumption occurs, and concomitantly glucose metabolism is increased via the hexose monophosphate shunt. Reactive oxygen metabolites, various proteolytic enzymes, and prostanoids are effector molecules of stimulated macrophages. Hydrogen peroxide may have a significant function in the cytolytic and immunosuppressive function of macrophages, but reactive oxygen species are thought not to participate greatly in the monocyte-mediated rapid killing of drug-treated tumor cells (Colotta et al. 1985).

The production of reactive oxygen species in macrophages depends on the specific stimulus, site of macrophage isolation, and the extent of differentiation of the cells. Monocytes contain a myeloperoxidase and produce oxidizing halogens and oxidizing radicals. During conversion of monocytes to macrophages they release their myeloperoxidase, and microbial killing by macrophages may therefore be accomplished only through oxidizing oxygen species. The ability of cultured monocytes to produce reactive oxygen species decreases with their differentiation into macrophages, histiocytes, epitheloid cells, and giant cells (Nakagawara et al. 1981), in order of declining potency.

Differentiation of fresh or cultured human monocytes into macrophages leads to a decrease in the magnitude of stimulated chemiluminescence (Jungi and Peterhans 1988). This suggests that oxidant-mediated cell injury may play a significant role only during the early period of macrophage activation. Macrophages of highly active hypersensitivity granulomas produce clearly more superoxide anion radicals and prostanoids than those from older granulomas (Chensu et al. 1983). Cellular chemiluminescence was suggested to be a useful assay for indicating the biological activity of phagocytic cells and disase states (Van Dyke and van Dyke 1986). Patients with acute sarcoidosis produce more stimulated monocyte chemiluminescence than control monocytes. When patients with active sarcoidosis are treated with glucocorticosteroids, the enhanced monocyte chemiluminescence normalizes.

4. Reactive Oxidants and Protease Inhibitors

Leukocyte-derived oxidants, especially the myeloperoxidase system, may contribute to proteolytic tissue injury by oxidative inactivation of protective antiproteases (Clark et al. 1981). A system of antiproteases is present in the circulation and tissue fluids which functions to inactivate proteases released from inflammatory cells. Protease inhibitors inhibit the respiratory burst in human neutrophils and rat macrophages (Goldstein et al. 1979a). α_1-Proteinase inhibitor is an important component of this antiprotease system and is capable of inhibiting both neutrophil lysosomal elastase and collagenase. Human neutrophils phagocytizing opsonized antigen-antibody complexes produce dialyzable species of activated oxidants which are capable of partially suppressing the elastase-inhibiting capacity of whole human serum or purified human proteinase inhibitor. Oxidative inactivation in the microenvironment of neutrophils accumulating at sites of inflammation may allow proteases released from the cells to damage more readily adjacent connective tissue structures (Carp and Janoff 1979). α_1-Proteinase inhibitor can be inactivated by modification of its reactive site methionine by reactive oxygen species (Mathewson et al. 1981). Methionine oxidation is also the key pathobiochemical event in the oxidative inactivation of α_1-antitrypsin, chemotactic factors, and complement C5A.

By oxidatively inactivating key proteinase inhibitors and simultaneously activating latent proteinases, neutrophils can create an environment in which elastase, collagenase, and gelatinase are able to exert destructive effects more efficiently and with greater specificity than could even enormous concentrations of oxidants alone. The combined action of reactive oxygen species and lysosomal proteinases can subvert all the intrinsic barriers designed to protect host tissue from neutrophil injury (Weiss 1989). If the inflammatory stimulus is eliminated, the antiprotease system of the inflamed site can be regenerated, thereby inhibiting the released enzymes.

Inflammatory cells may also modulate the synthesis of collagenase in skin fibroblasts, which are the major source of skin collagenase under physiological conditions. Human skin collagenase activity can be inhibited by human serum (Eisen et al. 1970) and by the serum protein α_2-macroglobulin (Wooley et al. 1976), the major skin collagenase inhibitor in human serum. α_1-Antitrypsin is a very weak inhibitor of skin collagenase. The serum collagenases are restricted in their action to the intravascular compartment because of their large molecular weight and may only be released into the extravascular space during inflammatory states. α_2-Macroglobulin is a nonspecific serum inhibitor of skin collagenases; the existence of a specific collagenase inhibitor produced by skin fibroblasts that synthesize collagen has been reported (Welgus et al. 1979). This latter inhibitor may be of great significance in the physiological regulation of connective tissue degradation and formation, because it has direct excess to the extravascular space. Its susceptibility to oxidative damage is not known.

II. Immune Complexes and Endothelial Injury

Reactive oxygen species play a role in the chemical- or neutrophil-mediated injury of endothelial cells by directly attacking endothelial cells of blood vessels with resulting tissue injury (Ager 1982; Del Maestro 1982; Weiss et al. 1981). Vascular endothelial cells are highly susceptible to oxidative injury mediated by reactive oxygen species (Sacks et al. 1978; Hennig and Chow 1988), lipid peroxidation products (Yagi 1982), and decompartmentalization of transition metal ions. In comparison with human fibroblasts, human endothelial cells are much more susceptible to oxidative injury (Jongkind et al. 1989). Oxidative damage to vessels of the microcirculation can be inhibited by antioxidants and transition metal ion chelating agents.

Oxidative injury of blood vessel endothelium may also be caused by lipid-peroxide-mediated inactivation of prostacyclin synthetase. Endothelial prostacyclin synthetase is an enzyme producing prostacyclin (PGI_2), which is an arterial endothelial protectant factor. Prostacyclin synthetase is highly susceptible to oxidative damage, and enzyme inhibition would result in a diminished prostacyclin/thromboxane ratio, favoring the action of thromboxane, which includes vasoconstriction and thrombocyte aggregation.

Reactive oxygen species play a central role in the early phases of damage produced in dermal vasculature by the deposition of immune complexes (McCormick et al. 1981). Immune-complex-mediated skin injury is complement and neutrophil dependent. Immune-complex-induced dermal vasculitis is linked to hydrogen peroxide generation from activated neutrophils and the subsequent conversion to the hydroxyl radical (Fliegiel et al. 1984). Hydrogen peroxide is a potent mediator of tissue injury in immune complex diseases, both by directly injuring the tissue components and by indirectly modifying the inflammatory process through complement activation in the serum (Shingu 1987). In human umbilicated vein endothelial cells, complement-C5a-activated neutrophils induce endothelial damage by production of reactive oxygen species (Sacks et al. 1978).

It must be stressed that the skin and lung vasculatures are fundamentally different with respect to mechanisms of acute immune-complex-mediated injury in the rat (Warren et al. 1990). Similar organ differences may also exist in humans. In immunological acute dermal vasculitis and alveolitis, catalase and desferrioxamine (an iron chelator) are highly protective in the rat lung, while SOD is most protective in rat skin. This indicates that the effector molecules causing tissue injury are different. In the lung, the effector cell ultimately causing tissue damage is different in IgA and IgG immune complex disease. Acute lung injury in the rat induced by IgA immune complex is mediated by reactive oxygen species, and macrophages may be the principal effector cell. IgG immune-complex-induced lung injury is also mediated by reactive oxygen species, but neutrophils are the effector cells (Johnson et al. 1986).

Reactive oxidants may not only have destructive effects on the vasculature but also regulatory properties. They can modulate the vascular tone by direct interaction with vascular smooth muscle and indirectly by reaction with vasoactive mediators (EDRF, norepinephrine). They can evoke both vasoconstriction and vasodilation (Rubanyi 1988).

In conclusion, it seems reasonable to suggest that several inflammatory skin diseases with endothelial cellular damage and elevated circulating immune complexes are mediated in part by the enhanced production of reactive oxygen species. Circulating immune complexes probably play an important role in the pathogenesis of SLE and cutaneous necrotizing vasculitis. The pathogenetic role of circulating immune complexes and consequently reactive oxidants is less well understood but presumably of similar significance in pemphigus vulgaris, bullous pemphigoid, pyoderma gangrenosum, IgA linear dermatosis, dermatitis herpetiformis Duhring, progressive systemic sclerosis, mixed connective tissue disease, erythema multiforme, and Behçet's disease (Yancey and Lawley 1984).

III. Clastogenic Products

Prevalence. Clastogenic products (chromosome breakage factors) were first described in plasma from patients treated with ionizing radiation (Faguet et al. 1983). These factors are also found in the plasma of patients with spontaneous chromosomal instability. The diseases comprise congenital breakage syndromes associated with a high incidence of malignancies, such as Bloom's syndrome (Emerit and Cerutti 1981 b) and ataxia telangiectasia (Shaham et al. 1980), as well as chronic inflammatory and autoimmune diseases: SLE progressive systemic sclerosis (Emerit 1976; Bernstein et al. 1980; Sherer et al. 1981), dermatomyositis, periarteriitis nodosa, rheumatoid arthritis and Crohn's disease (Emerit et al. 1972, 1974 a; Emerit 1976, 1983; Kong-ooGoh 1982; Konstantinova and Nedkova-Bratanova 1969). Recently, clastogenic factors were isolated from blood of patients during myocardial ischemia/reperfusion injury (Emerit et al. 1988). We have recently analyzed the plasma of AIDS patients for clastogenic activity. Although 4-hydroxynonenal was significantly elevated, clastogenic activity was similar to in control sera. Further studies are required to residue this apparent discrepancy (Fuchs 1991 c).

Generation by Reactive Oxidants. Chromosome breakage factors can be produced in vitro by exposing cells to a system generating superoxide radical (Emerit et al. 1985) or the tumor promotor phorbol myristate acetate, which induces a respiratory burst in neutrophils (Khan and Emerit 1985; Kozumbo et al. 1987; Emerit and Cerutti 1981 a; Emerit 1982). The chromosomal aberrations induced by the clastogenic factor can be counteracted by the antioxidant L-cysteine in lymphocyte cell cultures (Emerit et al. 1974b). Clastogenic activity can be transferred from one cell culture to the other, and the acceptor culture is protected from the clastogenic effect by pretreating it with SOD.

Chemical Composition. Clastogenic factors produced by treating lymphocyte cultures with an oxidant generating system (Emerit et al. 1985) are heat-labile, low molecular weight compounds considered to be reaction products of reactive oxygen species with polyunsaturated fatty acids. Preparations of clastogenic factors contain thiobarbituric acid reactive substances and conjugated dienes (Emerit et al. 1985; Khan and Emerit 1985). The lipid peroxide degradation product 4-hydroxynonenal is clastogenic in the nanomolar range, whereas peroxidized products of arachidonic acid and malondialdehyde were only slightly or not at all clastogenic (Emerit et al. 1991). Other clastogenic material was recently detected by fractionation of clastogenic factor preparations and was identified as inosine nucleotide (Auclair et al. 1990).

Biological Significance. Reactive oxygen species can be mutagenic in mammalian systems (Weitzmann and Stossel 1981, 1984; Phillips et al. 1984) by causing DNA base damage and strand breakage. Since reactive oxygen species have a short half-life, their hazardous potential seems to be limited to the ultimate vicinity in which they arise. Formation of secondary, longer-lived clastogenic material by reactive oxygen species, which can damage macromolecules at a distance from their site of formation, may be an important mechanism of oxidative injury. These factors are produced in free radical reactions and exert their clastogenic effect via free radical intermediates. Formation of clastogenic factors by reactive oxygen species may result in autocatalytic, self-sustaining tissue injury. Clastogenic factors may contribute to the chronicity of the inflammatory state in autoimmune diseases and are potential candidates for tumor promotion.

IV. Lipid Peroxidation Products

Lipid peroxidation products are presumably involved in sustaining the chronicity of inflammatory skin disorders. Once formed during autooxidative skin injury, they may induce further lipid peroxidation, irritate the skin, and attract leukocytes.

Skin Inflammation. Several decades ago lipid peroxidation products were suspected to be elevated in patients with rheumatoid arthritis, and increased levels of these products were later found in their synovial fluid (Lunec and Dormandy 1979; Rowley et al. 1984) and plasma (Muus et al. 1979). In inflammatory skin lesions, lipid peroxidation products were also hypothesized to be increased.

Lesional Skin. Lipid peroxide levels in skin lesions of patients with active inflammatory disease are markedly elevated in comparison with noninflammatory skin lesions (Niwa 1987a; Niwa et al. 1987a). A direct correlation of the concentration of skin lipid peroxidation products with the clinical severity of contact dermatitis has been noted (Hayakawa 1971; Izawa 1968). Lipid peroxidation products were measured in skin biopsies of patients with active inflammatory skin diseases, such as atopic dermatitis, contact dermatitis, and radiodermatitis, and in skin lesions of noninflammatory skin diseases, such as pityriasis versicolor, vitiligo, and verruca vulgaris. Their content was markedly elevated in lesions of inflammatory skin diseases compared with noninflammatory lesions (Niwa et al. 1987b; Sugiura et al. 1986b; Ueda et al. 1971; Fedorov 1990). In noninflammatory skin lesions, e.g., scars, hypertrophic scars, and keloids, the malondialdehyde concentration was in a similar range as in normal human skin (Sugiura et al. 1986a).

Sebum. Peroxidation of sebum lipids was analyzed by mass spectroscopy of skin surface lipids in patients with inflammatory acne vulgaris. In the sebum of healthy control persons, no lipid peroxidation products were detected, and peroxidation of sebum lipids is suggested to be linked to the inflammatory skin reaction (Fournier et al. 1980). An increased content of lipid peroxidation products in serum and sebum lipids of female patients with melanosis facie (melasma) decreased after oral and topical treatment with α-tocopherol, and hyperpigmentation improved clinically (Hayakawa and Okada 1977).

Lipid Peroxide/SOD Ratio: A Prognostic Marker. In patients with inflammatory skin diseases which were severe and highly resistant to therapy, SOD activity was only slightly increased. In patients with mild disease that responded well to therapy, SOD activity was markedly elevated. The ratio of lipid peroxide content to SOD activity was considered an accurate prognostic indicator, being raised only in the group not responding to treatment. The severity of inflammatory skin diseases or the response to treatment may in part be governed by the degree to which the patient's SOD activity is up-regulated in response to the generation of tissue-damaging substances such as lipid peroxidation products (Niwa et al. 1987a).

The lipid peroxide/SOD ratio was analyzed at three different sites in patients with superficial trichophyton infection. A high ratio was found at the raised erythematous margin of the lesion, and a low ratio both outside and within the margin. At the inflamed site of dermatophyte infection, there is an increase in cytotoxic lipid peroxides, and SOD activity is not sufficiently elevated to prevent this effect. Where inflammation has subsided, the SOD activity is augmented (Niwa, personal communication).

An increase in lipid peroxide levels with severe inflammatory skin lesions is accompanied by a stimulation of skin SOD activity in patients whose clinical course is better. Neutrophils from patients with cement dermatitis generate markedly increased levels of reactive oxygen species. In spite of the greater generation of oxidants by neutrophils, the SOD activities of the patients are comparable with those in healthy controls. In skin tissues from cement workers without cement dermatitis, enhanced SOD activity as well as increased neutrophil generation of reactive oxygen species are noted. Severe skin manifestations in some patients with cement dermatitis is partly explained by their defective capacity for inducing protective skin antioxidant activity (Miyachi et al. 1985).

Serum. Lipid peroxidation products (malondialdehyde and lipid hydroperoxides) were examined in the serum of patients with various dermatoses such as psoriasis, eczema, prurigo, bullous skin diseases, urticaria, acne vulgaris, seborrheic dermatitis, atopic dermatitis, SLE, progressive systemic sclerosis, generalized morphea, herpes zoster, and alopecia. Their concentration is raised in patients with psoriasis, eczema, prurigo, SLE, generalized morphea, atopic dermatitis, and alopecia (Demitsu et al. 1988). In patients with panniculitis, increased blood levels of malondialdehyde and decreased serum antioxidant activity have been seen (Ezhova et al. 1987).

Chemotactic Activity. Many factors with in vitro chemotactic activity for leukocytes have been described in the past. They can be divided into those derived from plasma (complement-derived factors), those that are cell associated, and those derived from microorganisms. Cell-derived chemotactic factors comprise various compounds such as peptides, amines, lymphokines, monokines, and specific lipid-derived products such as LTB4 (leukotriene), 5-HETE (hydroxyeicosatetraenoic acid), 12-HETE, and 15-HPETE (hydroperoxyeicosatetraenoic acid).

It was demonstrated in vitro that the oxidation of polyunsaturated fatty acids generates chemotactic stimuli (Turner et al. 1975; Bragt et al. 1980). These substances are heat-labile lipid derivatives and are also formed from arachidonic acid following incubation with a system generating reactive oxygen species (Perez and Goldstein 1980). A relationship between lipid peroxidation and the immune response was suggested based on the fact that hydroxyalkenals (degradation products of lipid peroxidation) have chemotactic activity (Curzio et al. 1982). 4-Hydroxynonenal and homologous aldehydes possess chemotactic activity in vitro at micro- to picomolar concentrations. At higher levels they exert strong toxic effects on neutrophils (Curzio 1988). 4-Hydroxynonenal is found in significant amounts under inflammatory and autoxidizing conditions in biological material. Reactive oxidants induce the formation of chemotactic factors that attract inflammatory cells and stimulate an oxygen burst (McCord et al. 1980; Petrone et al. 1980). By this mechanism, they may augment neutrophil accumulation and autotoxicity.

V. Prostanoid Metabolism

Prostaglandins and leukotrienes are important mediators of skin inflammation (Ford-Hutchinson and Rackman 1983; Ford-Hutchinson and Chan 1985; Hensby et al. 1983; Greaves 1982; Voorhees 1983). The control of prostaglandin synthesis can be regulated (Lands 1982) by modulating the availability of precursor fatty acids, by the presence of competing inhibitory fatty acid substrates (Lands 1989; Lands and Pendleton 1989) and other drugs, and by the tissue peroxide concentration (Lands 1984, Lands and Marshall 1985; Lands et al. 1985).

Hydroperoxides and Prostanoid Metabolism. Cyclooxygenase, lipoxygenase, and presumably phospholipase activities are regulated by the tissue hydroperoxide (ROOH) concentration. A variety of organic hydroperoxides rapidly accelerate cyclooxygenase activity to produce prostaglandins. Hydrogen peroxide also stimulates cyclooxygenase, but only at 0.001 of the efficacy of lipid hydroperoxides (Lands 1985). Inhibition of cyclooxygenase by glutathione peroxidase indicates that cyclooxygenase catalysis requires some hydroperoxide which apparently could be generated by cyclooxygenase itself. 5-Lipoxygenase is also inhibited by glutathione peroxidase and tocopherol and presumably requires a lipid hydroperoxide activator. This would mean that leukotriene formation could occur under similar control mechanisms. Cyclooxygenase contains a synthase and a peroxidase activity. With the synthase activity, peroxide serves as a trigger for an amplifying production of more peroxide. With the peroxidase activity, the peroxide is reduced to an alcohol. The affinity of the peroxide removing peroxidase for lipid hydroperoxides is in the micromolar range, whereas the peroxide-generating cyclooxygenase activity has a half-maximum activity at 20 n*M*. Therefore, the peroxidase activity will only be effective at higher hydroperoxide concentrations. Any process which generates peroxides may activate cyclooxygenase (Hemmler et al. 1979). Activated human leukocytes produce sufficient quantities of hydroperoxides to enhance prostaglandin synthesis in cells (Marshall and Lands 1986) (Fig. 3.30).

Prostacyclin and Thromboxane Synthetase. Concomitant with studies showing that peroxides can activate prostaglandin cyclooxygenase have come reports indicating that a radical species generated during the cyclooxygenase reaction inactivates the enzyme. Prostaglandin cyclooxygenase is irreversibly self-deactivated during the oxygenation of arachidonic acid. This might provide a negative feedback mechanism in peroxide-triggered cyclooxygenase activation. Inactivation is probably mediated by oxygen centered radicals formed as a result of the reductive breakdown of the hydroperoxide on PGG_2 (Egan et al. 1976). This radical destroys the action not only of the peroxidase and cyclooxygenase components of the prostaglandin synthetase enzyme but is also toxic to prostacyclin synthetase while having no effect on thromboxane synthetase. Hydroperoxides may therefore influence the ratio of prostacyclin and thromboxane and trigger pathological processes (Ham et al. 1979).

n-3 and n-6 Fatty Acids. Reaction of cyclooxygenase with eicosapentanoate (20:5 n-3) requires a much higher peroxide tone than with arachidonate (20:4 n-6) to form prostaglandins. The low peroxide concentration in normal tissues may permit 20:5 n-3 to be an effective antagonist of prostaglandin formation from 20:4 n-6 and may contribute to the health-related effects of n-3 fatty acids.

Prostanoid Second Messenger Hypothesis. A mechanism for regulating tissue antioxidant activities has been described (Cutler 1984). The rate of antioxidant synthesis is proposed to be determined by the ratio of intracellular levels of cGMP to those of cAMP. The validity of this concept is not proven in skin and experimental evidence is required before it can be generally accepted. The prostanoid-second messenger hypothesis is based upon the high sensitivity of arachidonic acid, cyclooxygenase, and guanylate cyclase to their redox microenvironments. The amount of arachidonic acid produced and/or the activity of cyclooxygenase determine the rate of PGH_2 synthesis, which then dictates the rate of prostaglandin biosynthesis and in turn the cellular levels of cAMP.

Active oxygen species inactivate cyclooxygenase and lower cAMP levels. In contrast, antioxidants reduce the concentration of reactive oxygen species and thus increase the overall arachidonic acid availability and/or enhance the activity of cyclooxygenase, resulting in a greater rate of PHG_2 synthesis and in turn a higher cellular level of cAMP. A negative feedback process stabilizes

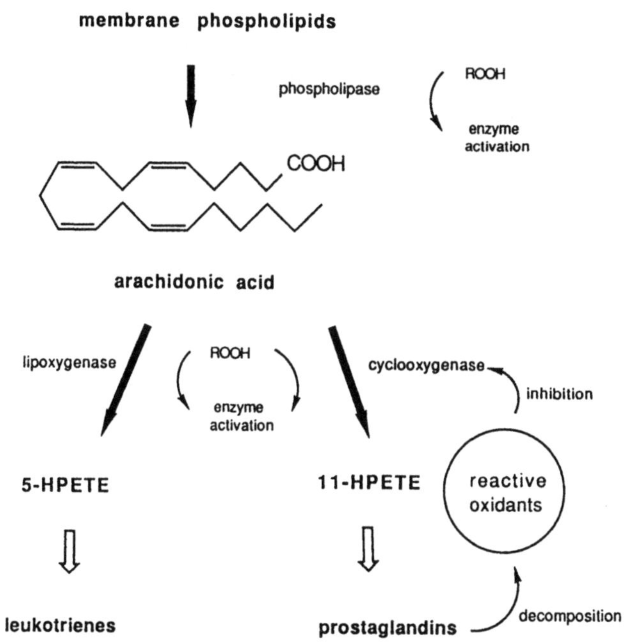

Fig. 3.30. Prostanoids and reactive oxidants

the overall prostaglandin synthesis pathways by being sensitive to the intracellular levels of active oxygen species.

cGMP and cGMP/cAMP Ratio. Guanylate cyclase activity is stimulated by a variety of reactive oxidants, such as hydrogen peroxide (White et al. 1976), superoxide anion (Vesely et al. 1979), hydroxyl radical (Mittal and Murad 1977a, b), nitrous oxide (Palmer et al. 1987), and stable nitroxide radicals (Lad et al. 1985). Activation of guanylate cyclase by free radicals and reactive oxygen species will elevate the cGMP level, and a net increase in the cGMP/cAMP ratio will result. If antioxidant levels are in excess or the normal level of reactive oxygen species is lowered, then a net decrease in the cGMP/cAMP ratio will arise.

VI. Reactive Oxidants as Modulators of Inflammation

The role of reactive oxidants as mediators of inflammation has been outlined several times (Del Maestro et al. 1980; Bragt et al. 1980; Schalwijk et al. 1986). In particular, they are involved significantly in the destruction of ocular tissue and amplification of the inflammatory process in experimental uveitis (Rao et al. 1987). They may also be important mediators in skin inflammation (Merk 1987). Reactive oxidants are thought to participate in the pathogenesis of allergic cutaneous vasculitis (Tosca and Stratigos 1988), and their role in inflammatory processes, with particular reference to skin, has recently been reviewed by Black (1989) and Kapp (1990).

Acute skin inflammation can be induced by intradermal injection of a superoxide anion radical generating system (Trenam 1991; Yoshioka et al. 1987b) or by attractors of inflammatory cells, e.g., dextran (Kamimura 1972). The inflammatory skin response due to the former is inhibited by concomitant injection of SOD or catalase. In oxidant-induced skin inflammation, histopathology study reveals vasodilation, hemorrhage, and early (3–24 h) infiltration of the upper dermis with neutrophils, monocytes becoming predominant after 48 h (Yoshioka et al. 1987b). In a similar study, skin inflammation was induced in mice by intracutaneous injection of reactive oxidant generating systems (xanthine oxidase/hypoxanthine, tertiary butylhydroperoxide, doxorubicin). The inflammatory response was inhibited by concomitant application of the antioxidant dihydrolipoate (Fuchs et al. 1990d).

Reactive oxygen species modulate the inflammatory response via complex mechanisms. Generation of reactive oxidants may provide a *negative feedback mechanism* to self-limit the inflammatory process by inactivation of neutrophils, inflammatory mediators, and chemo-attractants (Gruner et al. 1987a; Clark et al. 1980), as well as by direct down-regulation of the oxygen burst.

Reactive oxygen species attack the neutrophil itself, and this may be responsible for the death of these cells after phagocytosis (Badwey and Karnovsky 1980). The intracellular myeloperoxidase halide system of neutrophils was proposed to regulate the duration of superoxide anion radical production during the respiratory burst, presumably by inhibition of the plasmalemmal NADPH oxidase system (Edwards and Swan 1986). These mechanisms can help to terminate the respiratory burst after sufficient time for microbial killing and may serve in vivo as an important negative feedback mechanism to prevent excessive oxidant generation with resultant autooxidative damage to surrounding tissue (Speier and Newburger 1986). Phagocytozing neutrophil-derived reactive oxygen species can also deactivate proinflammatory lysosomal enzymes and chemotactic mediators.

Certain leukoattractants, including complement-derived and synthetic peptic agents, can be inactivated by the myeloperoxidase/hydrogen peroxide/halide system of neutrophils. Inactivation of peptide chemoattractants by the myeloperoxidase system is probably a consequence of methionine oxidation, and this biochemical alteration decreases the affinity of the peptide for neutrophil peptide receptors.

Leukotrienes are readily susceptible to free radical attack produced during flavin-mediated photooxidation of *o*-diansidine. In contrast, prostaglandins are rather unreactive. This relative susceptibility may play a role in the self-deactivation of chemotactic leukotrienes during the inflammatory process (Chopra et al. 1988).

Complement C5a Receptor Defect. Superoxide anion radical production and chemotaxis of neutrophils were studied in 135 patients with inflammatory skin disease (erysipela, acne conglobata, impetigo contagiosa, psoriasis). In 25 patients, a complement-C5a-dependent defect of neutrophil activation was found, while the neutrophil responsiveness for the tripeptide fMet-Leu-Phe (FMLP) was normal or elevated. The C5a-dependent defect was disease-nonspecific but correlated with disease activity (Schröder and Christophers 1985). It was suggested that it reflects down-regulation of C5a receptor functions in neutrophils, thereby counteracting tissue destruction by C5a-activated neutrophils (Christophers and Schröder 1987).

H. Oxidative Injury in Skin Diseases

Reactive oxidants are involved in the pathogenesis of skin diseases associated with inflammatory cells, immune complexes, hyperproliferation, and autoimmunopathies. Several skin diseases can be characterized histologically by the predominating inflammatory cell infiltrate (Table 3.2). This categorization may help in elucidating the pathophysiological mechanisms and the best therapeutic approach. In particular, reactive oxygen species have been suggested to participate in the pathophysiology of various inflammatory skin diseases (Miyachi et al. 1985, 1987).

Table 3.2. Skin diseases with characteristic inflammatory cell infiltrates

Micromorphology	Skin disease
Tissue neutrophilia	Psoriasis vulgaris Sweet's syndrome Dermatitis herpetiformis Duhring
Tissue eosinophilia	Bullous pemphigoid Pemphigus herpetiformis
Tissue lymphocytosis	Atopic dermatitis

I. Skin Diseases with Vasculitis

Cutaneous vasculitis is classified into different forms which share a common pathogenetic mechanism but can be manifested clinically in a large spectrum of cutaneous lesions. The clinical features and the histopathology of the individual diseases are markedly different. According to the predominating cellular element present in skin lesions, cutaneous vasculitis can be classified histopathologically into neutrophilic, lymphocytic, macrophage-derived (granulomatous), and eosinophilic vasculitis (Table 3.3).

Neutrophilic and lymphocytic vasculitis comprise various distinct clinical diseases listed in Table 3.3. The predominating inflammatory cells in *granulomatous vasculitis* (Table 3.3) are monocyte-derived cells (epitheloid cells) and giant cells (with syncytium formation). Although epitheloid cells are found in

Table 3.3. Types of cutaneous vasculitis

Micromorphology	Skin disease
Neutrophilic vasculitis	Behcet's disease Pyoderma gangrenosum Erythema elevatum et diutinum Urticaria vasculitis Vasculitis allergica Nodular fasciitis Granuloma faciale
Lymphocytic vasculitis	Erythema multiforme Systemic lupus erythematosus Dermatomyositis Pityriasis lichenoides Purpura pigmentosa et progressiva Venous stasis dermatitis
Granulomatous vasculitis	Granuloma annulare Rheumatoid nodule Necrobiosis lipoidica Wegener's granulomatosis
Eosinophilic vasculitis	Churg-Strauss syndrome

granuloma faciale, the predominating cell infiltrate is the neutrophil. Systemic diseases with skin involvement and giant cell vasculitis (arteriitis) are: temporal arteriitis, polymyalgia rheumatica, and Takayasu's disease. *Eosinophils* are significantly involved in the *vasculitis* occurring in Churg-Strauss syndrome.

1. Neutrophilic Vasculitis

Necrotizing vasculitis is defined by neutrophilic infiltration of blood vessel walls, neutrophilic karyohexis (leukocytoclasia), extravasation of erythrocytes, and fibrinoid necrosis of the vessel wall. In cases of less prominent erythrocyte extravasation, leukocytoclasia, and fibrinoid necrosis, the designation neutrophilic vascular reaction is preferred. Necrotizing vasculitis and neutrophilic vascular reactions are two variants of neutrophilic vasculitis. Necrotizing vasculitis is very likely an immune complex disease, in which the origin of the causative antigen is heterogeneous. Sera from patients with leukocytoclastic vasculitis induce an increased generation of reactive oxygen species in normal neutrophils. Circulating immune complexes are considered to be one of the factors responsible for this enhanced generation (Miyachi et al. 1982b).

Behçet's disease is a chronic, multisystem disorder involving the skin, mucous membranes, and gastrointestinal, cardiovascular, and central nervous systems. It is characterized by a relapsing inflammatory process and oral, genital, and ocular abnormalities. Vasculitis of small blood vessels is the basic histopathological feature of the disease (Arbesfeld and Kurban 1988). Histology of early lesions shows either leukocytoclastic vasculitis or a neutrophilic vascular reaction (Jorizzo et al. 1988). Numerous etiologies have been proposed; an autoimmune basis characterized by circulating immune complexes and complement activation are widely accepted. The presence of circulating immune complexes in patients has been confirmed (Levinsky and Lehner 1978; Gupta 1978; Burton-Kee et al. 1981). Increased chemotaxis of neutrophils is also reported by several authors (Sobel et al. 1977); Matsumura and Mizushima 1975; Fordhan et al. 1982), and neutrophil derived reactive oxygen species and lysosomal enzyme activities were elevated (Takeuchi et al. 1978). It was suggested that these neutrophils might have increased phagocytic activity and elevated production of reactive oxygen species (Jorizzo 1986). Indeed, it was demonstrated that activated neutrophils from patients with Behçet's disease generate higher levels of reactive oxygen species than normal neutrophils, which is believed to be partially responsible for the endothelial tissue damage (Niwa et al. 1982). Immunosuppressive agents (glucocorticosteroids) and colchicine are effective in the clinical therapy.

Pyoderma gangrenosum (dermatitis ulcerosa) is an ulcerating disease of the skin and is considered to be a neutrophilic vascular reaction (Jorizzo et al. 1988). Its histopathology is not diagnostic, the findings in the lesions are controversial, and necrotizing vasculitis is often described. Prominent features are massive neutrophil invasion, thrombosis of medium-sized and small blood vessels, and vascular necrosis. Intramural deposition of immunoglobulins and complement indicates a pathogenic role of immune complexes. In 50% of the cases, pyoderma gangrenosum is associated with internal diseases, such as colitis ulcerosa, Crohn's disease, seropositive and seronegative polyarthritis, lymphoreticular malignancies, and gammopathies. Besides immunosuppressive agents (glucocorticosteroids, azathioprine, chlorambucil, cyclophosphamide), dapsone, colchicine, and clofazimine are effective in the treatment. A case report of a patient with pyoderma gangrenosum and a neutrophil myeloperoxidase deficiency suggests that the myeloperoxidase enzyme is not essential in the pathogenesis (Disdier et al. 1991).

2. Lymphocytic Vasculitis

Erythema multiforme is an inflammatory disease of skin and mucous membranes. It may be idiopathic or may also occur as a drug reaction or associated with microbial infections. It is characterized histologically by dermal edema, lymphohistiocytic infiltration around vessel walls and at the dermal-epidermal interface, and necrotic keratinocytes. Neutrophilic infiltrates are not prominent. Recent observations indicate that circulating immune complexes are involved in the pathogenesis (Wuepper et al. 1980; Imamura et al. 1980). Deposition of circulating immune complexes in the superficial cutaneous vasculature may trigger lymphohistiocytic infiltration around the dermal blood vessels. It was demonstrated that neutrophils incubated with sera from patients with erythema multiforme are activated to produce oxygen radicals, indicating that the pathway of oxygen burst is amply accelerated by some unknown serum factor. This is not a phenomenon general to immunologically mediated and/or inflammatory skin diseases, since no significant increase in reactive oxygen species generation is noticed with sera from patients with bullous pemphigoid or inflammatory acne vulgaris. The reactive oxygen species could be involved in the tissue destruction leading to the formation of cutaneous lesions of immune complex type of erythema multiforme (Miyachi et al. 1986b). However, since neutrophils are usually not abundant in erythema multiforme lesions, the significance of this finding for the clinical situation has not been established.

II. Mesenchymal Autoimmune Disorders

SLE and progressive systemic sclerosis are autoimmune diseases, both exhibiting mainly lymphocytic vasculitis during inflammatory disease states. Although they share common histopathological and pathophysiological characteristics, their clinical pictures are quite distinct.

1. Systemic Lupus Erythematosus

SLE is an autoimmune disease of unknown origin, characterized by major alterations of both humoral and cellular immunity. Tissue injury in the course of the disease is discussed to be mediated by immune complexes, activated neutrophils, and clastogenic factors.

In the pathogenesis of SLE, a polygenetic or exogenously induced defect of the cellular microdebridement has been considered (Stone 1988). A consequence of this defect could be the reduced clearance of immune complexes. Although lymphocytes are the predominating inflammatory cells associated with the vasculitis, neutrophils may be of significant importance in the inflammation. The generation of activated oxidants by normal neutrophils after challenge with SLE and rheumatoid arthritis sera was studied, and strikingly different biological effects were induced. SLE sera stimulated generally a very strong generation of reactive oxygen species, while rheumatoid arthritis sera produced a rather weak respiratory burst of the neutrophils. The differing potencies of sera of SLE and rheumatoid arthritis patients to induce a respiratory burst might reflect altered clearance of circulating immune complexes by neutrophils in SLE and rheumatoid arthritis.

Serum-stimulated generation of reactive oxygen species in neutrophils was studied with serum from healthy volunteers, patients with bacterial infections, and patients with SLE (Niwa et al. 1985a). It was highest in the neutrophils stimulated with SLE-derived serum. That from patients with bacterial infections also showed a greater generation of reactive oxygen species compared with healthy controls (Niwa et al. 1984c). The activity of oxidant production by neutrophils may be an indicator of disease activity. It is suggested that tissue damage in SLE may be partially due to excessive generation of activated oxygen species and that both neutrophils themselves and a serum factor that activates them are involved in the mechanism of vascular injury (Niwa et al. 1985a). A serum factor has been identified as constituting immune complexes of intermediate size (Shingu et al. 1983). The high production of reactive oxygen species by sera of SLE patients may be causally related to the vasculitis and extensive tissue damage observed in this disease (D'Onofrio et al. 1984).

Patients with SLE have increased numbers of chromosome breaks (Tuschl et al. 1984), and this correlates with a low molecular weight chromosomal damaging agent that is released from the lymphocytes into the serum. This clastogenic factor also produces chromosome breaks in lymphocytes from healthy people when incubated in the presence of serum of SLE patients or lymphocytes or purified factor (Emerit and Michelson 1981). The increased risk of patients with SLE for the development of malignant lymphoma, however, is probably related to the intake of immunosuppressive drugs rather than to a direct effect of the clastogenic factor.

Chromosome breakage rates in lymphocytes are influenced by exogenous SOD in vitro and were also reduced in vivo by the injection of SOD in patients with lupus erythematosus and New Zealand Black (NZB) mice (Emerit and

Michelson 1980). The NZB mouse is an animal model for SLE; it develops spontaneously an autoimmune disease with antibody-positive hemolytic anemia, immune complex nephritis, production of nuclear-specific antibodies, and formation of clastogenic factors in the plasma (Emerit et al. 1980a). A high frequency of lymphoreticular neoplasms is found. In two NZB sublines with high and low chromosomal breakage rates, the incidence of pathological manifestations significantly correlated with the breakage frequency, indicating that chromosome damage is not a simple epiphenomenon (Emerit et al. 1980b). Increased formation of reactive oxidants in the sera and lymphocytes of SLE patients is suggested to inactivate preferentially cells in the suppressor T-lymphocyte lineage. Since it is generally believed that autoantibody production by B lymphocytes is due to a deficiency of suppressor T-lymphocyte activity, reactive oxidants may thereby participate in the pathogenesis of this autoimmune disease (Peller 1982).

Antioxidant Treatment. There are several studies concerning the possibility of insufficient free radical scavenging in patients with connective tissue diseases. Decreased SOD activity in the eryhtrocytes of patients with rheumatoid arthritis has been reported (Rister et al. 1978; Banford et al. 1982). Other authors, however, do not find significant differences in erythrocyte enzyme activities (Scudder et al. 1976) or even report increased SOD activity in neutrophils from patients with rheumatoid arthritis (Youssef and Baron 1983). The serum concentrations of α-tocopherol and β-carotene in lupus erythematosus patients were found to be low in a study by Welsh (1954), but the large physiological variability of serum concentrations was not considered in this study. Chronic cutaneous lupus erythematosus is a special monosymptomatic form of SLE which affects the skin only (Gilliam and Sontheimer 1981; Meigel 1987). In general, pathological alterations of blood vessels and immunological disturbances are less expressed in the case of chronic cutaneous lupus erythematosus than with SLE. Membrane lipid peroxidation may play a role in the pathogenesis of chronic cutaneous lupus erythematosus (Akimov 1987), and antioxidants are suggested to of therapeutical benefit (Kalamkarian et al. 1990).

Burgess first reported successful treatment of patients with chronic cutaneous lupus erythematosus using oral or systemic administration of vitamin E (Burgess 1948; Burgess and Pritchard 1948). These results were later confirmed by other authors (Bazex 1949; Doucas 1949; Silvers 1950; Sterzi 1950; Grubb 1952; Welsh 1954; Shinskii et al. 1962; Ayres and Mihan 1978c). However, less successful therapy outcome has been noted by several other authors (Sawicky 1950; Pascher et al. 1951; Sweet 1948; Jordan and Wulf 1950; Morgan 1951). Alimentary supplementation of NZB mice, a murine autoimmune disease model mimicking SLE, with α-tocopherol acetate results in a significant increase in their average lifespan (Harman 1980a).

α-Tocopherol may intervene as an antioxidant at several steps of the pathophysiological scenario. It may protect from reactive oxygen species produced by activated leukocytes, scavenge reaction intermediates produced by clastogenic factors, and inhibit the formation of clastogenic factors.

Photosensitivity. A prominent feature in patients with lupus erythematosus is high sensitivity to UVB and high-dose UVA irradiation. The lymphocytes of patients with SLE are sensitive to near-UV, and this photosensitivity seems to be related to the presence of the clastogenic factor in these cells. Superoxide anion radicals are involved in this photosensitivity reaction (Emerit and Michelson 1981). Lymphocytes of healthy subjects exposed to the factor also become sensitive to light of the same wavelengths. The presence of this photoactivated agent was thought to explain why the condition is aggravated after exposure to sunlight with the appearance of typical skin lesions. Tocopherol provides significant systemic photoprotection in patients with cutaneous lupus erythematosus (Jordan and Wulf 1950).

2. Progressive Systemic Sclerosis

Progressive systemic sclerosis (PSS) is a systemic, mesenchymal/autoimmune disease with sclerosis of the skin and internal organs. Immunological disturbances, endothelial cellular damage, elevated circulating immune complexes (Yancey and Lawley 1984), and the presence of clastogenic factors in plasma are described. Three pathways are proposed to be involved in the early stages.

The initial events include vascular alterations, dysfunction in the immune system, and dysregulation of connective tissue metabolism. Activation of connective tissue metabolism represents a final step in the pathogenesis of systemic scleroderma, leading to massive skin sclerosis (Krieg and Meurer 1988). Several mechanisms, including immunocompetent cells, are involved during the starting phase of fibrosis development. Vasculopathies in PSS are associated with a cytotoxic serum factor (Shanahan and Korn 1982; Cohen et al. 1983; Khaleh and Leroy 1983). Increased collagen production and fibrosis are attributed to a fibroblast growth factor (Leroy 1982; Leroy et al. 1983). Mediators secreted by macrophages, lymphocytes and platelets and growth factors of various origins can attract and induce proliferation of fibroblasts and stimulate collagen synthesis. Others such as γ-interferon inhibit fibroblast chemotaxis and collagen synthesis (Hein et al. 1988). In addition, reactive oxygen species may play a role of unknown significance in all three pathways. Reactive oxidants may be a secondary phenomenon in PSS, as was recently suggested for autoantibodies (Stone 1991).

Generally, in chronic inflammatory processes and after hyperoxygenation and radiation therapy, an enhanced production of collagen is observed, which may be mediated by reactive oxygen species. Idiopathic pulmonary fibrosis, for example, is a chronic, inflammatory condition of the lung interstitium characterized by the accumulation of alveolar macrophages and neutrophils.

Lung inflammatory cells (macrophages, neutrophils, lymphocytes) of patients with idiopathic pulmonary fibrosis show an increased production of reactive oxygen species (Cantin et al. 1987). Parenchymal cell injury finally results in fibrosis of the alveolar wall. In PSS, infiltration by inflammatory cells occurs significantly only during the early inflammatory phase and is not a characteristic feature of the chronic fibrotic phase.

In the late stage of PSS, the oxygen supply in sclerodermatous skin becomes impaired (Hiller et al. 1986; Silverstein et al. 1988), establishing a hypoxic cellular environment. Since the degree of oxygenation modulates fibroblast metabolism, the cutaneous hypoxia demonstrated in patients with systemic sclerosis may play a significant role in the pathogenesis of skin fibrosis (Silverstein et al. 1988). Cultures of human fibroblasts exhibit prolonged growth and an extended lifespan at decreased oxygen concentrations (Packer and Fuehr 1977). The extended lifespan may be due to decreased formation of reactive oxygen species during normal fibroblast cell metabolism. This phenomenon could participate in the late phase pathophysiology.

Although preexisting collagen may be destroyed in the initial phase of inflammation, its biosynthesis can be stimulated in chronic inflammation. Besides a direct effect, reactive oxygen species influence cellular second messengers and could thereby modulate collagen synthesis. In fibroblasts from patients with PSS, changes in cellular metabolism have been reported. The cGMP concentration is increased and that of cAMP decreased (Grozdova et al. 1984). Fibroblast proliferation, growth, and differentiation depend upon changes in cGMP and cAMP levels and on calcium ions (Chaplowski et al. 1975). Increased quantities of cAMP inhibit the labeling of collagen with radioactive precursors in fibroblasts (Baum et al. 1978, 1980). Also, cAMP can stimulate the incorporation of radioactively labeled sulfur into glycosaminoglycans and proteoglycans (Koyama et al. 1976; Schonhofer et al. 1974). It has been proposed that reactive oxygen species decrease cAMP levels (Cutler 1984; Vesely et al. 1979) and increase cGMP levels (White et al. 1976). Consequently, a net increase in the cGMP/cAMP ratio will result, which may have effects on the transcriptional control of collagen synthesis.

In patients with PSS, chromosomal instability and the presence of a chromosomal breakage factor in plasma have been described (Emerit et al. 1971, 1973, 1974a; Emerit 1976; Emerit and Michelson 1980; Emerit and Cerutti 1982; Pan et al. 1975; Sherer et al. 1981; Niwa et al. 1987a). Clastogenic activity is detected only in patients with PSS and not in those with morphea (circumscribed scleroderma) (Emerit, personal communication).

This clastogenic factor induces chromosome breaks in blood cell cultures of PSS patients (Emerit et al. 1974a) as well as of healthy people (Emerit et al. 1976). Increased chromosome damage is not only observed in lymphocyte and fibroblast cultures of PSS patients, but also in direct bone marrow preparations, indicating that it is a systemic phenomenon (Emerit et al. 1973). In healthy relatives of PSS patients increased chromosomal breakage is also observed (Emerit et al. 1976) and is interpreted in favor of the concept of a clonal evolution of genetically abnormal cells that could be at the origin of immune reactions in human connective tissue disease.

Clastogenic factors are probably composed of different substances, including lipid peroxidation products. Recently, a clastogenic inosine nucleotide was identified as a component of the chromosome breakage factor in PSS patients. Commercial inosine tri- and diphosphate added to lymphocyte cultures induced a clastogenic effect which was prevented by SOD (Auclair et al. 1990). This indicates that inosine nucleotides cause the formation of reactive oxidants in lymphocytes, which subsequently mediate chromosome breakage. An ultrafiltrate of PSS patients containing clastogenic factors stimulates thymidine uptake by fibroblasts and endothelial cells and triggers the respiratory burst of neutrophils (Emerit et al. 1989). This is considered additional indirect evidence for a significant role of reactive oxygen species in the pathogenesis of PSS (Emerit 1988; Emerit et al. 1989).

Although clastogenic products are mutagens, no increased cancer rate is reported in PSS patients. The incidence of lung cancer is raised only on the basis of lung fibrosis in these patients (Roumm and Medsger 1985). Other diseases associated with clastogenic products, such as dermatomyositis, periarteriitis nodosa, rheumatoid arthritis, and Sjögren's syndrome, are associated with a higher incidence of malignant neoplasms. However, dermatomyositis, for example, is considered a paraneoplastia.

III. Skin Diseases with Tissue Neutrophilia

Neutrophil-mediated skin diseases constitute a heterogeneous group of dermatoses with regard to etiology and clinical morphology. Dermatoses with significant tissue neutrophilia include dermatitis herpetiformis Duhring, Sweet's syndrome, and psoriasis vulgaris, especially the pustular variant. There are clearly some common aspects in their pathogenesis (Christophers 1987; Jorizzo et al. 1988).

1. Psoriasis Vulgaris

Psoriasis vulgaris is a common skin disease which is characterized by increased epidermal proliferation and dermal inflammation. There is a genetically controlled defect in the regulation of skin metabolism. Overresponsiveness of the dermal microvasculature to trauma and epidermal hyperproliferation may be the consequence of disturbed metabolism. Environmental and genetic factors do significantly modulate the molecular expression of these dysfunctions. The cellular defect is present in keratinocytes, fibroblasts, immune cells, and endothelial cells. The first histopathological features appearing in lesions are capillary dilatation and neutrophil and mononuclear leukocyte infiltration. A unknown serum factor has been described in patients with psoriasis that stimulates endothelial cell proliferation (Majewski et al. 1987) and fibroblast proliferation (Priestly and Adams 1985; Holzmann et al. 1988). This factor could also contribute to keratinocyte hyperproliferation in psoriais. However, it was recently demonstrated that the mitogenic effect of sera from psoriatic patients was not correlated with the clinical disease activity (Laubscher et al. 1991).

Neutrophil Infiltration. Oxidative injury may occur at various pathophysiological levels and can contribute to sustaining the inflammatory process and hyperproliferation. A significant component of the epidermal infiltrate in psoriasis is formed by neutrophils. Their function is altered in patients with psoriasis; their in vitro adherence (Sedgwick et al. 1980), phagocytosis, and chemotaxis are increased (Fräki et al. 1983b). This is believed to be partially due to increased tissue concentration of LTB4. 5-Lipoxygenase, the enzyme synthesizing leukotrienes, and phospholipase A_2, providing the substrate for the 5-lipoxygenase, are activated in psoriasis lesions. The pathophysiological link between prostanoid synthesis and tissue peroxide content, leading to the formation of potent chemoattractants, may contribute to the tissue neutrophilia.

Sera from patients with psoriasis stimulate an increased production of superoxide anion radicals in normal neutrophils (Sedgwick et al. 1981; Miyachi and Niwa 1983). Neutrophils of psoriasis patients also show higher basal and stimulated (opsonized zymosan) chemiluminescence levels than neutrophils of healthy controls (Larussa et al. 1986). The basal activities of NADPH oxidase and myeloperoxidase in leukocytes of patients with psoriasis are significantly raised over those of healthy controls. A definite relationship between the percentage of the surface area involved and the level of the two enzyme activities has been elucidated (Jain et al. 1985). After 4 weeks of therapy with PUVA, retinoids, or anthralin, neutrophil chemiluminescence returns to control values. Oscillations between high and low neutrophil chemiluminescence correspond to clinical relapses and remissions (Larussa et al. 1986).

Lipid Peroxidation. A membrane defect in the pathogenesis of psoriasis was suggested (Mahrle and Orfanos 1977). Functional and morphological alterations of the plasma membrane were discovered in keratinocytes (Gommans et al. 1979a, b; Roelfzema et al. 1981) and also in erythrocytes derived from psoriatic patients (Kumar et al. 1983a, b). This may indicate that a defective plasma membrane is a generalized phenomenon in psoriasis. It was suggested that perturbation of the keratinocyte lipid pattern results in elevated concentrations of polyunsaturated fatty acids. This increase should expose the keratinocyte cell membrane to lipid peroxidation by reactive oxidants (Corrocher et al. 1989). In platelets of psoriatic patients, the glutathione peroxidase activity, malonic dialdehyde concentration, and content of polyunsaturated fatty acids are significantly increased (Corrocher et al. 1989).

Antioxidants and Oxidants. Reactive oxidants were proposed to play a significant role in the pathogenesis of psoriasis (Popov et al. 1989; Goncharenko 1983; Ivanova 1987; Polkanov 1987). Regulation of mammalian cell proliferation is quite complex. In this scenario reactive oxidants acting as modulators was hypothesized. Very low concentrations of hydrogen peroxide or tertiary butyl-hydroperoxide (10^{-8} *M*) enhance the proliferation of monolayer fibroblast cultures. At higher concentrations, they inhibit cell growth by causing oxidative stress, leading to cell damage. Antioxidants generally enhance cell growth. It was suggested that at low concentrations the reactive oxidants directly activate cellular signal transduction mechanisms (Burdon and Gill 1989). The relevance of these findings for keratinocyte proliferation is not yet proven.

Serum Antioxidant Capacity. SOD and catalase activities in serum of psoriasis patients are comparable with those of healthy controls (Miyachi and Niwa 1983), and serum copper and ceruloplasmin levels of psoriasis patients are clearly higher than in healthy controls (Stratigos et al. 1975; Zackheim and Wolf 1972; Lipkin et al. 1962; Dogan et al. 1989). The serum antioxidant capacity of patients with active psoriasis is significantly elevated. This is believed to be due to compensatory up-regulation of the antioxidant potential in psoriasis patients (Popov et al. 1990).

Neutrophil Antioxidants. SOD activity in neutrophils of patients with psoriasis is significantly lower than that in healthy controls (Dogan et al. 1989). The cytoplasmic enzyme SOD protects neutrophils against reactive oxygen species. In the extracellular space, ceruloplasmin scavenges superoxide anion radicals, protecting the surrounding tissue from autooxidation. Decreased intracellular SOD and increased extracellular ceruloplasmin activity may indicate a higher susceptibility of the neutrophils to oxidative autoinactivation and some up-regulation of the extracellular antioxidant potential.

Skin Superoxide Dismutase. In clinically involved skin of psoriatic patients, the SOD activity is about half that of healthy controls. The SOD content in clinically uninvolved skin is not significantly different from that of healthy controls (Kamide et al. 1990). It is interesting to note that arsenite, an inducer of SOD activity, was used as Fowler's solution in the clinical treatment of psoriatic patients. Treatment of healthy skin with LTB_4 results in a significant suppression of SOD activity. LTB_4, a potent neutrophil chemoattractant, is present in high concentrations in psoriatic lesions. A causal relationship between high LTB_4 and low SOD activities in clinically involved skin was postulated (Van Baar et al. 1987 a).

Depressed SOD activity in psoriatic skin is, however, an unspecific phenomenon. In humans, epidermal SOD activity is decreased in various hyperproliferative conditions (psoriasis, squamous cell carcinoma, basal cell carcinoma). Tape stripping of the epidermis (removing the oxygen diffusion barrier) also results in decreased SOD activity, probably by increasing the epidermal oxygen tension and endogenous oxygen radical production. It has been suggested that decreased SOD activity might be induced by reactive oxygen species and that this is related to the hyperproliferative state (Ohkuma et al. 1987).

Xanthine Oxidase and Urate. In hyperproliferative skin diseases, e.g., psoriasis vulgaris, increased turnover of nucleic acids results in elevated production of urate. This can become clinically significant in a number of patients, resulting in hyperuricemia. Raised skin concentration of urate in patients with psoriasis may serve to protect skin from oxidative injury caused by the increased formation of reactive oxidants due to high metabolic turnover, e.g., increased mitochondrial respiration. On the other hand, augmented urate formation is an indicator of greater hypoxanthine turnover. Xanthine oxidase activity is stimulated in psoriatic epidermis; the average activity is about five times higher than that in normal epidermis (Kizaki et al. 1977).

Mitochondria. The stratum basale in the epidermis possesses ceaseless mitotic activity depending on mitochondrial production of high energy phosphates. Epidermal mitotic activity is increased following skin injury by UV, ionizing irradiation, wounding, toxic chemicals, and endogenous inflammatory processes. This is an ultimate defense, leading to desquamation of damaged cells and cell renewal. In cases of increased localized hyperproliferation, as in psoriasis, this mechanism is stimulated.

The number of epidermal mitochondria is greatly increased in psoriatic lesions (Dubertret et al. 1982), and skin oxygen consumption is elevated (Gans 1923; Buhmann 1936; Stüttgen 1955; Leonhardi 1961). Amplified mitochondrial density and activity could be a source of increased free radical flux in psoriatic lesions. The steady-state concentration of free radicals in involved skin is raised and declines after successful therapy. The rise is correlated with greater activities of mitochondrial succinate dehydrogenase and cytochrome *c* oxidase (Zadorozhny et al. 1973).

Cyclic Nucleotides. An important role in psoriatic epidermal metabolism is played by cyclic nucleotides. It was suspected that in response to increased superoxide generation they effect the regulation of the cellular metabolism of epidermal cells. The significant role of reactive oxidants in the modulation of cyclic nucleotide metabolism may contribute to dysregulation of epidermal turnover and membrane dysfunction in psoriasis (Miyachi and Niwa 1983).

Psoriasis Treatment. An unifying concept in the reaction mechanism of antipsoriatic drugs such as anthralin, pyrogallol, tar, UVB, and PUVA therapy has been proposed. They may act via a common mechanism, e.g., generation of reactive oxygen species in skin (Gruner et al. 1987b). This may explain the cytotoxic effects on inflammatory cells, e.g., neutrophils and rapidly proliferating keratinocytes.

2. Sweet's Syndrome

Sweet's syndrome, acute febrile neutrophilic dermatosis, is characterized by inflammatory cutaneous plaques, arthralgia, fever, leukocytosis, and infiltration of dermal vessels by neutrophils. However, in contrast to necrotizing vasculitis, fibrinoid necrosis of blood vessels is not seen (Jorizzo et al. 1988), and the term neutrophilic vascular reaction is used to describe the vasculopathy. Sweet's syndrome can be scussessfully treated with potassium iodine or colchicine.

The bowel-associated dermatosis-arthritis syndrome is a clinically related disease, and histology study reveals a micromorphological picture similar to Sweet's syndrome. It can develop in patients after jejunoileal bypass surgery for treatment of morbid obesity. Erythemas, papules, and subsequently pustular lesions develop on the trunk and arms. Bacterial overgrowth in a blind loop of the bowel may lead to the formation and circulation of immune complexes and, subsequently, massive skin infiltration with neutrophils. Antibacterial therapy is the treatment of choice.

3. Dermatitis Herpetiformis Duhring

Dermatitis herpetiformis is a polymorphous skin disease with pruritic papulovesicular lesions. There is a genetic predisposition, and the disease is influenced by environmental factors. Vesicle formation occurs at the dermoepidermal junction, and the dermal papillae are densely infiltrated with neutrophils and eosinophils. Usually neutrophils dominate over eosinophils in skin lesions. Two immunopathological patterns are described: granular immunoglobulin deposits (IgA) in the dermal papillae (85 % of cases) and linear IgA deposits (15 % of cases) along the basement membrane. The latter subtype is termed linear IgA bullous dermatoses. A prominent histopathological feature in dermatitis herpetiformis is infiltration of the epidermodermal junction with neutrophil and eosinophil leukocytes, which cause tissue injury. The deposits of immunoglobulin IgA at the dermal papillae and sometimes deposition of complement C3 may reflect an altered antigenic structure.

Tissue damage observed in dermatitis herpetiformis Duhring and linear IgA bullous dermatoses may be partially due to excessive production of activated oxygen species by neutrophils and a characteristic serum factor (Niwa et al. 1985b). Increased levels of IgA-containing circulating immune complexes have been detected in patients with dermatitis herpetiformis (Hall et al. 1980; Lancaster-Schmith et al. 1975). Dermatitis herpetiformis can be induced and/or aggravated by iodine. The pathogenetic role of iodine is believed to be due to a disease-specific chemotactic effect and stimulation of myeloperoxidase-mediated iodination in neutrophils, thereby generating potent oxidants. The specific chemotactic effect must be different from iodine's antichemotactic effect in other diseases (Honma et al. 1990). The treatment of choice in dermatitis herpetiformis involves dapsone, which is also clinically effective in some patients with linear IgA bullous disease. Lipid peroxidation products are elevated in the blood of Dermatitis herpetiformis patients (Mirzoev et al. 1986) and antioxidants (e.g., lipoate) have been successfully used in therapy (Mirzoev et al. 1988). A role of sulfhydryl groups in the pathogenesis was discussed (Mirzoev et al. 1985).

IV. Skin Diseases with Tissue Eosinophilia

Blood and tissue eosinophilia are frequently observed in parasitosis, hypereosinophilia syndrome, Kimura's disease, Ofuji's syndrome, Well's syndrome, granuloma faciale, herpes gestationis, and pemphigus herpetiformis. Skin diseases with occasional tissue eosinophilia include bullous pemphigoid, pemphigus vegetans, dermatitis herpetiformis Duhring, pemphigus seborrheicus, chronic urticaria, atopic eczema, sarcoidosis, and drug reactions (Czarnetzki 1987; Lever 1987; Aberer et al. 1988).

1. Bullous Pemphigoid

Bullous pemphigoid is characterized by dermal inflammation preceding subepidermal blister formation, infiltration of the basement membrane zone with activated leukocytes, and subsequently subepidermal blister formation within the basement membrane zone. Basement membrane zone-specific antibodies, dermal mast cells, and complement activation by membrane-deposited immune complexes are suggested to be the cause of inflammation. Bullous pemphigoid is an inflammatory bullous disease, and a dense dermal infiltrate composed of neutrophils and eosinophils is usually present. In contrast to neutrophils isolated from patients with dermatitis herpetiformis and linear IgA bullous dermatoses, no significant increase in the generation of reactive oxygen species by stimulated neutrophils was noticed in patients with bullous pemphigoid (Miyachi et al. 1987b). This negative result may be due to a higher clearance rate of immune complexes. Treatment of choice emplays glucocorticosteroids and azathoprine; sulfapyridine and dapsone are alternatives.

2. Pemphigus Herpetiformis

Pemphigus refers to a group of bullous skin diseases characterized by IgG autoantibodies specific for epithelial cells, acantholysis, and intraepidermal blister formation. The individual diseases have distinct histological and clinical features; they comprise pemphigus vulgaris, pemphigus vegetans, pemphigus foliaceus, Brazilian pemphigus, and pemphigus erythematosus. Neutrophil leukocytes can infiltrate acantholytic lesions of pemphigus vulgaris to varying degrees; however, acantholysis can occur in the absence of inflammatory cells (Emerson and Wilson-Jones 1968).

Acantholysis and tissue damage are believed to be caused by autoantibody-mediated activation of tissue proteinases (Schiltz et al. 1979) and complement. Eosinophils are the prominent inflammatory cells in pemphigus herpetiformis but not in pemphigus vulgaris. The role of reactive oxidants in the pathogenesis is not well characterized at the moment. Dapsone and sulfapyridine are used in the treatment of pemphigus herpetiformis.

V. Skin Diseases with Tissue Lymphocytosis

Lymphocytes are the predominating cell type in most chronic inflammations of the skin, particularly atopic dermatitis.

1. Atopic Dermatitis

Atopic dermatitis (endogenous eczema) is a chronically recurring skin disease with various clinical variants and has an as yet unresolved pathogenesis. The dermal perivascular infiltrate consists mainly of lymphocytes and also contains fairly numerous monocytes. Mast cells are present in increased numbers, but although blood eosinophilia is frequently observed, eosinophils are found only occasionally in skin tissue. Successful treatment regimens for atopic dermatitis patients have employed UVA phototherapy, glucocorticosteroids, and tar preparations.

In inflammatory skin lesions of patients with atopic dermatitis, products of lipid peroxidation are significantly increased (Sugiura et al. 1986 b; Ueda et al. 1971). However, in sun-exposed skin lesions they are found to be reduced in comparison with unexposed skin lesions (Sugiura et al. 1986 b). Their concentration is also elevated in serum (Demitsu et al. 1988).

In atopic dermatitis skin lesions, Mn-SOD activity is increased while Cu,Zn-SOD activity remains normal (Sugiura et al. 1986 b). Augmented SOD activity could be an adaptive response to an elevated concentration of superoxide anion radical.

It was proposed that the red cell antioxidant defense status, especially glutathione peroxidase activity, may be a reliable indicator of the overall status of the body's antioxidant defences (Levine and Kidd 1985 a). Erythrocyte glutathione peroxidase was suggested to be useful for monitoring the cutaneous antioxidant status in skin diseases (Juhlin et al. 1982). For reasons outlined in Chap 1, Sect. E. XI, the validity of this conclusion must be questioned, however, and assertions about the "cutaneous antioxidant status" require careful reevaluation. Erythrocyte glutathione peroxidase activity is markedly reduced in various skin disorders, including atopic dermatitis (Michaelson and Edquist 1984; Juhlin et al. 1982). In patients with atopic dermatitis, reduced concentrations of selenium in whole blood, plasma, and white cells and reduced activity of selenium-dependent glutathione peroxidase in red cells are found. Although selenium supplementation significantly increased glutathione peroxidase activity in the blood, no increase in the concentration of selenium in skin and no improvement of the dermatitis are observed. This indicates impaired skin bioavailability of dietary supplemented, selenium-enriched yeast (Fairris et al. 1989).

VI. Skin Diseases with Deficiency in Nutritional Antioxidants

1. Kwashiorkor Dermatitis

Kwashiorkor is a disease of infancy and childhood caused by inadequate protein intake. It is characterized by skin and hair changes, mental and gastrointestinal abnormalities, and developmental disturbances. Skin lesions in Kwashiorkor patients may be mild or severe and are characterized by depigmentation, erosive dermatitis, and increased skin fragility. In contrast to lesions of pellagra, the dorsal areas of hand and feet are seldomly affected, and sensitivity to sunlight is not a characteristic finding, although in some cases increases photosensitivity is present. Golden proposed that kwashiorkor results from an imbalance between decreased tissue antioxidants and increased production of reactive oxidants. In kwashiorkor patients, hepatic and bone marrow iron is increased, and serum ferritin is elevated. Infections, which are known to cause oxidative stress, often occur in kwashiorkor patients. Plasma tocopherol, β-carotene and retinol are reduced in children with kwashiorkor, and plasma zinc plasma, selenium, erythrocyte Cu,Zn-SOD, and erythrocyte glutathione peroxidase activity are low. Serum NADPH concentration is reduced and erythrocyte glutathione is very low in kwashiorkor patients, but normal in marasmus patients and controls. Glutathione reductase and glucose-6-phosphate dehydrogenase activity is elevated in kwashiorkor, indicating that erythrocyte glutathione is reduced in vivo predominantly because of increased consumption rather than decreased production. Golden and Ramdath (1987) suggested that an iron-overload syndrome coupled with a deficiency in nutritional antioxidants is involved in the pathogenesis of the kwashiorkor. Antioxidant therapy as well as supplementation with zinc and selenium, avoidance of polyunsaturated fatty acids in the diet, and early treatment of infections are recommended treatment modalities (Golden and Ramdath 1987). The clinical signifcance of reactive oxidants and antioxidants in the pathophysiology of protein-energy malnutrition awaits further characterization, and further studies are required to improve our understanding of the disease (Ahmed et al. 1990; Leichsenring, personal communication).

I. Skin Carcinogenesis

The molecular mechanisms of carcinogenesis have been extensively studied in skin as a model system. In 1761, John Hill speculated about a relationship between mucocutaneous carcinoma of the nose and the use of snuff tobacco. A correlation between scrotal skin cancer in humans and exposure to chimney soot was established by Sir Percivall Pott in 1775. The first experimental cancer in mouse skin was produced by topical application of coal tar condensate in 1915 (Yamagiwa and Ichikawa 1915), and subsequently croton oil was found to be a very effective agent to augment the skin carcinogenicity of benzpyrene (Berenblum 1941). The terms initiation and promotion were first applied to a stepwise model for eliciting tumors of the skin of rabbit ears (Rous and Kidd 1941). The multistep carcinogenesis model (Fig. 3.31) has been developed for many epithelial tissues including liver, lung, stomach, pancreas, and colon and is best studied in the epidermis. In mouse skin, three distinct stages of carcinogenesis can be defined: initiation, promotion, and progression.

Initiation is a phenomenon of gene alteration. Gene alterations can be caused directly by electromagnetic irradiation, carcinogenic chemicals, and oncogenic viruses. Skin tumor initiation probably occurs in dark, basal keratinocytes, which are keratinocyte stem cells and play an important role in the first stage of skin tumor promotion. The dark, basal keratinocytes are present in large numbers in animal skin during embryogenesis, in moderate numbers in newborns, in low numbers in young adults, and in very low numbers in older adults (Klein-Szanto and Slaga 1981). The initiating potential of mouse

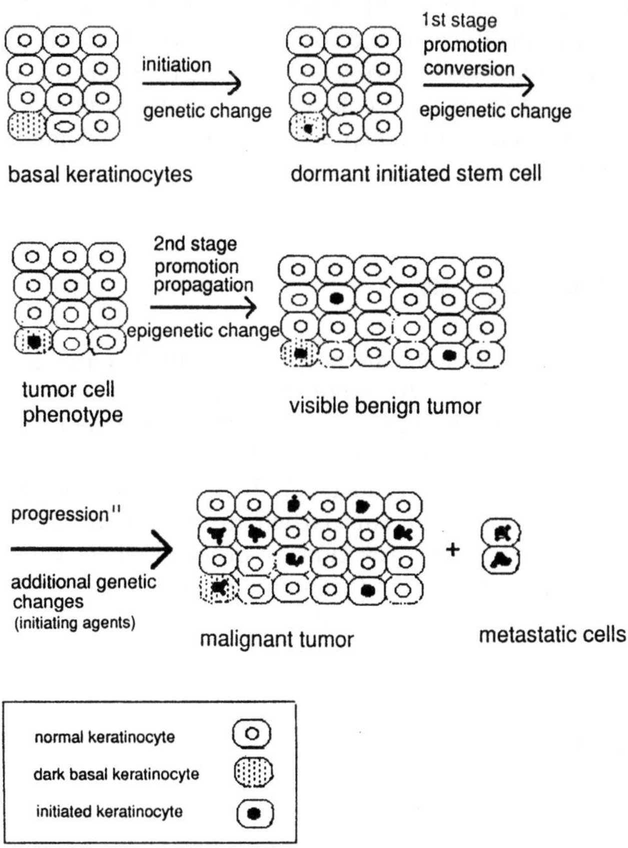

Fig. 3.31. Multistep model of carcinogenesis

skin decreases with age to a point at which it is very difficult to initiate mice over 1 year old as the dark, basal keratinocytes are very rare.

Promotion is a phenomenon of gene activation. Tumor-promoting agents such as phorbol esters, the active ingredients of croton oil, and nonpromoting agents, such as calcium ionophores, as well as wounding, increase the number of dark, basal keratinocytes. Promotion itself is divided into two steps, conversion and propagation (Slaga et al. 1980).

In stage I of tumor promotion, dark, basal keratinocytes multiply by stimulation of existing dark cells to divide and by the conversion of basal cells into dark cells. In stage II of tumor promotion, the increased number of dark, basal cells is maintained by the continuation of dark cell proliferation, and cellular metabolism (nucleic acid and protein synthesis) is modulated and activated. As a consequence, a benign tumor, a papilloma, is formed. Stage I is partially reversible, and stage II is initially irreversible.

Complete carcinogens possess both initiating and promoting properties; incomplete ones can only initiate. The promotion step is more readily reversible than the other two steps (initiation, progression) and has its greatest role in the development and maintenance of the benign state of neoplastic development. The promotion step can be blocked by pharmacological manipulations having little or no untoward effects on the host animals (Boutwell 1985).

The final stage of carcinogenesis is the *progression* of the transformed lesion into a malignant neoplasm (squamous cell carcinoma), involving additional genetic changes which may arise either spontaneously or through mediation by certain carcinogens.

I. Reactive Oxidants in Carcinogenesis

It is firmly established that most ultimate carcinogenic metabolites are electrophilic and specifically bind to DNA. The metabolic activation of xenobiotics to a proximate carcinogen may involve either one- or two-electron redox reactions catalyzed by the mixed function oxidase (MFO) system (cytochrome P 450, NADPH cytochrome P 450 reductase). The MFO system also produces reactive oxidants directly which can participate in non-enzymatic xenobiotic activation. One-electron redox reactions of procarcinogens yielding free radical intermediates may play a significant role in activation of carcinogens to DNA binding and damaging compounds.

II. Reactive Oxygen Species in Tumor Promotion

The hypothesis that reactive oxygen species might be involved in tumor promotion is based on the following observations (Kensler and Taffe 1986; Fischer et al. 1988):

1. Compounds generating free radicals such as various organic hydroperoxides are tumor promotors.
2. Systems generating reactive oxygen species can mimic the biochemical action of tumor promotors.
3. Some tumor promotors stimulate the production of reactive oxygen species from endogenous sources in a variety of cell types.
4. Tumor promotors modulate cellular antioxidant defense systems.
5. Free radical scavengers inhibit tumor promotion.

It is believed that tumor promotion is due to epigenetic rather than genetic effects. However, tumor promotors have been found to cause gene amplifications (Varshavsky 1981), chromosomal aberrations (Emerit and Cerutti 1981 a, 1982), and DNA strand breakage (Birnboim 1982), as well as to increase sister chromatid exchange in certain cell types (Kinsella and Radman 1978; Nagasawa and Little 1979). Cellular mediators, including reactive oxygen species, are thought to be released upon promotor stimulation since tumor promotors have never been found to interact directly with DNA.

Different classes of tumor promotors can evoke the production of reactive oxygen species by direct chemical generation or through the indirect activation or alteration of cellular metabolic sources (Fischer and Adams 1985; Fischer et al. 1986; Cerutti 1985; Kensler and Trush 1984). In some cell types, the production of reactive oxidants appears to arise through the *protein kinase C* pathways (Patriarca et al. 1970; Di Virgilio et al. 1984). Signal transduction for the pleiotropic actions of the phorbolester type and indolalkaloid class tumor promotors appear to be mediated through interaction with a calcium- and phospholipid-dependent receptor protein kinase C (Blumberg et al. 1984; Slaga 1984). Phosphorylation of proteins by this kinase may modulate many regulatory functions including gene activation (Nishizuka 1984). However, other classes of epidermal tumor promotors do not interact with this receptor molecule (Couturier et al. 1984).

Cyclooxygenase and *lipoxygenase products* as well as *polyamines* also play a significant and complex role in tumor promotion processes. PGE_2 participates in stage I of promotion and increases the number of dark cells induced by 12-O-tetradecanoylphorbol-13-acetate (TPA) (Slage et al. 1980; 1981 a).

Levels of cGMP and guanylate cyclase are increased in many tumors (Mittal and Murad 1977 b; DeRubertis and Cràven 1977; Veseley et al. 1977; Oberley and Buettner 1979). The former has been proposed as a regulator of cell proliferation, and guanylate cyclase is activated by reactive oxygen species. Lipid hydroperoxides and hydroxy radicals formed in the course of arachidonic acid metabolism (Kalyanaraman 1984) may be associated with an increased production of reactive oxidants in TPA promotion.

Oxidant stress can modify the regulation of genes controlling tumor promotion (Cerutti 1985; Kensler and Trush 1984). *Prooxidant states* may modulate the expression of prooxidant genes which are related to cell growth and differentiation by inducing alterations in the DNA structure or by epigenetic mechanisms. Cellular prooxidant states, e.g., increased concentrations of reactive oxygen species and other types of free radicals, can promote initiated cells to neoplastic growth.

The elaboration of *reactive oxygen species* may represent an additional mechanism for mediating the epigenetic actions of tumor promotors. Reactive oxygen species such as superoxide anion radical and hydrogen peroxide can act in a promotional manner that mimics the effects of the mouse skin tumor promotor TPA in mouse fibroblasts (Zimmerman and Cerutti 1984). Chromosomal aberrations seen after TPA treatment of lymphocytes and keratinocytes are supposed to be due to the action of a clastogenic factor which is thought to be generated by the cells in the course of oxygen activation (Emerit and Cerutti 1981 c; Emerit 1982). This factor of low molecular weight, which is released from TPA-stimulated granulocytes and monocytes, probably consists of lipid hydroperoxides, aldehyde compounds, and free arachidonic acid (Emerit 1982) and exerts its toxicity via formation of reactive oxygen species. Tumor-promotor-induced chromosome breaks are inhibited by antioxidants in vitro (Emerit et al. 1983).

The current theories concerning the mechanisms of multistage skin carcinogenesis have been reviewed by Perchellet and Perchellet (1989). It was suggested that different sequences and levels of reactive-oxidant-mediated cell damage contribute to the evolution of premalignant to malignant epidermal cells. Evidence is not convincing that reactive oxygen species play an essential role in stage I of skin tumor promotion. The concept that superoxide anion or related radicals are responsible for TPA-induced chromosomal aberrations has been criticized (Kinsella et al. 1983). Generation of reactive oxygen species is probably not unique for promotion but rather is involved in unspecific pleiotropic responses. Experimental evidence, however, suggests that reactive oxygen species are involved in stage II of tumor promotion (Marks and Fürstenberger 1985; Schwarz et al. 1984), tumor initiation and progression to cancer.

III. Peroxides as Tumor Promotors

Skin tumorigenic properties of various hydroperoxides, acylperoxides, peracids, and epoxides have been described (Kotin and Falk 1963). However, hydrogen peroxide, lauryl peroxide, and benzoyl peroxide are not carcinogenic when applied topically on mouse skin (Van Duuren et al. 1963).

In the two-stage carcinogenesis model, lauryl peroxide and benzoyl peroxide are tumor promotors (Klein-Szanto and Slaga 1982); hydrogen peroxide exhibits only a weak tumor-promoting potential. Repetitive treatment of benign mouse skin tumors with benzoyl peroxide increases the number of cumulative carcinomas significantly (O'Connell et al. 1986). Benzoyl peroxide is an effective promotor of both papillomas and squamous cell carcinomas (Slaga et al. 1981 b) and is also able to promote the formation of melanotic tumors in the skin of hamsters (Schweizer et al. 1987).

The promotor activity of peroxides is ascribed to their ability to generate free radicals, although they are quite stable to unimolecular homolysis at body temperature (Pryor 1975). Hydroperoxides are subject to decomposition by transition metal ions, and hematin can catalyze the breakdown into radical species (Kalyanaraman 1983). Butyl hydroperoxide and cumene hydroperoxide rapidly produce alkyl radicals when incubated with mouse keratinocytes (Taffe and Kensler 1986; Taffe et al. 1987). The nature of the system that presumably catalyzes peroxide cleavage to free radical species is unknown, but the activity is probably localized in the cytosolic fraction.

The fact that various organic peroxides display tumor-promoting activity is often cited as evidence that reactive oxidants are involved in the promotion step. It must be pointed out, however, that oxidants can act as promotors either by 1- or 2-electron oxidation pathways. Both types of mechanisms can be inhibited by antioxidants, which scavenge either free radicals or electrophils (Kennedy et al. 1989).

IV. Phorbol Ester Type Tumor Promotors

Generation of reactive oxygen species has been proposed as an important event in mouse epidermal tumor promotion by the phorbol ester TPA (Fig. 3.32), one of the active components of croton oil (Copeland 1983; Slaga et al. 1982).

Mouse skin tumor promotors of the phorbol ester type stimulate the production of reactive oxygen species in neutrophils and macrophages (Ohkawa et al. 1984; Witz et al. 1980; Goldstein et al. 1981; Czerniecky et al. 1985). In addition, TPA also induces the formation of reactive oxygen species in isolated murine epidermal cells (Fischer and Adams 1985). In keratinocytes of TPA-treated mouse epidermis, a 50% increase in the cellular hydrogen peroxide level has been reported (Robertson et al. 1988). TPA-mediated oxidant production in mouse epidermal cells can be suppressed by many, but not all, known inhibitors of mouse skin tumor promotion. The extent of the oxidant response directly correlates with the sensitivity to TPA tumor promotion among certain strains of mice.

The occurrence of breaks in the DNA of cells cocultured with phagocytes and TPA suggests indirect as well as direct mechanisms in tumor promotion

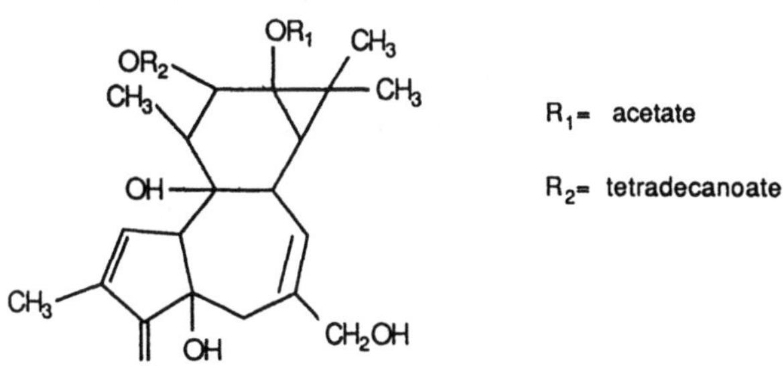

Fig. 3.32. 12-O-Tetradecanoylphorbol-13-acetate (TPA)

(Weitberg et al. 1983; Fanton and Ward 1982). Whether the source of reactive oxygen species in mouse skin is the epidermal cells or the usually associated inflammatory cells remains an unresolved question (Nakamura et al. 1985). It was suggested that free radical production by inflammatory cells does not necessarily need to be evoked when proposing mechanisms of tumor promotion (Fischer et al. 1986). However, other results are consistent with the hypothesis that production of reactive oxidants by inflammatory cells may be involved (Czerniecky et al. 1986).

TPA-induced chemiluminescence in mouse epidermal cells is diminished by agents which are lipoxygenase inhibitors, like benoxaprofen. Cyclooxygenase inhibitors such as indomethacin have no enhancing effects at low doses. This indicates that a major part of the TPA-induced chemiluminescence response is due to the metabolism of arachidonic acid, most probably by the lipoxygenase pathway (Fischer and Adams 1985; Fischer et al. 1988). Phospholipase C also induces chemiluminescence in mouse epidermis cells. TPA and phospholipase C may produce oxidant species through a common mechanism. It was suggested that increased phospholipid turnover induces protein kinase C activation (Fischer et al. 1988).

V. Modulation of Pro- and Antioxidant Skin Enzymes by Tumor Promotors

Treatment of isolated mouse epidermal cells with the tumor promotor (TPA) results in a rapid and transient increase in *glutathione peroxidase* activity, followed by a prolonged decrease. Multiple treatments maintain an elevated glutathione peroxidase activity. Glutathione reductase activity is depressed throughout a multidose exposure (Perchellet et al. 1985b; Taffe and Kensler 1986).

Epidermal SOD and *catalase* activity are reported to decrease following treatment with TPA. This suppression is maintained over a 12-h time course, but the activity rises above control levels several days after a single exposure (Perchellet et al. 1985b; Taffe and Kensler 1986). The treatment of adult mouse skin with TPA results in a sustained decrease in the basal activities of both SOD and catalase in the epidermis. The decline occurs within 3 h after application, and the maximum effect is seen after 16–17 h. The alterations in both enzymes arise against a background of enhanced protein synthesis, which indicates that the effect of TPA is selective for SOD and catalase. Oxidative injury which favors neoplastic progression can occur in TPA-treated mouse skin due to low levels of SOD and catalase activity. Accumulation of reactive oxygen species in mouse skin with TPA treatment may follow the decrease in SOD activity (Solanki et al. 1982; Kinsella et al. 1983) accompanied by inactivation of catalase (Kono and Fridowich 1982) and an increase in glutathione disulfide content (Kozumbo et al. 1983).

Tumor promotors induce de novo synthesis and conversion of both newly synthesized and preexisting *xanthine dehydrogenase* to the *oxidase* form. This effect is probably related to cell hyperplasia and terminal differentiation of keratinocytes (Reiners and Rupp 1989) and could contribute to the formation of reactive oxygen species.

Although production of reactive oxidants is increased and activity of antioxidant enzymes decreased following TPA treatment, a decline in skin peroxide content is observed. The decrease of skin *lipid peroxidation products* by TPA may possibly be related to their promitotic acitivity. Skin papilloma and carcinoma which also have high mitotic activity achieve lower levels of lipid peroxides compared with healthy epidermis of the same mice. Reducing the level of lipid peroxidation is apparently related to the high mitotic activity induced by TPA (Logani et al. 1982). The mechanism of this process is not known.

VI. Antioxidants as Antipromotors and Antiinitiators

Scavengers of reactive oxygen species, such as enzymic and nonenzymic antioxidants, inhibit biochemical and biological actions of tumor promotors in vitro and in vivo (Kensler et al. 1983; Kensler and Trush 1983; Kozumbo et al. 1983; Nakamura et al. 1985; Borek and Troll 1985; Zimmerman and Cerutti 1984). Some antioxidants also effectively counteract skin tumor initiation; these agents have an even greater inhibitor effect on skin tumor promotion by TPA and benzoyl peroxide. They specifically inhibit the second stage of promotion (Slaga 1984). The inhibition of polycyclic hydrocarbon tumorigenesis by antioxidants may be related to their ability to prevent the in vivo activation of hydrocarbons to carcinogenic epoxides and/or other electrophilic intermediates or may be related to their ability to increase detoxification of the reactive intermediates, which requires intact cells to be operational. It was suggested that antioxidants have an indirect effect on the epidermal metabolizing system which leads to a decrease in covalent binding of carcinogens to DNA (Slaga and Brocken 1977). A combination of inhibitors of the various stages of carcinogenesis (initiation, stages I and II of promotion) would be the most effective means of counteracting the induction of skin cancer in man (Slaga 1984).

Human cancer risks are inversely correlated with *blood retinol* and *dietary β-carotene* (Peto et al. 1981). Intake of dietary provitamin A, β-carotene, were inversely related to the 19-year incidence of lung cancer in a prospective epidermiological study (Shekelle et al. 1981). Some synthetic retinoids have been particularly useful for the prevention of bladder and breast cancer in experimental animals (Sporn and Newton 1979). Retinoic acid exerts a prophylactic and a therapeutic effect on chemically induced benign and malignant epithelial tumors in mice (Mayer 1978). β-carotene and canthaxanthin can prevent the development of UVB-induced skin cancer in hairless mice (Mathews-Roth and Krinsky 1985).

Tocopherol is a very promising inhibitor since it has a potent effect on both stages of promotion and a more moderate one on initiation. Tocopherol and BHT are effective inhibitors of 7,12-dimethylbenz[*a*]anthracene tumor initiation in a two-stage system of tumorigenesis (Slaga 1984). The croton oil dermatitis of the rabbit is inhibited by topical application of α-tocopherol (Kamimura 1972). In a previous study, topical treatment of acute croton oil dermatitis in the rabbit with tocopherol did not inhibit the inflammatory response; however, systemic therapy was successful (Yano 1959). It is not known whether tocopherol also inhibits phorbol ester-mediated skin tumorigenesis.

Antioxidants such as *butylated hydroxytoluene* (BHT) and *butylated hydroxyanisol* (BHA) suppress skin tumor promotion induced by UV and phorbol esters (Kozumbo et al. 1983; Black and Tigges 1986). Recently, it has been shown that phorbol ester-induced activation of ornithine decarboxylase (ODC) activity, which is an indicator of skin tumor promotion, is significantly reduced by topical application of BHA via suppression of ODC gene expression (Taniguchi et al. 1991). It is not known to what extent the antioxidant properties of BHA contribute to this effect.

Modulation of skin SOD activity is a possible way of inhibiting skin tumorigenesis. Inhibition of tumor promotion by retinoic acid, BHT, and glucocorticosteroids such as fluocinolone acetonide is, however, not mediated through alterations in SOD levels (Solanki et al. 1982). The glucocorticosteroid dexamethasone inhibits superoxide anion radical production in neutrophils stimulated by phorbol ester (Belman and Troll 1972). *SOD mimetics* inhibit multistage chemical carcinogenesis in mice (Egner and Kensler 1983). They are low molecular weight lipophilic copper coordination complexes that catalyze the disproportionation of superoxide anion at rates comparable with that of native Cu,Zn-SOD.

Glutathione level raising agents may inhibit by diverse means the effects of tumor promotors such as TPA on glutathione metabolism and skin tumor promotion. However, clonal expansion of tumor-initiated cells stimulated by TPA depends upon the availability of glutathione in the tissue. The mechanism by which depletion of glutathione could result in inhibition of skin tumor promotion is not known. Inactivation of glutathione may provide an effective control over proliferation of tumor cells (Kumar et al. 1988).

Glutathione level raising agents increase remarkably basal glutathione peroxidase activity and abolish the prolonged inhibitory effect of TPA on this enzyme. It is suggested that stimulators of the glutathione-dependent antioxidant protective system of the epidermis may inhibit the oxidative challenge linked to skin tumor promotion by TPA (Perchellet et al. 1985a, 1986b).

Tocopherol prevented completely the decrease in the GSH/GSSG ratio caused by TPA (Perchellet et al. 1985a). The anticarcinogenicity of sodium selenite, glutathione, and tocopherol may be related to their ability to facilitate or enhance the activity of the glutathione-dependent antioxidant protective system of the epidermis during the later stages of skin tumor promotion (Perchellet et al. 1987).

Antioxidants such as nordihydroguaiaretic acid and diallyl sulfide inhibit benzoyl peroxide tumor promotion in mice (Athar et al. 1989). Garlic oil, onion oil, and one of its constituents, dipropenylsulfide, increase glutathione peroxidase activity in isolated epidermal cells. Garlic and onion oil inhibit tumor promotion in skin of mice, which may result from their enhancement of the glutathione-dependent antioxidant protective system of the epidermis. In addition, onion and garlic oil are potent inhibitors of lipoxygenase (Perchellet et al. 1986a).

Protease inhibitors interfere with tumor promotion by modulating the cellular response to reactive oxygen species. Protease inhibitors are widely distributed in plants which are a major source of protein in vegetarian diets. A lower incidence of cancer is observed in populations eating vegetarian diets. It was suggested that diets rich in protease inhibitors may play an important role in preventing cancer (Troll and Wiesner 1985). Protease inhibitors have been shown to block tumor promotion in mouse skin. They suppress the production of superoxide anion radicals in neutrophils stimulated by tumor promotors. They also antagonize the stimulatory effect of a variety of activating agents and inhibit the respiratory burst in rat neutrophils as well as in alveolar macrophages (Goldstein et al. 1979).

VII. Endogenous Antioxidants in Skin Neoplasms

Animal Studies. Although tissues with a high metabolic rate appear to possess more SOD activity than do others, many animal tumor cell lines are low in SOD and catalase activities. It was proposed that decreased SOD activity together with increased oxidant generation might explain many of the properties of cancer cells (Oberley and Buettner 1979). The total SOD as well as the Mn-SOD activities decrease in hyperproliferative neoplastic tissue. Mn-SOD activity is diminished (Oberley and Buettner 1979, Peskin et al. 1977; Van Balgooy and Roberts 1979) or completely absent in various animal tumors (Oberley et al. 1978; Sahu et al. 1977).

Human Studies. SOD activities (Mn-SOD and Cu,Zn-SOD) in different human tissues and human malignant tumors do not correlate with tissue sensitivity to ionizing radiation. The Cu,Zn-SOD, Mn-SOD, catalase, and glutathione peroxidase activities among various human tumors and normal tissues are also not significantly different. Human tumors are thus considered different from those of other species which have very low Mn-SOD activity (Marklund et al. 1982; Westman and Marklund 1981). However, the total SOD content in human epidermal squamous cell epithelioma is significantly lower than in healthy epidermis (Galeotti et al. 1980).

Abnormal SOD activity is not pathognomonic for the neoplastic state; SOD activities are altered in various pathological conditions. Areas within a large tumor mass often have a limited blood supply and are anoxic. SOD is induced by oxygen, and thus a low SOD activity could be merely a consequence of tissue anoxia. Indeed, in the center of rat mammary carcinoma the SOD concentration is about half of the value found at the margin of the tumor (Petkau et al. 1977).

The present evidence does not support defects in protection against radicals as a general mechanism for the origin of human skin cancer, although increased free oxygen radical production can be carcinogenic (Halliwell and Gutteridge 1984).

Chapter 4

Dermatopharmacology

Sulfur, mercury, tar, arsenic, and complex products of plant origin were used as therapeutic agents in dermatology for several hundred years by physicians. In the past decade, many new drugs have been introduced in dermatological therapy and old drugs are being used in new ways. Some of them have antioxidant or prooxidant effects in vitro, others act in a dualistic mode. The clinical relevance of oxidative imbalance in the pathogenesis of skin diseases can be studied by analyzing the reaction mechanisms of antioxidant or prooxidant drugs used in the treatment of specific disorders. Conversely, pharmacological aspects of antioxidant/prooxidant drugs can be investigated in human or animal skin, which may be more relevant for pathophysiology. Roberfroid et al. postulated that drugs which specifically rebalance free-radical-mediated pathological processes may be useful in the treatment of diseases whose pathogenesis significantly involves reactive oxidants and free radicals (1987). There are certain limitations, however, in elucidating a drug reaction mechanism in complex biological systems. The in vivo effects of drugs can be different from their in vitro effects. Furthermore, the in vivo effect may result from a number of actions on different cell populations implicated in the various diseases, and no single unifying mechanism of action may be apparent. It is, therefore, difficult to extrapolate from in vitro studies to in vivo drug pharmacodynamics.

A. Chemotherapy

I. Tocopherol

The therapeutic uses of tocopherol in medicine have been reviewed by Horwitt (1980). Vitamin E is used clinically to correct a deficiency state and its associated pathophysiological disturbances. In contrast to the characteristic sequelae of tocopherol deficiency in certain animal species and premature infants, most vitamin E deficiency symptoms in humans are rather unspecific. Examples of specific deficiency diseases include respiratory distress syndrome and hemolytic anemia of low-birth weight infants. In the case of adults, tocopherol hypovitaminoses is observed only very rarely and are then mainly a consequence of malabsorption and alipoproteinemia. Reduced erythrocyte lifespan in patients with cystic fibrosis, hyperaggregability of platelets in patients with biliary atresia, and myopathy, cerebellar dysfunction, and retinitis pigmentosa in patients with hereditary abetalipoproteinemia can clinically be treated well with adequate tocopherol substitution (Oski 1980).

Vitamin E has also been reported to improve pathological conditions mediated by a deficiency of physiological antioxidants and/or increased formation of reactive oxidants. Examples include retrolental fibroblasia and bronchopulmonary dysplasia in premature infants exposed to prolonged oxygen administration in the treatment of respiratory distress syndrome. Vitamin E was employed successfuly in treatment of patients with hereditary hemolytic anemia due to deficiencies in glutathione synthase and glucose-6-phosphate dehydrogenase, and leukocyte dysfunction in patients with a deficiency in glutathione synthase (Oski 1980).

The designation of vitamin E as a fertility vitamin initially led to its use in treatment of male, fertility disturbances. According to the clinical experiences of many authors (Nikolowski 1950; Lindner 1958; Bayer 1959; Nikolowski 1960, 1962, 1970, 1974; Milbradt 1966; Bartak 1972; Glander 1973), the use of vitamin E in treatment of male subfertility is justified as a therapeutic trial. The therapeutical benefit of male subfertility treatment with α-tocopherol is, however, not clearly established scientifically. Recently, it has been recognized that oxidative damage of spermatozoa may contribute to male infertility disorders (Aitken and Clarkson 1987, Mann et al. 1980), and attempts are being made to investigate the clinical efficacy of antioxidants in male infertility (Ochsendorf et al. 1991).

Parallel to the use of vitamin E in treatment of male subfertility, the substance was also used as a drug for treatment of skin diseases. In 1936, Reffy and Leranth published reports on the clinical use of the substance for treatment of leg ulcers and other skin diseases. In 1948, Burgess and Prichard reported successful treatment of patients diagnosed as having lupus erythematosus with tocopherol. Subsequently, wider application of tocopherol in the treatment of various skin diseases ensued. Literature reviews (Sterzi 1950; Lohel 1951; Fegeler and Beckmann 1955; Yano 1959; Nikolowski 1959, 1960, 1985; Wulf 1962; Ayres and Mihan 1975; Fuchs and Milbradt 1989c) of clinical trials with vitamin E in dermatology indicate a broad spectrum of skin diseases in which clinical improvement has been attained, with varying success rates. The broad spectrum of clinical applications reflects an enthusiastic therapeutic approach, and a lack of knowledge with regard to the pharmacokinetic and pharmacodynamic properties of the vitamin. Consequently, therapeutic expectations have not always been fulfilled. Tocopherol is also used as an ingredient in cosmetic products, and has been suggested to increase skin smoothness and skin tension, have an effect as a deodorant, suppress sebum production, stimulate hair growth, prevent skin aging, and protect skin from UV damage (Djerassi et al. 1986; Möller et al. 1987).

The clinical use of tocopherol in dermatological therapy is not based on a hypothetical conception of a hypovitaminosis, as tocopherol deficiency dermatoses in the human are unknown. However, tocopherol deficiency in epidermal or dermal microcompartments may be a consequence of endogenously or exogenously induced oxidative injury in the skin. Dermatoses with significant involvement of reactive oxygen species in their pathogenesis include skin conditions mediated by electromagnetic radiation, inflammatory cells, prooxidant xenobiotics, and lipid peroxidation products. Furthermore, increased cellular metabolism during hyperproliferation, induction of prooxidant skin enzymes, and inhibition of skin antioxidants also contribute to prooxidant conditions. Dermatological disorders with significant involvement of oxidative damage may benefit from local and systemic antioxidant therapy. Hence, tocopherol may be of potential use in treatment of distinct dermatoses, e.g., those which are associated with vasculitis or tissue infiltration by inflammatory cells, and which are aggravated or induced by UV light.

Most of the biological effects of vitamin E can be explained by its antioxidant properties, presumably including interactions with prostanoid metabolism. Membrane-modulating effects may also contribute ot its biological activity to an unknown extent. The antioxidant and membrane-modulating effects are the rationale used to explain its antiinflammatory, photoprotective, and mesenchymal metabolism-modulating properties. Other pharmacodynamic properties are anticarcinogenic, antiarteriosclerotic, and immunomodulating effects.

The *antiinflammatory effect* of α-tocopherol has been demonstrated in several animal and clinical studies. Particularily in rheumatic diseases its clinical efficacy is well documented (Klein and Blankenhorn 1987; Blankenhorn 1986; Blankenhorn et al. 1985; Machtey and Quaknine 1978). Topical application of α-tocopherol inhibits anthralin dermatitis in humans (Finnen et al. 1984a) and reduces the enhancement of anthralin erythema produced by topical arachidonic acid (Lawrence and Shuster 1987). Tocopherol also inhibits the formation of persistent organic radicals in hairless mice skin after topical treatment with anthralin (Fuchs and Milbradt 1989b). Systemically administered α-tocopherol causes a significant antiinflammatory effect in dextran-mediated rat paw edema (Kamimura 1972). Croton acid dermatitis in the rabbit is inhibited by topical application of α-tocopherol (Kamimura 1972). In an earlier study, topical treatment of croton acid dermatitis in the rabbit with tocopherol was not successful, in contrast to systemic therapy (Yano 1959). Ayres and Mihan (1978b) hypothesized that free radicals and lipid peroxidation participate in the pathophysiology of inflammation occuring in acne vulgaris, and suggested that antioxidants such as tocopherol may be useful in treatment of acne vulgaris.

Topical application of *d*-α-tocopherol in rat skin prior to UV exposure inhibits long-term UVA- and UVB-induced hydroxyproline loss, but does not influence UVA-induced decrease in collagen solubility. It has been suggested that tocopherol is involved in the regulation of collagen metabolism following long-term exposure of hairless mouse skin to UV light (Trautinger et al. 1990b). Tocopherol deficiency in rat skin results in an increase in soluble collagen, probably via a defect of the inter- and intramolecular collagen cross-linking (Brown 1967). A single subcutaneous injection of tocopherol in rats inhibits *mesenchyma metabolism* in dermis, aorta, and cartilage for many weeks (inhibition of sulfate insertion rate; Rave et al. 1971). The growth of foreign body granulomas and the unspecific mesenchymal reaction (increased sulfate incorporation into mucopolysaccharides) induced by intraperitoneal injection of staphylococcus toxin is inhibited by tocopherol (Rave et al. 1971). In rats, tocopherol inhibits collagen synthesis and delays wound healing (Ehrlich et al. 1972). However, in rats exposed to ionizing radiation, an improvement in wound healing by tocopherol is observed (Taren et al. 1987), and may be caused by stimulation of epithelial regeneration. Indeed, Melkumian et al. (1978) reported that epithelial regeneration of burn wounds is improved by topical therapy with tocopherol. The data indicate that tocopherol is essentially involved in mesenchymal metabolism. In animals, high doses may cause inhibition of mesenchymal metabolism but low (physiological) concentrations are required for well-balanced collagen synthesis.

Tocopherol Serum Concentrations

To our knowledge there are no reports on tocopherol skin concentration in different dermatoses. Although it is not known whether there is a direct correlation between tocopherol serum concentration and tocopherol tissue content, particularily in skin, serum concentrations have been analyzed in several skin diseases.

Lupus Erythematosus. Welsh (1954) reported the average tocopherol blood concentration in a group of patients with lupus erythematosus was 567 µg/100 ml (µg %), while the blood level of a control group was 2370 µg %. The blood concentration of *β*-carotene was also decreased in the lupus patients. However, in this study standard deviation was not calculated (Welsh 1954). After oral administration of different tocopherol derivates (*dl-α-*, *d-α-*, and solubilized *d-α*-tocopherol acetate), tocopherol blood concentration was measured in these patients. Oral doses of *d-* and *dl-α*-tocopherol acetate and solubilized tocopherol (500–2000 mg/daily) resulted in an increase in blood tocopherol of 2500–5800 µg % for *d-* and *dl-α*-tocopherol acetate and of 3100–7100 µg % for the solubilized vitamin. In this study, bioavailability of the natural tocopherol derivative (*d-α*-tocopherol acetate) was not higher than that of the synthetic tocopherol derivative (*dl-α*-tocopherol acetate) (Welsh 1954).

Keratosis Follicularis. In Indian children belonging to a low socioeconomic population group who were diagnosed as having keratosis follicularis, serum tocopherol concentration was significantly lower (330 µg %) than in healthy children (660 µg %) of a comparable socioeconomic group (Nadiger 1980). Oral supplementation of the children with keratotis follicularis with daily doses of 300 mg *dl-α*-tocopherol acetate over 4 weeks resulted in a normalization of serum tocopherol concentration (1090 µg %) and to significant improvement of the skin condition. In this study, hypovitaminosis E was evident (< 500 μg% = 5 mg/l). The average plasma concentration of *α*-tocopherol in a healthy adult varies from 6 to 20 mg/l (10 mg/l = 22 μmol/l); *α*-tocopherol is found at concentrations of 1–2 mg/l. Other tocopherol stereoisomers and tocotrienols are only present in trace amounts. It was suggested that in addition to hypovitaminosis E, hypovitaminosis A may be present in these children, and that tocopherol inhibits oxidative inactivation of vitamin A and thereby improves keratotis follicularus via an indirect sparing effect on vitamin A.

Miscellaneous Skin Diseases. Determination of serum tocopherol levels in dermatology patients (acne vulgaris, neurodermitis, other eczemas, rosacea, psoriasis, scleroderma, ulcus cruris, urticaria) shows a mean value of 850 ± 162 μg% (μg/100 ml). This is below the mean value in healthy persons of about 1000 μg% (Fegeler and Beckmann 1955). The variability in the physiological serum concentration, however, is large (600–2000 μg%), so conclusions should be drawn from this study with caution. Fegeler could not demonstrate a clear correlation of tocopherol plasma concentration with severity of the skin diseases. In a different study (neurodermitis, other eczemas, ichthyosis vulgaris, pityriasis rubra pilaris, dermatitis herpetiformis Duhring, and others), no significant differences in tocopherol serum concentrations were noted in comparison to healthy controls (Rietschel 1963). A high variability of serum tocopherol concentration (530–1560 μg%) was evident in this study, too.

*Tocopherol-Supplemented Patients and Volunteers. dl-α-*Tocopherol was given orally (600 IU/day) to atopic dermatitis patients. Their plasma tocopherol concentration reached a maximum (2-fold increase), and this was followed by a plateau during the 6th–8th weeks of therapy (Fairris et al. 1989). Oral administration of 600 mg *d-α*-tocopherol in young healthy male adults resulted in a 2.5–3-fold increase in the levels in plasma and red and white blood cells, with a maximum occurring in the 4th week. The maximum was reached in thrombocytes after 8 weeks and in oral mucous membrane epithelia after 12 weeks. There was no further increase after the maximum was reached, although oral intake of tocopherol was continued. Within 1 week after termination of therapy, tissue concentrations declined to basal values. This indicates that after saturation of the membranes either resorption is reduced or elimination is increased (Kitagawa and Mino 1989). Substitution with 200 or 400 mg *dl-α*-tocopherol acetate in young healthy men resulted in an increase in the plasma tocopherol concentration. There was a tendency for the group treated with 400 mg to have higher plasma levels than the men taking 200 mg (Böhlers 1988). Supplementation of volunteers with all-rac-*α*-tocopherol doses of 150, 225, 800, and 1000 IU/day over 3 weeks results in increased oxidation resistance of their low density lipoprotein (LDL) of 1.17, 1.56, 1.35, and 1.75, respectively (relative values) (Esterbauer et al. 1990). This indicates that oral supplementation with increasing tocopherol doses results in increased protection from oxidative injury.

Treatment with Tocopherol

Lupus Erythematosus. Burgess and coworker first reported, promising results of treatment of lupus erythematosus (CDLE) with oral or systemic vitamin E (Burgess 1948; Burgess and Pritchard 1948). His reports were later confirmed by other authors (Bazex 1949; Doucas 1949; Silvers 1950; Sterzi 1951; Grubb 1952; Welsh 1954; Shinskii et al. 1962; Ayres and Mihan 1978a). Less successful results of treatment of lupus erythematosus (CDLE) with vitamin E have been reported by several other authors (Sawicky 1950; Pascher et al. 1951; Sweet 1948; Jordan and Wulf 1950; Kaminsky 1950; Morgan 1951). Ayres and Mihan (1971, 1975, 1978a) suggested that tocopherol therapy failures in lupus erythematosus may be due to a low-dose tocopherol regimen. Kaminsky (1950) attributed a large proportion of therapy failures to insufficiency of the dosage utilized and Welsh (1954) concluded that a megadose therapy with tocopherol is justified to obtain rapid therapeutic success in patients.

Although several investigators have used tocopherol for treatment of lupus erythematosus, most of these studies have not yielded consistent results. Some clinical investigations have shown beneficial effects, while others have suggested that tocopherol was more or less ineffective. These controversial results could partially be explained by clinical subsets of the disease and lack of adequate disease classification and heterogeneity of the patient populations with regard to duration, severity, and dynamics of progression or regression. Multicenter, prospective, long-term therapeutic trials with tocopherol in carefully selected cohorts of lupus erythematosus patients with comparable disease activity and duration are needed to resolve these differences.

Skin Disorders Associated with Sclerosis/Fibrosis. The second largest patient group reported to benefit from tocopherol therapy includes those suffering from disorders associated with skin sclerosis and fibrosis. Tocopherol is used clinically with varying success rates in treatment of various fibrotic skin diseases. In particular, tocopherol is used in treatment of:

- Fibromatosis palmaris et plantaris (Pambor et al. 1985; Reifferscheid and Matis 1951; Hanfstaengel 1951)
- Induratio penis plastica (Menzel 1977; Gartmann 1953; Pambor et al. 1985; Kunstmann 1954; Pult 1954; Marazzini 1955)
- Keloids (Pascher et al. 1951)
- Lichen sclerosus et atrophicus (Ayres and Mihan 1971, 1975; Martin 1952; Pascher et al. 1951)
- Scleroderma, systemic or localized (Ayres and Mihan 1971, 1978a; Pascher et al. 1951).

Miscellaneous Skin Diseases. Other skin diseases which are reported to benefit from tocopherol therapy include ones associated with vasculitis (acrodermatitis chronica atrophicans, necrobiosis lipoidica, granuloma annulare), other types of inflammatory injury (acne vulgaris, psoriasis vulgaris), and radiation and UV damage:

- Acne vulgaris (Michaelsson and Edquist 1984; Dainow 1953)
- Acrodermatitis chronica atrophicans (Grubb 1952; Hagerman 1951)
- Atopic dermatitis (Fairris et al. 1989)
- Epidermolysis bullosa dystrophicans (Sehgal et al. 1972; Michelson et al. 1974; Degreef and Fluor 1974; Smith and Michener 1973; Wilson 1964)
- Epidermolysis bullosa simplex Weber-Cockayne (Degreef and Fluor 1974; Ayres and Mirhan 1969)
- Granuloma annulare (Cochrane 1950; Ayres and Mihan 1971; Dainow 1958; Cornbleet et al. 1951)
- Keratosis follicularis (Nadiger 1980; Ayres and Mihan 1972).
- Lichen ruber (Lohel 1951)
- Necrobiosis lipodica diabeticorum (Pascher et al. 1951; Ayres and Mihan 1971)
- Neuralgia, postzoster (Ayres 1973a)
- Pemphigus benignus familiaris (Ayres and Mihan 1975)
- Pigment disturbances (chloasma and postinflammatory hyperpigmentation) (Hayakawa et al. 1981)
- Porphyria, hepatic and erythropoietic (Mustajoki 1972; Nair et al. 1971; Murty et al. 1969; Conley and Chisholm 1979)
- Pseudoxanthoma elasticum (Kahan et al. 1963; Hamilton-Gibbs 1964; Ayres and Mirhan 1969; Stout 1951)
- Psoriasis vulgaris (Schade 1952; Lohel 1952)
- Purpura of various etiopathologies (Gerloczy et al. 1966; Fujii 1972; Comi and Nesi 1950)
- Radiodermatitis and radiation ulcers (Frey 1954; Hütter 1954; Kemmerer-Görlitz 1952; Pascher et al. 1951)
- Subcorneal pustulosis Sneddon-Wilkinson (Ayres and Mirhan 1974)
- Ulcera cruris (Bijdendijk 1951; Reffy 1936; Leranth 1936; Siedentopf 1949)
- Yellow nail syndrome (Norton 1985; Ayres 1973a; Hazelrigg and McElroy 1980).

Biochemical and pharmacological studies indicate that α-tocopherol is a highly effective antioxidant with antiinflammatory, antifibrotic, photoprotective, and other properties. However, the evaluation of the clinical case studies and reports about treatment of various dermatoses with tocopherol (Fuchs and Packer 1991a) do not yet provide convincing evidence of its clinical efficiency in several cases. This discrepancy may be explained by false indications for drug therapy, insufficient dosage and/or duration of therapy, inadequate study protocols, low numbers of patients, heterogeneity of patient populations investigated, and distinct clinical subsets of skin diseases. Nonresponders and remissions caused by self-healing are further complicating factors in the analysis. The uncritical and widespread use of tocopherol in treatment of various skin diseases from the late 1940s to the 1960s reflected an ignorance of its pharmacodynamic properties and of the pathophysiology of the specific skin disease. Furthermore, it was the result of a therapeutic dilemma in the pre-glucocorticosteroid and pre-retinoid era. Today, tocopherol plays only a minor role in conventional dermatological therapy. Developments in the last decade, however, have made it possible to understand the pharmacokinetics and pharmacodynamics of the substance and to learn about oxidative injury mechanisms in dermatopathology. Now is the time to use tocopherol, logically and selectively, in dermatological therapy. Prospective, long-term, multicenter, double-blind studies in carefully selected patients are needed to unequivocally demonstrate the clinical efficacy of the antioxidant vitamin in distinct skin diseases.

II. Superoxide Dismutase

SOD has been used in the experimental and clinical therapy of chronic inflammatory and fibrotic processes for several years. Although the great majority of reports clearly indicate that it is effective in the treatment of these disorders, it was recently pointed out that most reports on the use of SOD are uncontrolled, and some even anecdotal (Greenwald 1990). In the future, well-controlled, mutlicenter, double-blind studies are needed to prove unequivocally the concept of SOD as a clinically effective antioxidant.

In chronic inflammatory and fibrotic processes, oxidant generation and fibroblast chemotaxis play a major role in the pathogenesis. Liposomal SOD suppresses luminol-amplified chemiluminescence in activated human neutrophils (Yanai et al. 1987), and SOD inhibits in a dose-dependent manner the chemotactic response of human skin fibroblasts. Suppression of the oxygen burst and inhibition of fibroblast chemotaxis can, in part, explain the beneficial results which are reported when SOD is used for the treatment of chronic inflammatory and fibrotic processes (Martin et al. 1985; Wach et al. 1987).

Animal Studies. The data derived from animal studies clearly indicate antioxidant protection by SOD: Cu,Zn-SOD prevents ischemia/reperfusion syndrome in mammalian skin (Manson et al. 1983), suppresses immune-complex-mediated dermal inflammation (McCormick et al. 1981), and is effectively used in treatment of severe dermatitis in the dog (Lowrey 1976). Intramuscular administration of Cu,Zn-SOD to NZB mice over several weeks prevented the development of lupus erythematosus nephritis (Sato et al. 1989).

Human Studies. Intravenous and subcutaneous application of liposomal SOD is clinically effective in patients with PSS, dermatomyositis, Behçet's disease, mucocutaneous lymph node disease, dermatitis herpetiformis, IgA linear bullous dermatosis, cement dermatitis, burn wounds, and postradiotherapy fibrosis and necrosis (Michelson 1982, 1986 b; Niwa et al. 1985 d; Niwa 1989; Somiya et al. 1987; Hoshi 1988). In humans, radiation-induced fibrosis of the skin and mucous membranes was treated with intramuscular application of liposomal SOD. A marked improvement was seen in about 80% of the patients after a short treatment period of 3 weeks (Michelson 1986 b). Hypertrophic scars and keloids (Grösser et al. 1984; Landthaler 1984 a), subcutaneous scarred indurations (Eckert and Koch 1987; Eckert 1988), and induratio penis plastica (Pyronie's disease) (Bartsch et al. 1981; Gustafson et al. 1981; Landthaler 1984 b; Chaussay et al. 1983) have successfully been treated by intralesional application of SOD.

Analysis of SOD content in normal skin, keloid, hypertrophic scar, and scar tissues reveal that total SOD, Mn-SOD and Cu,Zn-SOD are significantly lower in keloids than in healthy skin. No significant differences are found for hypertrophic scar tissue. In scar tissue, the total SOD content is significantly higher, that of Mn-SOD lies in a normal range, and Cu,Zn-SOD activity is higher than in normal skin tissue (Sugiura et al. 1986 a). The data do not support the hypothesis that SOD activity is generally diminished in fibrotic skin lesions.

Pretreatment of the epidermal surface of guinea pigs with liposomal SOD before exposure to UVB irradiation prevents an increase in skin lipid peroxidation and inhibits skin inflammation (Ogura et al. 1987). Exogenous SOD inhibits UVB phototoxic effects in cultured fibroblasts. Since intracellular penetration of nonliposomal SOD is slow and highly ineffective, it is suggested that membrane-bound SOD inhibits early phototoxic processes at the fibroblast plasma membrane (Emerit et al. 1981). A transitory stabilization of the tumor development in patients with xeroderma pigmentosum was achieved by local application of catalase (Vuillaume et al. 1983).

III. Retinoids

Vitamin A acid (retinoic acid), aldehyde (retinal), and alcohol (retinol) were recognized as endogenous regulators of epithelial and mesenchymal tissue differentiation (Wolbach and Howe 1925, 1928, 1933). By chemical modification of the parent compounds, three generations of retinoids are now available. The first generation includes 13-*cis*-retinoic acid and all-*trans*-retinoic acid (Fig. 4.1), the second generation etretinate, an aromatic retinoic acid ester (Fig. 4.2), and the third generation the arotinoids (polyaromatic retinoids; Fig. 4.3). Retinoids regulate epidermal differentiation and proliferation and have antiinflammatory, antioxidant, antineoplastic, and immunomodulatory effects.

First generation retinoids are used topically in the treatment of acne vulgaris and rosacea. Topical retinoids can be used for the prevention and repair of cutaneous photoaging (Kligman 1989; Goldfarb et al. 1989).

Second-generation retinoids are used systemically (retinoic acid ethylamide = motretinide is used locally only) in the treatment of various derma-

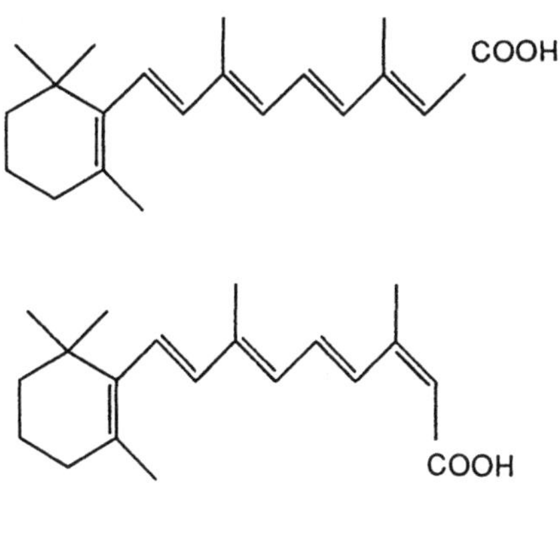

all-*trans*-retinoic acid

13-*cis*-retinoic acid

Fig. 4.1. First-generation retinoids

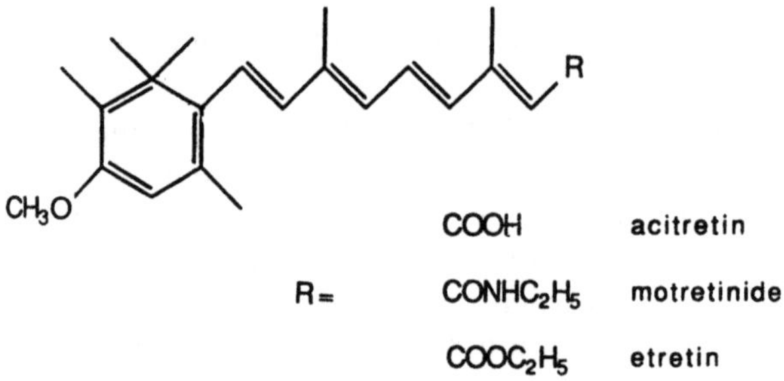

Fig. 4.2. Second-generation retinoids

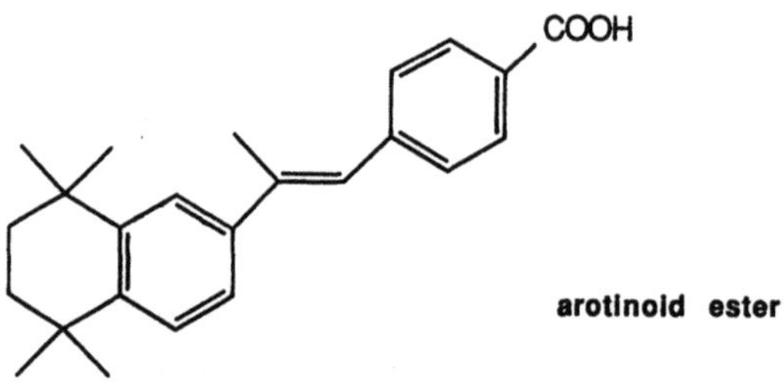

Fig. 4.3. Third-generation retinoids

toses, such as pustular psoriasis, lichen ruber, and keratinizing genodermatosis (Orfanos 1980).

The third-generation retinoids are currently undergoing experimental trials and clinical pilot studies; their pharmacodynamic actions occur at hormonelike plasma concentrations.

The lipophilic retinoids are bound in blood to serum proteins, and cellular uptake is mediated via a specific membrane receptor as well as by unspecific endocytosis. In 1987, a retinoic acid receptor similar to the steroid/thyroid receptor was identified (Petkovich et al. 1987; Gigure et al. 1987). Intracellular transport from the plasma membrane to the nucleus occurs via specific binding proteins.

Mechanism of Action. Retinoids trigger complex biochemical effects in different biological systems (Roberts and Sporn 1984), which cannot be explained by a single molecular mechanism (Elias and Williams 1981 b). At the nuclear level, retinol and retinoic acid affect genomic expression by activating and simultaneously repressing specific genes (Chytil and Sherman 1987), which may explain their multiple effects. The change in diverse patterns of phenotypic expression of cells in response to retinoids is characterized as a pleiotropic response. In the cytosol, posttranslational glycosylation of proteins, protein phosphorylation and dephosphorylation, and protein kinase are induced.

Retinoids and carotenoids exert antioxidant effects by different mechanisms. Retinoids such as tretinoin, retinol, retinal, and retinyl acetate decrease the production of superoxide radicals in activated neutrophils. The inhibitory effect of these retinoids is not exerted by directly scavenging superoxide anion radicals. This indicates that they might inhibit the generation of superoxide by affecting neutrophil functions (Yoshioka et al. 1986). Retinol, retinylacetate, and retinoic acid are also effective inhibitors of superoxide anion radical production in neutrophils activated by a tumor promoter (Witz et al. 1980). It is not known to what extent the antioxidant properties of retinoids may contribute to their different biological activities, e.g., antiinflammatory and antineoplastic effects.

The antioxidant action of various retinoids was examined in comparison with ascorbate, tocopherol, or glutathione using the DPPH radical assay. 13-*cis*-Retinoic acid and the furyl analogue of retinoic acid slightly quench the DPPH radical at high drug concentrations (2.5 m*M* retinoid, 100 μ*M* DPPH); acitretin, etretin, all-*trans*-retinoic acid, and arotinoid acid are not reactive (Hiramatsu, personal communication). This indicates that only some retinoids have a limited ability to participate directly in one-electron transfer reactions in vivo, e.g., scavenging of free radicals.

IV. Carotenoids

Carotenoids directly interact with reactive oxygen species. *β*-Carotene and canthaxanthin (Fig. 4.4) inhibit free-radical- and singlet-oxygen-induced lipid peroxidation in liposomes and biological systems (Krinsky and Deneke 1982; Krinsky 1989). Singlet oxygen is quenched in vitro very effectively by low concentrations of *β*-carotene (Foote and Denny 1968). The sequence of reactivity of carotenoids with singlet oxygen is in the order: lycopene > *γ*-carotene > canthaxanthin > *α*-carotene > *β*-carotene (DiMascio et al. 1989). The carotenoid pigments β-carotene and canthaxanthin quench photochemical reactions in the epidermis involving singlet oxygen and oxygen radicals that occur when epidermis is exposed to UVB (Mathews-Roth 1986).

β-Carotene and canthaxanthin were used in combination to produce a natural appearing sun tan. Oral administration of *β*-carotene significantly increased the minimal erythema dose of solar radiation in humans (Mathews-Roth et al. 1972). *β*-Carotene has been successfully used in the treatment of patients with various photodermatoses: erythropoietic protoporphyria (Mathews-Roth et al. 1970, 1974; Baart De La Faille et al. 1972; Pollit 1975), porphyria cutanea tarda (Ippen 1972), actinic reticuloid, solar urticaria

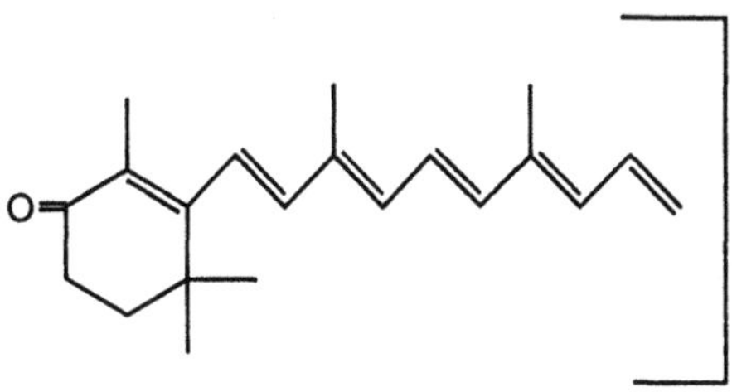

2

Fig. 4.4. Canthaxanthin

(Kobza et al. 1973), hydroa vacciniforme (Bickers et al. 1978), and polymorphous light eruption (Nordlund et al. 1973). Mentally handicapped patients medicated with photosensitizing psychopharmaceuticals were also given 18 mg β-carotene daily over 3 months. The patients showed significant clinical improvement of their photodermatological problems, and only mild to moderate skin discolouration was observed (Westermarck et al. 1990).

However, despite the many case reports suggesting that β-carotene is effective in treating several photodermatoses, a criticism is that controlled trials have not been undertaken (Pollitt 1975). Some points are noteworthy, though. The clinical efficacy of β-carotene in treatment of patients with erythropoietic protopoprhyria is documented by many studies. Despite some case reports, however, it is believed that β-carotene is of no clinical value in treatment of porphyria cutanea tarda, polymorphous light eruption, and solar urticaria. In these cases, reactive oxygen species, particularily singlet oxygen, may not play a predominant role (Hönigsmann, personal communication). An alternative explanation is that for these diseases the lipophilic β-carotene is not available in the specific microenvironments endangered by reactive oxidants. Differences in lipophilicity of porphyrin derivatives (Emiliani and Delmelle 1983) may account for contrasting signs of cutaneous photosensitivity in patients with porphyria cutanea tarda (predominance of hydrophilic uroporphyrin) and erythropoietic porphyria (predominance of lipophilic protoporphyrin). β-carotene (150 mg/day for 4 weeks) did not protect humans against UVA, UVB, and PUVA erythema, and gave no protection at the DNA level. It was suggested that β-carotene may not quench reactive oxidants efficiently in vivo, if these species play a role in these conditions at all (Wolf et al. 1988). Furthermore, it was hypothesized that carotenoid protection against UVB-induced skin carcinogenesis, which is documented in several studies (Epstein 1973; Mathews Roth 1982, 1983; Mathews-Roth and Krinsky 1984, 1985), does not operate by inhibition of mutagenic DNA lesions.

V. Anthralin

Anthralin (1,8-dihydroxy-9-anthrone; Fig. 4.5) was introduced into the external therapy of psoriasis in 1916 by Galewsky (1916). It superceded the natural compound chrysarobin (3-methylanthralin; Fig. 4.5), a constituent of the natural product Goa powder, which was used in the late nineteenth century for

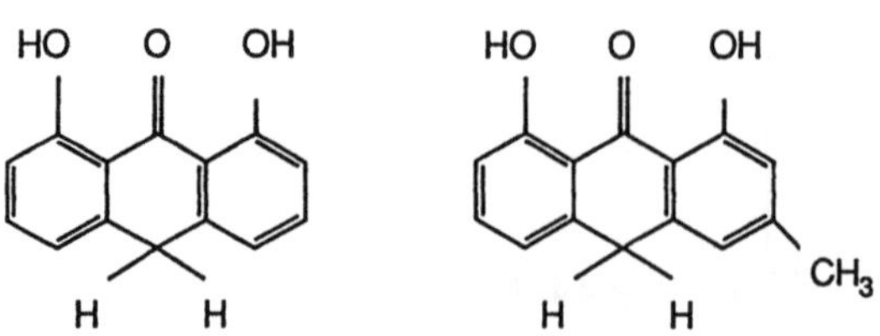

Fig. 4.5. Anthralin *(left)* and chrysarobin *(right)*

topical psoriasis treatment. It is also used effectively in the treatment of seborrheic dermatitis (Wölbling et al. 1987, 1989). Anthralin is a potent antipsoriatic compound, and its pharmacodynamic mode of action has been analyzed. However, it is difficult to study, because it readily decomposes under physiological conditions. Autoxidation of anthralin in skin is enhanced by proteins (Sa E Melo et al. 1983) and peroxidized lipids (Ducret et al. 1985).

Anthralin has a complex redox behavior; it may act both as an anti- and as a prooxidant (Fuchs et al. 1990g). It is a strong reductant that is readily oxidized by light, trace metal ions, and oxygen. The proposed 14 π electron structure (anthracentriol) is not proven, and anthralin presumably exists as a 2×6 π electron system. In physiological media the anthralin anion is formed immediately but rapidly decomposes (Kohen et al. 1986). Subsequently, reactive oxidants are generated which mediate the prooxidant effects (Fig. 4.6).

The anthralin and chrysarobin anions are photosensitizers of equal potency and generate singlet oxygen (Müller et al. 1986a, b; Joshi and Pathak 1984) and superoxide anion radical (Joshi and Pathak 1984). Anthralin possesses about 1/10 and chrysarobin 1/4 of the photosensitizing activity of the respective anions. The anthralin anion is oxidized by singlet oxygen, and anthraquinone is formed via anthralin endoperoxide (Fig. 4.7). Pulse radiolysis studies have revealed that the triplet state of anthralin rapidly reacts with oxygen, yielding the anthralin radical and singlet oxygen (Bruce et al. 1989).

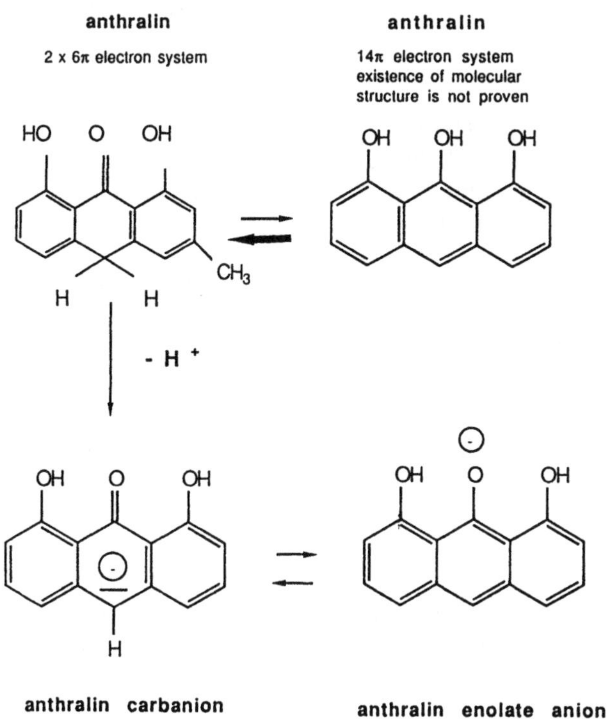

Fig. 4.6. Formation of anthralin anion

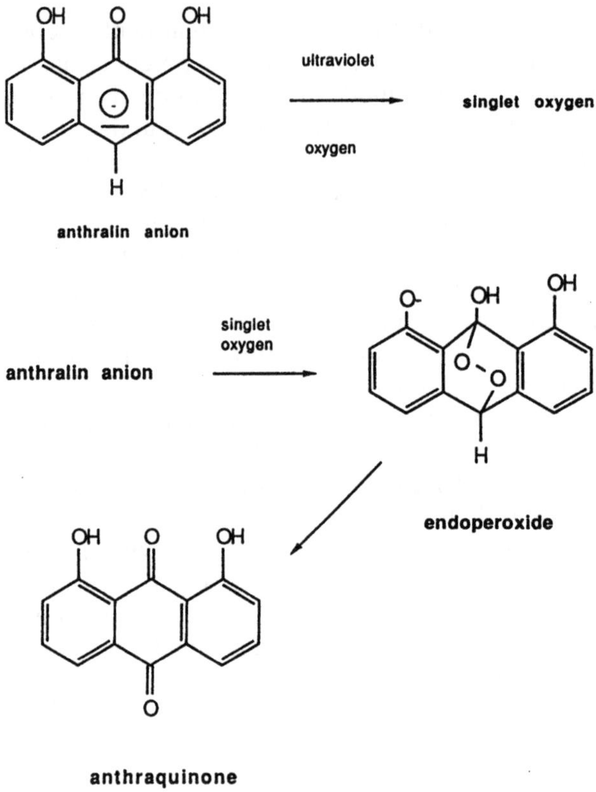

Fig. 4.7. Anthralin oxidation

The combined treatment of psoriasis patients with anthralin plus UVA irradiation was proposed to be of potential clinical interest (Hindson et al. 1983), but a significant therapeutical advantage could not be reproduced by others (Schauder and Mahrle 1982; Brun et al. 1984).

Anthralin Redox Chemistry. The initial oxidation product of anthralin is the 1,8-dihydroxy-9-anthrone-10-yl radical (anthralin radical) (D'Ischia and Prota 1985; Mustakallio et al. 1984), which is formed by reaction of anthralin anion with molecular oxygen, yielding reactive oxygen species (Fig. 4.8). Autoxidation of the anthralin anion involves the formation of reactive oxygen species, such as superoxide anion and hydrogen peroxide (Müller et al. 1986a, b; Unna 1916; Krebs 1969). Although the 1,8-dihydroxy-9-anthrone-10-yl radical is reported to be quite stable in benzene (Bruce et al. 1988), it is apparently very unstable in polar media such as water.

The radical reacts with molecular oxygen to form the highly reactive anthralin peroxyl radical, which decomposes via anthralin hydroperoxide into anthraquinone and may cause lipid peroxidation via hydrogen abstraction from unsaturated lipids (Fig. 4.9).

The anthralin radical may also react with itself to form anthralin dihydrodimer, which is further oxidized to a naphthodianthrone radical. Sub-

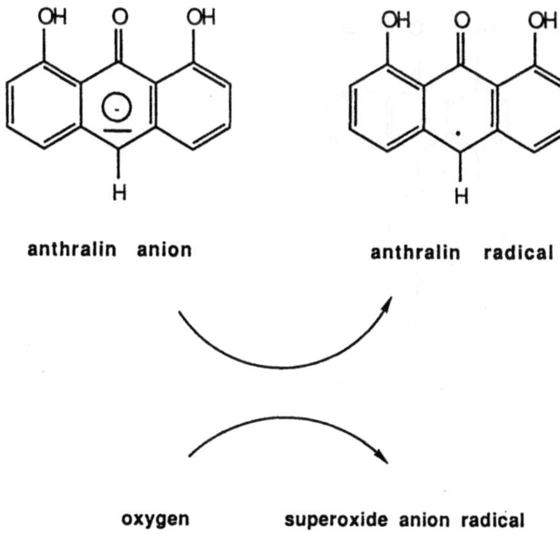

Fig. 4.8. Anthralin radical

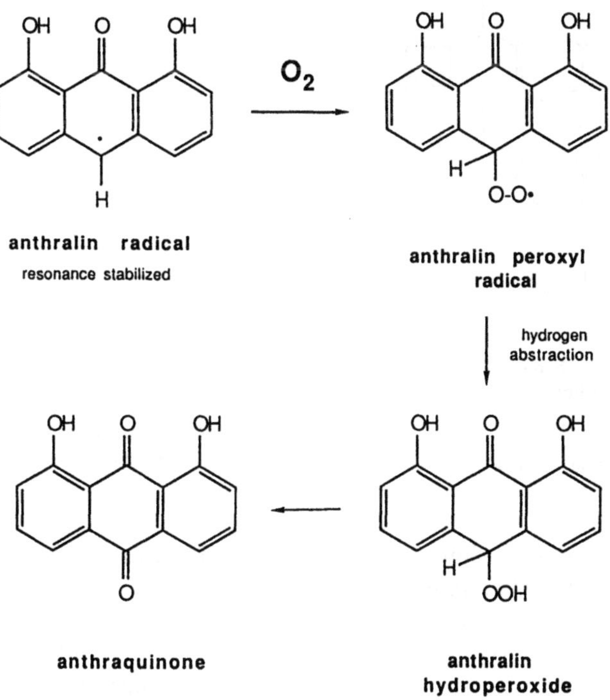

Fig. 4.9. Formation of anthraquinone

sequently, the chemically not well-characterized anthralin brown is formed (Fig. 4.10), which contains paramagnetic compounds (Mustakallio et al. 1984). In pig skin, persistent free radical species are detected after topical application of anthralin (Schroot and Brown 1986). In skin of hairless mice, persistent free radical species derived from anthralin were assigned to paramagnetic an-

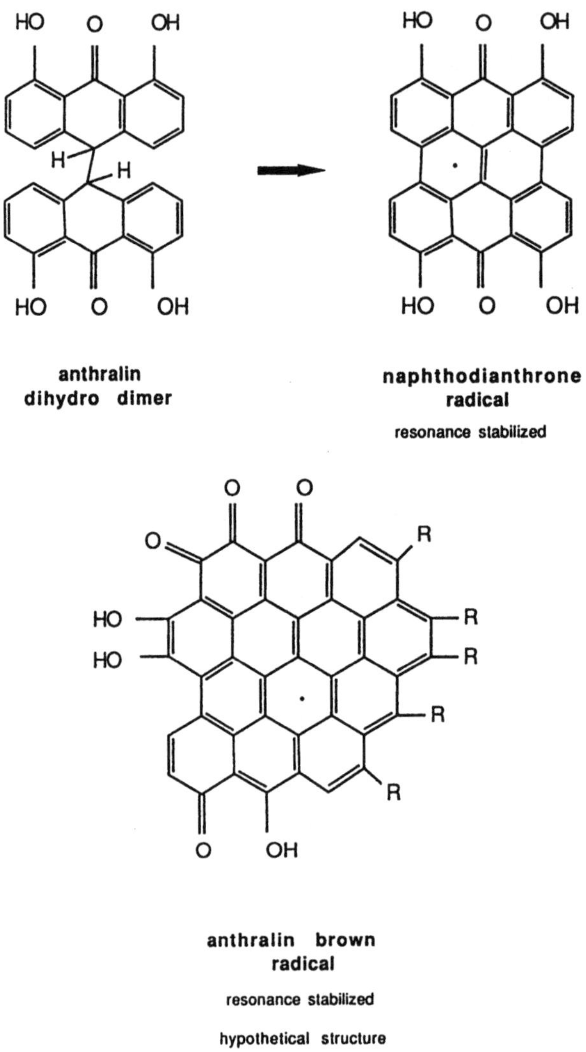

Fig. 4.10. Formation of anthralin brown

thralin brown products, and tocopherol inhibits formation of these persistent radicals in skin (Fuchs and Packer 1989).

Two different oxidation pathways of anthralin have been discussed. The one-electron mechanism leads to the formation of anthralin dimer and finally to anthralin brown products, whereas the two-electron mechanism results in the formation of anthraquinone (Schaltegger and Krebs 1984). Both reaction pathways involve the formation of free radical species. Under physiological conditions, the two-electron pathway may predominate, because oxidation of anthralin in heterogeneous systems (e.g., suspension) occurs mainly via the two-electron mechanism. Since anthralin has some solubility in the cytosol, the one-electron mechanism may also occur in vivo.

Oral ingestion of a laxative containing 1,8-dihydroxy-9-anthraquinone results in irritant toxic dermatitis in the perianal region, indicating in vivo reduction of the quinone to anthralin. This may indicate redox cycling of the drug in the strong reducing environment of the colon; however, the situation may be totally different at the rather oxidizing environment of the skin surface. Although some anthraquinones are redox cycling compounds (Kharasch and Novak 1981; Kappus 1986b), 1,8-dihydroxy-9,10-anthraquinone did not induce the formation of reactive oxygen species in isolated mitochondria and microsomes (Fuchs et al. 1990a).

In murine skin, epidermal NADPH-ubiquinone reductase is induced by anthralin in a dose-dependent manner (Merk et al. 1988). Anthralin-derived oxidation products such as 9,10-anthraquinone also induce NADPH-ubiquinone reductase in human keratinocytes (Bonnekoh et al. 1989). This mechanism was suggested to contribute to detoxification of redox cycling prooxidative quinone compounds in the skin. In vitro, however, NADPH ubiquinone reductase is inhibited in human keratinocytes by anthralin (Bonnekoh et al. 1989).

The capacity of anthralin to form free radicals and reactive oxygen species can provoke lipid peroxidation. Unna pointed out in 1916 that anthralin may induce the formation of hydrogen peroxide and cause lipid peroxidation in skin. Anthralin stimulates the UV-induced foramtion of lipid peroxides in skin surface extracts of psoriatic patients (Meffert and Reich 1969b). Topical application of anthralin augments ethane exhalation in animals, indicating increased skin lipid peroxidation (Müller et al. 1987). Anthralin causes lipid peroxidation in human skin (Finnen et al. 1984b), and anthralin dermatitis is inhibited by tocopherol, tertiary-butylhydroxyanisol, and retinol esters (Finnen et al. 1984a). Enhancement of anthralin dermatitis by arachidonic acid is also inhibited by tocopherol (Lawrence and Shuster 1987). Since arachidonic acid enhances the inflammatory response of anthralin, membrane phospholipids are suggested to be the substrate for anthralin free-radical-mediated inflammation (Lawrence and Shuster 1987). Intraperitoneal administration of SOD prior to elicitation of dithranol-induced toxic dermatitis significantly reduces the inflammatory response in mice (Kemeny et al. 1989).

Anthralin is a tumor promoter (McCann et al. 1975; Bock and Burns 1963; Segal et al. 1971); however an antimutagenic potential was also described (De Young et al. 1981; Bernd et al. 1987). Activated oxygen species are generated during oxidation of anthralin and cause inhibition of SOD and catalase activity in keratinocytes (Solanki et al. 1982). This may cause a prooxidant state, which can promote initiated cells to neoplastic growth (Cerutti 1985). There is, however, no clinical evidence indicating an increased incidence of skin cancer in anthralin-treated patients (Bernd et al. 1989).

Mechanisms of Action. Redox interactions of anthralin with biological membrane components are suggested to be an important feature in the molecular mechanism of anthralins action (Fuchs et al. 1990a).

Anthralin binds to the plasma membrane of keratinocytes and is subsequently accumulatetd in mitochondria (Reichert et al. 1985), which are considered target organelles of drug action (Morlier et al. 1985). Oxygen consumption is inhibited in isolated mitochondria (Morlier et al. 1985; Fuchs et al. 1986), keratinocytes (Reichert 1986), fibroblasts (Pätel et al. 1981), yeast and ascites cells (Raab and Patermann 1961), and skin (Leonhardi 1961). In keratinocytes, respiration is inhibited at lower anthralin concentrations than is thymidine incorporation. Thus, the antirespiratory effect of anthralin is suggested to be the main reason for its antipsoriatic activity (Reichert 1986). Anthralin-mediated oxidant injury of the mitochondria (Fuchs et al. 1986, 1990e) may be involved in its mode of action. The mitochondrial ATPase/synthetase complex is highly susceptible to oxidative injury (Hyslop et al. 1988). Electron microscopic studies of typical plaques from patients with psoriasis have revealed that keratinocyte mitochondria show an abnormal pathomorphological reaction in response to local treatment with anthralin (Klug and Schulze 1987; Klug et al. 1988).

Modulation of skin phagocyte activity and arachidonic acid metabolism may represent further important features that contribute to its clinical efficiency. Anthralin enhances the extracellular release of reactive oxidants by activated neutrophils in vitro. Intracellular generation of reactive oxygen species in activated neutrophils, however, is inhibited by anthralin (Anderson et al. 1987), probably due to autoinactivation of the neutrophil respiratory burst by high concentrations of extracellular oxidants (Edwards and Swan 1986). Two distinct dose-related prooxidative interactions of anthralin with human neutrophils are found, a low-dose priming and a high-dose activation of oxidant generation (Anderson 1989b). It is suggested that anthralin does not directly activate the membrane-associated oxidative metabolism of neutrophils but induces hyperreactivity of these cells to receptor-mediated stimuli. Contrary to these results, anthralin is reported to inhibit in a dose-dependent manner superoxide anion generation in stimulated neutrophils in vitro (Schröder et al. 1985).

VI. Organic Gold Compounds

Gold in elemental form has been used for centuries as an antipruritic to relieve the itching palm. In more modern times, the observation by Robert Koch in 1890 that gold inhibited *Mycobacterium tuberculosis* in vitro led to trials with arthritis and lupus erythematosus, thought by some to be tuberculous manifestations. However, the clinical efficacy of gold compounds, such as auranofin in the treatment of patients with SLE was only modest (Weisman et al. 1983). In contrast, results of treatment of patients with psoriatic arthritis (Dorwart et al. 1978) and pemphigus vulgaris with gold thiomalate and gold thioglucose (Penneys et al. 1976; Poulin et al. 1984) were rather promising. The topic of chrysotherapy in dermatology has been reviewed by Penneys (1979).

Three organic gold compounds are of interest in chrysotherapy: the hydrophilic aurothiomalate and aurothioglucose and the lipophilic, orally active auranofin (Fig. 4.11).

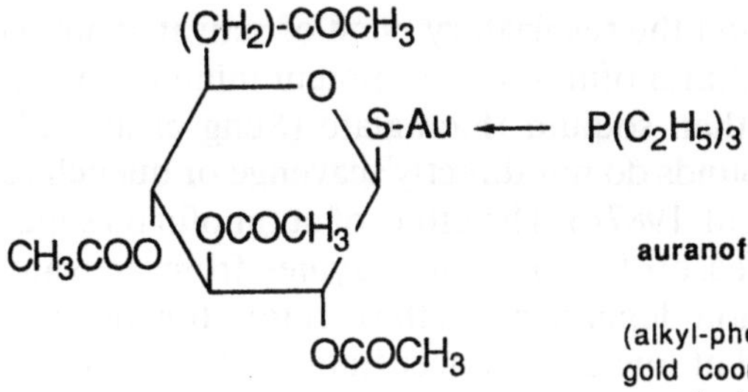

auranofin

(alkyl-phosphin gold coordination complex)

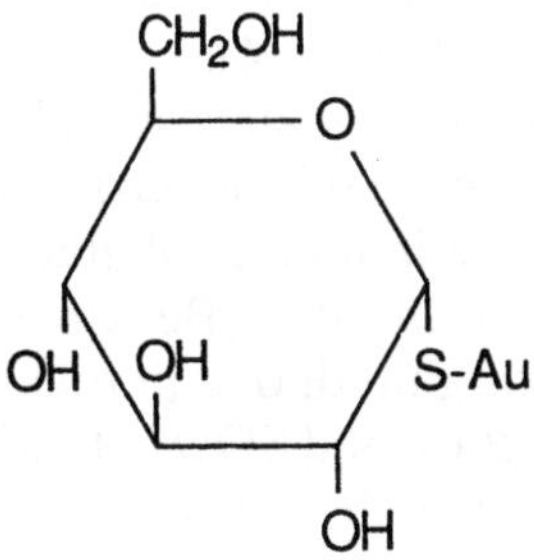

1-auro-D-thioglucose

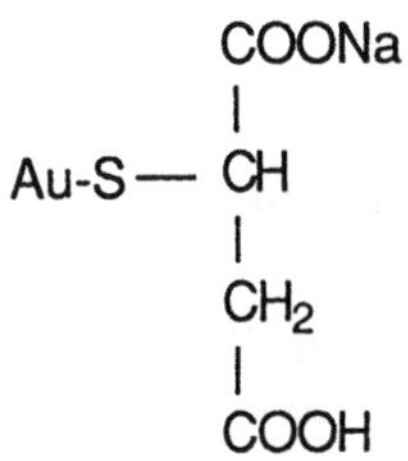

sodium aurothiomalate

Fig. 4.11. Organic gold compounds

Multiple studies have been conducted to characterize the biological activity of gold compounds. Qualitative and quantitative differences in the preclinical pharmacological profiles of these compounds have been analyzed (Walz et al. 1983).

Auranofin is more effective than gold thiomalate and gold thioglucose in suppressing inflammation and humoral as well as cell-mediated immunity. Auranofin has anti-inflammatory activity as evidenced by the inhibition of carrageenan-induced paw edema in rats and UV-induced erythema in guinea pigs. It possesses cytostatic activity as evidenced by the decreased synthesis of nucleic acids and proteins, partially inhibits antibody production, reduces elevated serum IgA, IgM, and IgG levels, and is a potent inhibitor of antibody-dependent, cell-mediated cytotoxicity. Furthermore, it reduces the chemotactic activity of monocytes, is a potent inhibitor of lysosomal enzyme release, and suppresses the generation of superoxide anion radicals of activated neutrophils (Crooke and Mirabelli 1983; Chaffman et al. 1984).

Organic gold compounds inhibit the respiratory burst of activated animal and human neutrophils in vitro. Auranofin is a more potent inhibitor of oxidant generation in neutrophils than sodium thiomalate (Sung et al. 1984; Roisman et al. 1983). The compounds do not directly scavenge or quench reactive oxygen species (Miyachi et al. 1987 c). The effect of auranofin on superoxide anion production and release of lysosomal enzymes from activated human neutrophils is concentration-dependent; both secretory functions are suppressed at high and increased at low drug concentrations. In neutrophils from patients treated with auranofin for several weeks, the chemiluminescence is enhanced, thus casting doubt on the clinical significance of the decreased superoxide production in vitro (Chaffman et al. 1984).

Gold thioglucose increases the toxicity of lipid peroxidative stimuli by the inhibition of microsomal and cytosolic defense mechanisms. Organic gold compounds like auranofin and gold thioglucose are inhibitors of glutathione peroxidase, glutathione-S transferase and catalase in vitro (Beckmann and Greene 1988; Mercurio and Combs 1985) and of glutathione peroxidase in vivo (Baker and Tappel 1986). Glutathione reductase and SOD activities are not influenced by gold thioglucose (Mercurio and Combs 1985).

In view of the multiple effects of gold compounds on enzyme activities and other cellular functions, it is not possible to draw definitive conclusions about their in vivo mechanism of action.

VII. Glucocorticosteroids and Nonsteroidal Antiphlogistic Drugs

Kendall, Hench, and Reichstein were awarded the Nobel Prize in 1950 for discovering the hormones of the adrenal cortex. Hydrocortisone was first used as a topical agent by Sulzberger and Witten in 1952 (Sulzberger and Witten 1952) and can be considered a major breakthrough in dermatologic therapy. Glucocorticosteroids modify the functions of epidermal and dermal cells as well as inflammatory cells. Their action via binding to cytoplasmic steroid receptors and subsequent modulation of nuclear transcription (protein synthesis) results in actions on arachidonic acid metabolism and IL-1 formation. These reactions contribute to its immunosuppressive, antimitogenic, and antiinflammatory effects but do not explain all their effects.

Glucocorticosteroids inhibit superoxide anion radical production in neutrophils stimulated by phorbolesters (Belman and Troll 1972) and induce a suppression of reactive oxygen species generation and release of lysosomal enzymes from neutrophils of patients with SLE (Rister et al. 1984). The inhibiting effect of glucocorticosteroids on the respiratory burst may be related to their inhibitory effect on phospholipase A_2 (Stacey and Klaassen 1982). Glucocorticosteroids are considered to possess an antioxidant potential and inhibit free radical reactions (Hess and Manson 1983; Probskina and Matyuskin 1982). However, their antiinflammatory activity is probably not related to their weak in vitro free radical scavenging properties (Puig-Parellada and Planas 1978).

Several nonsteroidal antiphlogistic drugs inhibit the respiratory burst of activated neutrophils; however, an opposite stimulating effect has also been described (Spisani et al. 1984; Sung et al. 1984). Non-steroidal antiphlogistic agents such as indomethacin and niflumic acid possess free radical scavenging activity in an in vitro assay (Puig-Parellada and Planas 1978); a SOD-like effect of their copper complexes has been proposed (Lengfelder 1984). The copper(II) complex of nonsteroidal antiphlogistic drugs like D-penicillamine, indomethacin, acetylsalicylic acid, and piroxicam has SOD-like activity (Betts et al. 1984; Lengfelder 1984; Sorensen 1984), but the clinical significance is not known.

VIII. Tetracyclines

Tetracyclines have a broad use in topical as well as in internal dermatologic therapy. They are used in the treatment of acne vulgaris, perioral dermatitis, rosacea, cheilitis granulomatosa, and other skin diseases. Their therapeutic efficacy in dermatoses not associated with microorganisms indicates that the compounds also have a paraantibiotic effect. It was suggested that the mode of action of many antibiotics mimics some of the natural defenses of phagocytizing cells in which reactive oxygen species are produced during the respiratory burst (Ames et al. 1986). Reactive oxygen species produced by activated neutrophils, eosinophils, and monocytes are involved in the killing of parasites and bacteria. The pharmacodynamic mechanism of various antimicrobial and antiparasitic drugs involves the formation of reactive oxygen species.

Prooxidative effects of tetracyclines may be mediated via the formation of charge transfer complexes with target molecules (e.g., DNA), (Ames et al. 1985). Tetracyclines may also initiate lipid peroxidation, DNA damage, and carbohydrate degradation, presumably by chelating transition metal ions (Rees and Slater 1988; Quinlan and Gutteridge 1988).

However, there is ample evidence that they also have *antioxidant properties.* The antibiotics tetracycline (Fig. 4.12), oxytetracycline, minocycline, and erythromycin reduce the generation of reactive oxygen species from activated neutrophils. Antibiotics such as cephalexin, penicillin G, streptomycin, and chloramphenicol do not have any inhibitory effect. It was suggested that the antioxidant effect of antibiotics does not stem from their capability to scavenge reactive oxygen species directly but rather from their effect on neutrophil function.

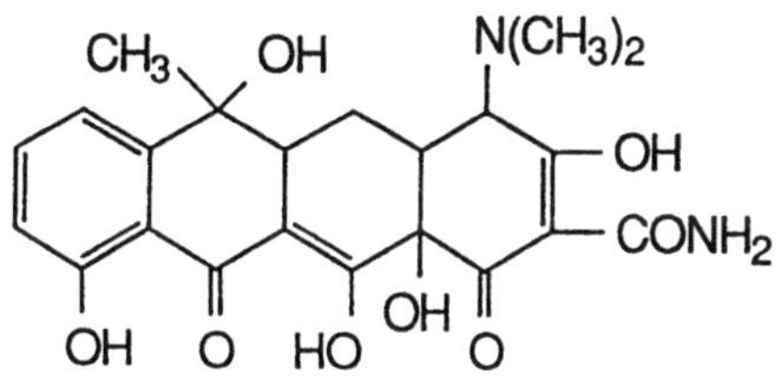

Fig. 4.12. Tetracycline

However, tetracycline, oxytetracycline, and minocycline inhibit the formation of hydrogen peroxide in the xanthine oxidase/hypoxanthine system, indicating some quenching activity (Miyachi et al. 1986d). Tetracyclines are suggested to be potent scavengers of superoxide anion radicals (Van Baar et al. 1987b).

IX. Metronidazole

Metronidazole, a nitroimidazole derivative, is used in the topical treatment of rosacea and acne vulgaris and as an antimicrobial drug in dermatology (Fig. 4.13).

Electrochemical data on antiprotozoal and antihelminthic agents including metronidazole indicate that production of reactive oxygen species via electron transfer processes may be significantly involved in the microbial killing (Ames et al. 1987a, b; Kovacic et al. 1989).

Protozoal Metabolism. Reductive metabolism of nitro heterocyclic drugs such as metronidazole involves the formation of nitro anion radicals as the first reduction product. This radical is readily oxidized by molecular oxygen with regeneration of the nitro group and formation of superoxide anion radical (Fig. 4.14). Generation of reactive oxygen species by metronidazole is therefore a possibility (Docampo and Moreno 1984). The nitro anion radicals of metronidazole and nifurtimox have been identified by EPR spectroscopy in protozoan parasites. Whether the nitro anion radical species contribute to the parasiticidal activity of these drugs is not known. Further reduction via a nitroso and nitroxide compound results in the formation of a hydroxylamine

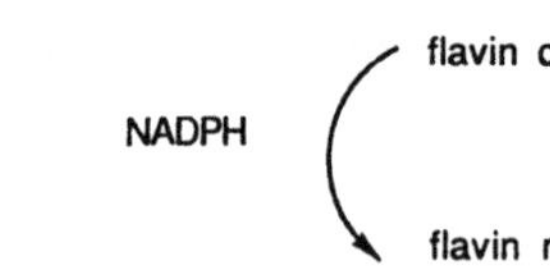

Fig. 4.13. Metronidazole

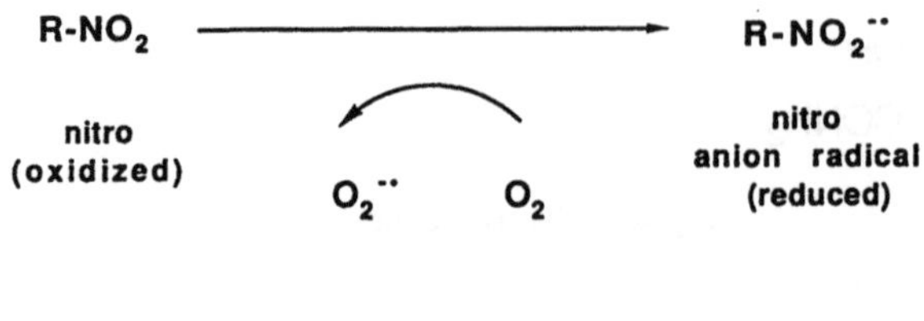

Fig. 4.14. Nitro anion radical formation

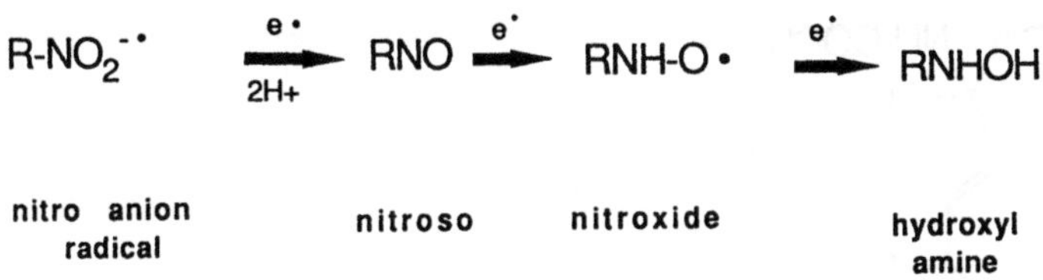

Fig. 4.15. Formation of a hydroxylamine

(Fig. 4.15). The hydroxylamine may readily autoxidize back to the nitroxide, resulting in redox cycling, thereby depleting cellular reducing equivalents.

Mammalian Cell Metabolism. The mode of action of metronidazole in normoxic mammalian cells is not well understood, but it seems feasible that it also undergoes reductive metabolism (Baines and McFadzean 1985). Certain hypoxic cell sensitizers used in chemotherapy, e. g., the nitroimidazole misonidazole, are reductively activated in hypoxic mammalian cells to a nitro anion radical. The nitro anion radical of misonidazole undergoes one-electron redox cycling, thereby generating reactive oxidants, which may be responsible for cytotoxicity. Mechanisms of action other than the generation of reactive oxidants have also been proposed. It was suspected that the hydroxylamine metabolite of metronidazole forms a complex with DNA, which results in the inhibition of nucleic acid polymerases.

Antioxidant properties of metronidazole may derive from its inhibitory effect on the generation of reactive oxygen species in activated neutrophils. Metronidazole does not scavenge reactive oxidants directly (Miyachi et al. 1986c). In the presence of palmitoleic acid (C16:1, cis 9), metronidazole inhibits anaerobic growth of *Propionebacterium acnes* and reduces neutrophil-generated reactive oxidants (Akamatsu et al. 1990a). Metronidazole alone does not affect either neutrophil phagocytosis or chemotaxis, and does not inhibit growth of *Propionebacterium acnes* in vitro. The slight inhibition of neutrophil-derived reactive oxidants (Miyachi et al 1986c) could not explain the marked effectiveness of metronidazole in treatment of acne vulgaris and rosacea. The synergistic effect of skin free fatty acids may, however, explain, at least in part, the antiinflammatory action of metronidazole in vivo (Akamatsu et al. 1990a).

X. Colchicine

Colchicine is an alkaloid with antiinflammatory effects (Fig. 4.16). It is used in the treatment of several diseases, particularly in Behçet's disease, recurrent aphthosis, and vasculitis allergica, and is very effective (Sterry 1987). Other diseases treated include: acute familial Mediterranean fever, acute neutrophil dermatosis (Sweet's syndrome), pustulosis subcornealis (Sneddon-Wilkinson

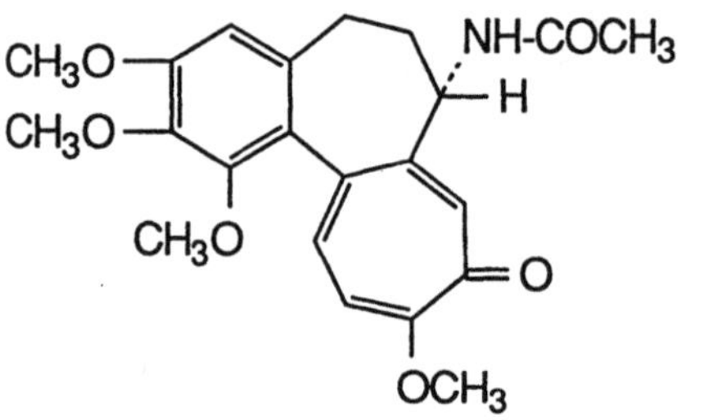

Fig. 4.16. Colchicine

syndrome), pustular psoriasis, and pyoderma gangrenosum. Colchicine suppresses the generation of reactive oxygen species in activated neutrophils (Miyachi and Niwa 1982; Niwa et al. 1984c) but does not scavenge reactive oxidants directly (Niwa et al. 1984c). Further mechanisms of action comprise inhibition of neutrophil and monocyte chemotaxis, inhibition of secretion of neutrophil lysosomal enzymes, and suppression of excretion of procollagen from fibroblasts (Sterry 1987). It is suggested that colchicine exerts its antiinflammatory effect through these different effects, including antioxidant properties, thus conferring protection from autooxidative tissue injury.

XI. Dapsone

Fromm and Wittmann discovered dapsone in 1908; its antiinflammatory potency was reported in 1937. In the early 1940s, its antimycobacterial activity was discovered, which finally led to its clinical application in the treatment of leprosy. Dapsone (4,4′-diaminodiphenylsulfone) is still used in the treatment of leprosy and malaria, and it probably acts in a similar way to sulfonamides by inhibiting folic acid synthesis. It is the drug of choice for dermatitis herpetiformis. The effectiveness of sulfones and sulfonamides in the treatment of dermatitis herpetiformis is not based on their antibacterial effects (Lorincz and Pearson 1962). The therapeutic activity of dapsone in dermatitis herpetiformis does not extend to other sulfonamides, except sulfapyridine. In dermatitis herpetiformis, polymorphonuclear leukocytes infiltrate the papillary dermis, and it is believed that dapsone may act through its modulation of neutrophil function.

Dapsone is successfully used in the treatment of various skin diseases involving vasculitis, chronic inflammatory attack on connective tissue, and infiltration with neutrophil granulocytes: pustular psoriasis, herpes gestationis, bullous pemphigoid, linear IgA disease, cutaneous vasculitis, pyoderma gangrenosum, relapsing polychondritis, erythema elevatum diutinum, pustulosis subcornealis (Sneddon-Wilkinson's syndrome), and acute febrile neutrophilic dermatosis (Sweet's syndrome). However, in these cases the response to dapsone is not as complete and rapid as in dermatitis herpetiformis. In bullous pemphigoid and pemphigus vulgaris, neutrophils do not form a prominent feature, and the two diseases do not generally respond well to dapsone. Cases which do respond may be attributed to an overlap syndrome with greater neutrophil involvement (Person and Rogers 1977). Sulfones are also used effectively in treating infantile acropustulosis (Duvanel and Harms 1988) and urticarial vasculitis (Wozel and Thiele 1987). It is feasible to suggest that the clinical efficiency of dapsone in the treatment of these distinct skin diseases is based on a common pathogenetic mechanism which is influenced by the drug. The neutrophil is of significant pathogenetic importance in many of these cutaneous disorders, and dapsone's efficacy in these clinically distinct and diverse skin diseases may reflect an action on this particular cell type.

Dapsone inhibits neutrophil-mediated oxidative inactivation of α_1-antiprotease (Theron and Anderson 1985), phagocytosis-linked chemiluminescence (Webster et al. 1984), and leukocyte-mediated cytotoxicity (Molin and Stendahl 1980). It inhibits the formation of reactive oxygen species in activated neutrophils (Miyachi and Niwa 1982; Niwa et al. 1984c). A concentration-dependent reduction of hydroxyl radical and hydrogen peroxide formation in neutrophils is observed, but no effect has been noted on superoxide anion radical generation. Dapsone directly scavenges singlet oxygen, hydrogen peroxide, and hydroxyl radical, but not the superoxide anion radical (Niwa et al. 1984c). However, no interaction with hydrogen peroxide generation was reported in a different study (Stendahl et al. 1978). Dapsone also interferes with the myeloperoxidase-halide system and may thereby decrease neutrophil generation of reactive oxygen species (Wozel 1987). Neutrophil myeloperoxidase is competitively inhibited by dapsone (Stendahl et al. 1978).

The adverse effects of dapsone, hemolytic anemia and methemoglobin formation, are thought to be caused by its hydroxylamine derivative (Kramer and Glader 1972). It is suggested that dapsone or its metabolite oxidizes hemoglobin and glutathione, thereby inducing hemolysis (Rasbridge and Scott 1973). Its erythrocyte toxicity is believed to be due to metabolites formed during its microsomal oxidation (Israili et al. 1973). Dapsone is metabolized to *N*-acetyldapsone and *N*-acetyldapsonehydroxylamine (Fig. 4.17).

The hydroxylamine form may account for up to one-third of the excreted drug and spontaneously oxidizes to a nitroxide radical. The nitroxide radical can easily be reduced by various endogenous reductants, and as a consequence redox cycling and consumption of reducing equivalents can occur. Therefore, the hydroxylamine metabolite of dapsone can produce oxidative injury in catalytically small amounts. Lipid peroxidation of erythrocyte cell membranes was detected in patients treated with dapsone (Goldstein and McDonagh 1976). The hydroxylamine metabolite induces the generation of superoxide anion and hydrogen peroxide in vitro (Weetman et al. 1980) and causes intracellular hydrogen peroxide formation in erythrocytes (Glader and Conrad 1973), which is associated with methemoglobin formation and red cell lysis. Hydrogen peroxide can form a complex with methemoglobin (peroxide-ferriheme complex) which has potent oxidant activity. Hemoglobin oxidation and red cell lysis induced by the intracellular formation of hydrogen peroxide

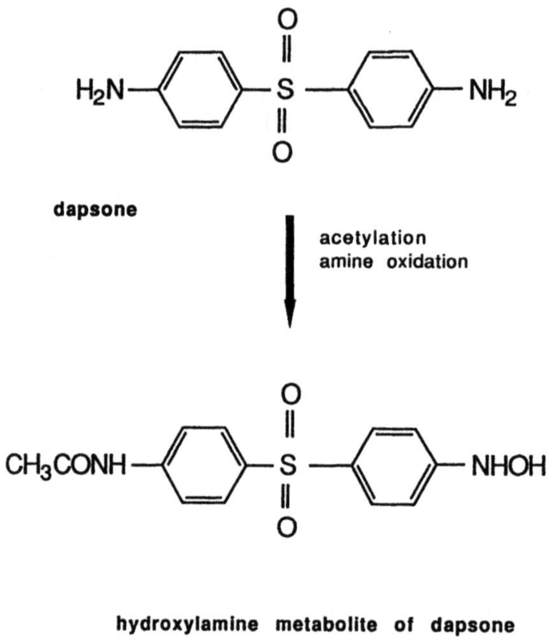

Fig. 4.17. Metabolism of dapsone

is also reported with the 8-aminoquinoline derivate primaquine, but not from the nonhemolytic 4-aminoquinoline derivate chloroquine (Cohen and Hochstein 1964; Summerfield and Tudhope 1978). Prooxidative effects of dapsone may also be induced via the formation of charge transfer complexes of the sulfone with target molecules (e.g., hemoglobin, DNA) (Ames et al. 1985).

The pro- and antioxidant properties of dapsone and its metabolites make it difficult to extrapolate to its in vivo mechanism of action. It was recently pointed out, however, that the mode of action of dapsone may not only be based on its effects on neutrophil function but also on certain cellular and humoral immune mechanisms, membrane phospholipid metabolism, and the complement cascade (Wozel 1988).

XII. Clofazimine

Clofazimine is a phenazine iminoquinone and has antibacterial, immunomodulating, and antiinflammatory properties (Fig. 4.18). It is used in the treatment of dermatosis such as chronic cutaneous lupus erythematosus, lichen ruber, granuloma annulare, pyoderma gangrenosum, pustular psoriasis, multibacillar and dapsone-resistant paucibacillar leprosy, leprosy reaction type I, and Melkerson-Rosenthal syndrome.

Its antiinflammatory properties are attributed to the inhibition of lymphocyte transformation and neutrophil migration.

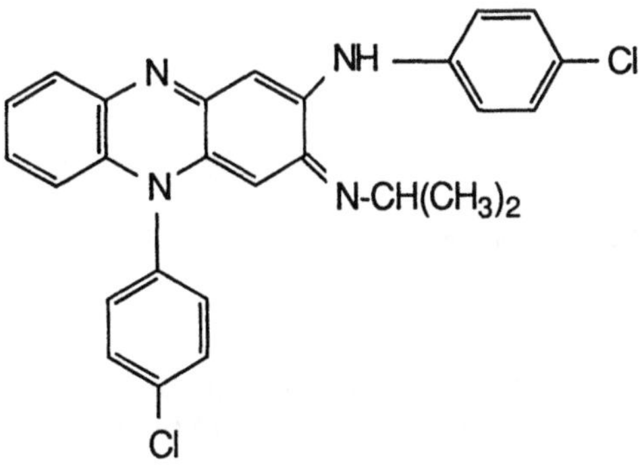

Fig. 4.18. Clofazimine

It readily binds to DNA (Morrison and Marly 1976) and possibly facilitates charge transfer reactions, thereby enhancing the generation of reactive oxygen species. It stimulates activated human neutrophils to generate increased amounts of reactive oxygen species. It does not activate the membrane-associated oxidative metabolism of neutrophils but rather primes the cells to hyperreact to various stimuli. Oral clofazimine administration in patients with lichen ruber planus and discoid lupus erythematosus augments the generation of reactive oxygen species in activated neutrophils (Anderson and Zeis 1988). Clofazimine also increases the formation of reactive oxidants in neutrophils from patients with leprosy (Niwa et al. 1984b). Impaired neutrophil phagocytic activity of patients with pustulosis palmaris et plantaris is improved in about 80% upon oral administration of clofazimine. This correlates with improvement of the clinical situation (Molin 1975). In cases of successfully treated patients with generalized pustular psoriasis, enhancement of phagocytic activity as well as clofazimine-DNA binding are discussed as possible reaction mechanisms (Chuaprapaislip and Piamphongsant 1978).

XIII. Thalidomide

Thalidomide is a glutamic acid derivative and is used in the treatment of leprosy reaction type II (erythema nodosum leprosum), prurigo nodularis Hyde, actinic prurigo, discoid lupus erythematosus, polymorphous light eruption, and recurrent aphthosis (Fig. 4.19). Favorable effects of the drug in the treatment of erythema nodosum leprosum and other dermatosis may be partly attributable to the same mechanism (Miyachi et al. 1982a). Pharmacodynamic qualities of action include immunomodulating and antiinflammatory properties (Hunziker and Krebs 1983; Peyron and Meynadier 1986). Phagocytosis of human macrophages is increased at low drug concentrations, which has no effect upon macrophage chemiluminescence. At high concentrations, the chemiluminescence of human macrophages is significantly reduced, while the chemiluminescence of human neutrophils is not influenced (Barnhill et al.

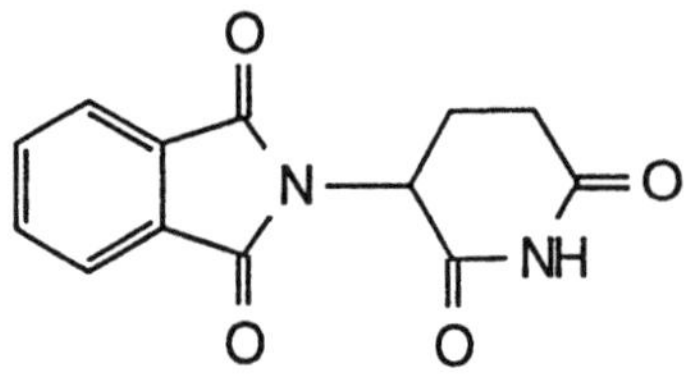

Fig. 4.19. Thalidomide

1984). However, in another study, thalidomide is reported to suppress the generation of reactive oxygen species in activated neutrophils (Miyachi et al. 1982a).

XIV. Iodine

The nonendocrine pharmacological effects of iodine have been analyzed (Stone 1971). Iodine is not a general antiinflammatory agent. It can intensify the acute inflammatory response in experimental animals (Stone and Willis 1967), but it inhibits granuloma formation (Mielens et al. 1968). It is used clinically in the treatment of erythema nodosum and Sweet's syndrome.

Potassium iodine suppressed the generation of reactive oxygen species in activate neutrophils. It might exert its antiinflammatory effects by interfering with the neutrophil-dependent production of activated oxidants, thus conferring protection from autooxidative tissue injury (Miyachi and Niwa 1982). During reductive metabolism of PGG_2 to PGH_2, oxidizing equivalents are released that inhibit the cyclooxygenase reaction (Egan et al. 1976). Iodine is thought to scavenge the reactive oxidant released during the reduction of PGG_2 to PGH_2, thereby modulating prostaglandin biosynthesis (Egan et al. 1978). Iodine stimulates the enzymatic oxygenation of arachidonic acid, increases the reduction of PGG_2 to PGH_2, quenches the electron paramagnetic resonance signal during oxidation of arachidonic acid, and depresses mouse ear edema (Egan et al. 1978). DNA damage caused by lipid peroxidation products is inhibited by potassium iodine, through its reaction with organic hydroperoxide and the subsequent formation of elemental iodine and an organic alcohol (Ueda et al. 1985). In the treatment of erythema nodosum, iodine-mediated modulation of heparin release from mast cells is proposed as contributing to the antiinflammatory action of the drug (Schulz and Whiting 1976). Recently, it was demonstrated by the modified Boyden chamber method that iodide inhibits neutrophil chemotaxis in human subjects. It was suggested that the therapeutic effect of potassium iodide on erythema nodosum and Sweet's syndrome might be mediated through the inhibition of neutrophil chemotaxis (Honma et al. 1990).

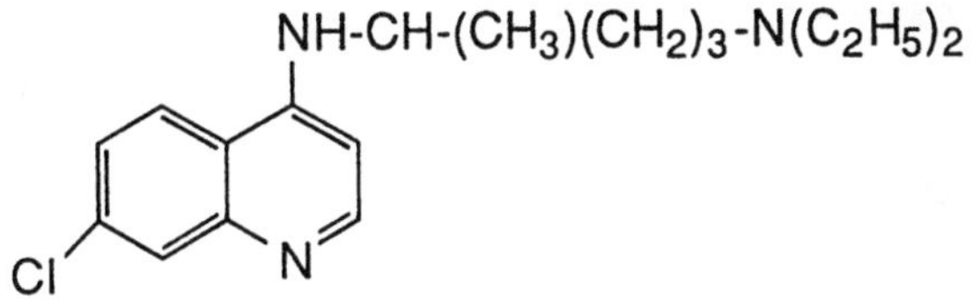

Fig. 4.20. Chloroquine

XV. Chloroquine

Atebrin (quinacrine), a 9-aminoacridine derivative, was synthesized in 1928 and is considered the first clinically useful synthetic antimalarial agent. Later, it was replaced by the less toxic chloroquine, a 4-aminoquinoline derivative (Fig. 4.20). Antimalarial drugs have antiinflammatory and immunosuppressive effects and are used in the treatment of a variety of photodermatoses, such as polymorphic light eruption, discoid lupus erythematosus, and porphyria cutanea tarda (Peyron and Meynadier 1986).

4-Aminoquinolines may work via at least three different modes of action. At low concentrations (10^{-7} M) they form a plasmodium cytotoxic complex with ferriprotoporphyrin IX, a degradation product of plasmodium-digested hemoglobin. This complex readily promotes the peroxidation of microsomal phospholipids, and it is suggested that this could explain the antimalarial action of chloroquine (Sugioka et al. 1987). Malaria parasites are considered to be quite vulnerable to oxidative stress, and many antimalarial compounds are believed to act through oxidative injury of the parasite (Clark and Cowden 1985). At concentrations of 10^{-6} M, the drug is lysosomotropic and may influence lysosomal function. At even higher concentrations (10^{-4} M), unspecifically binds it to various cellular constituents (melanin, enzymes, phospholipids, nucleic acids) (McChesney and Fitch 1984). At this last level, the pharmacological effects of 4-aminoquinolines may be mediated by the generation of reactive oxidants that arise via a charge transfer complex with target molecules (e.g., DNA) (Ames et al. 1985).

XVI. Flavonoids

Flavonoids are benzo-y-pyrone derivatives, and for centuries plant preparations containing them have been used to treat various human diseases. They comprise a large group of naturally occurring drugs with broad pharmacological activities (Havsteen 1983). In the early 1930s their vitamin C sparing activity was discovered (Clemetson and Andersen 1966), and since then the antioxidant effects have been of interest. In dermatology, plant products containing flavonoids are used as antiinflammatory drugs. The topical antiphlogistic properties of *Chamomilla recutita* extracts can be attributed to lipophilic compounds such as (–)-α-bisabolol and to hydrophilic flavones, e.g.,

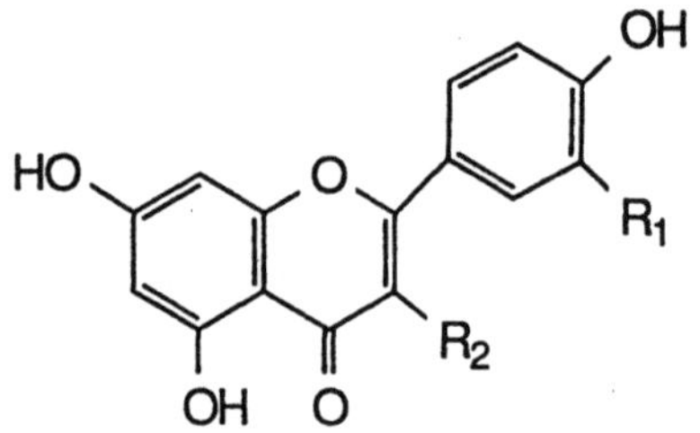

	R_1	R_2
apigenin	H	H
quercetin	OH	OH

Fig. 4.21. Flavones

apigenin, the hydrophilic flavones contributing most to the antiphlogistic effect, because the lipophilic compounds are present only in minute amounts and are less potent. The efficacy of the different flavones in inhibiting croton oil dermatitis is in the following order: apigenin > quercetin > myricetin > rutin (Fig. 4.21). *Apigenin* was even more effective than indomethacin and phenylbutazone (Della Loggia 1985).

Several distinct effects may contribute to the *antiphlogistic properties* of flavones: inhibition of arachidonic acid metabolism (inhibition of phospholipase, cyclooxygenase, and lipoxygenase), inhibition of histamine release (inhibition of calcium influx into mast cells, membrane stabilisation), and scavenging of reactive oxidants are possible mechanisms of action. Flavonoids can directly scavenge reactive oxygen species and also inhibit their production in neutrophils. Furthermore, they modulate the release of lysosomal enzymes. Apigenin inhibits the peak neutrophil chemiluminescence response to opsonized zymosan and suppresses the release of β-glucuronidase (Busse et al. 1984). It was shown that the scavenging properties of flavones did not correlate with their influence on cyclooxygenase activity (Baumann et al. 1980). The flavolignan complex *silymarin,* which is composed of silybin, silidianin and silichristin, is a hepatoprotective drug used in the treatment of alcohol and mushroom (phalloidin) hepatotoxicity. The antioxidant activity of flavonoids may be one of the important factors in the hepatoprotective activity.

Chromoglycinic acid, a synthetic flavonoid derivative (Fig. 4.22), is used in the prophylactic treatment of allergic mucocutaneous reactions. It is suggested that inhibition of the proton ATPase of mast cells is the key mechanism by which flavonoids decrease mast cell secretion. A number of flavonoids, e.g., quercetin, inhibit antigen-induced histamine release from sensitized mast cells (Fewtrell and Gomperts 1977a), and it was assumed that quercetin inhibits calcium influx into mast cells (Fewtrell and Gomperts 1977b). The inhibitory effect of flavonoids on neutrophil secretion is sup-

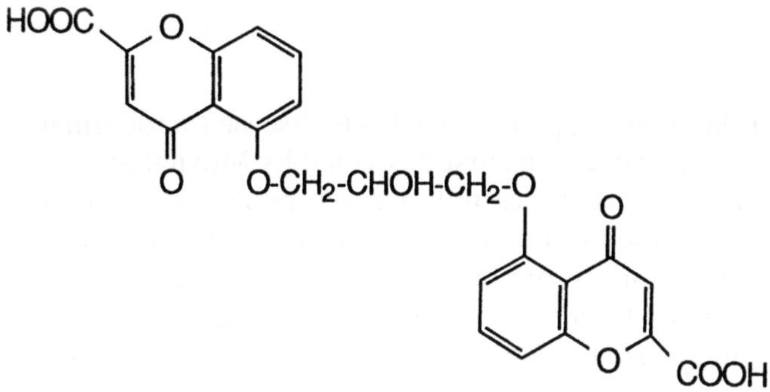

Fig. 4.22. Cromoglycinic acid

posed to be due to a mechanism different from inhibition of proton ATPase or calcium influx (Bennett et al. 1981), and it was hypothesized to be related to its antioxidant activity. As a corollary, this conclusion could be partially extrapolated to chromoglycinic acid, but no data on its antioxidant activity are available.

Quercetin is an antioxidant and inhibits lipid peroxidation of mitochondria and microsomes in vitro (Bindoli et al. 1977; Cavallini et al. 1978). It was proposed that flavonoids are effective scavengers of reactive oxygen species (Baumann et al. 1980; Takahama 1985, 1987). However, the specific scavenging activity of hydroxyl (Husain et al. 1987b), superoxide anion (Robak and Gryglewski 1988), and peroxyl (Torel et al. 1986) radicals may have been an overinterpretation of the data. Flavonoids react rapidly with hydroxyl radicals, although the very effective flavonol radical scavenger quercetin has only low rate constants (Bors et al. 1990), with superoxide anion radical. In contrast to the hydroxyl radical, superoxide is, however, not ultimately responsible for severe biological damage, due to its generally low chemical reactivity. The biological significance of free radical scavenging by flavonoids needs further evaluation, and may be a promising concept in understanding molecular mechanisms of phytopharmacology.

Curcumin, a phenolic antioxidant product of plant origin, has been used in India for centuries (Chandra 1989) to protect skin from UV damage and to prevent skin aging (Fig. 4.23).

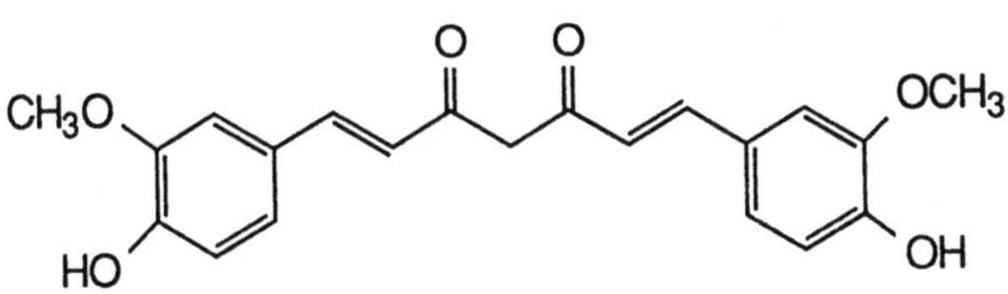

Fig. 4.23. Curcumin

XVII. Zinc

The link between an inflammatory, pluriorificial skin disease (acrodermatitis enterohepatica) and zinc deficiency was first described by Moynahan (1974). Acrodermatitis enterohepatica is successfully treated by zinc supplementation. The role of zinc in skin diseases has been reviewed recently (Dreno et al. 1988). Skin inflammation of various origins, e.g., sunburn reaction, anthralin dermatitis, and perioral dermatitis, is frequently treated topically with zinc preparations. It permeates intact human epidermis and dermis from topically applied zinc oxide (Ågren 1990).

Zinc is a cofactor in over 100 metalloenzymes in different species. It is an essential cofactor in RNA and DNA polymerase, alcohol dehydrogenase, and SOD (Riordan and Valle 1976). Human epidermal and dermal SOD activities are markedly enhanced by copper and zinc ions in vitro (Kim and Lee 1987). Besides its role as an enzyme cofactor, zinc is of significant importance in the immune system (Norris 1985; Fakter et al. 1986).

Skin Zinc and Plasma Zinc: A Correlation? In acrodermatitis enterohepatica, there is a good correlation between depressed plasma zinc levels and low alkaline phosphatase activity but a poor correlation with erythrocyte carbonic anhydrase activity (Neldner and Hambidge 1975); the correlation with SOD activity is not known. This and other reports (Huber and Gershoff 1973) indicate that correlation of zinc deficiency with enzyme activities depends on the particular enzyme studies. Plasma zinc values are not a good index of the tissue zinc level. Plasma zinc is almost entirely bound to transport proteins, and its tissue availability may be regulated by multiple factors. In patients with acne vulgaris, psoriasis, and Darier's disease there is a significant decrease in the epidermal zinc concentration; however, serum zinc levels are normal. Only in patients with dermatitis herpetiformis was a correlation found of decreased epidermal zinc level with decreased serum zinc concentrations (Michaelson and Ljunghall 1990).

Zinc protects the functional integrity of biomembranes (Chvapil 1976), and a particular role for it in the metabolism of free radical species has recently been discussed (Anonymus 1978; Koukay et al. 1987; Wilson 1988). In rats intoxicated with carbon tetrachloride, treatment with zinc reduces lipid peroxidation (Chvapil et al. 1973) and inhibits oxidative metabolism of drugs in liver microsomes (Chvapil et al. 1975). Carbon tetrachloride-induced collagen accumulation in rat livers is also decreased by zinc (Anttinen et al. 1984). The physiological role of zinc as an antioxidant by protecting thiol groups against oxidation and by the inhibition of metal-ion-mediated generation of reactive oxygen species.

Interaction with Transition Metal Ions. Redox inactive metals, such as zinc, which possess a high similarity in their ligand chemistry to that of copper or iron but are of relative low toxicity by themselves, may be useful in the intervention of transition-metal-mediated, free-radical-induced pathological processes. The endogenous free radical production is increased in lung microsomes as a function of dietary zinc deficiency but is not influenced by copper status (Kubow et al. 1986). In *Escherichia coli,* zinc protects against copper-mediated paraquat toxicity. The mechanism presumably involves the displacement of the redox active copper from its binding site, thus diverting the site of repeated production of free radicals (Korbashi et al. 1988). Zinc inhibits the Fenton reaction in vitro, but this effect does not occur in plasma (Favier et al. 1988). Apparently, this does not play a fundamental role in the extracellular space in vivo.

Interaction with the Glutathione System. An connection between the glutathione system and zinc metabolism has been established in rats (Hsu 1982). Zinc deficiency can have significant effects on the glutathione concentration in various tissues and the activity of glutathione-related enzymes. In zinc-deficient animals the glutathione concentration in erythrocytes and adrenal glands was significantly decreased, while in the liver and pancreas it was increased. In other soft tissues investigated, however, no differences were found. The activity of liver glutathione peroxidase and glutathione reductase were unaffected by zinc deficiency. The results of this study indicate that zinc deficiency causes a compensatory enhancement of glutathione biosynthesis (Hsu 1982).

XVIII. Benzoyl Peroxide

As early as 1905, benzoyl peroxide (Fig. 4.24) was suggested to possess wound healing properties (Loevenhart 1905). It is a strong oxidant that is topically used in the treatment of acne vulgaris, venous ulcers, and burn wounds. It has a broad spectrum antimicrobial activity (Kligman et al. 1977) and lowers the epidermal content of free fatty acids (Anderson et al. 1975).

It has been suggested that in the presence of copper ions, benzoyl peroxide can cause lipid peroxidation in rabbit dental pulp (Tsuyimoto et al. 1983; Terakado et al. 1984) and erythrocyte membranes (Tsuyimoto et al. 1988). Treatment of cultured human keratinocytes with benzoylperoxide causes lipid peroxidation only in the presence of metals ions or UVA (Artue and Kappus 1986). Therefore, it cannot be excluded that benzoyl peroxide treatment of patients causes skin lipid peroxidation. Lipid peroxidation may partially explain its antimicrobial activity and potential to induce skin irritation. Although benzoyl peroxide is a weak allergen, it is a strong irritant (Haustein et al. 1985). The irritant properties of commerical preparations impair their clinical use in some cases.

The low systemic toxicity of benzoyl peroxide is due to quick conversion of the peroxide in skin into benzoeic acid (Holzmann et al. 1979; Sahut et al. 1985). After topical application, the greatest amount of penetrating benzoyl peroxide is found in the stratum corneum, where biotransformation occurs; the stratum corneum also serves as a drug reservoir (Wepierre et al. 1986).

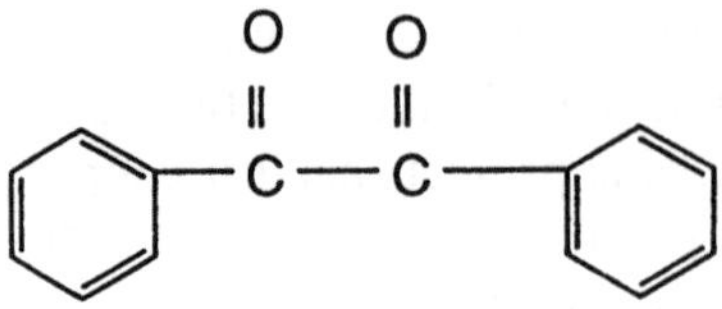

Fig. 4.24. Benzoyl peroxide

The drug was suggested to act in *acne vulgaris* though its slow release of oxygen, which acts on oxygen-sensitive anaerobic and microaerophilic organisms, suppressing the formation of free fatty acids. Its comedolytic effect may also be mediated via a direct sebosuppressive effect, by reducing the mitotic activity of sebaceous glands and gland secretion (Burkhart 1982). Inhibition of sebum production by benzoyl peroxide was, however, not confirmed in a another study (Puschmann 1982). Due to its tumor-promoting activity in animals, benzoyl peroxide therapy of acne was thought to be hardly justified in humans (Jones 1985). However, clinical reports of tumor induction during benzoyl peroxide treatment are not found in the literature (Gloor 1990).

Ulcer healing seems not to be primarily mediated through its antibacterial activity but rather by diret effects on wound healing, e.g., via activation of macrophages and angioneogenesis (Pace 1976).

XIX. Tetrachlorodecaoxide

Tetrachlorodecaoxide [chlor(I) oxide oxygen complex (4:1) hydrate] ($[Cl_4O_{10}] \times H_2O$) is clinically used to improve wound healing (Hinz et al. 1984, 1986), e.g., in the treatment of venous ulcers (Von Seebach and Stahl 1987). Being a chlorite oxygen complex, it is a weak oxidant, which is slowly decomposed by heme compounds to molecular oxygen and chloride ions. Heme-catalyzed decomposition of tetrachlorodecaoxide does not result in the formation of hydroxyl and superoxide anion radicals or singlet oxygen (Youngman et al. 1985). The oxidative property of tetrachlorodecaoxide is enhanced by reaction with heme compounds, forming unstable heme peroxides. These products are bactericidal and act as macrophage activators and vasodilators (Elstner 1988). However, it has been recently noted that tetrachlorodecaoxide may act as sodium chlorite ($NaOCl_2$) (Habermann and Müller 1989). Bactericidal activity, macrophage activation, vasodilation, and increased wound oxygen supply finally result in improvement of wound healing (Niinkoski et al. 1972; Knighton et al. 1983; Clark 1985; Reed and Clark 1985), although markedly reduced tissue oxygen concentration in venous ulcers are in general not incompatible with healing (Nemeth et al. 1989 a, b.

XX. Dimethylsulfoxide

The pharmacodynamic properties of dimethylsulfoxide include antiinflammatory activity, vasodilation, modulation of mesenchymal metabolism, and alteration of membrane permeability (Kligman 1965; Wood and Wood 1975). In addition, free radical scavenging properties are described (Ashwood-Smith 1975). Reports concerning topical dimethylsulfoxide therapy in systemic and localized scleroderma reveal contradictory results on its therapeutic effects (Baum 1975; Fleischmajer 1975). Topical dimethylsulfoxide in the therapy of cutaneous and systemic amyloidosis can be effective in some cases (Breathnach 1988). In veterinary medicine, it is used as a topical antiphlogistic drug.

XXI. Hyperbaric Oxygen

Hyperbaric oxygen has various biochemical effects on mammals, which are strongly dependent on oxygen tissue concentration. It affects the activity of several key enzymes and modulates acid-base balance and glucose, amino acid, and energy metabolism. Some of the physiological and most of the toxic effects of HBO are thought to be mediated by reactive oxygen species. Hyperbaric oxygen treatment is used in treatment of some dermatological conditions (Antonev 1986), including venous, arterial, diabetic, and decubitus ulcers, burns, other blister wounds, pyoderma gangrenosum, and toxic epidermal necrolysis. It is also applied clinically to improve the survival of free skin grafts and skin flaps (Fischer et al. 1988) and in treatment of scleroderma (Antonev et al. 1988). Antioxidants are frequently used in patients to prevent deleterious side effects of oxygen during hyperbaric oxygen treatment.

B. Photochemotherapy

Phototherapy of neonatal jaundice and uremic pruritus are well-established clinical modalities. Prolonged irradiation of jaundiced neonates with visible light results in the formation of hydrophilic bilirubin photoproducts which are more easily eliminated than the lipophilic bilirubin.

Photochemotherapy is defined as the combined use of UV or visible radiation and a drug, administered either topically or systemically. The history of photochemotherapy with psoralens may go back to 1550 B.C., as extracts of the plant *Ammi majus,* which contains furocoumarins, were used in the Nile delta for the treatment of leukoderma (Benedetto 1977). Some skin tumors were treated in the early twentieth century with eosin and light (Tanenier and Jesionek 1903). Photochemotherapy of herpes simplex in humans using neutral red was reported in 1973 (Felber et al. 1973). Parrish initiated PUVA (psoralen UVA) treatment of psoriasis (Parrish et al. 1974), starting a new area in dermatotherapy.

Phototoxicity is a central mechanism in the photochemotherapy of skin diseases. Epidermodermal lymphocytes and granulocytes are considered to be target cells in photo- and photochemotherapy of various dermatological disorders. Induction of normal protective mechanisms, alterations of metabolites, and killing of microorganisms are further possible mechanisms (Parrish 1981).

I. 8-Methoxypsoralen

Psoralens are extensively used in the photochemotherapy of cutaneous disorders like T-cell lmyphomas, mycosis fungoides, lichen ruber, psoriasis, vitiligo, urticaria pigmentosa, pityriasis lichenoides, and some photodermatoses (polymorphous light eruption, persistent light reaction).

The psoralens are benzopyran (coumarin) derivatives, and two types exist in nature: linear furocoumarins, in which a furan ring is linked to the 6,7 bond of benzopyron, and angular furocoumarins, in which the furan ring is joined to the 7,8 bond. The first furocoumarin, 5-methoxypsoralen, was isolated in 1834 from plant material. 8-Methoxypsoralen is a linear furocoumarin and was isolated from plants in the late 1940s; controlled clinical studies in the treatment of patients with vitiligo started in the 1950s (Fig. 4.25).

The major photobiological effects of psoralens were thought to be mediated through UVA-induced stable covalent photoaddition of psoralens to chromosomal DNA (reaction type I; Fig. 4.25) (Song and Tapley 1979). UVA-activated furocoumarin molecules react with the 5,6 double bond of pyrimidine bases of nucleic acids, thymines in particular, to produce C4 cycloaddition pro-

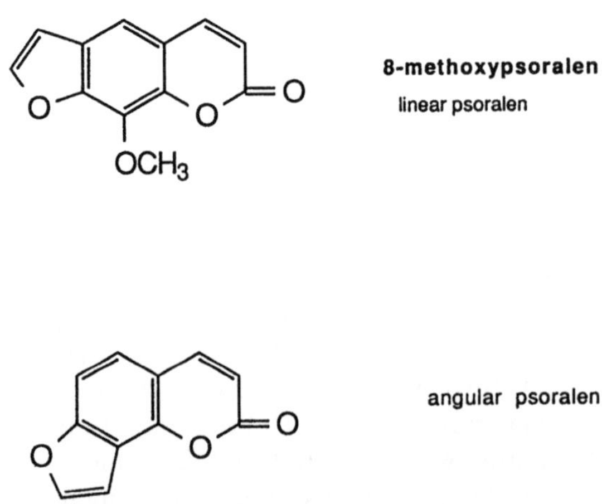

Fig. 4.25. Psoralens

ducts. This reaction does not require oxygen. With linear furocoumarins cycloaddition monoadducts and interstrand cross-links (bifunctional adduct) occur. With angular furocoumarins only monoadducts are obtained.

Psoralen-DNA cross-links were suggested to be essential for antipsoriatic effects. However, reactive oxygen species may also contribute to the antiproliferative effect.

Photodynamic Effects. In biological systems, photosensitization with psoralen was shown to be oxygen independent (Oginsky et al. 1959; Musajo and Rodighiero 1962). However, studies in the late 1960s indicated the involvement of free radical formation in psoralen photochemistry (Pathak et al. 1961). It is believed now that psoralens also undergo type II reactions in vivo (Fig. 4.26), in which the photoexcited psoralen reacts with oxygen and forms reactive oxygen species, e.g., singlet oxygen and superoxide anion (Joshi and Pathak 1983). The cell-killing effect of UVA-irradiated 8-methoxypsoralen is 2.5-fold increased in the presence of deuterated water, indicating the involvement of

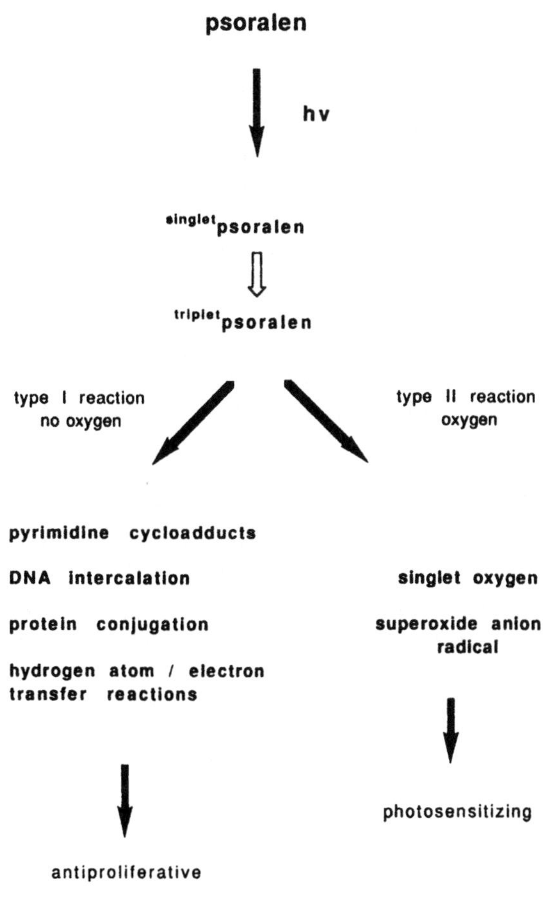

Fig. 4.26. Psoralen photochemistry

singlet oxygen (Meffert et al. 1982). Skin photosensitization by psoralens is significantly oxygen-dependent (Mobilio et al. 1990) and involves the generation of singlet oxygen, superoxide anion radical, and hydroxyl radicals (Pathak and Carraro 1987b). Singlet oxygen may cause lipid peroxidation (Wefers 1987). Indeed, PUVA therapy of guinea pigs increases the formation of lipid peroxidation products in skin and in serum; the changes are reversible after about 4 weeks (Akimov et al. 1984).

Membrane Effects. Furocoumarin/DNA photoadducts are detected in mammalian cell cultures and epidermis irradiated in vivo (Pathak et al. 1974; Musajo and Rodighiero 1972). It was suggested that UVA-activated 8-methoxypsoralen can bind to cellular proteins via intermediary free radicals and thus damage cellular membranes by induction of lipid peroxidation (Meffert et al. 1976b; Böhm and Meffert 1981). After topical application of 8-methoxypsoralen, membrane alterations of epidermal cells are noted without any signs of nuclear binding (Meffert et al. 1977). Cytotoxicity of PUVA in lymphocytes is observed in the absence of 8-methoxypsoralen binding to nucleic acids and is explained by lipid peroxidation of the lymphocyte plasma membrane (Böhm and Meffert 1981).

PUVA Erythema. Administration of 8-methoxypsoralen greatly lowers the minimal erythema dose (MED) of UVA. PUVA erythema and UVB or UVA erythema differ in time course and pathophysiology. PUVA erythema has its onset about 16–24 h after irradiation, peaks at 24–72 h, and persists longer than 7 days. No detectable changes in prostaglandin metabolism are shown in vivo with PUVA therapy, and cyclooxygenase inhibitors like indomethacin do not inhibit PUVA erythema. Neutrophils also do not appear to play a significant role in PUVA erythema (Anderson and Voorhees 1980). Lysosome destruction via initiation of free radical reactions in the lysosomal membrane was suggested to participate in PUVA-induced cytotoxicity (Meffert et al. 1976b).

Indirect evidence for the involvement of reactive oxidants in PUVA therapy stems from antioxidant treatment studies. α-Tocopherol protects rabbit skin against PUVA erythema caused by a medium UVA dose ($3 \times$ MED) only if applied topically before irradiation. If it is applied shortly after irradiation, its inhibitory effect is greatly weakened. α-Tocopherol application prior to PUVA treatment with a large UVA dose ($10 \times$ MED) has virtually no inhibitory action (Potapenko et al. 1980, 1984). The effect of PUVA on the mechanoelectrical properties of human skin was inhibited by tocopherol and it was concluded that the photochemical reactions of psoralen in the skin are inhibited (Potapenko et al. 1983).

Mechanisms of Action. Lymphocytes are suspected to be the target cells in PUVA therapy (Meffert et al. 1980, 1981; Lange et al. 1980). The *lymphocytotoxic* effect can explain the efficiency of PUVA treatment in such heterogeneous skin diseases.

Furthermore, singlet oxygen oxidizes psoralen to a compound that *con-*

jugates with complement C5a, thereby inhibiting its chemotactic activity to neutrophils. This is suggested to be one of the action mechanisms of PUVA treatment for the neutrophil-amplified dermatoses (Mizuno and Esaki 1987).

PUVA treatment of human neutrophils *inhibits acid hydrolase activity* and increases the formation of reactive oxygen species in zymosan-activated neutrophils. Excessive formation and release of lysosomal enzymes such as acid hydrolases can result in pronounced tissue damage. PUVA therapy may result in intracellular inactivation of acid hydrolase by increased formation of reactive oxygen species in neutrophils (Gruner et al. 1987a). It was suggested that lysosome destruction by PUVA treatment is a toxic reaction; however, lysosomal enzyme inactivation may be related to the antiinflammatory effects.

PUVA can *photoinactivate fibroblasts* in culture but stimulates synthesis of procollagen type III hin human skin. This mechanism may be involved in PUVA-induced skin sclerosis.

In healthy controls and patients with lupus erythematosus and PSS, extracorporeal photochemotherapy (photopheresis) with 8-methoxypsoralen *inhibits neutrophil NADPH oxidase activity,* resulting in decreased production of reactive oxygen species (Trautinger et al. 1990).

PUVA is a *mutagen* in human fibroblasts, causing chromosomal aberrations and tumor induction in mice. Several reports from the USA describe an increased incidence of squamous cell carcinoma in PUVA-treated patients (more than 260 treatments) have an 11-fold greater risk of developing squamous cell carcinoma than the general population (Stern and Lange 1988). A long-term follow-up study also revealed an increased incidence of skin cancer in patients with psoriasis treated by PUVA (Forman et al. 1989). However, this is in contrast with most European studies, which did not find an increase in the prevalence of non melanoma skin cancer in PUVA patients (Lassus et al. 1981; Roenick and Caro 1981). Differences in the PUVA schedule might explain different skin cancer risks found in Europe and in the USA. Recently, however, a European long-term follow-up study also showed an increased incidence of nonmelanoma skin cancer in Dutch patients treated with PUVA (Bruynzeel et al. 1991).

The *major photochemical reaction* of psoralen contributing to inhibition of DNA synthesis is presumably the formation of mono- and bifunctional adducts with pyrimidine bases (type I reaction). It is thought that this reaction may participate in the antipsoriatic effect. It is also believed that interstrand cross-linking is largely responsible for skin carcinogenesis and some photosensitization reactions. In addition, reactive oxygen species participate in mediating the cytotoxic effect. Different molecular mechanisms may contribute to the therapeutic effects in different diseases (e.g., vitiligo).

Monofunctional psoralens (angular structure) are mostly not phototoxic, their ability to generate reactive oxygen species, such as singlet oxygen, is of a lower order of magnitude to that of the *bifunctional psoralens* (linear structure). The capacity of monofunctional psoralens for DNA photobinding and their effectiveness in psoriasis treatment are less pronounced than those of the bifunctional compounds (Aubin and Manteaux 1990).

PUVA is also used in clinical prevention of solar urticaria, polymorphous light eruption, hydroa vacciniforme, persistent light reaction, and actinic reticuloid. If reactive oxygen species do participate in the pathophysiology of these photodermatoses, and PUVA therapy generates even more reactive oxygen species, why is PUVA an effective treatment? The mechanism of this "light hardening therapy" is not understood. Either reactive oxidants are an insignificant epiphenomenon in both cases, or a subtle regulation mechanism is triggered by excessive production through PUVA.

II. Hematoporphyrin

Hematoporphyrin derivatives are used in photodynamic cancer therapy (Moan 1986) and in the treatment of malignant cutaneous tumors (Dougherty et al. 1978; Dougherty 1981, 1986). This technique is based upon the tumor localizing ability of hematoporphyrin compounds and their photodynamic action. Exposure to activating visible light (red region of the spectrum) results in a relatively specific photosensitization of malignant tissues. Patients with psoriasis have also been successfully treated by topical application of hematoporphyrin derivative and visible/UVA irradiation (McCullough et al. 1983, 1987; Meffert et al. 1989; Pres et al. 1989). Recently, successful therapy of psoriasis unresponsive to other forms of therapy was described after systemic administration of thin protoporphyrin and UVA irradiation (Emtestam et al. 1989).

Mechanisms of Action. Skin damage in mice induced by porphyrin photosensitization occurs under oxic but not under anoxic conditions (Gomer and Razum 1984), indicating the involvement of reactive oxygen species in the mechanism of phototoxicity. *Photodynamic activation* of a hematoporphyrin derivative does not produce superoxide anion as the major oxygen radical (Gibson et al. 1984), as suggested by others (Buettner and Oberley 1980). Rather, it results in the generation of singlet oxygen (Keene et al. 1986; Blum and Fridovich 1985; Dougherty 1986), the cytotoxic agent apparently causing injury to cells. The sensitizing efficiency of porphyrins depends on the chemical nature of the side chains; chelation with metal ions strongly decreases it (Cauzzo et al. 1977). Another reactive species possibly involved in hematoporphyrin photochemotherapy is thiyl radicals. Thiyl radicals were detected during light activation of hematoporphyrin derivatives, and singlet oxygen as well as hydroxyl radicals were indicated to be the initiating species (Buettner 1984).

Microspectrofluorimetric studies of single living cells indicate that porphyrin photosensitization results in the production of lipofuscin-like pigments, implying *lipid photoperoxidation.* Chloroquine and the retinoid etretinate inhibit the formation (Reyftmann et al. 1986). No lipid peroxidation was, however, detectable in human keratinocytes exposed to hematoporphyrin derivatives and laser light (632 nm).

Activation of lysosomal enzymes is thought to be involved in hematoporphyrin-derivative-caused phototoxicity in these cells (Artuc et al. 1989). Porphyrin photosensitization of skin may occur due to released lysosomal enzymes, because in the presence of hematoporphyrin and 400 nm radiation lysosomes are easily destroyed, and lipid peroxides are formed in a process involving singlet oxygen (Torinuki et al. 1980). Damage to cutaneous lysosomes plays a part in the inflammatory reaction triggered by photosensitization (Allison et al. 1966), and porphyrin-derivative-mediated lysosomal rupture is hypothesized to be caused by a free radical reaction involving reactive oxygen species (Slater and Riley 1966). Other in vitro studies, though, indicate that lysosomal damage may not be the primary event of hematoporphyrin photosensitivity. These reports propose that lysosomal membrane damage may be the result rather than the cause of hematoporphyrin induced cell damage (Sandberg 1981, Sandberg and Romslo 1981).

III. Goeckerman Therapy

In 1923 Goeckerman introduced the combination of UV and coal tar in the treatment of psoriasis (Goeckerman 1925); numerous modifications have been introduced since. The combination of tar and UV irradiation is indicated to be superior to treatment with tar or UV alone (Marsico et al. 1976). However, it was also reported that the combined use of tar and UVB in Goeckerman therapy is not substantially superior to treatment with UVB and an emollient alone (Stern et al. 1986).

Crude coal tar is a mixture of about 10000 distinct chemical components and contains considerable amounts of carcinogenic polyaromatic hydrocarbons. It is the fractionated product of coal distillation in the absence of oxygen, and due to the complexity of the chemical changes that occur during carbonization, standardization of crude and purified coal tar is extremely difficult to achieve. The differences in tar efficiency in clinical studies may be due to its variable composition (Lowe et al. 1982).

UVB irradiation is identified as the light component primarily effective in Goeckerman therapy (Marsico and Eaglman 1973; Petrozzi et al. 1978). The UV component in tar photosensitivity in humans is UVA (Tanenbaum and Parrish 1975), and photosensitization is not the primary mechanism in Goeckerman photochemotherapy. The photosensitizing components of coal tar in humans are anthracene, pyrene, benzo[a]anthracene, fluoranthrene, and acridine (Wulf et al. 1963; Kochevar et al. 1982) and are identical with the carcinogenic and mutagenic tar components. In therapeutic, purified tar preparations benzo[*a*]pyrene content may vary from 0.05% to 0.9% (Bickers 1981). Many reports show that tar is an effective therapeutic agent in the treatment of dermatoses, such as eczema and psoriasis.

Mechanisms of Action. Tar inhibits DNA synthesis in the epidermis (Walter et al. 1978; Lowe et al. 1981) and has a cytostatic effect on skin (Lavker et al. 1981). The most potent phototoxic tar compounds are anthracene, pyrene, and fluoranthrene. The majority of tar components do not cause photosensitization. A photodynamic mechanism in tar-induced photosensitivity was suspected (Crow 1961), and participation of singlet oxygen was suggested (Kochevar et al. 1982). Crude coal tar used in Goeckerman therapy for psoriasis produces type I and type II photoreactions. The therapeutic effects of coal tar were suggested to be due to the photoconjugation of the photoreactive tar ingredients with DNA bases causing interstrand cross-links. Singlet oxygen is mainly responsible for tar photosensitization (Pathak and Joshi 1983); however, it cannot be excluded that reactive oxygen species also contribute to its antimitotic activity.

Although the carcinogenic and mutagenic properties of tar and its products are clearly established, a study indicates that in 719 tar-treated patients no increased incidence in skin carcinoma or other malignancies is found after a latency period of 10 years (Jones et al. 1985). Preaged skin in sun-exposed sites is reported in tar industry workers, and they still have increased photosensitivity several years after their last occupational exposure. The tendency towards the formation of tar keratoses, papillomata, and epitheliomata persists also for many years after leaving the factory (Götz 1976) and indicates that epidermal cells are transformed by tar.

IV. Ingram Therapy

Ingram proposed a combination of coal tar bath, UVB irradiation, and anthralin (dithranol) to be highly efficient in clearing psoriasis (Ingram 1953). At first, the patient is given a tar bath, then irradiated with UV, and finally treated topically with an anthralin ointment. Preformation of skin lipid peroxidation products by this regimen may contribute to acceleration of anthralin autoxidation in skin (Ducret et al. 1985) and the formation of reactive oxygen species. UV (>280 nm) photooxidizes anthralin and increases the generation of anthralin-derived free radicals in a chemical system (Fuchs and Packer 1989) and in skin of hairless mice (Fuchs, unpublished results). Although the Ingram regimen is highly successful, it lost popularity because it can only be carried out on inpatients. Furthermore, there are some clinical reports which indicate that its therapeutic advantage is rather small. The coal tar component of the Ingram regimen is thought to have no therapeutic advantage over an oil emulsion (Frödin et al. 1988).

Chapter 5
Dermatotoxicology

Adverse cutaneous effects of topically or internally administered drugs can in some cases be attributed to prooxidative properties of the compound. There are many chemically reactive species that can wreak havoc with living cells, but free radicals and reactive oxygen species are especially damaging because of their unique chemistry, e.g., their ability to participate in chain reactions which, once initiated, have the potential to spread rapidly among cellular membranes.

Free-radical-mediated reactions have been discussed as one of the basic mechanisms of drug toxicity (Roberfroid et al. 1987), and various xenobiotics are known to induce lipid peroxidation in vitro and invivo (Kappus and Sies 1981; Plaa and Witschi 1976). Redox cycling is a common mechanism of drug-induced cytotoxicity (Fig. 5.1). Various endogenous and exogenous compounds exert toxic effects via oxygen reduction. These compounds are first enzymatically reduced before they activate oxygen by electron transfer, resulting in the formation of reactive oxygen species and oxidized drug. Various enzymes, e.g., cytochrome P-450 reductase, unspecifically maintain the redox cycle by reducing the oxidized drug. As a consequence, cellular reducing equivalents (NADPH, NADH) are diminshed, and the steady-state concentration of reactive oxygen species is elevated. Drugs capable of redox cycling are azo, nitro, and nitroxide compounds, thiols/disulfides, metal chelates, quinones/hydroquinones and bipyridyls.

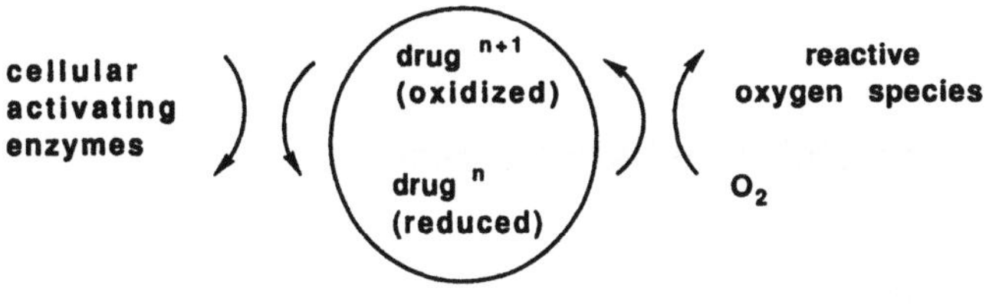

drug-oxidized	drug-reduced
quinone	semiquinone
semiquinone	hydroquinone
nitroxide	hydroxylamine
azo	azo anion radical
nitro	nitro anion radical
metal chelate^{ox}	metal chelate^{red}

Fig. 5.1. Redox cycling drugs

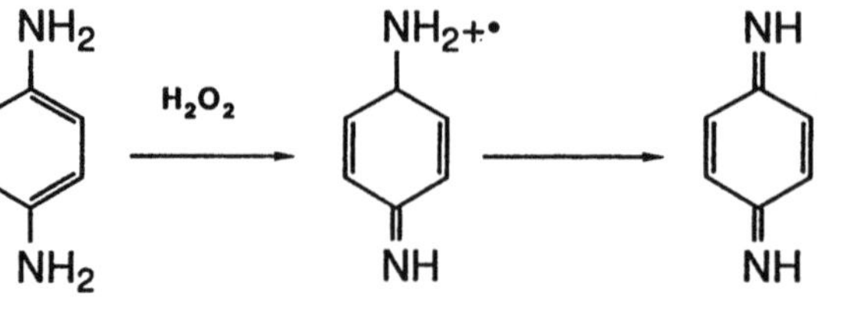

Fig. 5.2. *p*-Phenylenediamine oxidation

The role of free radicals and reactive oxygen species in dermatotoxicology has been a rather neglected field that deserves considerably more attention in the future. The study of cutaneous toxicology has important health implications (Marzulli and Maibach 1983). Occupational dermatoses include irritant and allergic contact dermatitis, skin cancer, chemical leukoderma, and chemical-induced comedogenesis as well as connective tissue disorders (Mathias 1988). It seems feasible to suggest that drug-induced peroxidation of cellular components may contribute significantly to toxic reactions of certain substances in the skin. However, the ultimate biological significance of free radical and reactive oxygen species in cutaneous drug toxicity is not known.

Skin can be a target organ for the toxic effects of systemic agents (Shupack 1976), as well as for topically applied substances. Cutaneous metabolism via cytochrome P-450, peroxidases, and flavoprotein dehydrogenases can lead to bioactivation of xenobiotics to prooxidant drugs and induce lipid peroxidation. Oxidation of hair dyes, e.g., *p*-phenylenediamine and 1,3-dihydroxybenzene, in the presence of hydrogen peroxide may exert skin toxicity via uncontrolled free radical reactions. A proposed intermediate of *p*-phenylenediamine oxidation is a semiquinone diimine which may cause skin toxicity via the generation of reactive oxygen species (Fig. 5.2). Additional reactive oxidants may derive from oxidation products of 1,3-dihydroxybenzenes (Spengler 1989).

A. Irritant Contact Dermatitis and Skin Necrosis

I. Lipid Peroxidation Products

Systemic toxicity of various lipid peroxidation products has been documented (Horgan et al. 1957; Holman and Greenberg 1960; Andrews 1960), e.g., in mice after intraperitoneal administration; LD_{50} values varying from 4 to 1000 μmol peroxide per mouse are reported (Horgan et al. 1957).

Organic hydroperoxides (Society of the Plastic Industry 1982) and lipid peroxidation products (Ohsawa et al. 1984; Tanaka 1979; Ogura 1981) also exert toxic effects in mammalian skin. Lipid peroxidation products can induce hyperkeratosis, hypergranulosis, dyskeratosis, intraepidermal edema, and spongiosis (Tanaka and Hayakawa 1986; Yagi 1987). Species differences in skin irritability by peroxidation products are noted, e.g., the skin of guinea pigs is usually more sensitive to irritation by lipid peroxidation products than human skin (Tanaka and Haykawa 1986).

Cytotoxic effects of lipid peroxidation products in skin may include: (a) perturbation of membrane fluidity, (b) inactivation of membrane receptors and membrane enzymes, (c) inactivation of proteins and nucleic acids in the hydrophilic cell compartments distant from the original site of lipid peroxidation, (d) inhibition of antioxidant systems, and (e) depletion of cellular reducing equivalents, e.g., NADPH.

It was proposed that epidermal damage due to epicutaneous application of lipid peroxidation products is mediated by the propagation of lipid peroxidation in situ (Hayakawa 1971; Yagi 1987). In mice, subcutaneous administration of an aqueous extract of UV-irradiated linoleic acid induces necrosis of the epidermis and dermis followed by an inflammatory response and disruption of collagen and elastic fibers (Waravdekar et al. 1965). The high eczematogenic activity of a certain turpentine is attributed to its high hydroperoxide content of d3 carene (Hellerström et al. 1955). Peroxidized squalene causes strong irritation of stripped human skin and growth inhibition of human keratinocytes in culture (Yamamoto et al. 1982). Oxidation products of linoleic acid release histamine from mast cells by a cytotoxic mechanism (Mannaioni and Masini 1988).

Respiration of fibroblasts and isolated epidermal cells is inhibited by degradation products of lipid peroxides (Meffert and Lohrisch 1971 a), probably by inhibition of mitochondrial succinate oxidation (Bernheim et al. 1952). Since polyunsaturated lipids of epidermal microsomes (Dixit et al. 1982, 1983 a) and mitochondria (Serri et al. 1979) are readily susceptible to peroxidation in vitro, energy production and metabolism of endogenous and exogenous molecules in skin may be significantly influenced by intracellular lipid peroxidation.

II. Anticancer Agents

Many reports concerning the skin toxicity of drugs which may induce lipid peroxidation in vivo involve anticancer agents. Skin necrosis following inadvertant extravasation of redox cycling antineoplastic drugs is a serious complication in cancer chemotherapy; the pathogenesis is based on a direct toxic effect and not on sensitization. Soft-tissue injuries by the anticancer drug doxorubicin (Svingen et al. 1981; Nobbs and Barr 1983; Daugherty and Khurana 1985), which is known to cause lipid peroxidation in vitro, are inhibited in vivo by antioxidants. Doxorubicin is reduced by several flavin-containing reductases to doxorubicin semiquinone, which subsequently reduces oxygen to reactive oxygen species (Fig. 5.3).

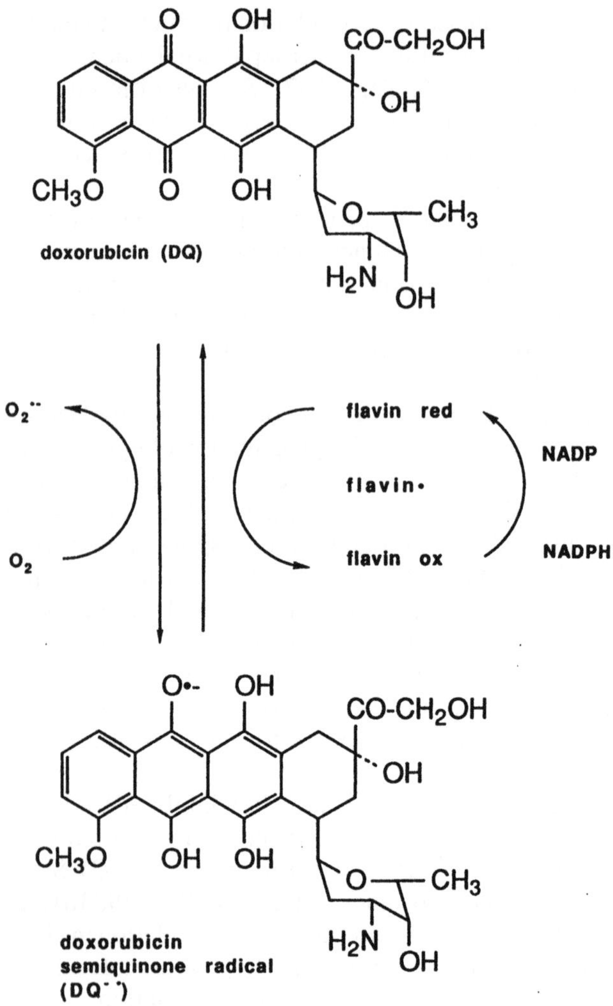

Fig. 5.3. Doxorubicin oxidation

III. Charge Transfer Mechanism

The *charge transfer mechanism* for the generation of reactive oxygen species is suggested as a unifying concept for the mechanism of action of anticancer agents and their toxicity (Kovacic et al. 1986), as well as for carcinogenesis by alkylating agents (Kovacic and Crawford 1986).

If an electron donor interacts with an electron acceptor, a complex may form in which the electronic charge passes from the donor to the acceptor. Such a complex is called a charge-transfer complex. The oxy radical charge transfer theory of pharmacodynamic/toxic actions of prooxidant drugs entails conversion of a target biomolecule by the drug to a conducting salt which can catalytically generate reactive oxygen species via electron transfer, and subsequently lipid peroxidation reactions may occur (Ames et al. 1986a, b).

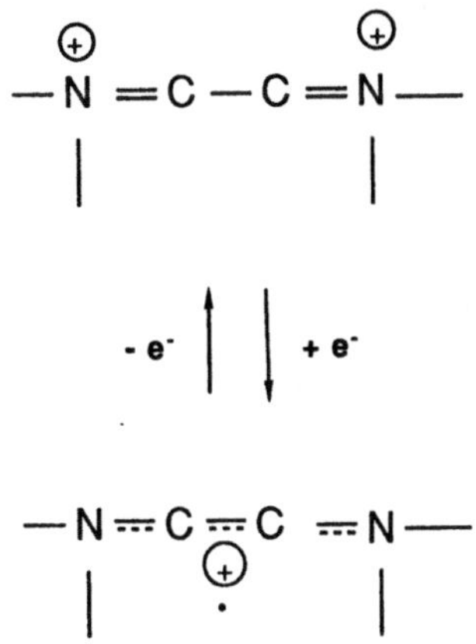

Fig. 5.4. Iminium ion charge transfer complex

Reactive oxygen species can be produced by charge transfer reaction from agents such as quinones, dapsone metabolites, metal complexes, peroxides, and iminium type ions. Particularly in the case of iminium ions, the positive charge enhances electron abstraction from donor molecules (Ames et al. 1985) (Fig. 5.4).

IV. Chemical Warfare Agents

Animals and plants use toxic chemicals for defense and to capture prey. Man has applied the same methods for similar purposes, and has used chemicals in warfare starting from early times and middle ages. Although the use of chemical weapons is outlawed under the Geneva Convention of 1925, they have been used since then in several occasions. The chemical weapons can be classified into: (1) lung irritants (chlorine, phosgene); (2) blood gases (hydrogen cyanide, cyanogen chloride); (3) vesicants (nitrogen and sulfur mustard, Lewisite); (4) nerve agents (e.g., organophosphates); (5) vomiting gases (e.g., diphenylchlorarsine); (6) tear gases (e.g., chloroacetophenone, 2-chlorobenzalmalonotrile); and (6) incapacitating agents (e.g., chinuklidinyl benzilate). Only vesicants and tear gases usually cause dermatological problems. Vesicants (skin irritants) are either alkylating agents (e.g., bis-2-chloroethyl sulfide ([sulfur mustard]; bis-2-chloroethylmethylamine [nitrogen mustard] or organic arsenicals (e.g., dichloro(2-chlorovinyl)arsine [Lewisite]). A detailed description, characterization, and overview of chemical weapons is given by Gillert (1944). The chemistry of nitrogen and sulphur mustard (Fox and Scott 1980) and their biological effects on the skin have been studied in detail (Sollmann 1918a, b; Lynch et al. 1918; Hopkins 1918; Marshall et al. 1918; Requena et al. 1988). In aqueous solution, nitrogen mustard cyclizes to an iminium ion (electrophile), which subsequently reacts with a nucleophile (RSH, RHN_2, ROH). Sulfur mustard forms the analogous sulfonium ion, but the bulk ionizes to the carbonium ion, which then reacts with a nucleophile (Fig. 5.5).

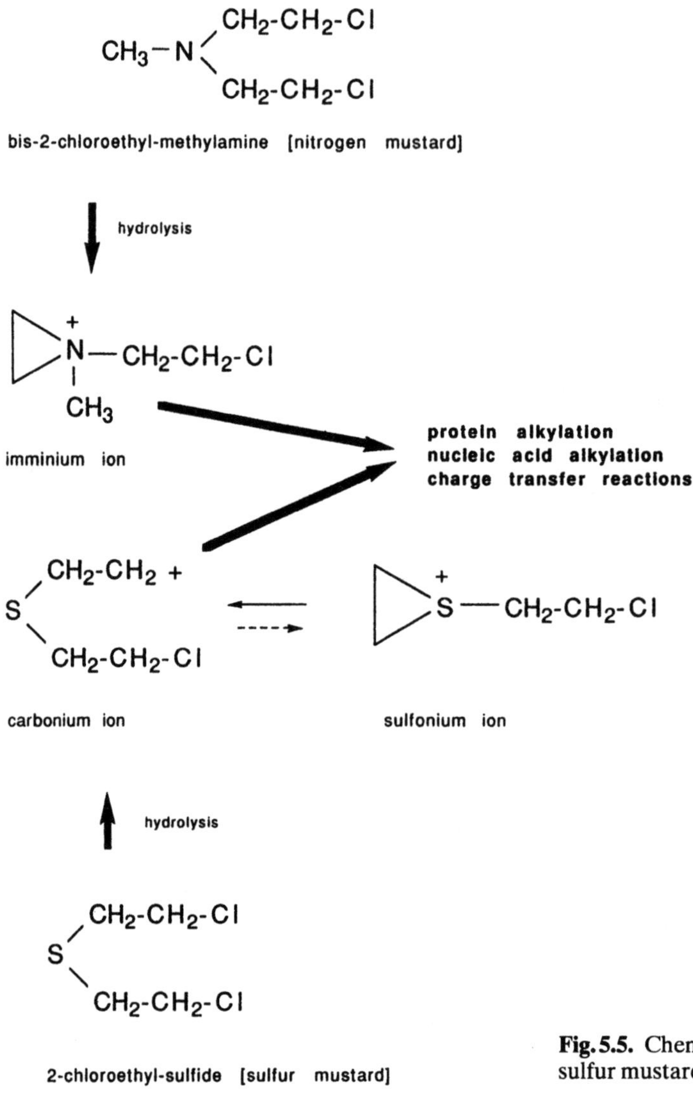

Fig. 5.5. Chemistry of nitrogen and sulfur mustard

The toxic effects of sulfur mustard include decreased tissue respiration, NAD and ATP depletion, inhibition of mitosis, and cell death. Most of the toxic effects are related to DNA alkylation. Mustard gas causes lipid peroxidation and causes depletion of ATP in the skin (Somani and Babu 1989). ATP depletion may result in oxygen radical formation (Nohl 1990). In addition, the oxy radical charge transfer mechanism (Ames et al. 1986a, b) may contribute to the pathogenesis of irritant contact dermatitis in humans caused by alkylating agents like sulfur mustard. It has been suggested that antioxidants might be of therapeutic benefit in patients suffering from chemical burns caused by these kind of agents. Indeed, tokopherol provides some protection against

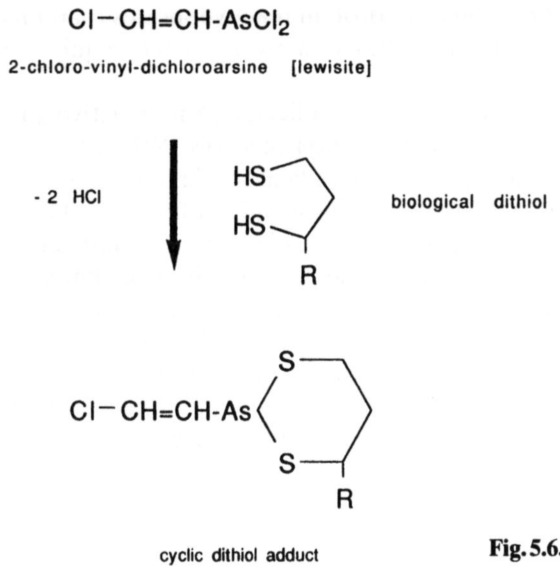

Fig. 5.6. Chemistry of Lewisite

sulfur and nitrogen mustard poisoning (Vojvodic et al. 1985). It was, however, suggested that there are no specific and very effective antidotes for sulfur and nitrogen mustard because the chemical reaction with biological tissue is so unspecific and rapid as to be irreversible for practical purposes (Somani and Babu 1989; Cernes 1989).

Lewisite (dichloro(2-chlorovinyl)arsine) decomposes in aqueous solution by a complex mechanism. It has a labile chlorine atom, trivalent arsenic, and an unsaturated bond, and so is very reactive. Its reactions are due to nucleophilic attack by thiols and water (Goldman and Dacre 1989). In Lewisite intoxication, Lewisite binds to dithiols (e. g., dihydrolipoic acid) to form a stable cyclic structure (in the case of dihydrolipoic acid, a six-membered ring) (Fig. 5.6).

Exposure to Lewisite can be effectively reversed by treatment with dithiols which chelate more effectively with Lewisite (e.g., British antilewisite = dimercaptopropanol). British antilewisite forms a more stable six-membered ring which is nontoxic and is excreted via the kidneys. There are no reports on the effectiveness of antioxidants as antidotes.

B. Allergic Contact Dermatitis

Poison ivy and poison oak (*Toxicodendron* species) dermatitis are the most common causes of allergic contact dermatitis to plants in humans. The induction is mediated by alkylcatechols (urushiols). The molecular mechanism of conversion of alkylcatechols (prohapten) to protein reactive species (hapten) involves oxidative activation and the formation of a protein reactive quinone (Byck and Dawson 1968; Dupuis 1979).

Recently, however, it was suggested that free radicals and not reactive quinones are the haptenic species derived from alkylcatechols (Schmidt et al. 1990). Furthermore, formation of haptenic free radical species and redox cycling processes were implicated in the molecular mechanism of contact allergy to halonitrobenzenes, alantolactones, and isoalantolactone (Schmidt et al. 1990). Schmidt suggested that free radical species are the ultimate skin haptenes in the case of dinitrohalobenzenes (with the exception of the fluoro derivative). Mouse skin exposed to dinitrohalobenzenes undergoes a classical oxidative stress response by decreasing glutathione und elevating glutathione disulfide (Schmidt and Chung 1990). Recently a possible role for superoxide anion production in the pathogenesis of nickel and chromate contact dermatitis was suggested (Sharkey et al. 1991).

C. Chemical-Induced Skin Sclerosis

Sclerosis is characterized by excessive deposition of connective tissue in a mesenchymal organ, and fibrosis is defined by pronounced accumulation of fibroblasts in connective tissue. Frequently, however, the terms sclerosis and fibrosis are used synonymously.

Drug-induced *organ fibrosis* is described for such chemically heterogenous compounds as oxygen (retrolental fibroplasia), phenytoin (gingival hypertrophy), methysergide (retroperitoneal fibrosis), nitrofurantoin, amiodarone, bleomycin (lung and skin fibrosis), and practolol (slerosing peritonitis). The herbicide paraquat serves as a model substance for producing fibrosis of the lung.

I. Paraquat-Induced Lung Fibrosis

Acute paraquat toxicity results in *severe alveolitis;* chronic toxicity is characterized by proliferative *lung fibrosis.* Specific toxicity to the lung by paraquat is thought to be caused by preferential uptake of the drug into the alveolar epithelial cells (Smith 1985). It is not known whether paraquat-induced lung fibrosis is related to superoxide radical production by paraquat directly, by production of reactive oxidants through the subsequently invading inflammatory cells, or just part of a repair process. It was suggested that the mechanism of paraquat-induced increase in lung collagen synthesis involves superoxide radicals because the same effect in organ cultures is abolished by SOD treatment (Hussain and Bhatnagar 1979).

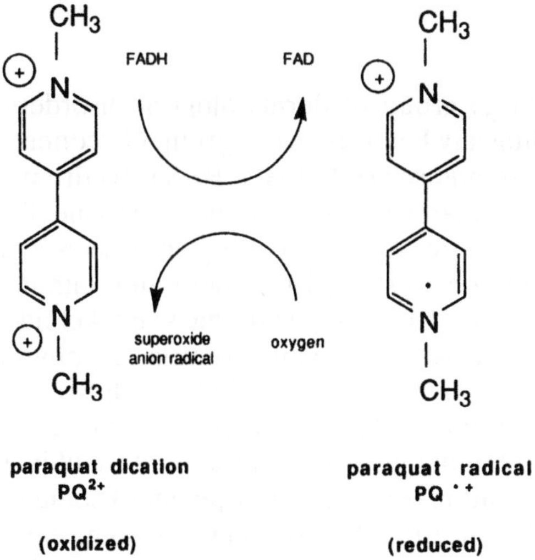

Fig. 5.7. Redox cycling of paraquat

The bipyridyl cation paraquat is rapidly reduced by flavin-containing oxidoreductases to the radical form which autooxidizes and yields the superoxide anion radical. By reaction of the paraquat radical with oxygen, oxidized paraquat is regenerated, thereby sustaining a redox cycling process (Fig. 5.7).

The common pathophysiological mechanism of bleomycin, paraquat, nitrofurantoin, and UV/ionizing-irradiation-mediated organ fibrosis may be the formation of reactive oxygen species. A similar mechanism has been suggested for chemical-induced skin sclerosis. This hypothesis is strengthened by the finding that collagen synthesis is stimulated by an elevated steady-state concentration of reactive oxygen species, e.g., in hyperoxic peridontal tissue culture of mice (Yen et al. 1979), hyperoxic rat lung organ culture (Bhatnagar et al. 1978), and rat lung by ozone (Last and Greenberg 1980). Although the molecular mechanism of collagen gene expression and synthesis is not yet known in detail, reactive oxidants are among the several mediators that affect these processes. Fibroblasts are not only the target of reactive oxidants but may also be the source of it. Primary cultures of human fibroblasts have the capacity to release superoxide anion radicals spontaneously or upon stimulation. Fibroblasts contain an NADPH oxidase system, similar to the NADPH oxidase of the phagocyte, but the nature of the physiological stimulants as well as the intracellular signal transduction are different (Meier and Sies 1990). It is important to emphasize, however, that fibroblast function and collagen synthesis are also regulated by various cytokines (Haustein et al. 1990). Reactive oxidants are not the only mediators of collagen metabolism. In addition to indirect effects via reactive oxidants and cytokines, fibroblasts may be directly affected by drugs causing organ fibrosis.

II. Skin Sclerosis

Fibrotic skin diseases comprise a large group of dermatological disorders. Sclerosis of the skin occurs in the clinically heterogeneous group of "connective tissue diseases": progressive systemic scleroderma, CREST syndrome, dermatomyositis, systemic lupus erythematosus, mixed connective tissue disease, and overlap syndromes. Furthermore, sclerosis is frequently seen in a variety of dermatopathological conditions distinct from these autoimmune diseases: localized scleroderma, lichen sclerosus, hypertrophic scars, keloids, dermatosclerosis due to venous insufficiency, phenylketonuria, porphyria cutanea tarda, carcinoid syndrome, and juvenile-onset diabetes mellitus. Although the etiopathology of the various forms of fibrotic skin diseases is largely unknown, cytokines such as transforming growth factor and interleukin-1 have been suggested to play a significant role in the development of some of the fibrotic skin diseases (Uito and Jimenez 1990). In recent years, some environmental substances have been implicated as inducing factors in progressive systemic scleroderma and scleroderma-like skin diseases. Occupational scleroderma is clinically indistinguishable from the idiopathic form.

Antioxidants in Treatment of Skin Fibrosis. Free radical scavengers and antioxidants were proposed to be of therapeutic value in the treatment of fibrotic processes (Borel 1983). Tocopherol and SOD are used clinically with varying success rates in the treatment of various fibrotic skin diseases such as induratio penis plastica, keloids, and hypertrophic scars. Recently, allopurinol, an inhibitor of xanthine oxidase, was used effectively in the treatment of patients with Dupuytren's contracture. It has been suggested that in palmar fibrosis free radicals released from the endothelial cells of narrowed microvessels may damage the perivascular connective tissue and induce a proliferative response in surrounding fibroblasts. The benefit of allopurinol in the management of palmar fibrosis and the finding that hypoxanthine is 6-fold elevated in Dupytren's palmar fascia support this hypothesis (Murrell et al. 1990b). By inhibiting production of reactive oxygen species allopurinol may break the cycle of localized ischemia, microvascular damage, fibroblast proliferation and collagen deposition (Murrell and Pilowsky 1990; Murrell et al. 1990a, b).

III. Chemical-Induced Scleroderma

Several years ago, it was pointed out that reversible, scleroderma-like lesions can be induced in human skin by a variety of chemicals (Korting and Holzmann 1967). Subsequently, a large clinical data base on drug- and environmental-agent-induced skin sclerosis has been amassed. Scleroderma-like skin lesions may be induced in susceptible patients by various exogenous trigger factors (Haustein and Ziegler 1986; Schwegler and Holzmann 1990):

Drugs
Bleomycin (Finch et al. 1980; Bork and Korting 1983)
Carbidopa, L-5-hydroxytryptophan (Sternberg et al. 1980)
8-Methoxypsoralen/UVA therapy (Thürlimann and Harms 1982)
Cocaine (Trozak and Gould 1984)
Pentazocin (Palestine et al. 1980; Rodnan and Medsger 1981)
Vitamin K_1 (Texier et al. 1975)

Environmental Agents
Epoxides, epoxy hardener (bis-4-amino-3-cyclohexylmethane) (Yamakage et al. 1980)
Denaturated rapeseed oil ("toxic oil" syndrome) (Phelps and Fleischmajer 1988)
Organic solvents (Haustein and Ziegler 1985, 1986)
Aromatic hydrocarbons (benzene, toluene, xylene) (Walder 1965, 1981)
Petroleum-derived crude solvents (naphtha, trichlorethylene) (Czirjak et al. 1987)
Alicyclic and aliphatic hydrocarbons, chlorinated hydrocarbons (naphtha, *n*-hexane, hexachloroethane) (Yamakage and Ishikawa 1982)
Perchlorethylene (Sparrow 1977)
Trichlorethylene (Reinl 1957; Saihan et al. 1978; Schirren 1971; Flindt-Hansen and Isager 1987)
Trichlorethane (Flindt-Hansen and Isager 1987)
Pesticides (malathione, parathione, chlordane, heptachlor) (Rowell 1986)
Vinylchloride (Maricq 1985; Veltman 1980)

PSS serves as a model disease for studying the pathology of sclerosis. Its induction requires a certain genetic predisposition in susceptible patients (Sollberg and Holzmann 1987). Genetic susceptibility is presumably an additional significant determinant in the development of skin sclerosis triggered by some chemicals, e.g., halogenated hydrocarbons, because not everybody exposed to these chemicals will acquire the sclerosis. It has been established that vinyl chloride- and "toxic oil"-induced scleroderma are associated with a genetic susceptibility. Other chemicals, e.g., bleomycin, cause skin sclerosis in rats via an obligate toxic mechanism (Mountz et al. 1983).

It is clearly that quartz can cause PSS in some susceptible subjects. Petroleum-derived crude solvents were suggested to cause PSS (Czirjak et al. 1987). Various aliphatic hydrocarbons (*n*-hexane, naphtha) are reported to bring about generalized morphea, sclerodactyly, Raynaud's phenomenon, esophageal dysfunction, and lung fibrosis (Yamake and Ishikawa 1982). Trichlorethylene and trichlorethane are also found to cause PSS-like symptoms (according to the American Rheumatology Association) (Flindt-Hansen and Isager 1987).

1. Quartz

In 1914, Byrom Bramwell, a Scottish physician, described nine patients with diffuse scleroderma; five were working as stone masons. Since then, increasing evidence links the prevalance of PSS with silicosis, and it is suggested that silicosis may be a predisposing factor in the pathogenesis of this disease (Erasmus 1957; Rodnan et al. 1967; Ziegler et al. 1982, Haustein and Ziegler 1986). Silica-induced skin sclerosis fulfills the criteria of the American Rheumatism Association to establish the diagnosis of PSS. PSS induced by silica is distinct from drug-induced, scleroderma-like lesions, because it does not resolve when exposure to the causing factors is stopped.

The pathogenesis of quartz-induced PSS is as follows: Quartz particles activate macrophages, which release IL-1, which then stimulated fibroblast collagen production (Hepplestone and Styles 1967). Reactive oxygen species may be involved in this type of cell activation (Meier et al. 1989). IL-1 also affects T-helper lymphocytes, which then release IL-2, finally stimulating B lymphocytes. Quartz-related free radicals may participate in the biochemical mechanisms involved in acute silicosis and presumably in quartz-induced PSS. Fresh quartz dust generates hydroxyl and possibly superoxide anion radicals in a cell-free aqueous medium (Dalal et al. 1989). In addition, phagocytosis-derived reactive oxidants may contribute to the pathophysiological process of crystal-induced inflammation (McCarty 1979). Silicates can activate the respiratory burst of leukocytes (Montsko et al. 1984). Phagocytosis of quartz particles by macrophages, subsequent macrophage cell death and release of the quartz particles, and continuation of this process (Löblich 1961) can lead to a rise in steady state levels of reactive oxygen species. Silicates may also stimulate fibroblast metabolism via immunologic mechanisms (Ziegler et al. 1986). In silica-induced skin sclerosis, macrophage-derived lymphokines and monokines are suggested to activate fibroblast collagen and glycosaminoglycan metabolism (Haustein et al. 1990).

2. Bleomycin

Bleomycins are glycopeptides produced by *Streptomyces* species; bleomycins A2 and B2 are used as antitumor agents. More than 200 types have been prepared by microbial processes or chemical modification. Bleomycin-mediated skin and lung fibrosis may result from its redox cycling properties and the formation of lipid peroxidation products. Bleomycin is presumably toxic by interaction with cellular iron ions, reduction of the metal complex formed, oxygen binding, reactive oxygen species formation, and DNA cleavage (Fig. 5.8).

In interstitial pulmonary fibrosis, bleomycin increases the rates of transcription of the genes coding for procollagen, elastin, and fibronectin (Raghow et al. 1985). The predilection of bleomycin toxicity for skin and lung results in part from an increased accumulation in these tissues and in part

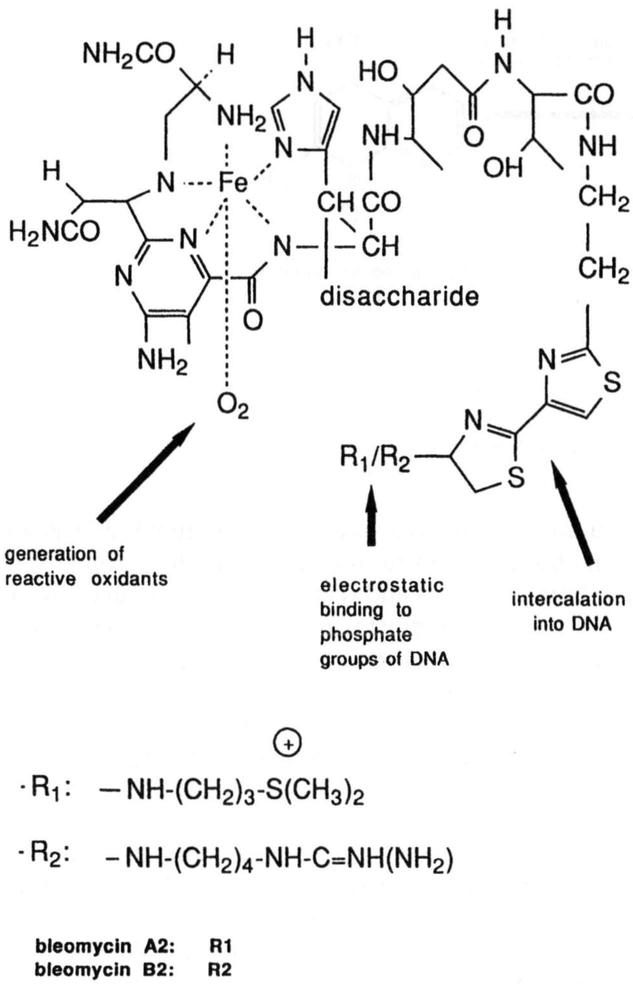

Fig. 5.8. Bleomycins

from decreased inactivation. Skin fibrosis in rats induced by bleomycin administration is an obligate toxic reaction and resembles skin changes in humans with PSS but without immunological alterations (Mountz et al. 1983).

3. Vitamin K_1

Although the redox chemistry of vitamin K derivatives is well-known, the biological significance of one-electron redox reactions in vitamin-K-induced skin sclerosis is purely hypothetical. Skin sclerosis after vitamin K_1 administration is observed very rarely (Texier et al. 1975), and only a few cases are reported in the literature. This rather indicates the involvement of multiple unknown variables.

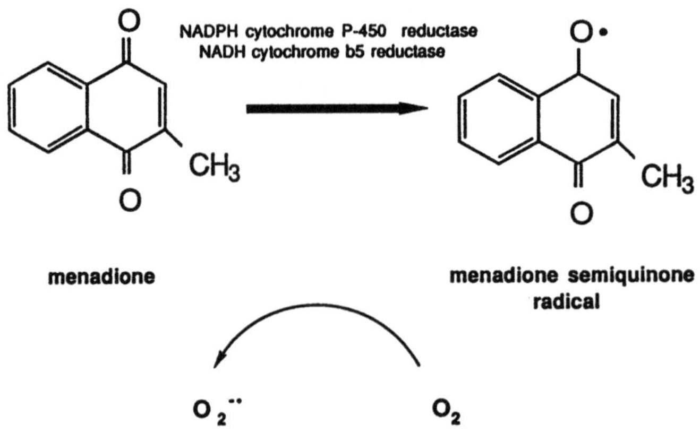

Fig. 5.9. Redox metabolism of menadione

Vitamin K_1 is a *p*-quinone derivative having a polyprenyl side chain, whereas vitamin K_3 (menadione) is substituted with a methyl group. Vitamin K is found in biomembrane domains and is involved in electron transport processes. Semiquinone radicals are readily formed during redox metabolism of vitamin K (Draw and Depew 1985) and may be a potential source of reactive oxygen species (Fig. 5.9). Menadione has been implicated in processes causing cell injury due to involvement in redox cycles producing reactive oxygen species and semiquinone radicals and depleting glutathione levels. Oxidation of intracellular glutathione is believed to be largely responsible for its cytotoxic effects (Thor et al. 1982).

Recently, however, it was pointed out that reactive oxidants generated during menadione-dependent redox cycling are not responsible for the mitochondrial damage. It was suggested that two-electron oxidation of intramitochondrial pyridine nucleotides via mitochondrial menadione reductases triggers calcium release and thereby causes impairment (Frei et al. 1986).

4. Toxic Oil Syndrome

In 1981 there were reports from Spain on the toxic oil syndrome. Rapeseed oil contaminated by aniline derivatives was used for preparing food. Rapeseed oil is normally used as an industrial lubricant and is supplemented with aniline to prevent its use as cooking oil. During an illegal attempt to refine the contaminated oil and sell it for human use, some chemical reactions occurred and various toxic substances were formed. The early stage of the syndrome was characterized by vasculitis in practically every organ clinically affected (lung, gastrointestinal tract, CNS). In the late stage, scleroderma-like skin lesions, Raynaud's phenomenon, sicca syndrome, dysphagia, myositis, and arthritis were observed. Mainly women with HLA DR3 and DR4 developed scleroderma-like lesions about 5–6 months after the intoxication, indicating again a genetic predisposition (Haustein and Ziegler 1986).

Free radicals and lipid peroxidation products generated by oil denaturants were suspected to participate in the pathogenesis of endothelial damage and skin fibrosis observed in the toxic oil syndrome (Rush et al. 1984; Haustein and Ziegler 1986; Spurzem and Lockey 1984; Gilsanz et al. 1984; Fournier et al. 1982; Tabuenca 1981). The aniline derivatives probably reacted with fatty acids and formed reactive oxidants, damaging endothelial cells and capillaries (Rush et al. 1984). Of course, this does not mean that free radicals and lipid peroxidation are the primary cause of the toxic oil syndrome; they may be just epiphenomena.

5. Biogenic Amines

Serotonin causes skin fibrosis in mice when injected intraperitoneally (McDonald et al. 1958). Several reports indicate that patients with carcinoid syndrome may develop scleroderma-like skin lesions (Zarafonetis et al. 1958; Hay 1964; Fries et al. 1973). Methysergide, which is structurally related to serotonin but acts pharmacologically as a serotonin antagonist, may cause retroperitoneal fibrosis in some patients (100 out of about 500000 patients treated; Graham 1967). Also, some patients treated with L-5-hydroxytryptophan and carbidopa (a 5-hydroxy-tryptophan decarboxylase inhibitor) may develop scleroderma-like skin lesions (Sternberg et al. 1980; Auffranc et al. 1985). Involvement of reactive oxygen species in tryptophan metabolism was suggested (Bus and Gibson 1982). High plasma serotonin levels and an abnormality with elevated plasma kynurenine may be involved in the pathogenesis. Plasma kynurenine was also found to be high in about 50% of patients with idiopathic morphea (Sternberg et al. 1980). Dysregulation of tryptophan metabolism was suggested to participate in development of sclerodermiform skin lesions (Jouve et al. 1987). Monoxygenase activity was suggested to be affected in patients with PSS or severe generalized morphea with vascular changes, resulting in accumulation of biogenic amines derived from tryptophan (Stachow et al. 1979). Patients with scleroderma have abnormally high urinary excretion of kynurenine, hydroxykynurenine, kynurenic acid, and *N*-acetylkynurenine after oral administration of tryptophan. Therapy with the metal-chelating agent EDTA improves the metabolic dysfunction significantly (Price et al. 1957). Disturbances in tryptophan metabolism are, however, not specific for scleroderma, but are also observed in patients with carcinomas of the bladder, prostate, kidney, and breast, Hodgkin's disease, rheumatoid arthritis, and endogenous depression (Rose 1972).

Tryptophan metabolism is associated with formation of reactive oxygen species (Fig. 5.10). Tryptophan is quite susceptible to enzymic and nonenzymic oxidative inactivation. Indolamine derivatives such as tryptophan, serotonin, and tryptamine are readily oxidized in vitro during cupric ion catalyzed peroxidation of arachidonic acid at physiological pH (Reyftmann et al. 1990). Tryptophan is also oxidized by the monoxygenase tryptophan-5-hydroxylase to 5-hydroxytryptophan, and by tryptophan dioxygenase and/or in-

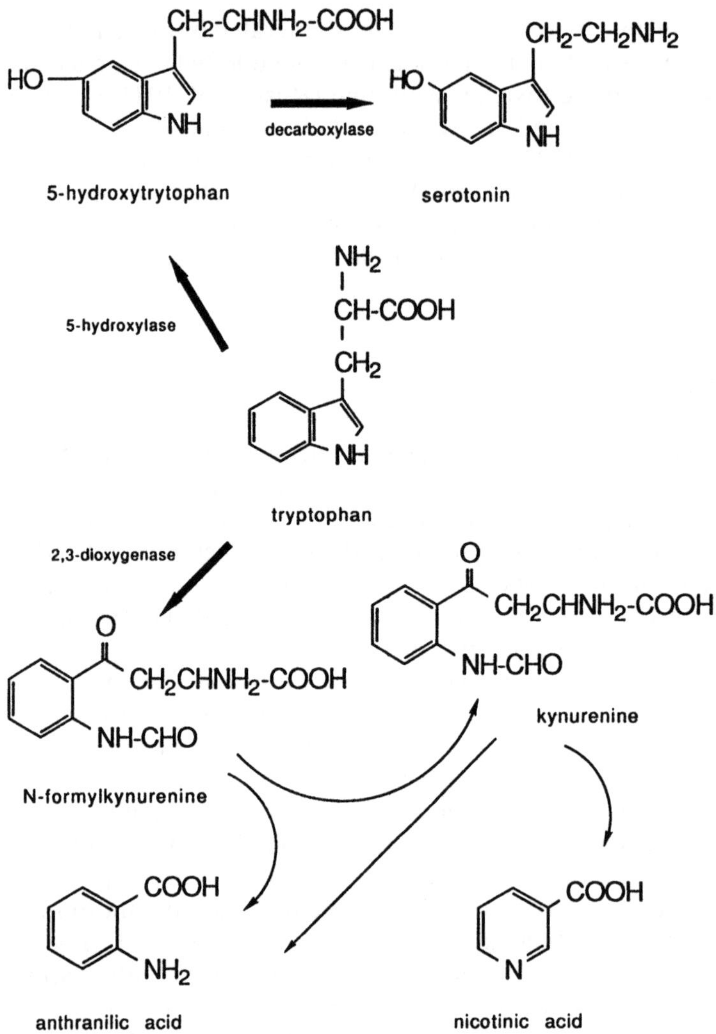

Fig. 5.10. Tryptophan metabolism

dolamine-2,3-dioxygenase to *N*-formylkynurenine. The enzymatic decarboxylation product of 5-hydroxytryptophan is serotonin. *N*-Formylkynurenine is converted to kynurenine in a reaction catalyzed by kynurenine formylase, or may be metabolized to anthranilic acid. Alternatively, anthranilic acid may arise from kynurenine. Kynurenine is then converted to 3-hydroxykynurenine by kynurenine-3-hydroxylase, employing molecular oxygen, and is finally metabolized via serveral steps to nicotin acid (Fig. 5.10).

During the various oxygenation and hydroxylation reactions, reactive oxygen species may be formed, presumably as transient intermediates on the active site of such enzymes. In contrast to tryptophan dioxygenase, indolamine-2,3-dioxygenase exhibits broad substrate specificity and acts upon various indolamine derivatives. It is an enzyme utilizing molecular oxygen and superoxide anion radical as cosubstrate (Hirata and Hayaishi 1975; Nishikimi 1975) for oxidation of indolamine derivatives, e.g., tryptophan. It is a hemoprotein and is induced in lung tissue by viral infections, interferon, and endotoxin (Hayaishi 1987). Tryptophan dioxygenase is inducible by administration of tryptophan. Indolamine-2,3-dioxygenase activity has been found in human skin (Schallreuter 1988).

The photochemistry of tryptophan and its degradation products also involves formation of reactive oxygen species. Tryptophan reacts rapidly with singlet oxygen to yield a dioxetane, which decomposes to *N*-formylkynurenine; excited carbonyls may be reaction intermediates. *N*-Formylkynurenine is also a photochemical (sunlight) breakdown product of tryptophan in proteins (Pirie and Dilley 1974), and a powerful photodynamic photosensitizer (wavelengths > 320 nm) for a great variety of biologically acitve compounds (Walrant and Santus 1974). Tryptophan and *N*-formylkynurenine are endogenous sources of hydrogen peroxide upon irradiation with near UV light (McCormick et al. 1976). Kynurenic acid is an endogenous photosensitizer; photohemolysis is oxygen-dependent and inhibited by β-carotene (Swanbeck and Wennersten 1974; Wennersten and Brunk 1977).

Although disturbances in tryptophan degradation may result in increased formation of reactive oxidants, direct experimental evidence is not available to support the hypothesis that this reaction is also linked pathophysiologically to induction of skin sclerosis. This is also the case for *tryptophan-associated eosinophilia-myalgia syndrome*. The eosinophilia-myalgia syndrome is a newly recognized illness that has been associated with the consumption of tryptophan products. The syndrome is characterized by eosinophilia, myalgia, arthritis, edema, polymorphous exanthema, and sclerodermiform skin lesions (Lehnert 1990); it may also been associated with diffuse fasciitis (Senff et al. 1990). Several studies suggest that eosinophilia-myalgia syndrome is not caused directly by tryptophan or tryptophan metabolites, but by an unidentified chemical contaminant (Belongia et al. 1990; Poesin et al. 1990). Further studies are required to solve the pathophysiology of this syndrome. However, it was suggested that tryptophan derived products may selectively influence connective tissue matrix protein synthesis either directly or indirectly (cytokines, growth factors) in susceptible patients (Kaufmann et al. 1990; Connolly et al. 1990)

6. Halogenated Aliphatic Hydrocarbons

Halogenated aliphatic hydrocarbons are ubiquitous environmental pollutants. Toxic metabolites of chlorinated alihatic hydrocarbons, e.g. trichlorethanol and/or metabolites of the dehalogenation, process, were thought to be involved in the mechanism of halogenated hydrocarbon skin sclerosis (Walder 1983).

Toxicity of halogenated aliphatic hydrocarbons is mediated by reactive metabolites formed during dehalogenation. The substances are subject to

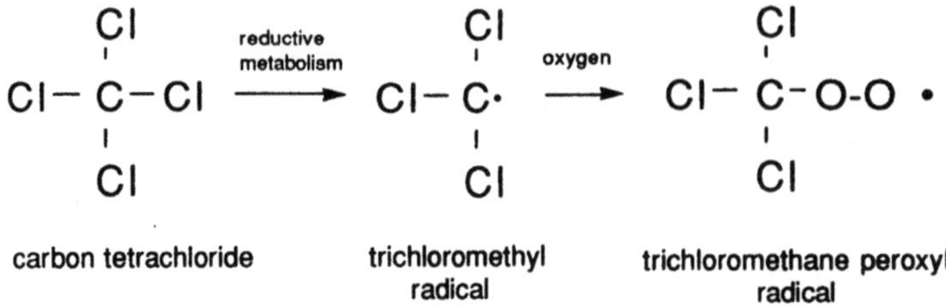

Fig. 5.11. Carbon tetrachloride metabolism

single-electron reduction by the mixed function oxidase system (reductive dehalogenation) to generate reactive alkyl and alkylperoxyl radicals. The more extensive the degree of chlorination, the easier the conversion to a reactive derivative, and the greater its toxicity. Carbon tetrachloride is metabolized by microsomal cytochrome P-450 to the trichloromethyl radical which, in the presence of oxygen, easily forms the trichloromethane peroxyl radical; this readily initiates lipid peroxidation in biological systems (Fig. 5.11).

Reductive metabolism of halogenated hydrocarbons can lead to the formation of lipid peroxidation products in cells and tissues. Free radical derivatives of halogenated aliphatic hydrocarbons are strongly alkylative and readily deplete tissue glutathione stores, causing a prooxidant cellular environment.

It is important to point out that the metabolic pathways of saturated and unsaturated halogenated aliphatic hydrocarbons are different. Generally, biotransformation of saturated halogenated hydrocarbons results in the formation of reactive free radicals (Fig. 5.12). Chloroform, methylene, and methyl chloride are less toxic than carbon tetrachloride, and each has a different metabolic degradation pathway. Metabolites of chloroform are trichloromethanol and phosgen, of methylene chloride chlormethanol, formaldehyde, and formic acid. Free radicals presumably do not play a major role in the mechanism of toxicity. Chlorinated ethane derivative metabolically form free radicals, and the tendency to free radical formation increases with the degree of halogenation, reaching a maximum at tetrachloroethane (Fig. 5.13).

In contrast to chlorinated alkanes, chlorinated alkenes are metabolized to

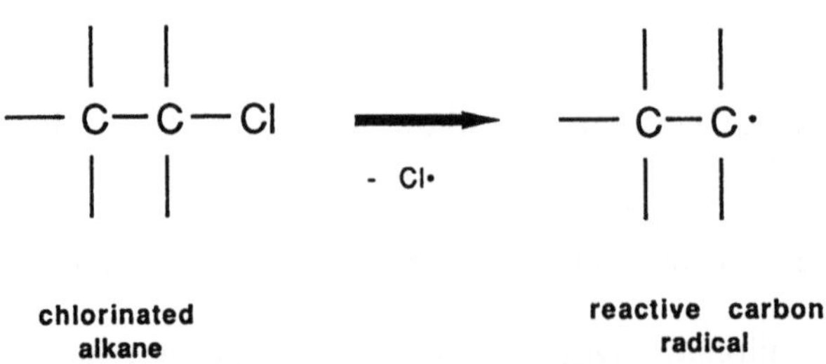

Fig. 5.12. Metabolism of chlorinated alkanes

$Cl_3C\text{-}CCl_3$	**hexachloroethane**
$Cl_2CH\text{-}CHCl_2$	**tetrachloroethane**
$Cl\text{-}CH_2\text{-}CH_2\text{-}Cl$	**dichloroethane**

Fig. 5.13. Chlorinated ethane derivatives

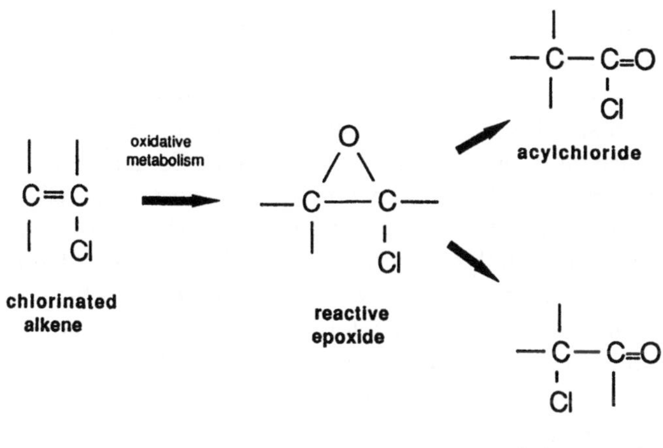

Fig. 5.14. Metabolism of chlorinated alkenes

$Cl_2C{=}CCl_2$	**tetrachloroethylene**
$Cl\text{-}CH{=}CCl_2$	**trichloroethylene**
$Cl\text{-}CH{=}CH\text{-}Cl$	**dichloroethylene**
$CH_2{=}CH\text{-}Cl$	**vinyl chloride**

Fig. 5.15. Chlorinated alkenes

highly reactive (nucleophilic) epoxides which are further converted to acylchlorides or aldehydes (Fig. 5.14).

Tetrachloroethylene is the most inert compound of this class, trichloroethylene the most reactive, and 1,1-dichloroethylene (vinylidenchloride) has intermediate reactivity (Fig. 5.15).

The "sclerotic potential" of different halogenated hydrocarbons is not known. This knowledge could help to improve our understanding of the mechanisms of halogenated-hydrocarbon-induced skin sclerosis. Since this is

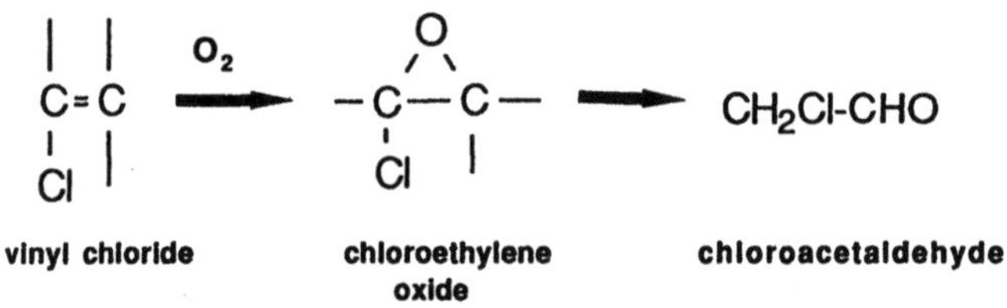

Fig. 5.16. Metabolism of vinyl chloride

not an obligate toxic reaction, it will be extremely difficult to evaluate the "sclerotic potential" of these compounds.

Vinyl chloride disease is characterized by pulmonary dysfunction, hepatic portal fibrosis, skin sclerosis, Raynaud's phenomenon, endothelial damage, and immunological disturbances. After first reports from the Soviet Union in 1949, the disease was described in detail as occupational acroosteolysis (Cordier et al. 1966; Markowitz et al. 1972). Industrial workers who clean autoclaves in which vinyl chloride is polymerized have a high incidence of skin sclerosis associated with other symptoms. Vinyl chloride monomer is bioactivated by the mixed function oxidase to form chloroethylene oxide and acyclic chloroethyl dioxide (Infante 1981; Ward et al. 1976) (Fig. 5.16). These reactive metabolites have been suggested to initiate lipid peroxidation and form covalent adducts, leading to the pathological sequelae. Involvement of vinylchloride-derived epoxides, peroxides, and free radicals is suggested in the pathogenesis (Haustein and Ziegler 1986). Vinyl chloride disease may be a representative model of free-radical-mediated inflammatory/immune disease resulting from exposure to halocarbons (Levine and Reinhardt 1983). It is reversible after exposure is stopped (Veltman 1980) and is associated with a genetic susceptibility of the patients, as shown by an increase in HLA DR5, B8, and DR3. There are clearly distinct differences between PSS and vinyl chloride disease (Mariqu 1981).

Silicone-induced skin sclerosis has been described in patients carrying silicone implants (Kondo et al. 1985; Kumagai et al. 1979, 1984; Spierra 1988). In some patients with silicone-induced skin sclerosis, the disease closely resembles initial cases of PSS. The compound may act as an immunologic adjuvant, and thereby trigger autoimmune diseases (Haustein and Ziegler 1986). It was demonstrated that dimethylpolysiloxane was released from the silicone implant into the tissue and macrophages. Although crystalline silica was not deteced, it may be speculated that the pathophysiology of silicone-induced skin sclerosis may resemble closey that of quartz-induced skin sclerosis.

The fact that *paraffin* and other fully saturated alkanes (Yamakage and Ishikawa 1982) can induce skin sclerosis argues against a common mechanism of xenobiotically induced skin sclerosis, e.g., free radical-induced sclerosis. However, it cannot be excluded that paraffin or alkanes contain trace amounts of unsaturated, peroxidizable compounds (alkenes) and lipid peroxidation products which trigger free radical reactions in susceptible patients. It has been pointed out that industrial hydrocarbons are frequently contaminated with chlorinated hydrocarbons (Walder 1983).

D. Chemical Leukoderma

Skin depigmentation (leukoderma) can be caused by various chemical agents, including azelaic acid (Nazzaro-Porro 1987), the glutathione reductase inhibitor carmustin (BCNU) (Hilger et al. 1974), bleomycin (Bork 1985) [bleomycin may also cause hyperpigmentation (Perrot and Ortonne 1978; Lowitz 1975)], certain mercaptoamines (Bleehen et al. 1968), Cer(IV) oxide (Rapaport 1982), alkyl- and arylphenols, and mercurials.

Topical application of *mercurials* was used for centuries for skin lightening. Because of nephrotoxic side effects, this kind of depigmenting therapy is obsolete. The mechanism is not known. Lipid peroxidation is observed in mercuric(II) chloride toxicity (Yonaha et al. 1982) and could be related to some effect of it on thiols (Gstrauenthaler et al. 1983). Depletion of cellular glutathione and other antioxidant thiols may then render the cells more susceptible to oxidative attack. Furthermore, mercuric(II) chloride inhibits cellular antioxidant systems, such as SOD, catalase, and glutathione peroxidase (Gstrauenthaler et al. 1983). The prooxidant state may then cause damage of the melanocytes, whose antioxidant capacity seems to be less effective than the antioxidant defense of other skin cell types (Muglia et al. 1986; Yohn et al. 1991).

The depigmenting effect of *phenolic compounds* (Fig. 5.17) was first noted in factory workers who whore rubber gloves containing hydroquinone monobenzyl ester as an antioxidant (Oliver et al. 1939, 1940). Phenolic reagents probably interfere with melanin biosynthesis by free radical type reactions. The following phenol derivatives have been reported to cause skin depigmentation: *p*-cresol (Shelley 1974), *p*-hydroquinone, *p*-hydroquinone monobenzyl ester (Spencer 1961; Snell 1964; Bentley-Phillips and Bayles 1975), *p*-tert-butylcatechol (Gellin et al. 1970), *p*-isopropylcatechol (Bleehen et al. 1968), *p*-hydroxyanisol, *p*-hydroxydiphenyl (Riley 1969 a, b), 1,3 di-tert-butyl-4-methylphenol (Vollum 1971), *p*-tert-amylphenol, *p*-tert-butylphenol (Kahn 1970).

Although other phenolic compounds such as epinephrine and norepinephrine are rapidly oxidized by tyrosinase in vitro, they do not cause skin depigmentation. Structure-activity analysis of depigmenting hydroquinones revealed that hydroxylation in the *para* position relative to a nonpolar side chain was an optimal requirement. Depigmenting compounds with an ether link in the side chain were more effective than those with hydrocarbon chains. 4-Isopropylcatechol is the most potent depigmenting agent, found by screening 33 catechol derivates, quinones, mercaptoamines, and thiols (Bleehen et al. 1968). The depigmenting effect of *m*-substituted catechols is weaker than that of *p*-substituted catechols (Kahn 1970).

Several mechanisms of melanocytotoxicity of phenolic compounds, including the formation of phenolic free radicals, have been discussed (Bolognial and Pawelek 1988). Tyrosinase is the critical enzyme that ultimately regulates melanogenesis (Hearing and Jimenez 1987), and the pigment cell

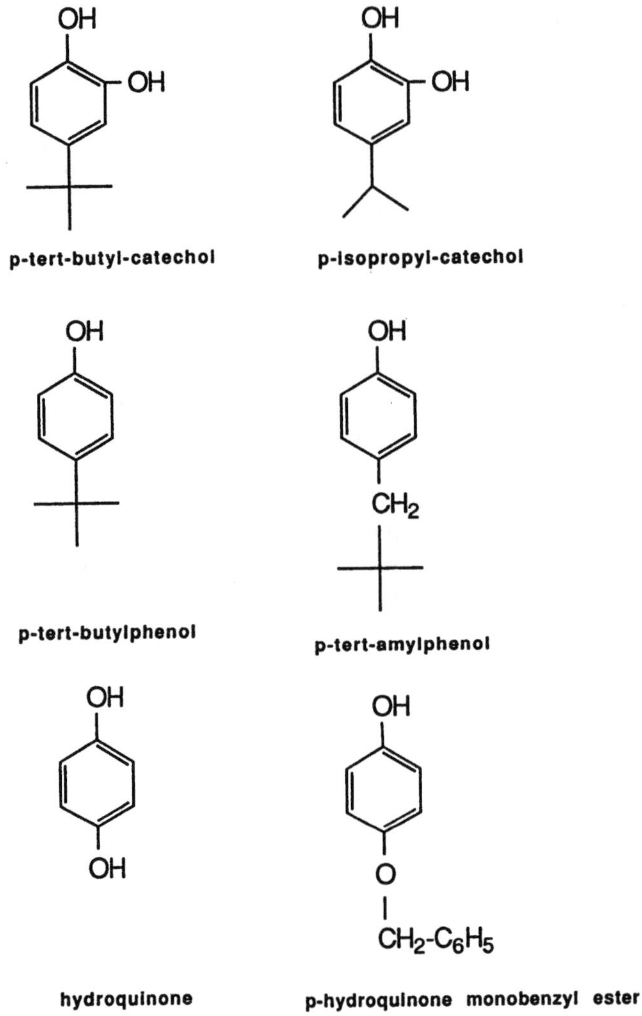

Fig. 5.17. Depigmenting phenolic compounds

specificity of the depigmenting chemicals can be explained by their selective uptake and metabolism by tyrosinase. Their reactive oxidation products, e.g., semiquinone radicals, may induce cytotoxicity via the induction of lipid peroxidation and redox cycling. The stability and reactivity of the semiquinone radical determine the extent of the depigmenting effect.

The cytotoxic effect of *p-hydroxyanisol* is selective towards melanocytes in mixed cultures containing keratinocytes and pigmented cells. This indicates that the melanocytotoxic effect is exerted by tyrosinase oxidation products (Riley 1969b). It was suggested that resonance-stabilized, semiquinone-type radicals of *p*-hydroxyanisol (anisyl semiquinone), which are formed inside the melanocytes via tyrosinase (Fig. 5.18), are melanocytotoxic and thus cause se-

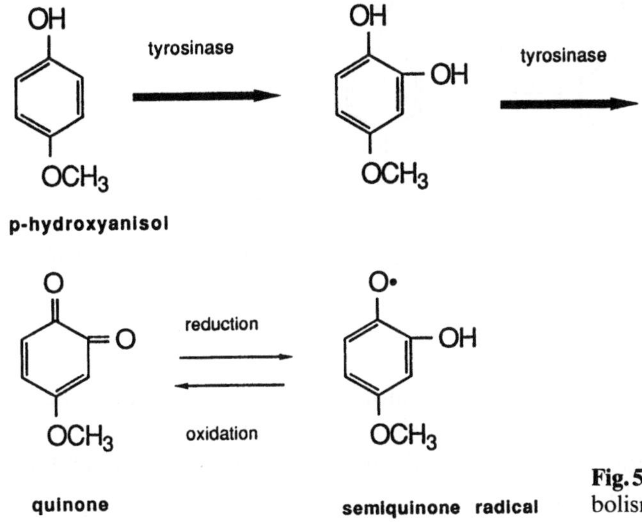

Fig. 5.18. Hydroxyanisol metabolism

lective destruction (Riley 1969b; Jimbow et al. 1974). Recently, however, the semiquinone radical of 4-hydroxyanisol was found to be quite unreactive with oxygen, indicating that superoxide anion radicals are not involved in the cytotoxic mechanism. The anisyl-3,4-quinone, which readily reacts with nucleophils, or less likely the anisyl semiquinone radical, are considered the cytotoxic species (Riley et al. 1989). Encouraging initial results have been obtained from clinical pilot studies using intraarterial infusion of *p*-hydroxyanisol in patients with localized recurrences of malignant melanoma (Riley 1985).

Treatment of guinea pig skin with *p*-hydroxyanisol results in the formation of a stable free radical species in the epidermis (Riley 1970). The lipohilic antioxidant α-tocopherol partially inhibits depigmentation of guinea pig skin induced by *p*-hydroxyanisol, wehreas the hydrophilic antioxidant ascorbate is not protective (Riley 1969a). In a melanocyte cell culture assay, ascorbate protects melanocytes from *p*-hydroxyanisol cytotoxicity, whereas α-tocopherolsuccinate does not (Riley et al. 1975). In vitro studies using guinea pig melanocytes revealed that the melanocytotoxic action of *p*-hdroxyanisol was prevented by antioxidants such as α-tocopherol, ubiquinone-6, cysteine, and ascorbate. These results clearly indicate that antioxidants can prevent the melanocytotoxic effects of *p*-hydroxyanisol. This favors the hypothesis of alkyl-, arylcatechol, free-radical-induced melanocytotoxicity.

As a corollary, the depigmenting effect of *hydroquinone* is believed to result from selective action on melanocyte metabolism and involves toxic tyrosinase-mediated oxidation products of hydroquinone (Penney et al. 1984). In the presence of a peroxidase and hydrogen peroxide, *p*-semiquinone radicals are immediately formed from hydroquinone and persist for several hours, whereas during nonenzymic autoxidation, only transient formation of small quantities of semiquinone radicals are observed (Yamazaki et al. 1960). Flavin-containing cytochrome b_5 and cytochrome *c* oxidoreductase can cata-

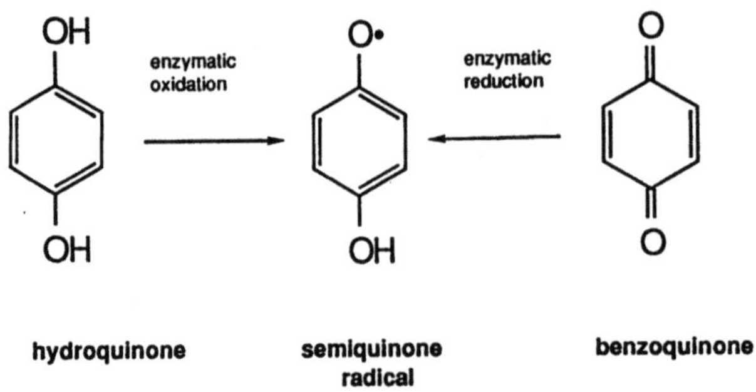

Fig. 5.19. Hydroquinone metabolism

lyze one-electron reduction of benzoquinone to the semiquinone radical (Iyanagi and Yamazai 1969). The semiquinone radical may also be formed by nonenzymatic comproportionation of hydroquinone and quinone in an electron transfer reaction (Fig. 5.19).

E. Chemical-Induced Comedogenesis

Chloracne was fist described by Von Bettmann in 1897 (Von Bettmann 1901) and 2 years later by Herxheimer. Herxheimer described acne among workers manufacturing hydrochloric acid and blamed chlorine gas for causing it; indeed, chlorinated aromatic hydrocarbons were responsible. Until the 1940s most cases of chloracne were due to exposure to chlorinated hydrocarbons. More recently, chlorinated herbicides and dioxins (Agent Orange = a 50:50 mixture of 2,4-dichlorophenoxyacetic acid and 2,3,7,8-tetrachlorodibenzo-*p*-dioxin) are a common cause. One of the key events of comedo formation is an increased proliferation and retention of horny cells in the infrainfundibulum of sebaceous glands. Decreased dehiscence of horny cells may be related to changes in the intercellular cement (lipids).

Comedogenesis is defined as the process of abnormal differentiation of the follicular epithelium that results in the formation of microcomedones. The microcomedones may evolve into noninflammatory comedos or inflammatory papules, nodules, or pustules. Comedo formation occurs in the course of various skin diseases, such as acne vulgaris, Favre-Racouchote disease and drug eruptions. Comedos can also be brought about by occupational hazards. Many kinds of topical substances, such as sebum, tar, oils, and chlorinated hydrocarbons, are known to produce acneiform eruptions in human skin. Physical agents that may cause acneigenic reactions are UV and ionizing irradiation.

All of the major components of sebum are comedogenic in the rabbit ear model (Kligman 1968; Kligman and Katz 1968). Glycerides are only weakly comedogenic. Comedogenicity of free fatty acids decreases with chain length from C12 to C20. Unsaturated fatty acids are more potent comedogenic agents than saturated fatty acids, and the hydroperoxide of oleic acid is more comedogenic than the parent compound itself. Squalene and its peroxides are the most potent comedogenic compounds in man (Mills et al. 1978; Motoyoshi 1983).

Ayres hypothesized that free radicals and lipid peroxidation participate in the pathophysiology of acne vulgaris and suggested that antioxidants such as tocopherol may useful in its treatment (Ayres and Mihan 1978c). It was suggested that the oxidation of squalene may induce comedogenesis and cause a microaerophilic condition in the pilosebaceous unit via reducing oxygen tension in the follicle. This may favor growth of anaerobic propionibacteria which synthesize porphyrins. Porphyrins catalyze oxidation of squalene, thereby self-perpetuating comedogenesis (Saint-Leger et al. 1986). The total absence of highly peroxidizable squalenes from monkey skin surface lipids can be related to the lack of acne in such animals (De Luca et al. 1989).

Peroxidized fatty acids, which cause epidermal hyperplasia in mice (Klein-Szanto and Slaga 1982), were suggested to be the species responsible for follicular hyperkeratosis and hyperplasia. However, another hypothesis suggests that high rates of sebum production result in low levels of linoleate in the sebum, subjecting the follicular epithelium to essential fatty acid deficiency and the characteristic follicular hyperkeratosis that results in comedo formation. The proportion of linoleic acid is markedly decreased in acne comedones (Wertz et al. 1985; Perisho et al. 1988). Deficiency of essential fatty acids containing linoleic acid in humans induces a scaly dermatosis, extensive transepidermal water loss, and epidermal hyperproliferation (Ziboh and Chapkin 1987). Suppression of sebum production by drugs elevates sebum linolate concentration and relieves follicular hyperkeratosis (Downing et al. 1987).

It was recently reported that linoleic acid inhibits phagocytosis and generation of reactive oxidants in human neutrophils, without affecting chemotaxis. The authors suggested that the decreased level of linolate in acne comedones contributes to the inflammatory process by a failure to inhibit neutrophil phagocytosis and reactive oxidant generation (Akamatsu et al. 1990a).

The role of chemicals in inducing comedogenesis is clearly established in the case of coal tar and halogenated hydrocarbons. These compounds induce comedos in the rabbit ear model and also produce acne in humans. Chlorinated aromatic hydrocarbons frequently produce chloracne in humans (Kimmig and Schulz 1957). 2,3,7,8-Tetrachlorodibenzo-*p*-dioxin (TCDD; Fig. 5.20) is comedogenic by topical contact or oral systemic administration in animals. The mechanism of comedo formation, however, remains obscure. In 1976, after an accident at a chemical plant in Seveso, Italy, chloracne was the most

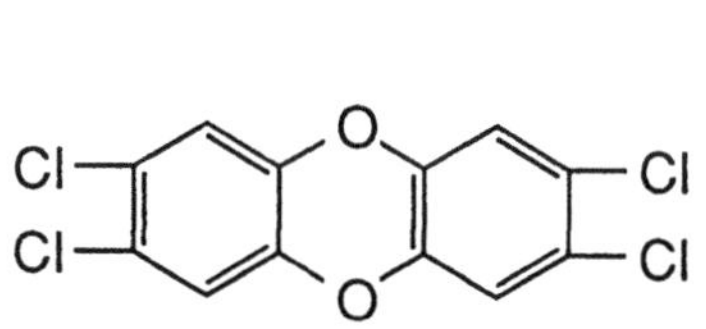

Fig. 5.20. 2,3,7,8-Tetrachlorodibenzo-p-dioxin

frequent cutaneous manifestation in 2,3,7,8-tetrachlorodibenzo-*p*-dioxin toxicity in humans (Goldmann 1973; Caputo et al. 1988), arising 30–60 days after exposure.

Although TCDD may affect epidermal differentiation and growth by influencing a specific receptor in epidermal basal cells, a role of increased lipid peroxidation in TCDD toxicity has been discussed (Albro et al. 1978; Sweeney et al. 1979). Furthermore, it was recently revealed that TCDD induces the formation of reactive oxygen species and inhibits oxidative phosphorylation in mitochondria (Nohl et al. 1989). TCDD probably stimulates the formation of pathophysiological redox cycling at the level of mitochondrial cytochrome *b* (Nohl 1989). It is speculated that lipid peroxidation induced by TCDD may contribute to comedo formation in skin. In hairless mice, vitamin A deficiency increases the sensitivity of the skin to TCDD-induced toxicity (Puhvel et al. 1990). However, more complex mechanisms than oxidative imbalance may contribute to TCDD comedogenicity.

Synopsis

Free radicals and reactive oxygen species play a considerable role in the pathophysiology of a wide spectrum of human diseases. In vitro free radical reactions are well understood and thoroughly analyzed. The biological consequences and clinical significance of free radical reactions in vivo, however, is only poorly understood. In most human diseases reactive oxidants are produced in excess as a consequence of the primary disease process, and not as a cause of it. It must be pointed out that activated oxygen species and other free radicals are essential components of many biochemical pathways. Pathology may also evolve from conditions in which there is decreased formation of reactive oxygen species or of supranormal levels of antioxidants are present. Reactive oxygen species can cause extensive tissue damage but are also required for the host's defense against infection. They can stimulate inflammation or may inhibit it by various mechanisms and are suggested to play a role as cellular messengers and immunomodultors. This duplicity of action will limit the feasibility of antioxidant therapy unless the precise molecular mechanisms of oxidant interaction with biological matter are elucidated. As a corollary, drugs may mediate pharmacological and a toxic effect via the same reactive oxidant species. Skin diseases where oxidative imbalance is involved in the pathogenesis may be treated effectively with antioxidant and/or prooxidant drugs. This indicates, again, that the biochemical fine tuning of the "antioxidant-prooxidant equilibrium" in a cellular microenvironment is an important factor. New methodological approaches and techniques are required to gain more detailed basic information on oxidative imbalance in skin diseases. Presently, we are only at the very beginning of understanding the clinical significance of oxidative imbalance. There is little direct evidence that free radicals and reactive oxygen species are *primarily* involved in cutaneous pathological processes. Most of the existing experimental evidence is indirect and based upon effects which appear to be abolished by free radical scavengers and on the detection of free radical reaction products, such as spin adducts and lipid peroxidation products of nucleic acids, lipids, and proteins.

Certain criteria must be considered for accepting a significant role of free radicals in a specific disease process (Procter and Reynolds 1984):

1. The disease must be known to be associated with an abnormal flux of free radicals or their intermediates.
2. Specific radical species or their characteristic reaction products must be demonstrated at the site of a lesion.

3. It must be demonstrated in vitro that radical species are involved in important mechanisms relevant to the specific disease in question.
4. Similar symptoms must be produced by otherwise dissimilar etiologic agents which generate free radical species in common or which inhibit or deplete components of the natural antioxidant defense system.
5. It must be possible modulate pathogenesis of a disease pharmacologically through the administration of antioxidants or free radical quenchers.

According to current knowledge, no dermatological disorder qualifies according to these criteria as an "oxidative injury disease." It seems unlikely that reactive oxidants are the only major significant determinant in skin inflammation, carcinogenesis, aging, ischemia, drug toxicity, and distinct dermatoses. Overemphasizing an all-encompassing role of oxidative injury in major pathological processes neglects their known complexity. However, numerous experimental and clinical studies do support the hypothesis that oxidative imbalance is *one* of the major determinants in general skin pathology and in the pathogenesis of certain skin diseases. The concept of oxidative injury in dermatopathology is a stimulating idea that needs to be tested extensively in future clinical and experimental work. It is a useful biomedical simplification that will help to improve our understanding of skin diseases. With the rapid advances occurring in this field of dermatology research, new diagnostic and therapeutic approaches may evolve and improve clinical patient management. My own view is that the future understanding of dermatoses will significantly advance if we continue to approach dermatology from the perspective of a "radical biologist."

References

Aberer W, Konrad K, Wolff K (1988) Well's syndrome is a distinctive disease entity and not a histologic diagnosis. J Am Acad Dermatol 18: 105–114

Ackermann Z, Seidenbaum M, Loewenthal E, Rubinow A (1988) Overload of iron in the skin of patients with varicose ulcers. Arch Dermatol 124: 1376–1378

Aebi H, Baggiolini M, Dewald B, Lauber E, Suter H, Micheli A, Frei J (1964) Observations in two Swiss families with acatalasia. II. Enzymol Biol Clin 4: 121–151

Ager A (1982) Effects of xanthine: xanthine oxidase on membrane function: an in-vitro model of endothelial damage. Agents Action 11 [Suppl]: 73–81

Aggarwal BB, Avi-Dor Y, Tinberg HM, Packer L (1976) Effect of visible light on the mitochondrial inner membrane. Biochem Biophys Res Commun 69: 362–368

Aggarwal BB, Quintanilha AT, Cammack R, Packer L (1978) Damage to mitochondrial electron transport and energy coupling by visible light. Biochim Biophys Acta 502: 367–382

Ågren MS (1990) Percutaneous absorption of zinc from zinc oxide applied topically to intact skin in man. Dermatologica 180: 36–39

Ahmad SI (1981) Synergistic killing of coliphage T7 by near ultraviolet radiation plus hydrogen peroxide: possible role of superoxide radicals. Photobiochem Photophys 2: 173–180

Ahmed HM, Laryea MD, El-Karib AO, El-Amin EO, Biggemann B, Leichsenring M, Mayatepek E, Bremer HJ (1990) Vitamin E status in Sudanese children with protein energy malnutrition. Z Ernährungswiss 29: 47–53

Aitken RJ, Clarkson JS (1987) Cellular basis of defective sperm function and its association with the genesis of reactive oxygen species by human spermatozoa. J Reprod Fertil 81: 459–469

Akamatsu H, Oguchi M, Nishijima S, Asada Y, Takahashi M, Ushijima T, Niwa Y (1990a) The inhibition of free radical generation hy human neutrophils through the synergistic effects of metronidazole with palmitoleic acid: a possible mechanism of action of metronidazole in rosacea and acne. Arch Dermatol Res 282: 449–454

Akamatsu H, Komura J, Miyachi Y, Asada Y, Niwa Y (1990b) Suppressive effects of linoleic acid on neutrophil oxygen metabolism and phagocytosis. J Invest Dermatol 95: 271–274

Akimov VG (1987) Role of peroxidation of membrane lipids in the pathogenesis of discoid lupus erythematosus. Vestn Dermatol Venerol 6: 7–9 (in Russian)

Akimov VG (1987) Contribution of iron to the pathogenesis of free radical processes induced by ultraviolet irradiation of the skin of guinea pigs. Vestn Dermatol Venerol 12: 7–10 (in Russian)

Akimov VG, Lashmanova AP, Marzeeva GI (1984) Peroxidation of membrane lipids in experimental photochemotherapy. Vestn Dermatol Venerol 8: 22–25 (in Russian)

Albro PW, Corbett JT, Harris M, Lawson LD (1978) Effects of 2,3,7,8-tetrachlorodibenzo-*p*-dioxin on lipid profiles in tissues of the Fischer rat. Chem Biol Interact 23: 315–320

Alexander H, Miller DL (1979) Determining skin thickness with pulsed ultrasound. J Invest Dermatol 72: 17–19

Al-Khalidi UAS, Chaglassian TH (1965) Species distribution of xanthine oxidase. Biochem J 97: 318–320

Allison AC, Magnus IA, Young MR (1966) Role of lysosomes and of cell membranes in photosensitization. Nature 209: 874–878

Ames BN, Cathcart R, Schwiers E, Hochstein P (1981) Uric acid provides an antioxidant defense in humans against oxidant- and radical-caused aging and cancer: a hypothesis. Proc Natl Acad Sci USA 78: 6858–6862

Ames BN (1983) Dietary carcinogenes and anticarcinogens. Science 221: 1256–1264

Ames BN (1989) Endogenous oxidative DNA damage, aging and cancer. Free Radic Res Commun 7: 121–128

Ames BN (1990) Oxidative DNA damage and aging. Free Radic Biol Med 9 (Suppl 1): 45

Ames BN, Hollstein MC, Cathcart R (1982) Lipid peroxidation and oxidative damage to DNA. In: Yagi K (ed) Lipid peroxides in biology and medicine. Academic, New York, pp 339–351

Ames JR, Ryan MD, Klayman DL, Kovacic P (1985) Charge transfer and oxy radicals in antimalarial action. Quinones, dapsone metabolites, metal complexes, iminium ions and peroxides. J Free Radic Biol Med 1: 353–361

Ames JR, Ryan MD, Kovacic P (1986) Mechanism of antibacterial action. Electron transfer and oxy radicals. J Free Radic Biol Med 2: 377–391

Ames JR, Ryan MD, Kovacic P (1987a) Mode of action of antiprotozoan agents. Electron transfer and oxy radicals. Life Sci 41: 1895–1902

Ames JR, Hollstein U, Gagneux AR, Ryan MD, Kovacic P (1987b) An integrated concept of amebicidal action: electron transfer and oxy radicals. Free Radic Biol Med 3: 85–96

Anderson AS, Galdys GJ, Green RC, Hohisel DW, Brown EP (1975) Improved reduction of cutaneous bacteria and free fatty acids with new benzoyl peroxide gel. Cutis 16: 307–310

Anderson RR (1989a) In-vivo fluorescence of human skin. Arch Dermatol 125: 999–1000

Anderson R (1989b) Dithranol mediated, dose dependent priming and activation of luminol enhanced chemiluminescence responses of human neutrophils in-vitro. Br J Dermatol 121: 1–9

Anderson R, Lukey PT, Dippenaar U, Eftychis HA, Findlay GH, Wooten MW, Nel AE (1987) Dithranol mediates pro-oxidative inhibition of polymorphonuclear leukocyte migration and lymphocyte proliferation. Br J Dermatol 117: 405–418

Anderson R, Zeis BM, Anderson IF (1988) Clofazimine mediated enhancement of reactive oxidant production by human phagocytes as a possible therapeutic mechanism. Dermatologica 176: 234–242

Anderson TF, Voorhees JJ (1980) Psoralen photochemotherapy of cutaneous disorders. Annu Rev Pharmacol Toxicol 20: 235–257

Andrews JS (1960) Toxicity of air oxidized soybean oil. J Nutr 70: 199–203

Angel MF, Narayanan K, Swartz WM, Ramasastry SS, Kuhns DB, Basford RE, Futrell JW (1986a) Deferoxamine increases skin flap survival: additional evidence of free radical involvement in ischemic flap surgery. Br J Plast Surg 39: 469–472

Angel MF, Narayanan K, Swartz WM, Ramasastry SS, Basford RE, Kuhns DB, Futrell JW (1986b) The etiologic role of free radicals in hematoma induced flap necrosis. Plast Reconstr Surg 77: 795–801

Angel MF, Ramasastry SS, Swartz WM, Basford RE, Futrell JW (1987) The cause of skin ulcerations associated with venous insufficiency: a unifying hypothesis. Plast Reconstr Surg 79: 289–297

Anneren KG, Epstein CJ (1987) Lipid peroxidation and superoxide dismutase-1 and glutathione peroxidase activities in trisomy 16 fetal mice and human trisomy 21 fibroblasts. Pediatr Res 21: 88–92

Annesley TM, Till GO, Ward PA (1985) Cutaneous thermal burn and oxidant mediated acute lung injury: appearance in serum of lung related LDH isoenzyme. J Free Radic Biol Med 1: 409–414

Anonymous (1978) A radical approach to zinc. Lancet i: 191–192 (editorial)

Anonymous (1981) Effect of vitamin E on prostanoid biosynthesis. Nutr Rev 39: 317–320

Anton'ev AA, Nomnoeva TN (1986) Use of hyperbaric oxygenation in dermatology. Vestn Dermatol Venerol 2: 27–31 (in Russian)

Anton'ev AA, Nemnoeva TN, Mel'nikova ON, Nifontova VI, Sidel'nikov VI (1988) Lipid

peroxidation in the foci of lesions in patients with scleroderma during treatment of hyperbaric oxygenation. Vestn Dermatol Venerol 1: 8–9 (in Russian)
Anttinen H, Ryhanen L, Oikarinen A (1980) Effects of divalent cations on collagen biosynthesis in isolated chick embryo tendon cells. Biochim Biophys Acta 609: 321–328
Anttinen H, Ryhanen L, Puistola V, Arranto A, Oikarinen A (1984) Decrease in liver collagen accumulation in carbon tetrachloride injured and normal growing rats upon administration of zinc. Gastroenterology 86: 532–539
Aoki T, Iriyama K (1986) Uric acid in burned skin excised from a person burnt to death. Jikeikai Med J 33: 209–215
Arbesfeld SJ, Kurban AK (1988) Behcet's disease. J Am Acad Dermatol 19: 767–779
Arduini A, Mezetti A, Porreca E, Lapenna D, DeJulia J, Marzio L, Polidoro G, Cuccurollo F (1988) Effect of ischemia and reperfusion on antioxidant enzymes and mitochondrial inner membrane proteins in perfused rat hearts. Biochim Biophys Acta 970: 113–121
Armstrong D (1984) Free radical involvement in the formation of lipopigments. In: Armstrong D, Sohal R, Cutler R, Slater T (eds) Free radicals in molecular biology, Aging and disease. Raven, New York, pp 129–141
Armstrong D, Koppang N, Rider J (1982) Ceroid lipofuscinosis (Batten's disease). Elsevier, Amsterdam
Aronoff S (1965) Catalase: kinetics of photooxidation. Science 150: 72–73
Artuc M, Ramshad M, Kappus (1989) Studies on acute toxic effects to keratinocytes induced by hematoporphyrin derivatives and laser light. Arch Dermatol Res 281: 491–494
Artue M, Kappus H (1986) Benzoyl peroxide induced lipid peroxidation in cultured human keratinocytes. J Invest Dermatol 86: 328
Asada K, Kanematsu S (1976) Reactivity of thiols with superoxide radicals. Agric Biol Chem 40: 1891–1892
Ashwood-Smith MJ (1975) Current concepts concerning radioprotective and cryoprotective properties of dimethyl sulfoxide in cellular systems. Ann NY Acad Sci 243: 246–256
Asokan P, Das M, Rosenkranz HS, Bickers DR, Mukhtar H (1985) Topically applied nitropyrenes are potent inducers of cutaneous and hepatic monooxygenase. Biochem Biophys Res Commun 129: 134–140
Athar M, Bickers DR, Mukhtar H (1989) Inhibition of benzoyl peroxide tumor promotion in 7,12-dimethylbenz[*a*]anthracene treated skin of Sencar mice by the antioxidants nordihydroguaiaretic acid and diallyl sulfide. J Invest Dermatol 92: 398
Au AM, Chan PH, Fishman RA (1985) Stimulation of phospholipase A2 activity by oxygen derived free radicals in isolated brain caillaries. J Cell Biochem 27: 449–453
Aubin F, Manteaux A (1990) Mecanismes de photoreaction des psoralenes. Ann Dermatol Venereol 117: 931–936
Auclair C, Gouyette A, Levy A, Emerit I (1990) Clastogenic inosine nucleotide as components of the chromosome breakage factor in scleroderma patients. Arch Biochem Biophys 278: 238–244
Auffranc IC, Berbis P, Fabre JF, Garnier JP, Privat Y (1985) Syndrome sclerodermiforme et poikilodermique observe au cours dun traitment par carbidopa et 5-hydroxy-tryptophan. Ann Dermatol Venereol 112: 691–692
Augusto O, Packer L (1981) Selective inactivation of microsomal drug metabolizing proteins by visible light. Photochem Photobiol 33: 765–767
Auletta M, Gange RW, Tan OT, Matzinger EM (1986) Effect of cutaneous hypoxia upon erythema and pigment response to UVA, UVB and PUVA (8-MOP + UVA) in human skin. J Invest Dermatol 86: 649–652
Aune TM, Pierce CW (1981) Conversion of soluble immune response suppressor to macrophage derived suppressor factor by peroxide. Proc Natl Acad Sci USA 78: 5099–5103
Aust SD, Morehousen LA, Thomas CE (1985) Role of metals in oxygen radical reactions. J Free Radic Biol Med 1: 3–25
Ayres S, Mirhan R (1969) Pseudoxanthoma elasticum and epidermolysis bullosa – response to vitamin E (tocopherol). Cutis 5: 287–294
Ayres S, Mihan R (1971) Vitamin E (tocopherol) – a reappraisal of its value in dermatoses of mesodermal tissues. Cutis 7: 35–45

Ayres S, Mihan R (1972) Keratosis follicularis. Arch Dermatol 106: 909–910
Ayres S, Mihan R (1973) Yellow nail syndrome. Arch Dermatol 108: 267–268
Ayres S, Mirhan R (1974) Subcorneal pustular dermatosis controlled by vitamin E. Arch Dermatol 109: 914
Ayres S, Mihan R (1975) Vitamin E and dermatology. Cutis 16: 1017–1021
Ayres S, Mihan R (1978b) Acne vulgaris and lipid peroxidation: new concepts in pathogenesis and treatment. Int J Dermatol 17: 305–307
Ayres S, Mihan R (1978a) Is vitamin E involved in the autoimmune mechanism? Cutis 21: 321–325
Baart De La Faille H, Suurmond D, Went LN, Stevenick J, Van Schothorst AA (1972) β-Carotene as a treatment for photohypersensitivity due to erythropoietic protoporphyria. Dermatologica 145: 389–394
Babior BM (1978a) Oxygen dependent microbial killing by phagocytes (first part). N Engl J Med 298: 659–668
Babior BM (1978b) Oxygen dependent microbial killig by phagocytes (second part). N Engl J Med 298: 721–725
Babior BM, Kipnes RS, Curnutte JT (1973) Biological defense mechanisms: the production by leukocytes of superoxide, a potential bactericidial agent. J Clin Invest 52: 741–744
Bada JL (1975) Amino acid racemisation reactions and their geochemical implications. Naturwissenschaften 62: 71–79
Badwey JA, Karnovsky ML (1980) Active oxygen species and the functions of phagocytic leukocytes. Annu Rev Biochem 49: 695–726
Bagatell FK, Dimitrov K, Dugan K (1967) An epidermal particulate fraction rich in mitochondria capable of independent respiration. J Invest Dermatol 48: 174–180
Bailey AJ, Ranta MH, Nicholls AC, Partidge SM, Elsden DF (1977) Isolation of α-amino adipic acid from mature dermal collagen and elastin. Evidence for an oxidative pathway in the maturation of collagen and elastin. Biochem Biophys Res Commun 78: 1403–1410
Baillet F, Housset M, Michelson AM, Puget K (1986) Treatment of radiofibrosis with liposomal superoxide dismutase. Preliminary result of 50 cases. Free Radic Res Commun 1: 387–394
Bain JA, Rusch HP, Kline BE (1943) Effects of temperature upon ultraviolet carcinogenesis with wavelenghts of 2800–3400 Å. Cancer Res 3: 610–612
Baines EJ, McFadzean JAM (1985) The action of metronidazole on anaerobic bacilli and similar organisms. Adv Pharmacol Chemother 18: 224–265
Baker DG, Leith JT (1975) Protection of the skin of mice against irradiation with cyclotron-accelerated helium ions by 2-mercaptoethylamine. Acta Radiol Ther Phys Biol 14: 561–571
Baker MA, Tappel AL (1986) Effects of ligands on gold inhibition of selenium glutathione peroxidase. Biochem Pharmacol 35: 2417–2422
Baldrige CW, Gerard RW (1933) The extra respiration of phagocytosis. Am J Physiol 103: 235–236
Balentine JD (1982) Pathology of oxygen toxicity. Academic, New York
Balin AK, Pratt LA (1989) Physiological consequences of human skin aging. Cutis 43: 431–436
Ballou D, Palmer G, Massey V (1969) Direct demonstration of superoxide anion production during the oxidation of reduced flavin and its catalytic decomposition by erythrocuprein. Biochem Biophys Res Commun 36: 898–904
Banford JC, Brown DH, Hazelton RA, McNeil CJ, Sturrock RD, Smith WE (1982) Serum copper and erythrocyte superoxide dismutase in rheumatoid arthritis. Ann Rheum Dis 41: 458–462
Bannister JV, Bannister WH (1985) Production of oxygen centered radicals by neutrophils and macrophages as studied by electron spin resonance (ESR). Environ Health Perspect 64: 37–43
Bannister JV, Bannister WH, Hill HAO, Mahood JF, Wilson RL, Wolfenden BS (1980) Does ceruloplasmin dismutate superoxide? FEBS Lett 118: 127–129
Bannister WH, Bannister JV (1987) Factor analysis of the activities of superoxide dismu-

tase, catalase and glutathione peroxidase in normal tissues and neoplastic cell lines. Free Radic Res Commun 4: 1–13

Barber AA (1977) Inhibiton of lipid peroxide formation by vertebrate blood serum. Arch Biochem Biophys 92: 38–43

Barnhill RL, Doll J, Millikan LE, Hastings RC (1984) Studies on the antiinflammatory properties of thalidomide: effects on polymorphonuclear leukocytes and monocytes. J Am Acad Dermatol 11: 814–819

Bartak V (1972) Zur Behandlung der Fertilitätsstörungen des Mannes mit Vitamin E und Methyltestosteron. Z Haut Geschl Kr 47: 919–924

Bartsch G, Menander-Huber KB, Huber W, Marberger H (1981) Orgoetin, a new drug for the treatment of Pyronie's disease. Eur J Rheumatol Inflamm 4: 250–259

Bast A (1986) Is formation of reactive oxygen by cytochrome P-450 perilious and predictable? Trends Pharmacol Sci 7: 266–270

Bast A, Haenen GRMM (1988) Interplay between lipoic acid and glutathione in the protection against microsomal lipid peroxidation. Biochim Biophys Acta 963: 558–569

Battino M, Bertoli E, Formiggini G, Gorini A, Villa RF, Lenaz G (1990) Comparative investigations on animal mitochondria ubiquinone content. A qualitative and quantitative study. In: Lenaz G, Barnabei O, Rabbi A, Battino M (eds) Highlights in ubiquinone research. Taylor and Francis, London, pp 178–181

Baum BJ, Moss J, Breul S, Crystal R (1978) Association in normal human fibroblasts of elevated levels of adenosine 3′,5′-monophosphate with a selective decrease in collagen production. J Biol Chem 253: 3391–3394

Baum BJ, Moss J, Breul S, Crystal R (1980) Effect of cyclic AMP on the intracellular degradation of newly synthesized collagen. J Biol Chem 255: 2843–2847

Baum J (1975) The present and potential role of dimethyl sulfoxide in connective tissue disorders. Ann NY Acad Sci 243: 391–392

Baumann J, Wurm G, Bruchhausen F (1980) Hemmung der Prostaglandinsynthetase durch Flavonoide and Phenolderivate im Vergleich mit deren $O_2^{-}\cdot$ Radikalfängereigenschaften. Arch Pharm (Weinheim) 313: 330–337

Baumgardner JE, Graves DJ, Neufeld GR, Quinn JA (1985) Gas flux through human skin: effect of temperature, stripping, and inspired tension. J Appl Physiol 58: 1536–1545

Bayer R (1959) Ergebnisse der Vitamin E Vorbehandlung zur Beeinflussung der primären und sekundären essentiellen Infertilität (Bericht über 100 Ehen). Wien Med Wschr 109: 271–275

Baxendale JH (1964) Effects of oxygen and pH in radiation chemistry of aqueous solutions. Radiat Res [Suppl] 4: 114–138

Bazex A (1949) La vitamine E dans le traitment des lupus erythemateux chroniques. Bull Soc Fr Dermatol Syphiligr 56: 56–57

Bech-Thomsen N, Wulf H, Poulsen T, Lundgreen K (1988) Pretreatment with long wave ultraviolet light inhibits ultraviolet induced skin tumor development in hairless mice. Arch Dermatol 124: 1215–1218

Beckmann JK, Greene HL (1988) Effects of aurothioglucose on iron-induced rat liver microsomal lipid peroxidation. Lipids 23: 899–903

Behrens WA, Madere R (1987) A procedure for the determination of ascorbic and dehydroascorbic acid in biological fluids, tissues, and foods. Ann NY Acad Sci 498: 458–459

Bellavite P (1988) The superoxide forming enzymatic system of phagocytes. Free Radic Biol Med 4: 225–261

Bellomo G, Mirabelli F, Finardi G (1988) Oxygen free radicals cause perturbations of the cytoskeletal structure and organization. International symposium on free radical in medicine, Vienna, 8–12 Nov 1988. (abstr Cs 11)

Belman S, Troll W (1972) The inhibition of croton oil promoted mouse skin tumorigenesis by steroid hormones. Cancer Res 32: 450–454

Belongia E, Hedberg CW, Gleich GJ, White KE, Mayeno AN, Loegering DA, Dunnette SL, Pirie PL, McDonald KL, Osterholm MT (1990) An investigation of the cause of the eosinophilia-myalgia syndrome associated with tryptophan use. N Engl J Med 323: 357–365

Bender JG, Van Epps DE (1983) Analysis of the bimodal chemiluminescence pattern stimulated in human neutrophils by chemotactic factors. Infect Immun 41: 1062–1070
Benedetto AV (1977) The psoralenes: an historical perspective. Cutis 20: 469–471
Benedetto JP, Ortonne JP, Voulot C, Khatchadourian C, Prota G, Thivolet J (1981) Role of thiol compounds in mammalian melanin pigmentation: part I. Reduced and oxidized glutathione. J Invest Dermatol 77: 402–405
Bennet JP, Gomperts BD, Wollenweber E (1981) Inhibitory effects of natural flavonoids on secretion from mast cells and neutrophils. Arzneimittelforschung Drug Res 31: 433–437
Bentley-Phillips B, Bayles MAH (1975) Cutaneous reactions to topical application of hydroquinone. S Afr Med J 9: 1391–1395
Bereiter-Hahn J (1986) Epidermal cell migration and wound repair. In: Bereiter-Hahn J, Matoltsky AG, Richards KS (eds) Biology of the integument, vol 2: Vertebrates. Springer, Berlin Heidelberg New York, pp 443–471
Bereiter-Hahn J, Matoltsy AG, Richards KS (eds) (1984) Biology of the integument, vol 1: Invertebrates. Springer, Berlin Heidelberg New York
Bereiter-Hahn J, Matoltsy AG, Richards KS (eds) (1986) Biology of the integument, vol 2: Vertebrates. Springer, Berlin Heidelberg New York
Berenblum I (1941) The cocarcinogenic action of croton resin. Cancer Res 1: 44–48
Bernd A, Holzmann H, Marsch WC, Kurulec B, Britvic S, Mueller WEG (1987) Antimutagenic potency of the cytotoxic and anti-psoriatic compound anthralin (cignolin). Pharmacol Res Commun 190: 367–379
Bernd A, Holzmann H, Kurelec B, Britivic S, Müller WEG (1989) Zur cytotoxischen und antimutagenen Potenz von Ditranol. Dermatol Monatsschr 175: 261–267
Bernheim F, Wilbur KM, Kenaston CB (1952) The effect of oxidized fatty acids on the activity of certain oxidative enzymes. Arch Biochem Biophys 38: 177–184
Bernstein R, Prinsloo I, Zwi S, Andrew MJA, Dawson B, Jenkins T (1980) Chromosomal aberrations in occupation associated progressive systemic sclerosis. S Afr Med J 58: 235–237
Betts WH, Cleland LG, Gee DJ, Whitehouse MW (1984) Effects of D penicillamine on a model of oxygen derived free radical mediated tissue damage. Agents Action 14: 283–290
Beyersdorf F, Fuchs J, Eberhardt B, Stauder M, Satter P, Zimmer G (1989) Myocardial protection by 2-mercaptopropionylglycine during global ischemia in dogs. Arzneimittelforsch/Drug Res 39: 6–9
Bhatnagar RS, Liu TZ (1972) Evidence for free radical involvement in the hydroxylation of proline: inhibiton by nitro blue tetrazolium. FEBS Lett 26: 32–34
Bhatnagar RS, Hussain MZ, Streifel JA, Tolentino M, Enriquez B (1978) Alteration of collagen synthesis in lung organ cultures by hyperoxic environments. Biochem Biophys Res Commun 83: 392–397
Biaglow JE, Lavik PS, Ferencz N Jr (1969) Modification of the radiation response through glucose controlled respiration. Radiat Res 39: 623–633
Bianchi E, Gasparini S (1955) L'aumentata resistenza tissulare cutanea alle radiazioni x in rapporto alla sommistrazione di betamercaptoetilamina al 0.5% per infiltrazione sottocutanea nel tratto da irradiare. Radiologia 11: 1137–1142
Bickers DR (1981) The carcinogenicity and mutagenicity of therapeutic coal tar – a perspective. J Invest Dermatol 77: 173–174
Bickers DR, Demar LK, DeLeo V, Poh-Fitzpatrick MB, Aronberg JM, Harber LC (1978) Hydroa vacciniforme. Arch Dermatol 114: 1193–1196
Biemond P, van Eijk HG, Swaak AJG, Koster JF (1984) Iron mobilization from ferritin by superoxide derived from stimulated polymorphonuclear leukocytes. J Clin Invest 73: 1576–1579
Bieri JG, Anderson AA (1960) Peroxidation of lipids in tissue homogenates as related to vitamin E. Arch Biochem Biophys 90: 105–110
Bijdendijk A, Noordhoeck FJ (1951) Die Behandlung von Ulcus cruris mit hohen Dosen Vitamin E. Nederl Tijdschr Geneesk 1039–1042
Bin OH (1991) Nitric oxide: a revolution of free radical theory of aging. Med Hypotheses 34: 60–61

Bindoli A (1988) Lipid peroxidation in mitochondria. Free Radic Biol Med 5: 247–261
Bindoli A, Cavallini L, Siliprandi N (1977) Inhibitory action of silymarin on lipid peroxid formation in rat liver mitochondria and microsomes. Biochem Pharmacol 26: 2405–2410
Bindoli A, Cavallini L, Jocelyn P (1982) Mitochondrial lipid peroxidation by cumene hydroperoxide and its prevention by succinate. Biochim Biophys Acta 681: 496–503
Bingham EL, Fenger TW, Sugar A, Smith JW (1985) Dependence on antibody for induction of chemiluminescence in polymorphonuclear leukocytes by herpes simplex virus. Invest Ophthalmol Vis Sci 26: 1236–1243
Birnboim HC (1982) DNA strand breakage in human leukocytes exposed to a tumor promotor, phorbol myristate acetate. Science 215: 1247–1249
Bissett DL, Hannon DP, Orr TV (1987) An animal model for solar aged skin: Histological, physical and visible changes in UV-irradiated hairless mouse skin. Photochem Photobiol 43: 367–378
Bissett DL, Chatterjee R, Hannon DP (1990) Photoprotective effect of superoxide-scavenging antioxidants against ultraviolet radiation induced chronic skin damage in the hairless mouse. Photodermatol Photoimmunol Photomed 7: 56–62
Bjørneboe A, Bjørneboe GEA, Bodd E, Hagen BF, Kveseth N, Drevon CA (1986) Transport and distribution of a-tocopherol in lymph, serum, and liver cells in rats. Biochim Biophys Acta 889: 310–315
Black HS (1987 a) Potential involvement of free radical reactions in ultraviolet light mediated cutaneous damage. Photochem Photobiol 46: 213–221
Black HS (1987 b) Systemic photoprotective agents. Photodermatology 4: 187–195
Black HS (1987 c) Photocarcinogenesis and diet. Fed Proc 46: 1901–1905
Black HS, Chan JT (1975) Suppression of ultraviolet light-induced tumor formation by dietary antioxidants. J Invest Dermatol 65: 412–414
Black HS (1989) Role of reactive oxygen species in inflammatory process. In: Schroot B, Schaefer H (eds) Pharmacology of the skin. Karger, Basel, vol 2, pp 1–20
Black HS, Douglas DR (1972) A model system for the evaluation of the role of cholesterol alpha-oxide in ultraviolet carcinogenesis. Cancer Res 32: 2630–2632
Black HS, Douglas DR (1973) Formation of a carcinogen of natural origin in the etiology of ultraviolet light induced carcinogenesis. Cancer Res 33: 2094–2095
Black HS, Lo WB (1971) Formation of a carcinogen in human skin irradiated with ultraviolet light. Nature 234: 306–307
Black HS, Tigges J (1986) Evaluation of structurally related phenols for anti photocarcinogenic and photoprotective properties. Photochem Photobiol 43: 403–408
Black HS, Lenger W, Gerguis J, McCann V (1984) Studies on the photoprotective mechanism of butylated hydroxytoluene. Photochem Photobiol 40: 69–75
Black HS, Lenger WA, Gerguis J, Thornby JI (1985) Relation of antioxidants and level of dietary lipid to epidermal lipid peroxidation and ultraviolett carcinogenesis. Cancer Res 45: 6254–6259
Blacker K, Olsen E, Boyer T (1989) Expression of glutathione S-transferase (GST) in cultured keratinocytes. J Invest Dermatol 92: 404
Blacker KL, Olson E, Vessey DA, Boyer TD (1991) Characterization of glutathione S-transferase in cultured human keratinocytes. J Invest Dermatol 97: 442–446
Blackett AD, Hall DA (1980) The action of vitamin E on the ageing of connective tissues in the mouse. Mech Ageing Dev 14: 305–316
Blakeslee JR Jr, Yamamoto N, Hinuma Y (1985) Human T-cell leukemia virus I induction by 5-iodo-2′-deoxyuridine and *N*-methyl-*N*′-nitro-*N*-nitrosoguanidine: inhibition by retinoids, L-ascorbic acid, and *dl*-α-tocopherol. Cancer Res 45: 3471–3476
Blankenhorn G (1986) Klinische Wirksamkeit von Spondyvit[R] (Vitamin E) bei aktivierten Arthrosen. Z Orthop 124: 340–343
Blankenhorn G, Fischer I, Seuß J (1985) Möglichkeiten und Grenzen der Antioxidans-Therapie entzündlicher rheumatischer Erkrankungen mit hochdosiertem Vitamin E. Akt Rheumatol 10: 125–128
Bleehen SS, Pathak MA, Hori Y, Fitzpatrick TB (1968) Depigmentation of skin with 4-isopropylcatechol, mercaptoamines and other compounds. J Invest Dermatol 50: 103–117

Bligard CA (1987) Kawasaki disease and its diagnosis. Pediatr Dermatol 4: 75–84
Blois MS, Zahlan AB, Maling JE (1964) Electron spin resonance studies on melanin. Biophys J 4: 471–490
Blokha VV, Kagan VY, Sitkovskii MV, Danilov VS, Kol's OR, Kozlov YP (1972) Peroxidation of lipids and conduction of excitation in frog muscles. Biofizika 17: 549–551
Blum J, Fridovich I (1985) Inactivation of glutathione peroxidase by superoxide radical. Arch Biochem Biophys 240: 500–508
Blum A, Grossweiner LI (1985) Singlet oxygen generation by hematoporphyrin IX, uroporphyrin I and hematoporphyrin derivative at 546 nm in phosphate buffer and in the presence of egg phosphatidylcholine liposomes. Photochem Photobiol 41: 27–32
Blumberg PM, Jaken S, Konig B, Sharkey NA, Leach KL, Jeng AY, Yet E (1984) Mechanism of action of the phorbol esters tumor promotors: Specific receptors for lipophilic agents. Biochem Pharmacol 33: 933–940
Blumenthaler M, Ambach W, Canaval H (1985) Seasonal variations of solar UV-radiation at a high mountain station. Photochem Photobiol 42: 1147–1152
Bock BG, Burns R (1963) Tumor promoting properties of anthralin (1.8.9 anthracenetriol). J Natl Cancer Inst 30: 393–397
Böhlers H (1988) Resorptionsdynamik unterschiedlicher Vitamin E Mengen bei gesunden jungen Männern. Vitaminspur 3: 134–136
Böhm F, Meffert H (1981) PUVA verursachte Membranschädigung menschlicher Lymphocyten. Dermatol Monatsschr 167: 119–121
Bohnert E, Jung EG (1989) Nevoid basal cell carcinoma syndrome. Arch Dermatol Res 281: 218
Bolognia LJ, Pawelek JM (1988) Biology of hypopigmentation. J Am Acad Dermatol 19: 217–255
Bonifacic M, Asmus KD (1984) Radical reactions in aqueous disulfide-thiol systems. Int J Radiat Biol 46: 35–45
Bonnekoh B, Jugert F, Merk H, Mahrle G (1989) NAD(P)H: Quinone reductase (NQR) in human keratinocytes, induction by naphtoflavone (β-NF), tert.-butyl-hydroxyanisole (BHA), danthrone (D) and inhibition by anthralin (A). J Invest Dermatol 92: 406
Bonomi F, Pagani S (1986) Removal of ferritin bound iron by DL-dihydrolipoate and DL-dihydrolipoamide. Eur J Biochem 155: 295–300
Bonomi F, Ceriolio A, Pagani S (1989) Molecular aspects of the removal of ferritin bound iron by DL-dihydrolipoate. Biochim Biophys Acta
Borek C, Troll W (1985) Modifiers of free radicals inhibit in-vitro the oncogenic actions of X-rays, bleomycin and the tumor promotor 12-O-tetradecanoyl-phorbol-13-acetate. Proc Natl Acad Sci USA 80: 1304–1307
Borel JP (1983) Radicaux libres oxygenes, tissu conjonctif et inflammation. Pathol Biol (Paris) 31: 8–10
Borello S, Seccia A, Galeotti T, Bartoli GM, Farallo E, Serri F (1984) Protective enzymes in human epidermal carcinomas and psoriasis. Arch Dermatol Res 276: 338–340
Borg DC, Schaich KM, Elmore JJ, Bell JA (1978) Cytotoxic reactions of free radical species of oxygen. Photochem Photobiol 28: 887–907
Bork K (1985) Medikamentöse Zahnverfärbungen. In: Bork K (ed) Kutane Arzneimittelnebenwirkungen. Schattauer, Stuttgart, pp 297–302
Bork K, Korting GW (1983) Symptomatische Sklerodermie durch Bleomycin. Hautarzt 34: 10–12
Bors W, Heller W, Michel C, Saran M (1990) Flavonoids as antioxidants. Determination of their radical scavenging efficiencies. Methods Enzymol 186: 343–355
Boutwell RK (1985) Tumor promotors in human carcinogenesis. In: DeVita VT, Hellman S, Rosenberg SA (eds) Important advances in oncology 1985. Lippincott, Philadelphia, pp 16–27
Boveris A, Cadenas E, Reiter R, Filipkowski M, Nakase Y, Chance B (1979) Organ chemiluminescence: noninvasive assay for oxidative radical reactions. Proc Natl Acad Sci USA 77: 347–351

Bragt PC, Bansberg JI, Bonta IL (1980) Antiinflammatory effect of free radical scavengers and antioxidants. Inflammation 4: 289–305

Brasch RC, London DA, Wesbey GE, Tozer TN, Nitecki DE, Williams RD, Doemeny J, Dallas Tuck L, Lallemand DP (1983) Work in progress: Nuclear magnetic resonance studies of a paramagnetic nitroxide contrast agent for enhancement of renal structures in experimental animals. Radiology 147: 773–779

Braun-Falco O (1961) Zur Histotopographie der Cytochromoxidase in normaler und pathologisch veränderter Haut sowie in Hauttumoren. Arch Klin Exp Dermatol 214: 176–224

Braverman IM, Fonferko E (1982a) Studies in cutaneous aging. I. The elastic fibre work. J Invest Dermatol 78: 434–443

Braverman IM, Fonferko E (1982b) Studies in cutaneous aging. II. The microvasculature. J Invest Dermatol 78: 444–448

Bray RC, Cockle SA (1974) Reduction and inactivation of superoxide dismutase by hydrogen peroxide. Biochem J 139: 42–48

Bray TM, Bettger WJ (1990) The physiological role of zinc as an antioxidant. Free Radic Biol Med 8: 281–291

Breathnach SM (1988) Amyloid and amyloidosis. J Am Acad Dermatol 18: 1–16

Breimer LH, Lindahl T (1985) Enzymatic excision of DNA bases damaged by exposure to ionizing radiation of oxidizing agents. Mutat Res 150: 85–89

Breimer LH (1991) Repair of DNA damage induced by reactive oxygen species. Free Radic Res Commun 14: 159–171

Briggs RT, Karnovsky ML, Karnovsky MJ (1977) Hydrogen peroxide production in chronic granulomatous disease. J Clin Invest 59: 1088–1098

Brinkman R, Lamberts HB (1958) Ozone as a possible radiomimetic gas. Nature 181: 1202–1203

Brody I (1968) The epidermis. In: Gans O, Steigleider GK (eds) Normale und pathologische Anatomie der Haut. Springer, Berlin Heidelberg New York, pp 1–142 (Handbuch der Haut- und Geschlechtskrankheiten, vol 1, pt 1)

Brooksbank BWL, Balazs R (1984) Superoxide dismutase, glutathione peroxidase and lipoperoxidation in Down's syndrom fetal brain. Dev Brain Res 16: 37–44

Brown RG, Button GM, Smith JT (1967) Effect of vitamin E deficiency on collagen metabolism in the rat's skin. J Nutr 91: 99–106

Bruce JM, Gorman AA, Hamblett I, Kerr CW, Lambert C, McNeeney SP (1988) Characterization of the anthralin radical by pulse radiolysis and laser photolysis. In: Moreno G et al. (eds) Photosensitisation. Springer, Berlin Heidelberg New York, pp 73–75 (NATO ASI series, vol H15)

Bruce JM, Gorman AA, Hamblett I, Kerr CW, Lambert C, McNeeney SP (1989) Anthralin derived transients. I. The triplet state and the products of its reaction with oxygen in benzene. Photochem Photobiol 49: 439–445

Brugh M (1977) Butylated hydroxytoluene protects chicken exposed to Newcastle disease virus. Science 197: 1291–1292

Brun P, Juhlin L, Schalla W (1984) Short contact anthralin therapy of psoriasis and maintenance schedule to prevent relapses. Acta Derm Venereol (Stockh) 64: 174–177

Brunmark A, Brunk UT (1990) Superoxide radical exposure of lysosomes causes rapid loss of membrane integrity. Antioxidants and degenerative diseases, satellite meeting of the Oxygen Society and the International Society for Free Radical Research, Berkeley, CA, 26–27 Jan 1990

Bruynzeel I, Bergmann W, Hartevelt HM, Kenter CCA, van de Velde EA, Schothorts AA, Suurmond D (1991) High single dose European regimen also causes an excess of nonmelanoma skin cancer. Br J Dermatol 124: 49–55

Bruze M, Emmett EA, Creasey J, Strickland PT (1989) Cyclobutane dithymidine induction by solar simulating UV radiation in human skin: II. Individual responses. J Invest Dermatol 93: 341–344

Buettner GR (1984) Thiyl free radical production with hematoporphyrin derivative, cystein and light: a spin trapping study. FEBS Lett 177: 295–299

Buettner GR, Oberley LW (1980) The apparent production of superoxide and hydroxyl radicals by hematoporphyrin and light as seen by spin trapping. FEBS Lett 121: 161–164
Buettner GR, Motten AG, Hall RD, Chignell CF (1986) Free radical production by chlorpromazine sulfoxide, an ESR spin trapping and flash photolysis study. Photochem Photobiol 46: 5–10
Buettner GR, Motten AG, Hall RD, Chignell CF (1987) ESR detection of endogenous ascorbate free radical in mouse skin: Enhancement of radical production during UV irradiation following topical application of chlorpromazine. Photochem Photobiol 46: 161–164
Buhl R, Holroyd KJ, Cantin AM, Jaffe HA, Wells FB, Saltini C, Crystal RG (1989) Systemic glutathione-deficiency in symptom-free HIV seropositive individuals. Lancett ii: 1294–1298
Buhmann A (1936) Atmung und Glycolyse in normaler und pathologisch veränderter Haut. Biochem Z 287: 145–152
Burdon RH, Gill V (1989) Cell proliferation and oxidative stress. Free Radic Res Commun 7: 149–159
Burdon RH, Gill VM, Rice-Evans C (1987) Oxidative stress and heat shock protein induction in human cells. Free Radic Res Commun 3: 129–139
Burgess JF (1948) Vitamin E (tocopherols) in the collagenoses. Lancet ii 215–217
Burgess JF, Pritchard JE (1948) Tocopherols (vitamin E). Treatment of lupus erythematosus: preliminary report. Arch Dermatol Syphilol 57: 953–964
Burkhart CG (1982) A reexamination of the function of benzoyl peroxide in acne. Dermatology 23
Burns FJ (1989) Cancer risk associated with therapeutic irradiation of skin. Arch Dermatol 125: 979–981
Burns JC, Geha RS, Schneeberger EE, Newburger JW, Rosen FS, Glezen LS, Huang AS, Natale J, Leung DYM (1986) Polymerase activity in lymphocyte culture supernatants from patients with Kawasaki disease. Nature 323: 814–816
Burton W, Cheeseman KH, Doba T, Ingold KU, Slater F (1983) Vitamin E as an antioxidant in-vitro and in-vivo. Ciba Found Symp 101: 4–14
Burton GW, Ingold KU (1984) β-Carotene: an unusual type of antioxidant. Science 224: 569–573
Burton GW, Ingold KU (1986) Vitamin E: application of the principles of physical organic chemistry to the exploration of its structure and function. Acc Chem Res 19: 194–201
Burton-Kee JE, Mowbray JF, Lehner T (1981) Different cross reacting circulating immune complexes in Behcet's syndrome and recurrent oral ulcers. J Lab Clin Med 97: 559–567
Bus JS, Gibson JE (1982) Mechanism of superoxide radical mediated toxicity. J Toxicol Clin Toxicol 19: 689–697
Busse WW, Kopp DE, Middleton E (1984) Flavonoid modulation of human neutrophil function. J Allergy Clin Immunol 73: 801–809
Byck JS, Dawson CR (1968) Assay of protein-quinone coupling involving compounds structurally related to the active principle of poison ivy. Anal Biochem 25: 123–135
Cabrera-Juarez E (1981) Near ultraviolet mutation of transforming DNA irradiated in-vivo. Mutat Res 83: 301–306
Cadenas E, Sies H (1984) Low-level chemiluminescence as an indicator of singlet molecular oxygen in biological systems. Methods Enzymol 105: 221–231
Cadenas E, Boveris A, Ragan CI, Stoppani AOM (1977) Production of superoxide radicals and hydrogen peroxide by NADH-ubiquinone reductase and ubiquinol cytochrome c reductase from beaf heart mitochondria. Arch Biochem Biophys 180: 248–257
Cadet J, Berger M, Decarboz C, Wagner JR, Van Lier JE, Ginot YM, Vigny P (1986) Photosensitized reactions of nucleic acids. Biochimie 68: 813–834
Cand F, Verdetti J (1989) Superoxide dismutase, glutathione peroxidase, catalase, and lipid peroxidation in the major organs of the aging rats. Free Radic Biol Med 7: 59–63
Cantin AM, North SL, Fells GA, Hubbard RC, Crystal RG (1987) Oxidant mediated epithelial cell injury in idiopathic pulmonary fibrosis. J Clin Invest 79: 1665–1673
Caprini L, Pasquali P, Tadolini B, Sechi AM, Landi L (1986) Antioxidant behaviour of ubi-

quinone and β-carotene incorporated in model membranes. Free Radic Res Commun 2: 85–92

Capron A, Ameisen JC, Joseph M, Auriault C, Tonnel AB, Caen J (1985) New function of platelets and their pathological implications. Int Arch Allergy Appl Immunol 64: 359–360

Caputo R, Monti M, Ermacora E, Carminati G, Gelmetti C, Gianotti R, Gianni E, Puccinelli V (1958) Cutaneous manifestations of tetrachlorodibenzo-p-dioxin in children and adolescents. J Am Acad Dermatol 19: 812–819

Carlsson J, Carpenter VS (1980) The recA + gene product is more important than catalase and superoxide dismutase in protecting Escherichia coli against hydrogen peroxide toxicity. J Bacteriol 142: 319–321

Carnes SA, Watson AP (1989) Disposing of the US chemical weapons stockpile. JAMA 262: 653–659

Carp H, Janoff A (1979) In-vitro suppression of serum elastase inhibitory capacity by reactive oxygen species generated by phagocytosing polymorphonuclear leukocytes. J Clin Invest 63: 793–797

Carraro C, Pathak MA (1988) Characterization of superoxide dismutase from mammalian skin epidermis. J Invest Dermatol 90: 31–36

Cascinelli N, Marchesini R (1989) Increasing incidence of cutaneous melanoma, ultraviolet radiation and the clinician. Photochem Photobiol 50: 497–505

Cathcart RF (1986) The vitamin C treatment of allergy and the normally unprimed state of antibodies. Med Hypotheses 21: 307–321

Cathcart R, Schwiers E, Saul RL, Ames BN (1984) Thymine glycol and thymidine glycol in human and rat urine: a possible assay for oxidative DNA damage. Proc Natl Acad Sci USA 81: 5633–5637

Cauzzo G, Gennari G, Jori G, Spikes JD (1977) The effect of chemical structure on the photosensitizing efficiencies of porphyrins. Photochem Photobiol 25: 389–395

Cavalier E, Rogan E (1985) Role of radical cations in aromatic hydrocarbon carcinogenesis. Environ Health Perspect 64: 69–84

Cavallini L, Bindoli A, Siliprandi N (1978) Comparative evaluation of antiperoxidative action of silymarin and other flavonoids. Pharmacol Res Commun 10: 133–136

Carnes SA, Watson AP (1989) Disposing of the US chemical weapons stockpile. JAMA 262: 653–659

Cerutti PA (1974) Effects of ionizing radiation on mammalian cells. Naturwissenschaften 61: 51–59

Cerutti PA (1985) Prooxidant states and promotion. Science 227: 375–381

Chaffman M, Brogden RN, Heel RC, Speight TM, Avery GS (1984) Auranofin. A preliminary review of its pharmacological properties and therapeutic use in rheumatoid arthritis. Drugs 27: 378–424

Chakrabarti B, Banerjee S (1955) Dehydroascorbic acid level in blood of patients suffering from various infectious diseases. Proc Soc Exp Biol Med 88: 581–583

Chan SW, Chan PC, Bielski BHJ (1974) Studies on the lipoic acid free radical. Biochim Biophys Acta 338: 213–223

Chance B, Sies H, Boveris A (1979) Hydroperoxide metabolism in mammalian organs. Physiol Rev 59: 527–605

Chandra T (1989) Effect of curcumin on lipid peroxidation in CCl4 induced fatty liver. International symposium on biological oxidation systems, Bangalore, 23–26 Oct 1989

Chaplowski FA, Kelly LA, Butcher RW (1975) Cyclic nucleotides in cultured cells. Adv Cyclic Nucleotide Res: 245–278

Chatterjee R, Benzinger MJ, Ritter JL, Bissett DL (1990) Chronic ultraviolet B radiation induced biochemical changes in the skin of hairless mice. Phochem Photobiol 51: 91–97

Chaussay SA, Staehler G, Brandl H (1983) Preliminary evaluation of superoxide dismutase (orgotein) in the treatment of Pyronie's disease. In: Progress in reproductive biology and medicine, vol 9. Karger, Basel, pp 54–56

Chedekel MR, Smith SK, Post PW, Pokora A, Vessel DL (1978) Photodestruction of phaeomelanin: role of oxygen. Proc Natl Acad Sci USA 75: 5395–5399

Chedekel MR, Zeise L (1988) Sunlight, melanogenesis and radicals in the skin. Lipids 23: 587–591

Cheng LY, Packer L (1979) Photodamage to hepatocytes by visible light. FEBS Lett 97: 124–128

Cheng SC, Burton GW, Ingold KU, Foster DO (1987) Chiral discrimination in the exchange of α-tocopherol stereoisomers between plasma and red blood cells. Lipids 22: 469–473

Chensue SW, Kunkel SL, Higashi GI, Ward PA, Boros DL (1983) Production of superoxide anion, prostaglandins and hydroxyeicosatetraenoic acids by macrophages from hypersensitivity type (Schistosoma mansoni egg) and foreign body granuloma. Infect Immun 42: 1116–1125

Chignell CF (1989) Free radicals in drug induced photosensitivity reactions: spin trapping study of the photolysis of photoallergenic halogenated salicylanilides. In: Hayaishi O, Niki E, Kondo M, Yoshokawa T (eds) Medical, biochemical and chemical aspects of free radicals. Elsevier, Amsterdam, pp 209–212

Chignell CF, Kalyanaraman B, Mason RP, Sik H (1980) Spectroscopic studies of cutaneous photosensitizing agents. I. Spin trapping of photolysis products from sulfanilamide. Dermatologica 32: 563–571

Chignell CF, Kalyanaraman B, Sik RH, Mason RP (1981) Spectroscopic studies of cutaneous photosensitizing agents. II. Spin trapping of photolysis products from sulfanilamide and 4-aminobenzoic acid using 5,5-dimethyl-1-pyrroline-1-oxide. Photochem Photobiol 34: 147–156

Chignell CF, Motten AG, Buettner GR (1985) Photoinduced free radicals from chlorpromazine and related phentothiazines: relationship to phenothiazine-induced photosensitization. Environ Health Perspect 64: 103–110

Chopra M, Belch JJF, Smith WE (1988) A comparison of the free radical scavenging activity of leukotrienes and prostaglandins. Free Radic Res Commun 5: 95–99

Chow CK (1982) Influence of dietary vitamin E on susceptibility to ozone exposure. In: Lee SD, Mustafa MG, Mehlman MA (eds) The biomedical effects of ozone and related photochemical oxidants. Princeton Scientific, Princeton, pp 75–94

Christophers E (1987) Neutrophile Dermatosen. In: Braun-Falco O, Schill WB (eds) Fortschritte der praktischen Dermatologie und Venerologie. Springer, Berlin Heidelberg New York, pp 195–202

Christophers E, Schröder JM (1987) Differential impairment of neutrophil function in inflammatory skin conditions. In: Hayaishi O, Imamura S, Miyachi Y (eds) The biological role of reactive oxygen species in skin. Elsevier, New York, pp 163–170

Chuaprapaislip T, Piamphongsant T (1978) Treatment of pustular psoriasis with clofazimine. Br J Dermatol 99: 303–305

Chvapil M (1976) Effects of zinc on cells and biomembranes. Med Clin North Am 60: 799–812

Chvapil M, Ryan JN, Elias SL, Peng YM (1973) Protective effect of zinc on carbon tetrachloride induced liver injury in rats. Exp Mol Pathol 19: 186–196

Chvapil M, Sipes IG, Ludwig JC, Halladay SC (1975) Inhibition of NADPH oxidase and oxidative metabolism of drugs in liver microsomes by zinc. Biochem Pharmacol 2: 917–919

Chytil F, Sherman DR (1987) How do retinoids work? Dermatologica 175 [Suppl I]: 8–12

Cillard J, Cillard P, Cormier M (1980) Effect of experimental factors on the prooxidant behavior of a-tocopherol. J Am Oil Chem Soc 57: 255–261

Cipolaro AC, Crossland PM (1967) Radiodermatitis and other clinical effects of ionizing radiation on the skin. In: X-rays and radium in the treatment of diseases of the skin. Lea and Febiger, Philadelphia, pp 357–410

Clare NT, Whitten LK, Filmer DB (1947) A photosensitized keratitis in young cattle following the use of phenothiazine as an anthelmintic. III Identification of the photosensitizing agent. Aust Vet J 23: 344–348

Clark A, Cowden WB (1985) Antimalarials. In: Sies H (ed) Oxidative stress. Academic, London, pp 131–149

Clark IA, Hunt NH, Cowden WB (1986) Oxygen derived free radicals in the pathogenesis of parasitic disease. Adv Parasitol 25: 1–44

Clark RAF (1985) Cutaneous tissue repair: basic biologic considerations. I. J Am Acad Dermatol 13: 701–725

Clark RA, Szot S, Venkatasubramanian K, Schiffmann E (1980) Chemotactic factor inactivation by myeloperoxidase mediated oxidation of methionine. J Immunol 124: 2020–2026

Clark RA, Stone PJ, Hag AE, Calore JD, Franzblau C (1981) Myeloperoxidase catalyzed inactivation of alpha-1 protease inhibitor by human neutrophils. J Biol Chem 256: 3348–3353

Clemetson CAB, Andersen L (1966) Plant polyphenols as antioxidants for ascorbic acid. Ann NY Acad Sci 136: 341–376

Cochrane T (1950) Granuloma annulare: treatment with vitamin E. Br J Dermatol Syph 62: 316–318

Cohen BE, Harmon CS, Phizackerley PJR (1983) Glucose metabolism in experimental skin flaps. Plast Reconstr Surg 71: 79–85

Cohen G, Hochstein P (1963) Glutathione peroxidase: the primary agent for the elimination of hydrogen peroxide in erythrocytes. Biochemistry 2: 1420–1428

Cohen G, Hochstein P (1964) Generation of hydrogen peroxide in erythrocytes by hemolytic agents. Biochemistry 3: 895–900

Cohen IK, McCoy BJ (1983) Wound healing. In: Goldsmith LA (ed) Biochemistry and physiology of the skin. Oxford University Press, New York, pp 462–470

Cohen S, Johnson AR, Hurd E (1983) Cytotoxicity of sera from patients with scleroderma. Arthritis Rheum 26: 170–178

Colotta F, Bersani L, Lazzarin A, Poli G, Mantovani A (1985) Rapid killing of actinomycin D treated tumor cells by human monocytes. J Immunol 134: 3524–3531

Comi G, Nesi G (1950) Contributo allo studio delle azioni esercitate dalla vitamina E. Riv Clin Med 50: 214–226

Commoner B, Townsend J, Pake GW (1954) Free radicals in biological materials. Nature 174: 689–691

Condell RA, Tappel AL (1983) Evidence for suitability of glutathione peroxidase as a protective enzyme: studies of oxidative damage, denaturation and proteolysis. Arch Biochem Biophys 223: 407–416

Conley CL, Chisholm JJ (1979) Ausheilung einer Leber-Dekompensation bei einer Protoporphyrie (EPP oder erythropoetische Protoporphyrie). John Hopkins Med J 145: 237–240

Connolly SM, Quimby SR, Griffing WL, Winkelmann RK (1990) Scleroderma and L-tryptophan: a possible explanation of the eosinophilia-myalgia syndrome. J Am Acad Dermatol 23: 451–457

Connor MJ, Wheeler LA (1987) Depletion of cutaneous glutathione by ultraviolet radiation. Photochem Photobiol 47: 239–245

Copeland ES (1983) Free radicals in promotion: a chemical pathology study section workshop. Cancer Res 43: 5631–5637

Cordier JH, Fieviez C, Lefevre MJ (1966) Acroosteolyse et lesions cutanees chez deux ouvriers affectees au nettoyage d'autoclaves. Can Med Travail 4: 3

Corey EJ, Mehrotra MM, Khan AU (1987) Water induced dismutation of superoxide anion generates singlet molecular oxygen. Biochem Biophys Res Commun 145: 842–846

Cornbleet T, Cohen D, Bielinsky S (1951) Granuloma annulare generalized. Arch Dermatol Syphilol 64: 249

Corrocher R, Ferrari S, de Gironcoli M, Bassi A, Olivieri O, Guarini P, Stanzial A, Barbara AL, Grgolini L (1989) Effect of fish oil supplementation on erythrocyte lipid pattern, malondialdehyde production and glutathione peroxidase activity in psoriasis. Clin Chim Acta 179: 121–132

Couet WR, Eriksson UG, Brasch RC, Tozer TN (1985) Pharmacokinetics of two prototype nitroxide spin labels for contrast enhancement in magnetic resonance imaging. Pharm Res 2: 6973

Coulomb B, Dubertret L, Bell E, Merrill C, Fosse M, Breton-Gorius J, Frost C, Touraine R (1983) Endogenous peroxidases in normal human dermis: a marker of fibroblast differentiation. J Invest Dermatol 81: 75–78

Couturier A, Bazgar S, Castagna M (1984) Further characterization of tumor promotor mediated activation of protein kinase C. Biochem Biophys Res Commun 121: 448–455

Crane FL, Roberst H, Linnane AW, Löw H (1982) Transmembrane ferricyanide reduction by cells of the yeast Sacharomyces cerevisiae. J Bioenerg Biomembr 14: 191–205

Crane FL, Löw H, Clark MG (1985 a) Plasma membrane redox enzymes. In: Martonosi AN (ed) The enzymes of biological membranes, vol 4. Plenum New York, pp 465–510

Crane FL, Sun IL, Clark MG, Grebing C, Löw H (1985 b) Transplasma membrane redox systems in growth and development. Biochim Biophys Acta 811: 233–264

Crapo JD (1986) Morphologic changes in pulmonary oxygen toxicity. Annu Rev Physiol 48: 721–731

Crapo JD, Tierney DF (1974) Superoxide dismutase and pulmonary oxygen toxicity. Am J Physiol 226: 1401–1407

Crapo JD, Sjodtrom K, Drew RT (1978) Tolerance and cross-tolerance using NO2 and O2. I. Toxicology and biochemistry. J Appl Physiol 44: 364–369

Craw MT, Depew MC (1985) Contributions of electron spin resonance spectroscopy to the study of vitamins C, E and K. Rev Chem Intermed 6: 1–31

Crooke S, Mirabelli CK (1983) Molecular mechanisms of action of auranofin and other gold complexes as related to their biological activities. Am J Med 30: 109–113

Crow KD (1961) Photosensitivity due to pitch. Br J Dermatol 73: 220–232

Cruickshank CND, Trotter MD (1956) The oxygen uptake, glucose utilization and lactic acid production of guinea pig skin in relation to oxygen tension. Biochem J 62: 57–61

Cruickshank CND, Trotter MD, Cooper JR (1962) A study of available substrates for the endogenous respiration of skin "in-vitro". J Invest Dermatol 39: 175–178

Cruz PD, Bergstresser PR (1989) Ultraviolet radiation, Langerhans cells and skin cancer. Arch Dermatol 125: 975–979

Crystal RG, Fulmer JD, Baum BJ (1958) Cells, collagen and idiopathic pulmonary fibrosis. Lung 155: 199

Culling-Berglund AJ, Newcomb SA, Gagne M, Morfitt WS, Davis TP (1989) J Micronutr Anal 5: 139–148

Cunningham ML, Krinsky NI, Giovanazzi SM, Peak MJ (1985) Superoxide anion is generated from cellular metabolites by solar radiation and its components. J Free Radic Biol Med 1: 381–385

Curzio M (1988) Interaction between neutrophils and 4-hydroxyalkenals and consequences on neutrophil motility. Free Radic Res Commun 5: 55–66

Curzio M, Torielli MV, Giroud JP, Esterbauer H, Dianzani MU (1982) Neutrophil chemotactic response to aldehydes. Res Commun Chem Pathol Pharmacol 36: 463–476

Cutler RG (1984) Antioxidants, aging and longevity. Free Radic Biol Med 4: 371–428

Czapski G, Goldstein S (1986) When do metal complexes protect the biological system from superoxide toxicity and when do they enhance it? Free Radic Res Commun 1: 157–161

Czarnetzki BM (1980) Die Rolle eosinophiler Leukozyten bei entzündlichen Dermatosen. Hautarzt 31: 291–296

Czarnetzki BM (1987) Les polynucleaires eosinophiles. Ann Dermatol Venerol 114: 1147–1156

Czerniecki B, Goldstein BD, Witz G (1985) Production of superoxide anion radicals by murine peritoneal exsudate cells stimulated with tumor promotors. Proc Am Assoc Cancer Res 76: 133–142

Czerniecki B, Gad SC, Reilly C, Smith AC, With G (1986) Phorbol diacetate inhibits superoxide anion radical production and tumor promotion by mezerein. Carcinogenesis 7: 1637–1641

Czirjak L, Danko K, Schlammadinger J, Suranyi P, Tamasi L, Szegedi GY (1987) Progressive systemic sclerosis occuring in patients exposed to chemicals. Int J Dermatol 26: 374–378

Dagadu JM (1967) Distribution of carotene and vitamin A in liver, pancreas and body fat of Ghanaians. Br J Nutr 21: 453–456

Dahlgren C (1989) Is lysosomal fusion required for the granulocyte chemiluminescence reactions? Free Radic Biol Med 6: 399–403

Dainow I (1958) Granuloma annulaire traite par les vitamines E et A. Dermatologica 116: 367–369

Dainow PI (1953) La vitamin E dans le traitment de l'acne. Dermatologica 106: 197–200

Dalal NS, Shi X, Vallyathan V (1989) Oxygenated radical formation by fresh quartz dust in a cell free aqueous medium and its inhibition by scavengers. 2nd International symposium on spin trapping and aminoxyl radical chemistry, University of Guelph, Guelph, Ontario, 2–7 July 1989

Dale WM (1954) Basic radiation biochemistry. In: Hollaender A (ed) Radiation biology. McGraw Hill, New York, vol 1, part 1, pp 255–281

Dalle-Carbonare M, Pathak MA (1989) Photoaging and the role of UV (290–400 nm) and reactive oxygen species in collagen cross linking. J Invest Dermatol 92: 410

Damas J, Lagneaux D, Lecompte T (1981) Liberation d'histamine par les mastocytes due rat en hyperoxie. CR Soc Biol 175: 266–271

Danno K, Horio T (1987) Sunburn cell: factors involved in its formation. Photochem Photobiol 43: 683–690

Danno KT, Horio M, Takigawa, Imamura S (1984) Role of oxygen intermediates in UV-induced epidermal cell injury. J Invest Dermatol 83: 166–168

Darr D, Combs S, Dunstan S, Manning T, Pinnell S (1991) Topical vitamin C treatment inhibits ultraviolet radiation-induced damage to porcine skin. J Invest Dermatol 96: 353

Das DK, Engelman RM, Rousou JA, Breyer RH, Otani H, Lemeshow S (1986) Pathophysiology of superoxide radical as potential mediator of reperfusion injury in pig heart. Basic Res Cardiol 81: 156–166

Das UN, Ells G, Begin ME, Horrobin DF (1986) Free radicals as possible mediators of the actions of interferon. J Free Radic Biol Med 2: 183–188

Daugherty JP, Khurana A (1985) Amelioration of doxorubicin induced skin necrosis in mice by butylated hydroxytoluene. Cancer Chemother Pharmacol 14: 243–246

Davies KJA, Quintanilha AT, Brooks GA, Packer L (1982) Free radicals and tissue damage produced by exercise. Biochem Biophys Res Commun 107: 1198–1205

Davies MJ, Slater TF (1987) Studies on the metal ion and lipoxygenase catalysed breakdown of hydroperoxides using electron spin resonance spectroscopy. Biochem J 245: 167–173

Decker RH (1971) Nature and regulation of energy metabolism in the epidermis. J Invest Dermatol 57: 351–363

Decker RH, Andersen RL (1965) Oxidative phosphorylation in the epidermis of man and the hairless mouse. J Invest Dermatol 45: 168–172

De Duve C, Baudhuin P (1966) Peroxisomes (Microbodies and related particles). Physiol Rev 46: 323–357

Degreef H, Fluor M (1974) Epidermolysis bullosa: treatment with vitamin E, preliminary results. Arch Belg Dermatol 30: 83–87

De Gruijl FR, Van der Meer JB, Van der Leun JC (1983) Dose-time dependency of tumor formation by chronic UV exposure. Photochem Photobiol 37: 53–62

Delahanty JN, Evans JC, Rowlands CC (1989) Radicals involved in photoallergen/protein interactions. Free Radic Biol Med 7: 231–236

De Leo VA, Horlick H, Hanson D, Eisinger M, Harber LC (1984) Ultraviolet radiation induces changes in membrane metabolism of human keratinocytes in culture. J Invest Dermatol 83: 323–326

De Leo VA, Hanson D, Weinstein IB, Harber LC (1985) Ultraviolet radiation stimulates the release of arachidonic acid from mammalian cells in culture. Photochem Photobiol 41: 51–56

Delfino M, Fabbrocini G, Di Vaia E (1990) Effects of hyperbaric oxygen in cryobiologic wounds. Arch Dermatol Res 282: 267–269

Della Loggia R (1985) Lokale antiphlogistische Wirkung der Kamillen-Falvone. Dtsch Apoth Ztg 43 (Suppl 1): 9–11

Deloria L, Abbott V, Gooderham N, Mannering GJ (1985) Induction of xanthine oxidase and depression of cytochrome P-450 by interferon inducers: genetic difference in the responses of mice. Biochem Biophys Res Commun 131: 109–114

Del Maestro RF (1982) Role of superoxide anion radicals in microvascular permeability and leukocyte behaviour. J Physiol Pharmacol 160: 1406–1414

Del Maestro RF, Thaw HH, Björk J, Planker M, Arfors KE (1980) Free radicals as mediators of tissue injury. Acta Physiol Scand 492 [Suppl]: 43–57

De Luca C, Fanfoni GB, Picardo M, Nazzaro-Porro M, Passi S (1989) The skin surface lipids of man compared with those of other different primates. J Invest Dermatol 92: 473A

De Marchenna O, Guarnieri M, McKhann G (1974) Glutathione peroxidase levels in brain. J Neurochem 22: 773–776

Demitsu T, Katayama H, Yaoita H (1988) Serum lipid peroxide levels in various dermatoses. J Dermatol 15: 325–329

Denq RY, Fridovich I (1989) Formation of endonuclease III-sensitive sites as a consequence of oxygen radical attack on DNA. Free Radic Biol Med 6: 123–129

De Rios G, Chan TJ, Black HS, Rudolph AH, Knox JM (1978) Systemic protection by antioxidants against UVL-induced erythema. J Invest Dermatol 70: 123–125

De Rubertis FR, Craven P (1977) Increased guanylate cyclase activity and guanosine-3′,5′-monophosphate content in ethionine induced hepatomas. Cancer Res 37: 15–21

De Simone C, Rusciani L, Venier A, Larussa FM, Littaruru GP, Oradei A, Lippa S, Serri F (1987) Vitamin E and coenzyme Q10 content in epidermis and basal cell epithelioma. J Invest Dermatol 89: 317

Devasagayam TPA, Mascio PDI, Kaiser S, Sies H (1989) Thiols as singlet oxygen quenchers. Free radicals and cancer, winter meeting, Society for Free Radical Research, Tübingen, 8–9 Dec 1989

De Young LM, Helmes CT, Chao WR, Young LM, Miller V (1981) Paradoxical effect of anthralin on 12-0-tetradecanoyl-phorbol-13-acetate induced mouse epidermal ornithine decarboxylase activity, proliferation and tumor promotion. Cancer Res 41: 204–208

Di Mascio P, Kaiser S, Sies H (1989) Lycopene as the most efficient biological carotenoid singlet oxygen quencher. Arch Biochem Biophys 274: 532–538

D'Ischia M, Prota G (1985) Generation and role of superoxide ion in the autoxidation of 1,8-dihydroxy-9-anthrone. Gazz Chim Ital 115: 511–514

Di Virgilio, Lew DP, Pozzan T (1984) Protein kinase C activation of physiological processes in human neutrophils at vanishingly small cytosolic Ca^{2+} levels. Nature 310: 691–693

Disdier P, Harle JR, Weiller-Merli C, Andrac L, Weiller PJ (1991) Neutrophilic dermatosis despite myeloperoxidase deficiency. J Am Acad Dermatol 24: 654–655

Dixit R, Mukhatar H, Bickers DR (1982) Evidence that lipid peroxidation in microsomal membranes of epidermis is associated with generation of hydrogene peroxide and singlet oxygen. Biochem Biophys Res Commun 105: 546–552

Dixit R, Mukhatar H, Bickers DR (1983 a) Studies on the role of reactive oxygen species in mediating lipid peroxide formation in epidermal microsomes of rat skin. J Invest Dermatol 81: 369–375

Dixit R, Mukhatar H, Bickers DR (1983 b) Destruction of cytochrome P-450 by reactive oxygen species generated during photosensitization of hematoporphyrin derivative. Photochem Photobiol 37: 173–176

Djerassi D, Machlin LJ, Nocka C (1986) Vitamin E: biochemical function and its role in cosmetics. Drug Cosmet: 29–34

Doba T, Burton GW, Ingold KU (1985) Antioxidant and co-antioxidant activity of vitamin C. The effect of vitamin C, either alone or in presence of vitamin E or a water soluble vitamin E analogue, upon peroxidation of aqueous multilamellar phospholipid liposomes. Biochim Biophys Acta 835: 298–303

Docampo R, Moreno SNJ (1984) Free radical intermediates in the antiparasitic action of drugs and phagocytic cells. Free Radic Biol 4: 243–287

Dogan P, Soyuer Ü, Tanrikulu G (1989) Superoxide dismutase and myeloperoxidase activity in polymorphonuclear leukocytes, and serum ceruloplasmin and copper levels, in psoriasis. Br J Dermatol 120: 239–244

Doniger J, Jacobson ED, Krell K, DiPaolo JA (1981) UV light action spectra for neoplastic transformation and lethality of Syrian hamster embryo cells correlate with that for pyrimidine dimer formation in cellular DNA. Proc Natl Acad Sci USA 78: 2378–2382

D'Onofrio C, Maly FE, Fischer H, Maas D (1984) Differential generation of chemiluminescence-detectable oxygen radicals by polymorphonuclear leukocytes challenged with sera from systemic lupus erythematosus and rheumatoid arthritis patients. Klin Wochenschr 62: 710–716

Dormandy TL (1969) Biological rancidification. Lancet ii: 684–688

Dorwart BB, Gall EP, Schumacher HR, Krauser RE (1978) Chrysotherapy in psoriatic arthritis. Arthritis Rheum 21: 513–515

Doucas C (1949) Vitamin E in case of Lupus erythematosus with poikiloderma and in Meleda disease: successful therapy. Bull Soc Fr Dermatol Syphiligr 56: 449–454

Dougherty TJ (1981) Photoradiation therapy for cutaneous and subcutaneous malignancies. J Invest Dermatol 77: 122–124

Dougherty TJ (1986) Photosensitization of malignant tumors. Semin Surg Oncol 2: 24–37

Dougherty TJ (1987) Photosensitizers: therapy and detection of malignant tumors. Photochem Photobiol 47: 879–889

Dougherty TJ, Kaufman JE, Goldfarb A, Weishaupt KR, Boyle D, Mittleman A (1978) Photoradiation therapy for the treatment of malignant tumors. Cancer Res 38: 2628–2635

Douglas DM, Bandyopadhyay D, Russell JC, Hoory S, Antar M, Das DK (1989) Free radical involvement in the wound healing process. In: Hayaishi O, Niki E, Kondo M, Yoshokawa T (eds) Medical, biochemical and chemical aspects of free radicals. Elsevier, Amsterdam, pp 1275–1279

Dover JS, Phillips TJ, Arndt KA (1989) Cutaneous effects and therapeutic uses of heat with emphasis on infrared radiation. J Am Acad Dermatol 20: 278–286

Downing DT, Stewart ME, Wertz PW, Colton SW, Abraham W, Strauss JS (1987) Skin lipids: an update. J Invest Dermatol 88: 2S

Dreno B, Vandermeeren MA, Rigous V (1988) Le zinc et al peau. Ann Dermatol Venereol 115: 741–746

Dröge W, Eck HP, Näher H, Pekar U, Daniel V (1988) Abnormal amino acid concentration in the blood of patients with acquired immunodeficiency syndrome (AIDS) may contribute to the immunological defect. Biol Chem Hoppe Seyler 369: 143–148

Dubbelmann TMAR, de Bruijne AW, van Steenick J (1977) Photodynamic effects of protoporphyrin on red blood cell deformability. Biochem Biophys Res Commun 77: 811–817

Dubbelmann TMAR, de Goeiji AFPM, van Stevenick J (1978) Protoporphyrin sensitized photodynamic modification of proteins in isolated human red blood cell membranes. Photochem Photobiol 28: 197–204

Dubertret L, Breton-Gorius J, Fosse M, Tourraine R (1982) Diaminobenzidine cytochemistry in unfixed human epidermis: a marker for epidermal differentiation and for mitochondria. J Invest Dermatol 78: 77–81

Duchesne J (1977) A unifying biochemical theory of cancer, senescence and maximal life span. J Theor Biol 66: 137–145

Ducret F, Lamblet P, Löliger J, Savoy MC (1985) Antipsoriatic drug action of anthralin oxidation reactions with peroxidizing lipids. J Free Radic Biol Med 1: 301–306

Dunlop JI, Nicholls CH (1965) Electron spin resonance studies of ultraviolet irradiated keratin and related proteins. Photochem Photobiol 4: 881–890

Dupuis G (1979) Studies on poison ivy. In-vitro lymphocyte transformation by urushiol-protein conjugates. Br J Dermatol 101: 617–624

Duvanel T, Harms M (1988) Infantile Akropustulose. Hautarzt 39: 1–4

Dworkin BM, Rosenthal WS, Wormser GP, Weiss L (1986) Selenium deficiency in the acquired immune deficiency syndrome. JPEN 10: 405–407

Eaglstein WH, Sakai M, Mizuno N (1979) Ultraviolet radiation induced inflammation and leukocytes. J Invest Dermatol 72: 59–63

Earle WR (1928) Studies upon the effect of light on blood and tissue cells. III. The action of light on fibroblasts in-vitro. J Exp Med 48: 683–693

Eck HP, Dröge W (1989) Influence of the extracellular glutamate concentration on the in-

tracellular cyst(e)ine concentration in macrophages and on the capacity to release cysteine. Biol Chem Hoppe Seyler 370: 109–113
Eck HP, Gmünder H, Hartmann M, Petzold D, Daniel V, Dröge W (1989) Low concentration of acid soluble thiol (cystein) in the blood plasma of HIV-1 infected patients. Biol Chem Hoppe Seyler 370: 101–108
Eckert P (1988) Orgotein beseitigt subkutane Narben. Selecta 36: 5
Eckert P, Koch A (1987) Subkutane narbige Indurationen. Münch Med Wochenschr 129: 342–344
Edwards SW, Swan TF (1986) Regulation of superoxide generation by myeloperoxidase during the respiratory burst of human neutrophils. Biochem J 237: 601–604
Egan RW, Paxton J, Kuehl FA (1976) Mechanism for irreversible self deactivation of prostaglandin synthetase. J Biol Chem 251: 7329–7335
Egan RW, Gale PH, Beveridge GC, Phillips GB (1978) Radical scavenging as the mechanisms for stimulation of prostaglandin cyclooxygenase and depression of inflammation by lipoic acid and sodium iodide. Prostaglandins 16: 860–869
Egner PA, Kensler TW (1985) Effects of a biomimetic superoxide dismutase on complete and multistage carcinogenesis in mouse skin. Carcinogenesis 6: 1167–1172
Ehrlich HP, Tarver H, Hunt T (1972) Inhibitory effects of vitamin E on collagen synthesis and wound repair. Ann Surg 17: 235–240
Ehrmann S (1909) Weitere Untersuchungen über die Lichtwirkung bei Hydroa aestivalis (Bazin), summer eruption (nach Hutchinson). Arch Dermatol 97: 75–86
Eisen AZ (1969) Human skin collagenase: relationship to the pathogenesis of epidermolysis bullosa dystrophica. J Invest Dermatol 52: 449–453
Eisen AZ, Bloch KJ, Sakai T (1970) Inhibition of human skin collagenase by human serum. J Lab Clin Med 75: 258–263
Eisenstark A (1987) Mutagenic and lethal effects of near ultraviolet radiation (290–400 nm) on bacteria and phage. Environ Mol Mutagen 10: 317–337
Eisenstark A, Perrot G (1987) Catalase has only minor role in protection against near ultraviolet radiation damage in bacteria. Mol Gen Genet 207: 68–72
El-Alfi OS, Smith PM, Biesele JJ (1965) Chromosomal breaks in human leukocyte cultures induced by an agent in the plasma of infectious hepatitis patients. Hereditas 52: 285–294
Eldred GE, Katz ML (1989) The autofluorescent products of lipid peroxidation may not be lipofuscin like. Free Radic Biol Med 7: 157–163
Elias PM (1981) Epidermal lipids, membranes, and keratinization (review). Int J Dermatol 20: 1–12
Elias PM, Williams MI (1981) Retinoids, cancer and the skin. Arch Dermatol 117: 160–180
Eling TE, Curtis JF, Harman LS, Mason RP (1986) Oxidation of glutathione to its thiyl free radical metabolite by prostaglandin H synthase. J Biol Chem 261: 5023–5028
Elliot J, Munk PL, Stevenson KJ, Armstrong DA (1980) Reactions of oxidizing and reducing radical probes with lipoamide dehydrogenase. Biochemistry 19: 4945–4950
Elstner EF (1988) Heme peroxides: possible initiators of cellular repair reactions induced by tetrachlorodecaoxide (TCDO). International symposium on free radicals in medicine, Vienna, 8–12 Nov 1988 (abstr RPC-16)
Emerit I (1976) Chromosomal breakage in systemic sclerosis and related disorders. Dermatologica 153: 145–156
Emerit I (1982) Chromosome breakage factors: origin and possible significance. Prog Mutat Res 4: 61–72
Emerit I (1983) Properties and action mechanism of clastogenic factors. Lymphokines 8: 413–424
Emerit I (1988) Oxidative reactions and connective tissue diseases. In: Chow CK (ed) Cellular antioxidant defense mechanism. CRC, Boca Raton, vol 3, pp 111–121
Emerit I (1989) Chromosomal instability in collagen diseases. J Rheumatol 39: 84–89
Emerit I, Cerutti P (1981 a) Tumor promotor phorbol-12-myristate-13-acetate induces chromosomal damage via indirect action. Nature 293: 144–146
Emerit I, Cerutti P (1981 b) Clastogenic activity from Bloom's syndrome fibroblast cultures. Proc Natl Acad Sci USA 78: 1866–1872

Emerit I, Cerutti PA (1982) Tumor promotor phorbol-12-myristate-13-acetate induces a clastogenic factor in human leucocytes. Proc Natl Acad Sci USA 79: 7509–7513

Emerit I, Housset E (1973) Chromosome studies on bone marrow from patients with systemic sclerosis. Evidence for chromosomal breakage in-vivo. Biomedicine 19: 550–558

Emerit I, Michelson AM (1980) Chromosome instability in human and murine autoimmune disease: anticlastogenic effect of superoxide dismutase. Acta Physiol Scand 492 [Suppl]: 59–65

Emerit I, Michelson AM (1981) Mechanism of photosensitivity in systemic lupus erythematosus patients. Proc Natl Acad Sci USA 78: 2537–2540

Emerit I, Housset E, de Grouchy J, Camus JP (1971) Chromosomal breakage in diffuse scleroderma. A study of 27 patients. Biomedicine 16: 684–691

Emerit I, Emerit J, Tosoni-Pittoni A, Bousquet O, Sarrazin A (1972) Chromosome studies in patients with ulcerative colitis. Humangenetik 16: 313–316

Emerit I, Levy A, Housset E (1973) Sclerodermie generalisiee et cassures chromosomiques. Mise en evidence d'un "facteur cassant" dans le serum de malades. Ann Genet (Paris) 16: 135–138

Emerit I, Feingold DJ, Cams JP, Houset E (1974 a) Etude chromosomique des maladies du collagene. Ann Genet (Paris) 17: 251–256

Emerit I, Levy A, Housset E (1974 b) Breakage factor in systemic sclerosis and protector effect of L-cysteine. Humangenetik 25: 221–226

Emerit I, Housset E, Feingold J (1976) Chromosomal breakage and scleroderma: Studies in family members. J Lab Clin Med 88: 81–86

Emerit I, Levy A, De Vaux Saint Cyr C (1980 a) Chromosome damaging agent of low molecular weight in the serum of New Zealand Black mice. Cytogenet Cell Genet 26: 41–47

Emerit I, Feingold J, Levy A, Martin E, Housset E (1980 b) Tumor incidence and development of autoimmune hemolytic anemia in two breeding lines of the NZB mouse strain, that differ in chromosome breakage. J Natl Cancer Inst 64: 513–518

Emerit I, Michelson AM, Martin E, Emerit J (1981) Perinuclear halo formation as an indication of phototoxic effects. Dermatologica 163: 295–299

Emerit I, Levy A, Cerutti P (1983) Suppression of tumor promotor phorbol myristate acetate induced chromosome breakage by antioxidants and inhibitors of arachidonic acid metabolism. Mutat Res 110: 327–335

Emerit I, Khan SH, Cerutti PA (1985) Treatment of lymphocyte cultures with a hypoxanthine xanthine oxidase system induces the formation of transferable clastogenic material. J Free Radic Biol Med 1: 51–57

Emerit I, Fabiani JN, Ponzio O, Murday A, Lunel F, Carpentier A (1988) Clastogenic factor in ischemia reperfusion injury during open heart surgery: protective effect of allopurinol. Ann Thorac Surg 46: 619–624

Emerit I, Autor A, Auclair C, Levy A, Ensworth S, Lahoud-Maghani M, Stein H, Camus JP (1989) Role of oxygen free radical in progressive systemic sclerosis (scleroderma). In: Hayaishi O, Niki E, Kondo M, Yoshikawa T (eds) Medical, biochemical and chemical aspects of free radicals. Elsevier, Amsterdam, pp 1211–1218

Emerit I, Khan SH, Esterbauer H (1991) Hydroxynonenal. A component of clastogenic factors. Free Radic Biol Med 10: 371–377

Emerson RW, Wilson-Jones E (1968) Eosinophilic spongiosis in pemphigus: a report of an unusual histologic change in pemphigus. Arch Dermatol 97: 252–257

Emiliani C, Delmelle M (1983) The lipid solubility of poprhyrins modulates their phototoxicity in membrane models. Photochem Photobiol 37: 487–492

Emtestam L, Berglund L, Angelin, Drummond GS, Kappas A (1989) Tin-protoporphyrin and long wave length ultraviolet light in treatment of psoriasis. Lancet i: 1231–1233

Engel A, Johnson ML, Haynes SG (1988) Health effects of sunlight exposure in the United States. Arch Dermatol 124: 72–79

Epstein CJ, Avraham KB, Lovett M, Smith S, Elroy-Stein O, Rotman G, Bry C, Groner A (1987) Transgenic mice with increased Cu/Zn-superoxide dismutase activity: animal model of dosage effects in Down syndrome. Proc Natl Acad Sci USA 84: 8044–8048

Epstein JH (1973) Effects of β-carotene on ultraviolet induced cancer formation in the hairless mouse skin. Photochem Photobiol 25: 211–213
Epstein JH (1977a) The pathological effects of light on the skin. Free Radic Biol 3: 219–249
Epstein JH (1977b) Effects of β-carotene on ultraviolet induced cancer formation in the hairless mouse skin. Photochem Photobiol 25: 211–213
Epstein JH, Wintroub BU (1985) Photosensitivity due to drugs. Drugs 30: 42–57
Epstein JH, Wuepper KD, Maibach HI (1968) Photocontact dermatitis to halogenated salicylanilides and related compounds. Arch Dermatol 97: 236–244
Epstein S (1939) Photoallergy and primary photosensitivity to sulfanilamide. J Invest Dermatol 2: 43–51
Erasmus LD (1957) Scleroderma in gold miners on the Witwatersrand with particular reference to pulmonary manifestations. S Afr J Lab Clin Med 3: 209–231
Ernst E, Matrai A (1985) Einfluß von alpha-Tocopherol (Vitamin E) auf die Fließeigenschaften des Blutes. Therapiewoche 35: 5701–5702
Erum ODL (1969) Oxygen consumption of basal and differentiating cells from hairless mouse epidermis. J Invest Dermatol 52: 204–211
Escribano LM, Gabriel LC, Villa E, Navarro JL (1987) Endogenous peroxidase activity in human cutaneous and adenoid mast cells. J Histochem Cytochem 35: 213–220
Esterbauer H, Dieber-Rotheneder M, Waeg G, Puhl H, Tatzber F (1990) Endogenous antioxidants and lipoprotein oxidation. Biochem Soc Trans 18: 1059–1061
Evans TNS, Naylor PFD, Rowlinson G (1981) Diffusion of oxygen through the mouse ear. Br J Dermatol 105: 45–56
Ezhova MN, Verbenko EV, Tishenina RS (1987) Studies of lipid peroxidation processes in the blood of patients with vasculitis or spontaneous panniculitis and antioxidant correction of these processes. Vestn Dermatol Venerol 6: 14–17 (in Russian)
Fabian J, Zahradnik R (1989) Auf der Suche nach tieffarbigen organischen Verbindungen. Angew Chemie 1101: 693–710
Faguet GB, Reichard SM, Welter DA (1983) Radiation induced clastogenic plasma factors. Cancer Genet Cytogenet 12: 73–83
Fairris GM, Perkins PJ, Lloyd B, Hinks L, Clayton BE (1989) The effect on atopic dermatitis of supplementation with selenium and vitamin E. Acta Dermatol Venereol (Stockh) 69: 359–362
Fantone JC, Ward PA (1982) Role of oxygen-derived free radicals and metabolites in leukocyte-dependent inflammatory reaction. Am J Pathol 107: 397–418
Faraggi M, Redpath JL, Tal Y (1975) Pulse radiolysis studies of electron transfer reaction in molecules of biological interest. I. The reduction of a disulfide bridge by peptide radicals. Radiat Res 64: 452–466
Faria JAF, Mukai MK (1983) Use of a gas chromatographic reactor to study lipid photooxidation. J Am Oil Chem Soc 60: 77–81
Farris G, Oliva L (1957) Prime osservazioni sull'incremento dell'azione radiobiologica in funzione delle variazioni indotte nelle tensioni superficiali del tessuto vivente. Radiol Clin 26: 223–228
Favier A, Koukay N, Richard MJ (1988) Zinc and protection against oxygen free radicals. Society of Free Radical Research, winter meeting on free radicals in medicine, Current status of antioxidant therapy, Paris 9–10 Dec 1988
Fedorov SM, Grikurov KG, Fedorova TN (1990) Lipid peroxidation in patients with true eczema. Vestn Dermatol Venerol 2: 32–33 (in Russian)
Fegeler F, Beckmann R (1955) Über den Vitamin E Gehalt in Serum bei verschiedenen Hautkrankheiten, mit Bemerkungen über den Wirkungsmechanismus und die klinische Anwendung des Vitamin E. Dermatol Wochenschr 131: 433–440
Felber TD, Smith EB, Knox JM, Wallis C, Melnick JL (1973) Photodynamic inactivation of herpes simplex: report of a clinical trial. JAMA 223: 289–292
Fewtrell CMS, Gomperts BD (1977a) Effect of flavone inhibitors of transport ATPases on histamine secretion from rat mast cells. Nature 265: 635–636
Fewtrell CMS, Gomperts BD (1977b) Quercetin: a novel inhibitor of Ca^{2+} influx and exocytosis in rat peritoneal mast cells. Biochim Biophys Acta 469: 52–60

Finch WR, Rodnan GP, Buckingham RB (1980) Bleomycin induced scleroderma. J Rheumatol 7: 651–659
Findlay GH (1964) Oxygen consumption and changes of colour in the human epidermis: their measurement in-vivo with the cyclops oximeter. S Afr J Lab Clin Med 19: 73–76
Fine AS, Egnor RW, Forrester E, Scopp IW, Stahl SS (1978) Oxidative phosphorylation in rat oral mucosal mitochondria. J Invest Dermatol 70: 16–20
Finnen MJ, Lawrence CM, Shuster S (1984a) Inhibition of dithranol inflammation by free radical scavengers. Lancet ii: 1129–1130
Finnen MJ, Lawrence CM, Shuster S (1984b) Anthralin increases lipid peroxide formation in skin and free radical scavengers reduce anthralin irritancy. Br J Dermatol 111: 717
Finnen MJ, Herdman ML, Shuster S (1985) Distribution and sub-cellular localization of drug metabolizing enzymes in the skin. Br J Dermatol 113: 713–721
Fischer B, Jain KK, Braun E, Lehrl S (1988) Handbook of hyperbaric oxygen therapy. Springer, Berlin Heidelberg New York
Fischer SM, Adams LM (1985) Suppression of tumor promotor induced chemiluminescence in mouse epidermal cells by several inhibitors of arachidonic acid metabolism. Cancer Res 45: 3130–3136
Fischer SM, Baldwin JK, Adams LM (1986) Effects of anti-promotors and strain of mouse on tumor promoter-induced oxidants in murine epidermal cells. Carcinogenesis 7: 915–918
Fischer SM, Cameron GS, Baldwin JK, Jasheway DW, Patrick KE (1988) Reactive oxygen in the tumor promotion stage of skin carcinogenesis. Lipids 23: 592–597
Fleishmajer R (1975) Methodology and techniques in the evaluation of dimethyl sulfoxide for connective tissue disorders. Ann NY Acad Sci 243: 393–394
Flesch P, Rothman S (1948) Role of sulfhydryl compounds in pigmentation. Science 180: 505–506
Flesch P, Satanoce A (1955) Sulfhydryl groups and disulfide linkages in human epidermal components. Br J Dermatol 67: 343–347
Fliegiel SEG, Ward PA, Johnson KJ, Till GO (1984) Evidence for a role of hydroxyl radical in immune complex induced vasculitis. Am J Pathol 115: 375–382
Flindt-Hansen H, Isager H (1987) Scleroderma after occupational exposure to trichlorethylene and trichlorethane. Acta Derm Venereol (Stockh) 67: 263–264
Flohe L (1982) Glutathione peroxidase brought into focus. Free Radic Biol 5: 223–254
Flohe L, Giertz H (1987) Endotoxins, arachidonic acid, and superoxide formation. Rev Infect Dis 9 [Suppl 5]: S553–S561
Floyd RA (1980) Free radicals and cancer. Dekker, New York
Fogh B (1960) Local chemical protection of the skin against roentgen radiation injury. Acta Radiol 53: 49–56
Folkers K, Langsjoen P, Nara Y, Muratsu K, Komorowski J, Richardson PC, Smith K (1988) Biochemical deficiencies of coenzyme Q10 in HIV infection and exploratory treatment. Biochem Biophys Res Commun 153: 888–896
Foote CS, Denny RW (1968) Chemistry of singlet oxygen. VII. Quenching by β-carotene. J Am Chem Soc 90: 6233–6235
Foote CS, Clough RL, Yee BG (1978) Photooxidation of tocopherols. In: de Duve C, Hayashi O (eds) Tocopherol, oxygen and biomembranes. Elsevier/North Holland, New York, pp 13–21
Forbes PD (1981) Photocarcinogenesis: an overview. J Invest Dermatol 77: 139–143
Forbes PD, Cole C (1982) Skin cancer by UV-A alone. 10th Annual meeting of the American Society for Photobiology, Vancouver
Fordhan JN, Daies PG, Kirk A, Currey HLF (1982) Polymorphonuclear function in Behcet's syndrome. Ann Rheum Dis 41: 421–425
Ford-Hutchinson AW, Chan CC (1985) Pharmacological actions of leukotrienes in the skin. Br J Dermatol 113 (Suppl 28): 95–97
Ford-Hutchinson AW, Rackman A (1983) Leukotrienes as mediators of skin inflammation. Br J Dermatol 109 (Suppl 25): 26–29

Forman AB, Roenigk HH, Caro WA, Magid ML (1989) Long term follow up of skin cancer in PUVA-48 cooperative study. Arch Dermatol 125: 515–519
Forman HF, York JL, Fisher AB (1980) Mechanism of the potentiation of oxygen toxicity by disulfiram. J Pharmacol Exp Ther 212: 452–455
Forman HJ (1986) Oxidant production and bacterial activity of phagocytes. Annu Rev Physiol 48: 669–680
Forman HJ, Rotman EI, Fisher AB (1983) Role of selenium and sulfur containing amino acids in protection against oxygen toxicity. Lab Invest 49: 148–153
Forman JH, Fischer AB (1981) Antioxidant enzymes of rat granules pneumocytes: constitutive levels and effect of hyperoxia. Lab Invest 45: 1–6
Forni LG, Willson RL (1986a) Thiyl free radicals and the oxidation of ferrocytochrome c. Biochem J 240: 905–907
Forni LG, Willson RL (1986b) Thiyl and phenoxyl free radicals and NADH. Biochem J 240: 897–903
Forrester AR, Hay JM, Thomson RH (1968) Organic chemistry of stable free radicals. Academic, New York
Forssberg A (1950) On the possibility of protecting the living organism against roentgen rays by chemical means. Acta Radiol 33: 296–304
Fournier J, Emerit J, Michelson AM (1980) Biochmimie Appliquee – la peroxdation des lipides dans les processus inflammatoires. CR Acad Sci (Paris) [D] 3: 753–755
Fournier E, Efthymiou ML, Lecorsier A (1982) Spanish adulterated oil matter: an important discovery by Spanish toxicologists: the toxicity of anilides of unsaturated fatty acids. Toxicol Eur Res 4: 107–112
Fox M, Scott D (1980) The genetic toxicology of nitrogen and sulphur mustard. Mutat Res 75: 131–168
Fraker PJ, Gershwin ME, Good RA, Prasad A (1986) Interrelationship between zinc and immune function. Fed Proc 45: 1474–1479
Fräki JE, Schechter NM, Lazarus GS (1983a) Human skin proteinases as inflammatory mediators. Br J Dermatol 109: 72–76
Fräki JE, Jakoi L, Davies AO (1983b) Polymorphonuclear leukocyte function in psoriasis: chemotaxis, chemokinesis, β-adrenergic receptors, and proteolytic enzymes of polymorphonuclear leucocytes in the peripheral blood from psoriatic patients. J Invest Dermatol 81: 254–257
Frank L, Massaro D (1980) Oxygen toxicity. Am J Med 69: 117–126
Frank L, Yam L, Roberst RJ (1978) The role of endotoxin in protection of adult rats from oxygen induced lung toxicity. J Clin Invest 61: 269–275
Frank L, Summerville J, Massaro D (1980) Protection from oxygen toxicity with endotoxin. J Clin Invest 65: 1104–1110
Franz E, Bosse K, Beck B (1978) Der Sauerstoffverbrauch des spontanen und künstlich induzierten Hautcyclus zu verschiedenen Tageszeiten. Arch Dermatol Res 262: 167–172
Frederick JE, Snell HE, Haywood EK (1989) Solar ultraviolet radiation at the earths surface. Photochem Photobiol 50: 443–450
Freeman BA, Crapo JD (1981) Hyperoxia increases oxygen radical production in rat lungs and lung mitochondria. J Biol Chem 256: 10986–10992
Freeman RG, Nox GM (1964) Influence of temperature on ultraviolet injury. Arch Dermatol 89: 858–864
Freeman SE, Gange RW, Matzinger EA, Sutherland BM (1986) Higher pyrimidine dimer yields in skin of normal humans with higher UVB sensitivity. J Invest Dermatol 86: 34–36
Freeman SE, Gange RW, Sutherland JC, Matzinger EA, Sutherland BM (1987) Production of pyrimidine dimers in DNA of human skin exposed in situ to UVA radiation. J Invest Dermatol 88: 430–433
Frei B, Winterhalter KH, Richter C (1986) Menadione-(2-methyl-1,4-naphtoquinone-) dependent enzymatic redox cycling and calcium release by mitochondria. Biochemistry 25: 4438–4443
Frei B, Stocker R, Ames BN (1988) Antioxidant defenses and lipid peroxidation in human blood plasma. Proc Natl Acad Sci USA 85: 9748–9752

Frei B, England L, Ames BN (1989) Ascorbate is an outstanding antioxidant in human blood plasma. Proc Natl Acad Sci USA 86: 6377–6381

Freierabend J, Engel S (1986) Photoinactiation of catalase in-vitro and in leaves. Arch Biochem Biophys 251: 567–576

Frey JG (1954) Über die Kombinationsbehandlung von Röntgenspätschäden der Haut mit Kurzwellen und Vitamin E. Strahlentherapie 95: 440–443

Frey PA, Ikeda BH, Gavino GR, Speckhard DC, Wong SS (1978) Escherichia coli pyruvate dehydrogenase complex. J Biol Chem 253: 7234–7241

Fridovich I (1978) The biology of oxygen radicals. Science 201: 875–880

Fridovich I (1976) Oxygen radicals, hydrogen peroxide and oxygen toxicity. Free Radic Biol 1: 239–277

Fries JF, Lindgren LA, Bull JM (1973) Scleroderma like lesions and the carcinoid syndrome. Arch Intern Med 131: 550–553

Frings MM, Kind PP, Goerz G, Abel J (1989) The effect of topically corticosteroids and zinc on the metallothionein content of skin in an experimental model. Clin Exp Dermatol 14: 434–436

Frödin T, Skogh M, Molin L (1988) The importance of the coal tar bath in the Ingram treatment of psoriasis. Evaluation by evaporimetry and laser Doppler flowmetry. Br J Dermatol 118: 429–434

Fromm E, Wittmann J (1908) Derivate des p-Nitrophenols. Ber Dtsch Chem Ges 41: 2264–2273

Fuchs J (1988a) Free radical reduction by thioredoxin reductase. (letter to the editor) J Invest Dermatol 91: 92–93

Fuchs J (1988b) Validity of the "bioassay" for thioredoxin reductase activity. (letter to the editor) Arch Dermatol 124: 849–851

Fuchs J, Milbradt R (1989a) Wirkung von topischer DL-Dihydroliponsäure bei durch UVB Bestrahlung induziertem Symptomenkomplex bei gesunden Probanden. In: Borbe H, Ulrich H (eds) Thioctsäure Neue biochemische, pharmakologische und klinische Erkenntnisse zur Thioctsäure. PMI, Frankfurt, pp 299–303

Fuchs J, Milbradt R (1989b) The redox couple dihydrolipoate/lipoate prevents acute photodamage in human skin. International symposium on biological oxidation systems, Bangalore, 23–26 Oct 1989

Fuchs J, Milbradt R (1989c) Vitamin E als Therapeutikum in der klinischen Dermatologie. Zentralbl Haut- Geschlechtskr 156: 407–425

Fuchs J, Milbradt R (1990) Anwendung der Elektronen Spin Resonanz Spektroskopie in der dermatologischen Forschung. 36th Meeting of the German Dermatological Society, Hannover, 29 Aug–2 Sept 1990

Fuchs J, Packer L (1987) EPR spectroscopic investigations on the interaction of alpha tocopherol with various antioxidants. Satellite meeting of the Society for Free Radical Research on oxygen radicals and antioxidants in cancer and aging, Berkeley, 2–7 Feb, 1987

Fuchs J, Packer L (1989) Investigations on the anthralin free radicals in model systems and in skin of hairless mice. J Invest Dermatol 92: 677–682

Fuchs J, Packer L (1990) Ultraviolet irradiation and the skin antioxidant system. Photodermatol Photoimmunol Photomed 7: 90–92

Fuchs J, Packer L (1991a) The antioxidant tocopherol in clinical dermatology. In: Packer L, Fuchs J (eds) Vitamin E: biochemistry and clinical applications. Dekker, New York (in press)

Fuchs J, Packer L (1991b) Electron paramagnetic resonance in dermatologic research with particular reference to photodermatology. Photoderm Photoimmunol Photomed 7: 229–232

Fuchs J, Schürer N (1991) Lipid peroxidation in dermatopathology. In: Vigo-Pelfrey C (ed) Membrane lipid oxidation, vol 3. CRC Press, Boca Raton, pp 165–189

Fuchs J, Zimmer G (1986) 31P-NMR spectroscopic investigations and mitochondrial studies on the cardioprotective efficiency of 2-mercaptopropionylglycine. Biochem Pharmacol 35: 4381–4385

Fuchs J, Veit P, Zimmer G (1985a) Uncoupler- and hypoxia-induced damage of the working rat heart and its treatment. II. Hypoxic reduction of aortic flow and its reversal. Basic Res Cardiol 80: 231–240

Fuchs J, Mainka L, Zimmer G (1985b) 2-Mercaptopropionylglycine and related compounds in treatment of mitochondrial dysfunction and postischemic myocardial damage. Arzneimittelforschung/Drug Res 35: 1394–1402

Fuchs J, Zimmer G, Wölbling RH, Milbradt R (1986) On the interaction between anthralin and mitochondria: a revision. Arch Dermatol Res 279: 59–65

Fuchs J, Beyersdorf F, Zimmer G (1987) A novel cardioprotective regimen for improvement of inner mitochondrial membrane function after ischemic stress. Arzneimittelforsch/Drug Res 37: 1030–1034

Fuchs J, Huflejt ME, Rothfuss LM, Wilson DB, Gerardo C, Packer L (1989a) Impairment of enzymic and non enzymic antioxidants in skin by photooxidative stress. J Invest Dermatol 93: 769–773

Fuchs J, Mehlhorn RJ, Packer L (1989b) Free radical reduction mechanisms in mouse epidermis and skin homogenates. J Invest Dermatol 93: 633–640

Fuchs J, Huflejt M, Rothfuss L, Carcamero G, Packer L (1989c) Acute effects of near ultraviolet and visible light on the cutaneous antioxidant defense system. Photochem Photobiol 50: 739–744

Fuchs J, Nitschmann WH, Packer L (1990a) The antipsoriatic compound anthralin influences bioenergetic parameters and redox properties of energy transducing membranes. J Invest Dermatol 94: 71–76

Fuchs J, Groth N, Herrling T (1990b) Electron paramagnetic resonance tomography in skin. J Invest Dermatol 94: 526

Fuchs J, Mehlhorn R, Packer L (1990c) Nitroxide radical reduction in nude mouse skin – reply (letter to the editor). J Invest Dermatol 95: 242–243

Fuchs J, Milbradt R, Zimmer G (1990d) Dihydrolipoate inhibits reactive oxygen species mediated skin inflammation. J Invest Dermatol 94: 526

Fuchs J, Milbradt R, Zimmer G (1990e) Multifunctional analysis of the interaction of anthralin and its metabolites anthraquinone, and anthralin dimer with the inner mitochondrial membrane. Arch Dermatol Res 282: 47–55

Fuchs J, Milbradt R, Zimmer G, Groth N, Herrling T (1990f) One- and two-dimensional EPR imaging studies in skin. 8th International symposium on bioengineering and the skin, Stresa, 13–16 June 1990

Fuchs J, Nitschmann W, Packer L (1990g) Antioxidant and prooxidant effects of the antipsoriatic compound anthralin in skin and subcellular fractions. In: Emerit I, Packer L, Auclair C (eds) Antioxidants in therapy and preventive medicine. Plenum, London, Advances in Experimental Medicine and Biology 264: 537–541

Fuchs J, Ochsendorf F, Schöfer H, Milbradt R, Rübsamen-Weigmann H (1991a) Oxidative imbalance in HIV infected patients. Med Hypotheses 36: 60–64

Fuchs J, Janka S, Ochsendorf F, Schöfer H, Milbradt R, Freisleben HJ, Buhl R, Unkelbach U, Oster O, Siems W, Esterbauer H, Emerit I (1991b) Oxidants and antioxidants in HIV infected patients. Seventh international conference on AIDS, Florence, Italy, 16–21 June, 1991

Fujii T (1972) The clinical effects of Vitamin E on purpuras due to vascular defects. J Vitaminol 18: 125–130

Fukuzumi K (1986) Relationship between lipoperoxides and disease. J Environ Pathol Toxicol Oncol 6: 25–56

Fuller BJ, Green CJ (1986) Oxidative stress in organs stored at low temperatures for transplantation. In: Rice-Evans (ed) Free radicals, cell damage and disease. Richelieu, London, pp 223–240

Furukawa F, Kashihara-Sawami M, Lyons MB, Norris DA (1990) Binding of antibodies to the extractable nuclear antigens SS-A/Ro and SS-B/La is induced on the surface of human keratinocytes by ultraviolet light (UVL): implications of the pathogenesis of photosensitive cutaneous lupus. J Invest Dermatol 94: 77–85

Galeotti T, Borello S, Seccia A, Farallo E, Bartoli GM, Serri F (1980) Superoxide dismutase content in human epidermis and squamous cell epithelioma. Arch Dermatol Res 267: 83–86

Galewsky E (1916) Cignolin, ein Ersatzpräparat des Chrysarobins. Dermatol Wochenschr 6: 113–115

Gans O (1923) Über die Gewebsatmung in der gesunden und kranken Haut. Dtsch Med Wochenschr 49 (I): 16

Ganther HE, Kraus RJ (1981) Oxidation states of erythrocyte glutathione peroxidase: isolation of an enzyme-glutathione complex. In: Spallholz JE, Martin JL, Ganther HE (eds) Selenium in biology and medicine. AVI, Westport, pp 54–69

Gartmann H, Sieler H (1953) Zur Vitamin E-Behandlung der Induratio penis plastica. Dermatol Wochenschr 128: 1213–1219

Gast W, Ryter M, Schippel G, Schippel K, Walther T, Haustein UF (1988) Ultrastrukturelle Befunde an UV bestrahlten Lymphocyten. Dermatol Monatsschr 174: 762–765

Geesin JC, Gordon JS, Berg RA (1989) Retinoids inhibit collagen synthesis through inhibition of ascorbate stimulated lipid peroxidation. J Invest Dermatol 92: 433

Geesin JC, Hendricks L, Gordon JS, Berg RA (1990) Lipid peroxidation stimulates collagen synthesis in human dermal fibroblasts. Free Radic Biol Med 9 (Suppl 1): 4

Gellin GA, Possick PA, Peronne VB (1970) Depigmentation from 4-tertiary butyl catechol – an experimental study. J Invest Dermatol 55: 190–197

Gemsa D, Leser HG, Seitz M, Debatin M, Bärlin E, Deimann W, Kramer W (1982) Cells in inflammation. The role of macrophages in inflammation. Agents Action 11 [Suppl]: 93–114

Gerloczy F, Lancos F, Szabo J (1966) Wirkung des Vitamin E auf die Kapillarresistenz bei Purpura im Kindesalter. Acta Paediatr Acad Sci Hung 7: 363–367

Gerschmann R, Gilbert DL, Nye SW, Dweyer P, Fenn WO (1954) Oxygen poisoning and x-irradiation: a mechanism in common. Science 119: 623–626

Ghezzi P, Bianchi M, Mantovani A, Spreafico F, Salmona M (1984) Enhanced xanthine oxidase activity in mice treated with interferon and interferon inducers. Biochem Biophys Res Commun 119: 144–149

Ghezzi P, Saccardo B, Bianchi M (1986) Induction of xanthine oxidase and heme oxygenase and depression of liver drug metabolism by interferon: a study with different recombinant interferons. J Interferon Res 6: 251–256

Giblin FJ, McCready JP, Reddan JR, Dziedzic DC, Reddy VN (1985) Detoxification of H_2O_2 by cultured rabbit lens epithelial cells: participation of the glutathione redox cycle. Exp Eye Res 40: 827–840

Gibson SL, Cohen HJ, Hilf R (1984) Evidence against the production of superoxide by photoirradiation of hematoporphyrin derivative. Photochem Photobiol 40: 441–448

Gigli I, Schothorst AA, Soter NA (1980) Erythropoietic protoporphyria: photoactivation of the complement system. J Clin Invest 66: 517–522

Gigure V, Ong ES, Segui P, Evans RM (1987) Identification of a receptor for the morphogen retinoic acid. Nature 330: 624–629

Gilchrest BA, Soter NA, Stoff JS, Mihm MC (1981) The human sunburn reaction: histologic and biochemical studies. J Am Acad Dermatol 5: 411–422

Gilchrest BA, Szabo G, Flynn E, Goldwyne RM (1983) Chronologic and actinic induced aging in human facial skin. J Invest Dermatol 80: 81 s–85 s

Gillert E (1944) Die Kampfstoffverletzung. Urban and Schwarzenberg, Vienna

Gillespie E, Lichtenstein LM (1972) Histamine release from human leukocytes: studies with deuterium oxide, colchicine and cytochalasin B. J Clin Invest 51: 2941–2947

Gillespie E, Levine RJ, Malawista SE (1968) Histamine release from rat peritoneal mast cells: inhibition by colchicine and potentiation by deuterium oxide. J Pharmacol Exp Ther 164: 158–165

Gilliam JN, Sontheimer RD (1981) Distinctive cutaneous subsets in the spectrum of lupus erythematosus. J Am Acad Dermatol 4: 471–475

Gilsanz V, Alvarez JP, Serrano S, Simon J (1984) Evolution of the alimentary toxic oil syndrome due to ingestion of denaturated rapseed oil. Arch Intern Med 144: 254–256

Giovannini L, Bertelli AAE, Scalori V, Dell'Osso L, Alessandri MG, Mian M (1988) Skin penetration of CoQ 10 in the rat. Int J Tissue Reac X: 103–105

Girotti AW (1979) Protoprophyrin sensitized photodamage in isolated membranes of human erythrocytes. Biochemistry 18: 4403–4411

Glader BE, Conrad ME (1973) Hemolysis by diphenylsulfones: Comparative effects of DDS and hydroxylamine DDS. J Lab Clin Med 81: 267–272

Glasenapp I, Leonhardi G (1953) Die biologische Oxidation in der menschlichen Haut. Arch Dermatol Syphilol 196: 319–324

Glass GA, Gershon D (1981) Enzymatic changes in rat erythrocytes with increasing cell and donor age: loss of superoxide dismutase activity associated with increases in catalytically defective forms. Biochem Biophys Res Commun 103: 1245–1253

Glavind J, Christensen F (1967) Influence of nutrition and light on the peroxide content of the skin surface lipids of rats. Acta Derm Venereol (Stockh) 47: 339–344

Glavind J, Christensen F (1969) Further studies on the peroxidation of the surface lipids of the skin of rodents. Acta Derm Venereol (Stockh) 49: 536–546

Glavind J, Faber M (1967) Antioxidants and lipoperoxides in the organs of irradiated rats. III. Importance of the dose rate. Int J Radiat Biol 12: 121–126

Glette J, Sandberg S (1986) Phototoxicity of tetracyclines as related to singlet oxygen production and uptake by polymorphonuclear leucocytes. Biochem Pharmacol 35: 2883–2885

Gloor M (1990) Benzoylperoxide in der dermatologischen Lokaltherapie. Zentralbl Haut-Geschlechtskr 157: 1010–1015

Goeckerman WH (1925) The treatment of psoriasis. Northwest Med 24: 229–231

Goerz G, Merk H (1983) Biochemische Aspekte der Entzündungsreaktion – unter besonderer Berücksichtung des Sauerstoffs. Z Hautkr 58: 1669–1686

Gohil K, Rothfuss L, Lang J, Packer L (1987) Effect of exercise training in tissue vitamin E and ubiquinone content. J Appl Physiol 63: 1638–1641

Gohil K, Viguie, Stanley WC, Packer L, Brooks GA (1988) Blood glutathione oxidation during human exercise. J Appl Physiol 64: 115–119

Golden MHN, Ramdath D (1987) Free radicals in the pathogenesis of kwashiorkor. Proc Nutr Soc 46: 53–68

Goldfarb MT, Ellis CN, Weiss JS, Voorhees JJ (1989) Topical tretinoin and photoaged skin. Cutis 43: 476–482

Goldfischer S, Reddy JK (1984) Peroxisomes (microbodies) in cell pathology. Int Rev Exp Pathol 26: 45–84

Goldman M, Dacre JC (1989) Lewisite: it's chemistry, toxicology, and biological effects. Rev Environ Contam Toxicol 110: 76–115

Goldman PJ (1973) Schwerste akute Chlorakne, eine Massenintoxikation durch 2,3,6,7-Tetrachlorodibenzodioxin. Hautarzt 24: 149

Goldstein BD, Harber LC (1972) Erythropoietic protoporphyria: lipid peroxidation and red cell membrane damage associated with photohemolysis. J Clin Invest 51: 892–902

Goldstein BD, McDonagh EM (1976) Spectrofluorescent detection of in-vivo red cell lipid peroxidation in patients treated with diaminodiphenylsulfone. J Clin Invest 57: 1302–1307

Goldstein BD, Witz G, Amoruso M, Troll W (1979) Protease inhibitors antagonize the activation of polymorphonuclear leukocytes oxygen consumption. Biochem Biophys Res Commun 88: 854–860

Goldstein BD, Witz G, Amoruso M, Stone DS, Troll W (1981) Stimulation of human polymorphonuclear leucocyte superoxide anion radical production by tumor promotors. Cancer Lett 11: 257–262

Goldstein IM, Kaplan HB, Edelson HS, Weisman G (1979) Ceruloplasmin: a scavenger of superoxide anion radicals. J Biol Chem 254: 4040–4045

Golumbic C, Mattill HA (1941) Antioxidants and the autoxidation of fats. XIII. The antioxygenic action of ascorbic acid with tocopherols, hydroquinones and related compounds. J Am Chem Soc 63: 1279–1280

Gomer CJ, Razum NJ (1984) Acute skin response in albino mice following porphyrin photosensitization under oxic and anoxic conditions. Photochem Photobiol 40: 435–439

Gommans JM, Bergers M, Van Erp PEJ, Van Den Hurk JMA, Mier PD, Roelfzema H

(1979a) Studies on the plasma membrane of normal and psoriatic keratinocytes. 1. Preparation of material and morphological characterization. Br J Dermatol 101: 407–412
Gommans JM, Bergers M, Van Erp PEJ, Van Den Hurk JMA, Van Den Kerkhof P, Mier PD, Roelfzema H (1979b) Studies on the plasma membrane of normal and psoriatic keratinocytes. 2. Cyclic AMP and its response to hormonal stimulation. Br J Dermatol 101: 413–418
Goncharenko MS (1983) Status and role of free radical processes in psoriasis patients. Vestn Dermatol Venerol 6: 7–11 (in Russian)
Goodchild NT, Kwock L, Lin PS (1981) Melanin: a possible cellular superoxide scavenger. In: Rodgers MAJ, Powers EL (eds) Oxygen and oxy-radicals in chemistry and biology. Academic, New York, pp 645–648
Gorgas K (1987) Morphogenesis of peroxisomes in lipid synthesizing epithelia. In: Fahimi HD, Sies H (eds) Peroxisomes in biology and medicine. Springer, Berlin Heidelberg New York, pp 3–17
Götz H (1976) The relationship between ultraviolet light sensitization and tar exposure of the skin. Aust J Dermatol 17: 57–60
Gower JD (1988) A role for dietary lipids and antioxidants in the activation of carcinogens. Free Radic Biol Med 5: 95–111
Graham JR (1967) Cardiac and pulmonary fibrosis during methysergide therapy for headache. Am J Med Sci 254: 23–34
Granger DN, Höllwarth ME, Parks DA (1986) Ischemia reperfusion injury: role of oxygen derived free radicals. Acta Physiol Scand Suppl 548: 47–63
Grant D, Long WF, Williamson FB (1989) A comparison of the antioxidant requirements of proteins with those of synthetic polymers suggests an antioxidant function for clusters of aromatic and bivalent sulphur containing amino acid residues. Med Hypotheses 28: 245–253
Gray GM, Yardley HJ (1975a) Mitochondria and nuclei of pig and human epidermis: isolation and lipid composition. J Invest Dermatol 64: 423–430
Gray GM, Yardley GM (1975b) Different populations of pig epidermal cells: isolation and lipid composition. J Lipid Res 16: 441–452
Greaves MW (1982) Prostaglandine, Leukotriene, antiinflammatorische Substanzen und deren Bedeutung bei entzündlichen Reaktionen der Haut. Hautarzt 33: 123–129
Green HA, Margolis R, Boll J, Kochevar IE, Parrish JA, Oseroff AR (1987) Unscheduled DNA synthesis in human skin after in-vitro ultraviolet-excimer laser ablatio. J Invest Dermatol 89: 201–204
Green SG, Stewart ME, Downing DT (1984) Variation in sebum fatty acid composition among adult humans. J Invest Dermatol 83: 114–117
Greene RS, Downing DT, Pochi PE, Strauss JS (1970) Anatomical variation in the amount and composition of human skin surface lipids. J Invest Dermatol 54: 240–247
Greenwald RA (1980) Effect of oxygen derived free radicals on connective tissue macromolecules. In: Bannister WH, Bannister JV (eds) Biological and clinical aspects of superoxide dismutase, developments in biochemistry, vol 11B. Elsevier, Amsterdam, pp 160–171
Greenwald RA (1990) Superoxide dismutase and catalase as therapeutic agents for human diseases. A critical review. Free Radic Biol Med 8: 201–209
Gregory EM, Fridovich I (1973) Induction of superoxide dismutase by molecular oxygen. J Bacteriol 114: 543–548
Gregory EM, Yost FJ, Fridovich I (1973) Superoxide dismutase of Escherichia coli: intracellular localization and functions. J Bacteriol 115: 987–991
Grootveld M, Halliwell B (1987) Measurement of allantoin and uric acid in human body fluids. A potential index of free radical reactions in-vivo? Biochem J 243: 803–808
Grösser A, Stark F, Landthaler M (1984) Behandlung von hypertrophen Narben und Keloiden mit Orgotein. Hautarzt 35: 377–378
Grozdova MD, Khokhlova JV, Panasyuk AF, Guseva NG, Trofimova TM (1984) Cyclic nucleotides and calcium transport in cultured dermal fibroblasts from progressive systemic sclerosis and rheumatoid arthritis patients. Arthritis Rheum 27: 1144–1149

Grubb E (1952) Our experiences of vitamin E treatment. Acta Derm Venereol (Stockh) 32: 256–258

Grum CM, Ragsdale RA, Ketai LH, Simon RH (1987) Plasma xanthine oxidase activity in patients with adult respiratory distress syndrome. J Crit Care 2: 22–26

Gruner S, Plötz M, Zwirner A, Müller GM, Paskaleva I, Sönnichsen N, Diezel W (1987a) Intrazelluläre Enzymhemmung durch PUVA. Hemmung saurer Hydrolasen in menschlichen Granulocyten. Dermatol Monatschr 173: 249–257

Gruner S, Strunk D, Diezel W (1987b) Der Einfluß topischer Antipsoriatika auf ATPase-positive epidermale Langerhans Zellen. Tierexperimentelle Untersuchungen. Dermatol Monatsschr 173: 436–440

Gruner S, Stoppe H, Eckert R, Sönnichsen N, Diezel W (1990) Verlängerung der Transplantatüberlebenszeit durch eine PUVA Behandlung des Transplantatempfängers. Dermatol Monatsschr 176: 49–54

Gstrautenthaler G, Pfaller W, Katanko P (1983) Glutathione depletion and in-vitro lipid peroxidation in mercury of maleate induced acute renal failure. Biochem Pharmacol 32: 2969–2972

Gupta RC (1978) Circulating immune complexes in active Behcet's disease. Clin Exp Immunol 34: 213–217

Gustafson H, Johansson B, Edsmyr F (1981) Pyronie's disease: experience of local treatment with orgotein. Eur Urol 7: 346–348

Gutterridge JMC (1978) The membrane effects of vitamin E, cholesterol and their acetates on peroxidative permeability. Res Commun Chem Pathol Pharmacol 22: 563–572

Gutteridge JMC (1986) Aspects to consider when detecting and measuring lipid peroxidation. Free Radic Res Commun 1: 173–184

Guzman-Barron ES (1954) The effect of x-rays on systems of biological importance. In: Hollaender A (ed) Radiation biology, vol 1, pt 1. McGraw Hill, New York, pp 283–314

Haber F, Weiss J (1932) Über die Katalyse des Hydroperoxides. Naturwissenschaften 20: 948–950

Haber F, Weiss J (1934) The catalytic decomposition of hydrogen peroxide by iron salts. Proc R Soc Lond [A] 147: 332–351

Habermann E, Müller B (1989) Oxoferin und Natriumchlorit – ein Vergleich. Klin Wochenschr 67: 20–24

Hadler WA, Silveira SR (1981) Identification of tocopherol (vitamin E) in the skin and its histochemical detection. Acta Histochem 68: 11–21

Haenen GRMM, Bast A (1983) Protection against lipid peroxidation by a microsomal glutathione dependent labile factor. FEBS Lett 159: 24–28

Haenen GRMM, Bast A (1989) Modulation of lipid peroxidation by dihydrolipoate in interplay with vitamin C, vitamin E and glutathione. Free radical and cancer, winter meeting, Society for Free Radical Research, Tübingen, 8–9 Dec 1989

Hagerman G (1951) On the use of vitamin E in dermatologic practice. Acta Derm Venereol (Stockh) 31: 225–226

Hall RP, Lawley TJ, Heck JA, Katz SI (1980) IgA containing circulating immune complexes in dermatitis herpetiformis, Hennoch Schönlein purpura, systemic lupus erythematosus and other diseases. Clin Exp Immunol 40: 431–438

Halliwell B, Cross CE (1991) Reactive oxygen species, antioxidants and acquired immunodeficiency syndrome. Sense or speculation? Arch Intern Med 151: 29–31

Halliwell B, Gutteridge MC (1984) Oxygen toxicity, oxygen radicals, transition metals and disease. Biochem J 219: 1–14

Halliwell B, Gutteridge JMC (1985) Free radicals in biology and medicine. Clarendon, Oxford

Halliwell B, Gutteridge JMC (1986a) Iron and free radical reactions: two aspects of antioxidant protection. Trends Biochem Sci 11: 372–375

Halliwell B, Gutteridge JMC (1986b) Oxygen free radicals and iron in relation to biology and medicine: some problems and concepts. Arch Biochem Biophys 246: 501–514

Halprin K, Ohkawara A (1967) The measurement of gluthathione in human epidermis using gluthatione reductase. J Invest Dermatol 48: 149–152

Ham EA, Egan RW, Soderman DD, Gale-Huehl FA (1979) Peroxidase dependent deactivation of prostacyclin synthetase. J Biol Chem 254: 2195–2198

Hamer DH (1986) Metallothionein. Annu Rev Biochem 55: 913–951

Hamilton-Gibbs JS (1964) Pseudoxanthoma elasticum. Aust J Dermatol 7: 156–163

Hanada K, Gange R, Connor MJ (1991) Effect of glutathione depletion on sun burn cell formation in the hairless mouse. J Invest Dermatol 96: 838–840

Hanfstaengel E (1951) Die Behandlung der Dupuytrenschen Palmarfascienkontraktur mit Vitamin E. Med Klin 373–375

Hanson DL, De Leo VA (1989) Long wave ultraviolet radiation stimulates arachidonic acid release and cyclooxygenase activity in mammalian cells in culture. Photochem Photobiol 49: 423–430

Hanson DL, Bodian A, De Leo V (1989) Ultraviolet A (320–400 nm) induces phospholipase activation in human epidermal keratinocytes in culture. J Invest Dermatol 92: 39

Hansson HA, Holmgren A, Rozell B, Stemme S (1986) Localization of thioredoxin, thioredoxin reductase and ribonucleotide reductase in cells: immunohistochemical aspects. In: Holmgren A (ed) Thioredoxin and glutaredoxin systems: structure and function. Raven, New York, pp 177–187

Hansson RB, Hansson HA, Jonsson O, Lindholm L, Pettersson S, Schersten T, Waldenstrom J (1987) Protection of renal cells against free radical damage in-vitro. J Clin Pathol 87: 601–607

Hanstein WG (1976) Uncoupling of oxidative phosphorylation. Biochim Biophys Acta 456: 129–148

Harman D (1956) Aging: a theory based on free radical and radiation chemistry. J Gerontol 11: 298–300

Harman D (1980 a) Free radical theory of aging: beneficial effect of antioxidants on the life span of male NZB mice; role of free radical reactions in the deterioration of the immune system with age and in the pathogenesis of systemic lupus erythematosus. Age 3: 64–73

Harman D (1980 b) Secondary amyloidosis and antioxidants. Lancet ii: 593

Harman D (1984) Free radical theory of aging: the "free radical" diseases. Age 7: 111–131

Harman D, Eddy D, Noffsinger J (1976) Free radical theory of aging: inhibition of amyloidosis in mice by antioxidants possible mechanisms. J Am Geriatr Soc 24: 203–210

Harman HH (1976) Modern factor analysis, 3rd edn. University of Chicago Press, Chicago

Harman LS, Mottley C, Mason RP (1984) Free radical metabolites of L-cysteine oxidation. J Biol Chem 259: 5606–5611

Harmon CS, Masser MR, Phizackerley PJR (1986) Effect of ischemia and reperfusion of pig skin flaps on epidermal glycogen metabolism. J Invest Dermatol 86: 69–73

Harris M, Herp A, Phelps R, Pigman W (1971 a) A study of the breakdown of carbohydrates by redox generated free radicals. Biochim Biophys Acta 244: 501–504

Harris MJ, Herp A, Pigman W (1971 b) Depolymerization of polysacharides through the generation of free radicals at a platinum surface: a novel procedure for the controlled production of free radical oxidations. Arch Biochem Biophys 142: 615–622

Harris RS, Bunker JM, Milas NA (1932 a) The germicidal activity of vapors from irradiated oils. J Bacteriol 23: 429–435

Harris RS, Bunker JM, Milas NA (1932 b) Chemical nature of germicidal vapors emanating from irradiated oils. Ind Eng Chem 24: 1181–1183

Hartz JW, Funakoshi S, Deutsch HF (1973) The levels of superoxide dismutase and catalase in human tissues as determined immunohistochemically. Clin Chim Acta 46: 125–132

Harwood R, Grant ME, Jackson DS (1974) Influence of ascorbic acid on ribosomal patterns and collagen biosynthesis in healing wounds of scorbutic guinea pigs. Biochem J 142: 641–651

Hassan HM (1988) Biosynthesis and regulation of superoxide dismutases. Free Radic Biol Med 5: 377–385

Hatzelmann A, Ullrich V (1987) Regulation of 5-lipoxygenase activity by the glutathione status in human polymorphonuclear leukocytes. Eur J Biochem 169: 175–184

Hausmann W (1908) Die sensibilisierende Wirkung tierischer Farbstoffe und ihre physiologische Bedeutung. Wien Klin Wochenschr 21: 1527–1529

Haustein UF, Ziegler V (1985) Environmentally induced systemic sclerosis like disorders. Int J Dermatol 24: 147–151
Haustein UF, Ziegler V (1986) Sklerodermie und Sklerodermie ähnliche Erkrankungen durch Umweltsubstanzen. Dermatosen 34: 61–67
Haustein UF, Tegtmeyer L, Ziegler V (1985) Allergic and irritant potential of benzoyl peroxide. Contact Dermatitis 13: 252–257
Haustein UF, Ziegler V, Herrmann K, Mehlhorn J, Schmidt C (1990) Silica induced scleroderma. J Am Acad Dermatol 22: 444–448
Havsteen B (1983) Flavonoids, a class of natural products of high pharmacological potency. Biochem Pharmacol 32: 1141–1148
Hay DR (1964) Malignant carcinoid syndrome with scleroderma. NZ Med J 63: 90–92
Hayaishi O (1987) Indolamine 2,3 dioxygenase: O 2-requiring enzyme and its role in interferon action. In: Hayaishi O, Imamura S, Miyachi Y (eds) The biological role of reactive oxygen species in skin. Elsevier, New York, pp 3–8
Hayakawa K, Oizumi J (1989) Determination of lipolysine derived from enzymes by liquid chromatography. J Chromatogr 490: 33–41
Hayakawa K, Matsuo I, Fujita H, Ohkido M (1987) Thiazide phototoxicity: evidence for participation of singlet oxygen in photohemolysis. In: Hayaishi O, Imamura S, Miyachi Y (eds) The biological role of reactive oxygen species in skin. Elsevier, New York, pp 101–106
Hayakawa R (1971) Relation with facial skin diseases and lipids and lipoperoxide of serum and sebum, pH on the face and buffering power. Jpn J Dermatol 81: 11–29 (in Japanese)
Hayakawa R, Okada Y (1977) Lipoperoxide in melanosis faciei feminus. Nishinihon J Dermatol 39: 70–74 (in Japanese)
Hayakawa R, Ueda H, Nozaki T, Izawa Y, Yokotake J, Yazaki K, Azumi T, Okada Y, Kobayashi M, Usuda T, Ishida J, Kondo T, Adachi A, Kawase A, Matsunaga K (1981) Effects of combination treatment with vitamins E and C on chloasma and pigmented contact dermatitis. A double blind controlled clinical trial. Acta Vitaminol Enzymol 3: 31–38
Hayden RE, Yeung CST, Paniello RC, Bello SL, Dawson SM (1987) The effect of glutathione and vitamins A, C and E on acute skin flap survival. Laryngoscope 97: 1176–1179
Hazelrigg DE, Mc Elroy RJ (1980) The yellow nail syndrome. J Assoc Milit Dermatol 6: 1–21
Hazelton GA, Lang CA (1985) Glutathione peroxidase and reductase activities in the aging mouse. Mech Ageing Dev 29: 71–88
Hearing VJ, Jimenez M (1987) Mammalian tyrosinase – the critical regulatory control point in melanocyte pigmentation. Int J Biochem 19: 1141–1147
Heckly RJ (1980) Free radicals in dry biological systems. Free Radic Biol 4: 135–158
Hein R, Mauch C, Braun-Falco O, Krieg T (1988) Zur Pathophysiologie von Fibrosen. Hautarzt 39: 65–71
Hellerström S, Thyresson N, Blohm SG, Widmark G (1955) On the nature of the eczematogenic component of oxidized d 3-carene. J Invest Dermatol 24: 217–224
Hemler ME, Cook HW, Lands WEM (1979) Prostaglandin biosynthesis can be triggered by lipid peroxides. Arch Biochem Biophys 193: 340–345
Hendrick SJ, Silverman AK, Solomon AR, Headington JT (1988) α_1-Antitrypsin deficiency associated with panniculitis. J Am Acad Dermatol 18: 684–692
Hennet T, Peterhans E, Stocker R (1988) Redox status of mouse tissues during influenza A infection. Free radicals in medicine, Society for Free Radical Research, winter meeting, Paris, 9–10 Dec 1988
Hennig B, Chow CK (1988) Lipid peroxidation and endothelial cell injury: implications in atherosclerosis. Free Radic Biol Med 4: 99–106
Henricks PAJ, Van der Tol ME, Verhoef J (1985) Interactions between human polymorphonuclear leukocytes and influenza virus. Scand J Immunol 22: 721–725
Hensby CN, Schroot B, Schaefer H, Juhlin L, Ortonne JP (1983) Prostaglandins in human skin disease. Br J Dermatol 109 (Suppl 25): 22–25
Henschler D (1975) Vergiftungen durch Halogen-Kohlenwasserstoffe. In: Henschler D (ed) Gesundheitsschädliche Arbeitsstoffe. Verlag Chemie, Weinheim, pp 181–189

Hepplestone AG, Styles FS (1967) Activity of macrophage factor in collagen formation. Nature 214: 521–522
Herndon JH, McGuire JS (1967) Oxidation of fatty acids in guinea pig epidermis. Arch Biochem Biophys 119: 583–585
Herrling T, Klimes N, Karthe W, Ewert U, Ewert B (1982) EPR zeugmatography using modulated magnetic field gradients. J Magn Reson 49: 203–211
Herzenberg LA, Raju PA, Staal F, Montano M, Herzenberg LA, Roeder M (1989) Thiols inhibit HIV replication in an in-vitro model system: potential therapeutic use of N-acetylcysteine. AIDS E Sindromi Correlate, III Convegno Nazionale, Naples, 10–12 Nov 1989
Hess ML, Manson NH (1983) The paradox of steroid therapy. Inhibition of oxygen free radicals. Circ Shock 10: 205–213
Heytler PG (1979) Uncouplers of oxidative phosphorylation. Methods Enzymol 55: 462–472
Hicks M, Gebicki JM (1981) Inhibition of peroxidation in linoleic acid membranes by nitroxide radicals, butylated hydroxytoluene and a-tocopherol. Arch Biochem Biophys 210: 56–63
Hidiroglou M (1986) Vitamin E response in sheep to various modes of administration. Int J Vitamin Nutr Res 56: 27–252
Higuchi K, Matsumura A, Honma A, Takeshita S, Hashimoto K, Hosokawa M, Yasuhira K, Takeda T (1983) Systemic senile amyloid in senescence accelerated mice. Lab Invest 48: 231–240
Hilger R, Fukuyama K, Zackheim HS, Epstein JH (1974) Increased melanocytes in hairless mice following topical treatment with carmustine (BCNU) and nitrogen mustard (NM). Clin Res 22: 159 A
Hill KE, Burk RF (1984) Influence of vitamin E and selenium on glutathione-dependent protection against microsomal lipid peroxidation. Biochem Pharmacol 33: 1065–1068
Hiller D, Kessler M, Hornstein OP (1986) Vergleichende kutane Sauerstoffmessung (p O 2) bei Gesunden und bei Patienten mit progressiver Sklerodermie. Hautarzt 37: 83–89
Hilton JG (1980) Effects of alterations of polyunsaturated fatty acid metabolism upon plasma volume loss induced by thermal trauma. J Trauma 20: 663–668
Hindson C, Diffey B, Lawlor F, Downey A (1983) Dithranol UVA phototherapy (DUVA) for psoriasis: a treatment without dressings. Br J Dermatol 108: 457–460
Hintz P, Kalyanaraman B (1986) Metal ion induced activation of molecular oxygen in pigmented polymers. Biochim Biophys Acta 883: 41–45
Hinz J, Kühne FW, Stahl KW (1984) Local tetrachlorodecaoxide treatment to improve oxygen supply to non-healing wounds. Lancet ii: 630
Hinz J, Hautzinger H, Stahl KW (1986) Rationale for and results from a randomized double blind clinical trial of tetrachlorodecaoxygen anion complex in wound healing. Lancet i: 823–828
Hiramatsu M, Izawa Y, Hagihara M, Nishigaki I, Yagi K (1984) Serum lipid peroxide levels of patients suffering from thermal injury. Burns 11: 111–116
Hirata F, Hayaishi O (1975) Studies on indolamine 2,3 dioxygenase. J Biol Chem 250: 5960–5966
Holman RT, Greenberg SI (1960) Note on toxicities of methyl oleate peroxide and ethyl linoleate peroxide. J Am Oil Chem Soc 35: 707–711
Holmes B, Page AR, Good RA (1967) Studies on the metabolic activity of leukocytes from patients with a genetic abnormality of phagocytic function. J Clin Invest 46: 1422–1432
Holmgren A (1979) Thioredoxin catalyzes the reduction of insulin disulfide by dithiothreitol and dihydrolipoamide. J Biol Chem 254: 9627–9632
Holmgren A (1985) Thioredoxin. Annu Rev Biochem 54: 237–271
Holmgren A, Luthman M (1978) Tissue distribution and subcellular localization of bovine thioredoxin determined by radioimmunoassay. Biochemistry 17: 4071–4077
Holthusen H (1921) Beiträge zur Biologie der Strahlenwirkung. Pflügers Arch Ges Physiol 187: 1–24
Holzmann H, Morsches B, Benes P (1979) The absorption of benzoyl peroxide from leg ulcers. Arzneimittelforsch/Drug Res 29: 1180–1183

Holzmann H, Bernd A, Hohlmaier K (1988) Proliferation of non-psoriatic human fibroblasts in vitro by serum from patients with psoriasis. Lancet ii: 1031
Honma K, Saga, Onodera H, Takahashi M (1990) Potassium iodide inhibits neutrophil chemotaxis. Acta Derm Venereol (Stockh) 70: 247–249
Hopkins EF (1918) On dichlorethylsulphide (mustard gas). III Solubility and hydrolysis of dichlor-ethylsulphide. With a new method for estimating small amounts of the same. J Pharmacol Exp Ther 12: 393–403
Hopsu-Havu VK, Fräki JE (1981) Proteinases and their inhibitors in skin disease. Int J Dermatol 20: 159–163
Horacek J, Cernikova-Brünn M (1961) Der Einfluß von Strahlung auf Lipoide des menschlichen Hauttalges. Arch Klin Exp Dermatol 213: 124–129
Horgan VJ, Philpot STL, Porter WA, Roodyn DB (1957) Toxicity of autoxidized squalene and linoleic acid, and of simpler peroxides, in relation to toxicity of radiation. Biochem J 67: 551–558
Horikoshi T, Balin AK, Carter DM (1986) Effect of oxygen on the growth of human epidermal keratinocytes. J Invest Dermatol 86: 424–427
Horikoshi T, Balin AK, Carter DM (1991) Effects of oxygen tension on the growth and pigmentation of normal human melanocytes. J Invest Dermatol 96: 841–844
Horio T, Okamoto H (1987) Oxygen intermediates are involved in ultraviolet radiation-induced damage of Langerhans cells. J Invest Dermatol 88: 699–702
Hornsby PJ, Crivello JF (1983) The role of lipid peroxidation and biological antioxidants in the function of the adrenal cortex, part 2. Mol Cell Endocrinol 30: 123–147
Horwitt MK (1980) Therapeutic uses of vitamin E in medicine. Nutr Rev 38: 105–113
Hoshi K, Oka H, Haramoto S, Takahashi H, Yanagawa A, Mizushima Y (1989) Clinical application of liposomal SOD in several intractable diseases. In: Hayaishi O, Niki E, Kondo M, Yoshokawa T (eds) Medical, biochemical and chemical aspects of free radicals. Elsevier, Amsterdam, pp 731–734
Housset B, Ody C, Rubin DB, Eleman G, Junod AF (1983) Oxygen toxicity in cultured aortic endothelium: selenium-induced partial protective effect. J Appl Physiol 55: 343–352
Hsu JM (1982) Zinc deficiency and glutathione linked enzymes in rat liver. Nutr Rep Int 25: 573–582
Huang L, Privalle CT, Serafin D, Klitzman B (1987) Increased survival of skin flaps by scavengers of superoxide radical. FASEB J 1: 129–132
Huber AM, Gershoff SN (1973) Effects of dietary zinc on zinc enzymes in the rat. J Nutr 103: 1173–1181
Huber W (1981) Orgotein (bovine Cu-Zn superoxide dismutase), an antiinflammatory protein drug: Discovery, toxicology and pharmacology. Eur J Rheumatol Inflamm 4: 173–182
Hug DH (1978) The activation of enzymes with light. Photochem Photobiol Rev 3: 1–33
Hug DH (1981) Photoactivation of enzymes. Photochem Photobiol Rev 6: 87–138
Hung SSO, Moon TW, Hilton JW, Slinger SJ (1982) Uptake, transport and distribution of *dl-α*-tocopherylacetate compared to *d-α*-tocopherol in rainbow trout. J Nutr 112: 1590–1596
Hunt TK (1980) Disorders of wound healing. World J Surg 4: 271–277
Hunter JA, Bhutani LK, Magnus IA (1970) Chlorpromazine photosensitivity in mice: its action spectrum and effect of antiinflammatory agents. Br J Dermatol 82: 157–168
Hunziker T, Krebs A (1983) Thalidomid in der Dermatologie. Hautarzt 34: 66–72
Husain SR, Cillard J, Cillard P (1987a) *α*-Tocopherol prooxidant effect and malondialdehyde production. J Am Oil Chem Soc 64: 109–111
Husain SR, Cillard J, Cillard P (1987b) Hydroxyl radical scavenging activity of flavonoids. Phytochemistry 26: 2489–2491
Hussain MZ, Bhatnagar RS (1979) Involvement of superoxide in the paraquat-induced enhancement of lung collagen synthesis in organ culture. Biochem Biophys Res Commun 89: 71–76
Hütter H (1954) Neue Anwendungsgebiete für die Vitamin E Behandlung. Dtsch Gesundheitswes: 277–279

Hyslop PA, Hinshaw DB, Halsey WA, Schraufstätter IU, Sauerheber RD, Spragg RG, Jackson JH, Cochrane CG (1988) Mechanisms of oxidant mediated cell injury. J Biol Chem 263: 1665–1675

Igarashi A, Uzuka M, Nakajima K (1989) The effects of vitamin E deficiency on rat skin. Br J Dermatol 121: 43–49

Ikeda S, Naito I, Imai R, Manabe M, Takamori K, Ogawa H (1985) Origin and properties of the blister formation factor in the blister fluids from recessive dystrophic epidermolysis bullosa. Br J Dermatol 113: 661–667

Ikeda S, Manabe M, Muramatsu T, Takamori K, Ogawa H (1988) Protease inhibitor therapy for recessive dystrophic epidermolysis bullosa. J Am Acad Dermatol 18: 1246–1251

Ilizuka H, Ohkawara A (1986) "Ischemic" rise of epidermal cyclic AMP is a beta adrenergic adenylate cyclase dependent process. J Invest Dermatol 86: 271–274

Im MJ, Hoopes JE (1984) Age dependent decrease in superoxide dismutase activity in rat skin. J Invest Dermatol 82: 437

Im MJ, Shen WH, Pak CJ, Manson PN, Bulkley GB, Hoopes JE (1984) Effect of allopurinol on the survival of hyperemic island skin flaps. Plast Reconstr Surg 73: 276–278

Imamura S, Yanase K, Taniguchi S, Ofugi S, Mangavil L (1980) Erythema multiforme: Demonstration of immune complexes in the sera and skin lesions. Br J Dermatol 102: 161–166

Imre S, Toth F, Fachet J (1984) Superoxide dismutase, catalase and lipid peroxidation in liver of young mice of different ages. Mech Ageing Dev 28: 297–304

Infante PF (1981) Observations of the site specific carcinogenicity of vinyl chloride to humans. Environ Health Perspect 41: 89–94

Ingold KU (1986) Antioxidants in normal and in tumor tissue. In: Annual scientific progress report, project of the National Foundation for Cancer Research, pp 243–247

Ingold KU (1991) Uptake and loss of vitamine E in animals and in man. German International Workshop on a-Tocopherol. Kluvensiek, Germany, Jan 31–Feb 5, 1991

Ingold KU, Burton GW, Foster DO, Hughes L, Lindsay DA, Webb A (1987) Biokinetics of and discrimination between dietary RRR- and SRR-α-tocopherols in the male rat. Lipids 22: 163–172

Ingram JT (1953) The approach to psoriasis. Br Med J 2: 591–594

Ippen H (1972) Beta-carotin als Lichtschutz bei Porphyria cutanea tarda. Hautarzt 23: 47–50

Israili ZH, Cucinelli SA, Vaught J, Davis E, Lesser M, Dayton PG (1973) Studies on the metabolism of dapsone in man and experimental animals: formation of n-hydroxy metabolites. J Pharmacol Exp Ther 187: 138–151

Ito T (1978) Cellular and subcellular mechanisms of photodynamic action: the 102 hypothesis as a driving force in recent research. Photochem Photobiol 28: 493–508

Ivanova IP, Mareeva TE (1987) Disorders of lipid peroxidation and of lysosomal hydrolase activity and their correction in psoriasis patients. Vestn Dermatol Venerol 4: 26–31 (in Russian)

Iwaoka T, Tabata F, Takahashi T (1987) Lipid peroxidation and lipid peroxide detected by chemiluminescence. Free Radic Biol Med 3: 329–333

Iyanagi T, Yamazaki I (1969) One electron transfer reactions in biochemical system. III. One electron reduction of quinones by microsomal flavin enzymes. Biochim Biophys Acta 172: 370–381

Iyer GYN, Islam DMF, Quastel JH (1961) Biochemical aspects of phagocytosis. Nature 192: 535–541

Izawa Y (1968) Lipid peroxides and skin diseases. Tokyo Tanabe Q 8: 55–67 (in Japanese)

Jain VK, Gupta V, Verman KC, Saini AS (1985) NADPH oxidase and myeloperoxidase activity in psoriasis leukocytes. J Dermatol 12: 425–428

Jamieson D, van den Brenk HAS (1965) Electrode size and tissue pO 2 measurement in rats exposed to air or high pressure oxygen. J Appl Physiol 20: 514–518

Jamieson D, Chance B, Cadenas E, Boveris A (1986) The relation of free radical production to hyperoxia. Annu Rev Physiol 48: 703–719

Januszkiewicz AJ, Faiman MD (1984) The effect of in-vivo hyperoxic exposure on the re-

lease of endogenous histamine from the rat isolated perfused lung. Toxicol Appl Pharmacol 72: 134–140
Januszkiewicz AJ, Huntrakoon M, Wilson PK, Faiman MD (1986) Isolated perfused lung histamine release, lipid peroxidation, and tissue superoxide dismutase from rats exposed to normobaric hyperoxia. Toxicology 39: 37–46
Janzen EG (1980) A critical review of spin trapping in biological systems. Free Radic Biol 4: 115–155
Jarasch ED, Bruder G, Heid HW (1986) Significance of xanthine oxidase in capillary endothelial cells. Acta Physiol Scand 126 [Suppl 548]: 39–46
Javier JJ, Fodyce-Baum MK, Beach RS, Gavancho M, Cabrejos C, Mantero-Atienza E (1990) Antioxidant micronutrients and immune function in HIV1 infection. FASEB Proc 1: 234–235
Jenkins FP, Welti D, Baines D (1964) Photochemical reactions of tetrachlorosalicylanilide. Nature 201: 827–828
Ji LL, Stratman FW, Lardy HA (1988) Antioxidant enzyme systems in rat liver and skeletal muscle. Arch Biochem Biophys 263: 150–160
Jimbow K, Obata H, Pathak M, Fitzpatrick TB (1974) Mechanism of depigmentation by hydroquinone. J Invest Dermatol 62: 436–449
Jimenez SA, Dehm P, Olsen BR, Prockop DJ (1973) Intracellular collagen and protocollagen from embryonic tendon cells. J Biol Chem 248: 720–729
Jocelyn PC (1972a) Thiols and disulfides in animal tissues. In: Jocelyn PC (ed) Biochemistry of the SH group. Academic, London, pp 261–278
Jocelyn PC (1972b) Naturally occuring thiols and disulfides. In: Jocelyn PC (ed) Biochemistry of the SH group. Academic, London, pp 1–46
Johansen I, Howard-Flanders P (1965) Macromolecular repair and free radical scavenging in the protection of bacteria against x-rays. Radiat Res 24: 184–200
Johnson BE (1968) Ultraviolet radiation and lysosomes in skin. Nature 219: 1258–1259
Johnson BE, Farrington D (1969) Lysosomes and the reactions of skin to ultraviolet radiation. J Invest Dermatol 53: 85–94
Johnson BE, Mier P (1962) Phospholipids and the reaction of skin to ultra-violet radiation. Nature 194: 101–102
Johnson KJ, Ward PA, Kunkel RG, Wilson BS (1986) Mediation of IgA induced lung injury in the rat. Lab Invest 54: 499–506
Jones GRN (1985) Skin cancer: risk to individuals using the tumor promotor benzoyl peroxide for acne treatment. Hum Toxicol 4: 75–78
Jones GRN (1987) Free radicals in immunological killing: the case of tumour necrotising factor (TNF). Med Hypotheses 21: 267–271
Jones SK, Mackie RM, Hole DJ, Gillis CR (1985) Further evidence of the safety of tar in the managment of psoriasis. Br J Dermatol 113: 97–101
Jongkind JF, Verkerk A, Baggen RGA (1989) Glutathione metabolism of human vascular endothelial cells under peroxidative stress. Free Radic Biol Med 7: 507–512
Jordan P, Wulf K (1950) Erythematodes Behandlung mit Vitamin E. Hautarzt 1: 233–234
Jorizzo JL (1986) Behcet's disease. Arch Dermatol 122: 556–558
Jorizzo JL, Solomon AR, Zanolli MD, Leshin B (1988) Neutrophilic vascular reactions. J Am Acad Dermatol 19: 983–1005
Joselow SA, Mannik M (1984) Localization of preformed circulating immune complexes in murine skin. J Invest Dermatol 82: 335–340
Joshi PC (1989) Flavins as endogenous chromophores for UV induced biological oxidation reactions in skin. International symposium on biological oxidation systems, Bangalore, 23–26 Oct 1989
Joshi PC, Pathak MA (1983) Production of singlet oxygen and superoxide radicals by psoralens and their biological significance. Biochem Biophys Res Commun 112: 638–646
Joshi PC, Pathak MA (1984) The role of active oxygen (102 and O2-) induced by crude coal tar and its ingredients used in photochemotherapy of skin diseases. J Invest Dermatol 82: 67–73
Joshi PC, Carraro C, Pathak MA (1987) Involvement of reactive oxygen species in the oxi-

dation of tyrosine and DOPA to melanin and in skin tanning. Biochem Biophys Res Commun 142: 265–274

Jouve I, Berbis P, Garnier JP, Privat Y (1987) Les troubles du metabolisme du tryptophane en dermatologie. Ann Dermatol Venereol 114: 1571–1582

Jugert F, Merk H (1990) NAD(P)H: quinone reductase: same enzymatic from present in murine liver and skin, and human keratinocytes. J Invest Dermatol 94: 539

Juhlin L, Edquist LE, Ekman LG, Ljunghall K, Olsson M (1982) Blood glutathione-peroxidase levels in skin diseases: effect of selenium and vitamin E treatment. Acta Derm Venereol (Stockh) 62: 211–214

Juhlin LA, Civier A, Shroot S, Hensby C (1983) Effect of infrared irradiation on the recoverable levels of free arachidonic acid and prostaglandins in human forearm skin. J Invest Dermatol 81: 297–300

Jung EG (1978) Xeroderma pigmentosum: heterogenous syndrome and model for UV carcinogenesis. Bull Cancer (Paris) 65: 315–322

Jung EG (1990) Lichtbiologie des malignen Melanoms. Zentralbl Haut- Geschlechtskr 157: 833

Jung EG, Bohnert E (1979) Lichtbiologie der Haut. In: Schwarz E, Spier HW, Stüttgen G (eds) Normale und pathologische Physiologie der Haut II. Springer, Berlin Heidelberg New York, pp 459–540 (Handbuch der Haut- und Geschlechtskrankheiten, puppl, vol 4, pt 4)

Jung EG, Bohnert E, Boonen H (1986) Dysplastic nevus syndrome: ultraviolet hypermutability confirmed in-vitro by elevated sister chromatid exchanges. Dermatologica 173: 297–300

Jungi TW, Peterhans E (1988) Change in the chemiluminescence reactivity pattern during in-vitro differentiation of human monocytes to macrophages. Blut 56: 213–220

Juva K, Prockop DJ, Cooper GW, Lash JW (1966) Hydroxylation of proline and the intracellular accumulation of a polypeptide precursor of collagen. Science 152: 92–94

Kagan V, Khan S, Swanson C, Shvedova A, Serbinova E, Packer L (1990) Antioxidant action of thioctic and dihydrolipoic acid. Free Radic Biol Med 9 (Suppl 1): 15

Kahl R, Weinke S (1988) Comparison of antioxidant and prooxidant activity of various synthetic antioxidants. Society for Free Radical Research, winter meeting on free radicals in medicine, Current status of antioxidant therapy, Paris, 9–10 Dec 1988

Kahan A, Kahan IL, Tapaszto I, Bencze G (1963) Therapieversuche beim Pseudoxanthoma elasticum Syndrom. Klin Monatsbl Augenheilkd 142: 1047–1058

Kahn G (1970) Depigmentation caused by phenolic detergent germicides. Arch Dermatol 102: 177–187

Kaidbey KH, Kurban AK (1976) The influence of corticosteroids and topical indomethacin on sunburn erythema. J Invest Dermatol 66: 153–156

Kalamkarian AA, Samsonov VA, Olisova MO, Lashmanova AP, Polianskaia NP (1989) Antioxidants in the treatment of patients with discoid lupus erythematosus. Vestn Dermatol Venerol 10: 29–31 (in Russian)

Kalyanaraman B (1982) Detection of toxic free radicals in biology and medicine. Rev Biochem Toxicol 4: 73–139

Kalyanaraman B (1984) The electron spin resonance study of free radicals formed during the arachidonic acid cascade and cooxidation of xenobiotics by prostaglandin synthase. Free Radic Biol 6: 149–198

Kalyanaraman B, Mottley C, Mason RP (1983) A direct electron spin resonance and spin trapping investigation of peroxy free radical formation by hematin/hydroperoxide systems. J Biol Chem 258: 3855–3858

Kalyanaraman B, Felix CC, Sealy RC (1984) Photoionization and photohomolysis of melanin: an electron spin resonance spin trapping study. J Am Chem Soc 106: 7327–7330

Kamide R, Hashimoto T, Yamagishi R, Niimura M (1990) Decreased superoxide dismutase activity in the lesional skin of psoriasis. J Invest Dermatol 94: 540

Kamimura M (1972) Antiinflammatory activity of vitamin E. J Vitaminol 18: 204–209

Kamimura M, Matsuzawa T (1968) Percutaneous absorption of α-tocopherol acetate. J Vitaminol 14: 150–159

Kaminsky A, Kaplan H, Viglioglia P (1950) Lupus eritematoso y tocoferoles. In: Separata de sesiones dermatologicas en homenjae al profesor Luis E. Pierini, Buenos Aires, Nov. 1949, Buenos Aires, Lopez & Etchegoyen SRL

Kapoor M, Sveenivasan GM (1988) The heat shock response of Neurospora crassa: stress induced thermotolerance in relation to peroxidase and superoxide dismutase levels. Biochem Biophys Res Commun 156: 1097–1102

Kapp A (1990) Aktivierte Granulocyten als Produzenten reaktiver Sauserstoff-Spezies – ihre Bedeutung für die Entzündungsreaktion. Hautarzt 41: 196–203

Kappus H (1986a) Lipid peroxidation: mechanism, analysis, enzymology and biological relevance. In: Sies H (ed) Oxidative stress. Academic, New York, pp 273–309

Kappus H (1986b) Overview of enzyme systems involved in bioreduction of drugs and in redox cycling. Biochem Pharmacol 35: 1–6

Kappus H, Sies H (1981) Toxic drug effects associated with oxygen metabolism: redox cycling and lipid peroxidation. Experientia 37: 1233–1241

Kasha M, Khan AU (1970) The physics, chemistry and biology of singlet oxygen. Ann NY Acad Sci 171: 1–23

Kaufman LD, Seidman RJ, Phillips ME, Gruber BL (1990) Cutaneous manifestations of the L-tryptophan-associated eosinophilia-mylagia syndrome: a spectrum of sclerodermatous skin disease. J Am Acad Dermatol 23: 1063–1069

Kawai S, Komura J, Asada Y, Niwa Y (1988) Experimental burn-induced changes in lipid peroxide levels, and activity of superoxide dismutase and glutathione peroxidase in skin lesions, serum, and liver of mice. Arch Dermatol Res 280: 171–175

Kazama T, Kihara I, Oite T, Shimizu F, Ito M, Sato Y (1985) Ultrastructural localization of anionic sites in the rat skin demonstrated with cationic probes. J Dermatol 12: 472–478

Kazimierczak W, Plaut M, Knauer KA, Meier HL, Lichtenstein LM (1984) Deuterium oxide induced histamine release from basophils of allergic subjects. Am Rev Respir Dis 129: 592–596

Kearns DR, Hollins RA, Khan AU, Radick P (1967) Evidence for the participation of $1\Sigma g+$ and $1\Delta g$ oxygen in dye sensitized photooxygenation reactions. J Am Chem Soc 82: 5456–5457

Kedar I, Greenwald M, Ravid M (1977) Treatment of experimental murine amyloidosis with dimethyl sulfoxide. Eur J Clin Invest 7: 149–150

Kedar I, Sohar E, Ravid M (1982) Degradation of amyloid by a serum component and inhibition of degradation. J Lab Clin Med 99: 693–700

Kedziora J, Bartosz G, Gromadzinska J, Sklodowska M, Wesowicz W, Scianowski J (1986) Lipid peroxides in blood plasma and enzymatic antioxidative defense of erythrocytes in Down's syndrome. Clin Chim Acta 154: 191–194

Keene JP, Kessel D, Land EJ, Redmond RW, Truscott TG (1986) Direct detection of singlet oxygen sensitized by hematoporphyrin and related compounds. Photochem Photobiol 43: 117–120

Kemeny L, Csato M, Dobozy A (1989) Pharmacological studies on dithranol induced irritative dermatitis in mice. Arch Dermatol Res 281: 362–365

Kemmer-Görlitz CH (1952) Vitamin E Behandlung des Röntgenulkus. Dermatol Wochenschr 126: 1209–1210

Kennedy CH, Dyer JM, Church DF, Winston GW, Pryor WA (1989) Radical production in liver mitochondria by peroxidic tumor promotors. Biochem Biophys Res Commun 160: 1067–1072

Kensler TW, Taffe BG (1986) Free radicals and tumor promotion. Adv Free Radic Biol Med 2: 347–387

Kensler TW, Trush MA (1983) Inhibition of oxygen radical metabolism in phorbol ester activated polymorphonuclear leukocytes by an antitumor promoting copper complex with superoxide dismutase mimetic activity. Biochem Pharmacol 32: 3485–3487

Kensler TW, Trush MA (1984) Role of oxygen radicals in tumor promotion. Environ Mutagen 6: 593–616

Kensler TW, Bush DM, Kozumbo WJ (1983) Inhibition of tumor promotion by a biomimetic superoxide dismutase. Science 221: 75–77

Kerimov RF, Goloshchapov AN, Burlakova EB, Dzhafarov AI (1987) Effect of the functional groups of tocopherol molecules on lipid viscosity of mitochondrial lipids. Biull Eksp Biol Med (USSR) 103: 540–543

Kessel D, Rossi E (1982) Determinants of porphyria induced photo oxidation characterized by fluorescence and absorption spectra. Photochem Photobiol 35: 37

Kettle AJ, Winterbourn CC (1988) Superoxide modulates the activity of myeloperoxidase and optimizes the production of hypochlorus acid. Biochem J 252: 529–536

Khaleh MB, Leroy EC (1983) Endothelial injury in scleroderma. J Exp Med 101: 553–560

Khan SH, Emerit I (1985) Lipid peroxidation products and clastogenic material in culture media of human leukocytes exposed to the tumor promotor phorbol myristate acetate. J Free Radic Biol Med 1: 443–449

Khan WA, Das M, Stick S, Javed S, Bickers DR, Mukhtar H (1987) Induction of epidermal NAD(P)H: quinone reductase by chemical carcinogens: a possible mechanism for the detoxification. Biochem Biophys Res Commun 146: 126–133

Kharasch ED, Novak RF (1981) Anthracendione activation by NADPH-cytochrome P-450 reductase; comparison with anthracyclins. Biochem Pharmacol 30: 2881–2884

Khettab N, Amory MC, Briand G, Bousquet B, Combre A, Forlot P, Barey M (1988) Photoprotective effect of vitamins A and E on polyamine and oxygenated free radical metabolism in hairless mouse epidermis. Biochimie 70: 1709–1713

Khrapova NG (1977) Kinetic aspects of the action of tocopherols as antioxidants. Biofizika 22: 36–442

Kihlström M, Salminen A (1988) Cardiac antioxidants in exercise. Society for Free Radical Research, winter meeting on free radicals in medicine, Current status of antioxidant therapy, Paris, 9–10 Dec 1988

Kim JA, Baker DG, Hahn SS, Goodchild NT, Constable WC (1983) Topical use of *N*-acetylcysteine for reduction of skin reaction to radiation therapy. Semin Oncol 10: 86–88

Kim K, Kim IH, Lee KY, Rhee SG, Stadtman ER (1988) The isolation and purification of a specific "protector" protein which inhibits enzyme inactivation by a thiol/Fe(III)/O2 mixed function oxidation system. J Biol Chem 263: 4704–4711

Kim TY, Kripke ML, Ullrich SE (1990) Immunosuppression by factors released from UV-irradiated epidermal cells: selective effects on the generation of contact and delayed hypersensitivity after exposure to UVA or UVB radiation. J Invest Dermatol 94: 26–32

Kim YP, Lee SC (1987) Superoxide dismutase activities in the human skin. In: Hayaishi O, Imamura S, Miyachi Y (eds) The biological role of reactive oxygen species in skin. Elsevier, New York, pp 225–230

Kimball RE, Reddy K, Pierce TH, Schwartz LW, Mustafa MG, Cross CE (1976) Oxygen toxicity: augmentation of antioxidant defense mechanisms in rat lung. Am J Physiol 230: 1425–1431

Kimmig J, Schulz KH (1957) Berufliche Akne (sog. Chlorakne) durch chlorierte aromatische zyclische Äther. Dermatologica 115: 540–542

Kind P, Lehmann P (1990) Photobiologie des Lupus erythematodes. Hautarzt 41: 66–71

Kingston TP, Connor MJ, Lowe NJ (1985) Anthralin and cutaneous inflammation. Br J Dermatol 113: 780–781

Kinsella AR, Radman M (1978) Tumor promotor induces sister chromatid exchanges: relevance to the mechanism of carcinogenesis. Proc Natl Acad Sci USA 75: 6149–6153

Kinsella AR, Gainer SCH, Butler J (1983) Investigations of a possible role for superoxide anion production in tumor promotion. Carcinogenesis 4: 717–719

Kitagawa M, Mino M (1989) Effects of elevated *d*-α (RRR)-tocopherol dosage in man. J Nutr Sci Vitaminol 35: 133–142

Kizaki H, Matsuo I, Sakurada T (1977) Xanthine oxidase and guanase activities in normal and psoriatic epidermis. Clin Chim Acta 75: 1–4

Klebanoff SJ (1967a) Iodination of bacteria: a bactericidal mechanism. J Exp Med 126: 1063–1078

Klebanoff SJ (1967b) A peroxidase mediated antimicrobial system in leucocytes. J Clin Invest 46: 1078

Klebanoff SJ (1968) Myeloperoxidase halide hydrogen peroxide antibacterial system. J Bacteriol 95: 2131–2138

Klein KG, Blankenhorn G (1987) Vergleich der klinischen Wirksamkeit von Vitamin E und Diclofenac-Natrium bei Spondylitis ankylosans (Morbus Bechterew). Vitaminspur 2: 133–138

Klein-Szanto AJP, Slaga TJ (1981) Numerical variation of dark cells in normal and chemically induced hyperplastic epidermis with age of animal and efficiency of tumor promotor. Cancer Res 41: 4437–4440

Klein-Szanto AJP, Slaga TJ (1982) Effects of peroxides on rodent skin: epidermal hyperplasia and tumor promotion. J Invest Dermatol 79: 30–34

Kligman AM (1965) Topical pharmacology and toxicology of dimethylsulfoxide – part 1. JAMA 193: 140–148

Kligman AM (1968) Pathogenesis of acne vulgaris II. Histopathology of comedones induced in the rabbit ear by human sebum. Arch Dermatol 98: 58–66

Kligman AM, Katz AG (1968) Pathogenesis of acne vulgaris I. Comedogenic properties of human sebum in external ear canal of the rabbit. Arch Dermatol 98: 53–57

Kligman AM, Leyden JJ, Stewart R (1977) New uses for benzoyl peroxide: a broad spectrum antimicrobial agent. Int J Dermatol 16: 413–417

Kligman LH (1982) Intensification of ultraviolet induced dermal damage by infrared radiation. Arch Dermatol Res 272: 229–238

Kligman LH (1989) Prevention and repair of photoaging: sunscreens and retinoids. Cutis 43: 458–465

Kligman LH, Kligman AM (1984) Reflections on heat. Br J Dermatol 110: 369–375

Kligman LH, Kligman AM (1985) Cutaneous photoaging by ultraviolet radiation. In: Maibach HJ, Lowe NJ (eds) Models in dermatology, vol 1. Karger, Basel, pp 59–69

Klug KH, Schulze P (1987) Mitochondrial ring phenomenon of keratinocytes in psoriasis vulgaris after dithranol therapy. J Cutan Pathol 14: 212–216

Klug KH, Schulze P, Sönnichsen N (1988) Abnorme Mitochondrien – ein spezifisches Phänomen psoriatischer Keratinocyten? Dermatol Monatsschr 174: 305–307

Knight KR, Mellow CG, Abbey PA, Angel MF, Lepore DA, Arnold LI, Gray KA, McC O'Brian B (1990) Interaction between thromboxane and free radical mechanisms in experimental ischaemic rabbit skin flaps. Res Exp Med 190: 423–433

Knighton DR, Hunt TK, Scheuenstuhl H, Halliday BJ, Werb Z, Banda MJ (1983) Oxygen tension regulates the expression of angiogenesis factor by macrophages. Science 221: 1283–1285

Knott EK (1948) Development of ultraviolet blood irradiation. Am J Surg 76: 165–171

Kobayashi T, Matsumoto M, Iizuka H, Suzuki K, Taniguchi N (1991) Superoxide dismutase in psoriasis, squamous cell carcinoma and basal cell epithelioma: an immunohistochemical study. Br J Dermatol 124: 555–559

Kobza A, Ramsay CA, Magnus IA (1973) Oral β-carotene therapy in actinic reticuloid and solar urticaria. Br J Dermatol 88: 157–161

Koch WH, Chedekel MR (1987) Photochemistry and photobiology of melanogenic metabolites: formation of free radicals. Photochem Photobiol 42: 229–238

Kochevar IE (1987) Mechanisms of drug photosensitization. Photochem Photobiol 46: 891–895

Kochevar IE, Armstrong RB, Einbinder J, Walther RR, Harber LC (1982) Coal tar phototoxicity: active compounds and action spectra. Photochem Photobiol 36: 65–69

Kochevar IE, Morison WL, Lamm JL, McAuliffe DJ, Western A, Hood AF (1986) Possible mechanism of piroxicam induced photosensitivity. Arch Dermatol 122: 1283–1287

Kohen E, Kohen C, Morlier P, Santus R, Reyftmann JP, Dubertret L, Hirschberg JG, Coulomb B (1986) A microspectrofluorimetric study of the effect of anthralin, and antipsoriatic drug, on cellular structures and metabolism. Cell Biochem Funct 4: 157–168

Komura S, Yoshino K, Kondo K, Yagi K (1988) Lipid peroxide levels in the skin of the senescence accelerated mouse. J Clin Biochem Nutr 5: 255–260

Kondo H, Kumagai Y, Shiokawa Y (1985) Scleroderma following cosmetic surgery ("adjuvant disease"): a review of nine cases reported in Japan. In: Black CM, Myers AR (eds) Current topics in rheumatology: systemic sclerosis (scleroderma). Gower, New York, pp 135–137

Kong-ooGoh JRF (1982) Chromosomes in rheumatoid arthritis. Ann Rheum Dis 41: 644–652

Konings AWT, Oosterloo SK (1980) Radiation effects on membranes. Radiat Res 81: 200–207

Kono K, Fridovich I (1982) Superoxide radical inhibits catalase. J Biol Chem 257: 5751–5754

Konstantinova B, Nedkova-Bratanova N (1969) Chromosomal aberrations in patients with ulcero-haemorrhagic colitis. Digestion 2: 329–333

Koone MD, Black HS (1986) A mode of action for butylated hydroxytoluene mediated photoprotection. J Invest Dermatol 87: 343–347

Korbashi P, Katzhendler J, Saltman P, Chevion M (1988) Zinc protects E. coli against copper mediated paraquat induced damage. Society of Free Radical Research, winter meeting on free radicals in medicine, Current status of antioxidant therapy, Paris, 9–10 Dec 1988

Koren A, Schara M, Sentjurc M (1980) EPR measurements of free radicals during tetanic contraction of frog skeletal mucle. Period Biol 82: 399–401

Koren A, Sauber C, Sentjurc M, Schara M (1983) Free radicals in tetanic activity of isolated skeletal muscle. Comp Biochem Physiol [B] 74: 633–635

Korting GW, Holzmann H (1967) Die Sklerodermie und ihr nahestehende Bindegewebsprobleme. Thieme, Stuttgart

Korytowski W, Hintz P, Sealy RC, Kalyanaraman B (1985) Mechanism of dismutation of superoxide produced during autoxication of melanin pigments. Biochem Biophys Res Commun 131: 659–665

Korytowski W, Kalyanaraman B, Menon IA, Sarna T, Sealy RC (1986) Reaction of superoxide anions with melanins: electron spin resonance and spin trapping studies. Biochim Biophys Acta 882: 145–153

Kotin P, Falk HL (1963) Organic peroxides, hydrogen peroxide, epoxides and neoplasia. Radiat Res 3 [Suppl]: 193–211

Koukay N, La Porte F, Favier A (1987) Zinc et radicaux libres. Diet Med 3: 147–153

Kovacic P, Crawford PW (1986) Charge transfer mechanism for carcinogenesis by alkylating and other agents. Bioelectrochem Bioenerg 15: 305–316

Kovacic P, Ames JR, Lumme P, Elo H, Cox O, Jackson H, Rivera LA, Ramirez L, Ryan MD (1986) Charge transfer oxy radical mechanism for anti-cancer agents. Anticancer Drug Des 1: 197–214

Kovacic P, Ames JR, Rector DL, Jawdosiuk M, Ryan MD (1989) Reduction potentials of antihelminthic drugs: possible relationship to activity. Free Radic Biol Med 6: 131–139

Koyama H, Tomida M, Ono T (1976) The enhanced production of hyaluronic acid by cultured rat fibroblast cells treated with cyclic AMP and its dibutyryl derivative. J Cell Physiol 87: 189–197

Kozumbo WJ, Seed JL, Kensler TW (1983) Inhibition by 2(3)-tert-butyl-4-hydroxyanisole and other antioxidants of epidermal ornithine decarboxylase activity induced by 12-O-tetradecanoylphorbol-1,3-acetate. Cancer Res 43: 2555–2559

Kozumbo WJ, Muehlematter D, Jörg A, Emerit I, Cerutti P (1987) Phorbol ester induced formation of castogenic factor from human monocytes. Carcinogenesis 8: 521–526

Kraemer KH, Lee MM, Scotto J (1984) DNA repair protects against cutaneous and internal neoplasia: evidence from xeroderma pigmentosum. Carcinogenesis 5: 511–514

Kralli A, Moss SH (1987) The sensitivity of an actinic reticuloid cell strain to near ultraviolet radiation and its modification by Trolox-C, a vitamin E analogue. Br J Dermatol 116: 761–772

Krämer GF, Ames BN (1987) Oxidative mechanisms of toxicity of low intensity near ultraviolet UV light. J Bacteriol 169: 2259–2266

Kramer PA, Glader BE, Li TK (1972) Mechanism of methemoglobin formation by diphenylsulfones. Effect of 4-amino-4′-hydroxyaminodiphenylsulfone and other p-substituted derivatives. Biochem Pharmacol 21: 1265–1274

Krämer R (1986) Die Beteiligung der Peroxisomen am Lipidstoffwechsel. J Clin Chem Clin Biochem 24: 109–118

Krebs A (1969) Untersuchungen zur Strukturspezifität der Psoriasisheilmittel Chrysarobin und Dithranol. Hautarzt 20: 204–209
Krieg T, Aumailley M (1990) Connective tissue diseases in the skin. J Dermatol 17: 67–84
Krieg T, Meurer M (1988) Systemic scleroderma. J Am Acad Dermatol 18: 457–481
Krinsky NI (1989) Antioxidant functions of carotenoids. Free Radic Biol Med 7: 617–635
Krinsky NI, Deneke SM (1982) Interaction of oxygen and oxy-radicals with carotenoids. J Natl Cancer Inst 69: 205–209
Kroegel C, Costabel U, Matthys H, Barnes PJ (1988a) Die pathogenetische Bedeutung des eosinophilen Granulocyten. I. Morphologie, Biochemie und Effektormechanismen. Dtsch Med Wochenschr 113: 1405–1411
Kroegel C, Costabel U, Barnes PJ, Matthys (1988b) Die pathogenetische Bedeutung des eosinophilen Granulozyten. II. Hypereosinophile Erkrankungen. Dtsch Med Wochenschr 113: 1446–1452
Krylenkov VA, Malygin AM (1982) UV action on the surface of the immunocompetent cells of mammalians. Citologia 24: 1411–1416 (in Russian)
Krylenkov VA, Roschupkin DI, Potapenko AY, Malygin AM (1984) Changes in the expression of lymphocyte surface antigens under UV irradiation effect of antioxidants. Photobiochem Photobiophys 8: 19–23
Krylov VI, Olekhnovich VM, Sorogin PS, Zhogin SV (1984) Activities of phospholipases and antioxidants in bronchial secretion products of children with bronchopulmonary diseases. Vopr Med Khim 30: 52–55 (in Russian)
Kubow S, Bray TM, Bettger WJ (1986) Effects of dietary zinc and copper on free radical production in rat lung and liver. Can J Physiol Pharmacol 64: 1281–1285
Kumagai Y, Abe C, Shiokawa Y (1979) Scleroderma after cosmetic surgery. Arthritis Rheum 22: 532–537
Kumagai Y, Shiokawa Y, Medsger TA Jr (1984) Clinical spectrum of connective tissue disease after cosmetic surgery. Arthritis Rheum 27: 1–12
Kumar R, Weiss VC, West DP, Chiero LA (1983a) Altered erythrocyte membrane phosphorylation in psoriasis. Br J Dermatol 109: 267–275
Kumar R, Weiss VC, West DP, Chiero LA (1983b) Erythrocyte membrane phosphorylation in untreated and in etretinate treated psoriatic patients. Br J Dermatol 109: 277–286
Kumar S, Antony M, Mehrotra NK (1988) Role of cutaneous GSH in 12-O-tetradecanoylphorbol-13-acetate induced mouse skin tumor promotion. Arch Dermatol Res 280: 371–374
Kunert KJ, Tappel AL (1983) The effect of vitamin C on in-vivo lipid peroxidation in guinea pigs as measured by pentane and ethane production. Lipids 18: 271–274
Kunstmann H (1954) Zur Behandlung der Induratio penis plastica mit Vitamin E. Die Medizinische: 125–126
Kupper TS (1989) Mechanisms of cutaneous inflammation. Arch Dermatol 125: 1406–1412
Lad PM, Glovsky MM, Richards JH, Smiley PA, Backstrom B (1985) Regulation of human neutrophil guanylate cyclase by metal ions, free radicals and muscarinic cholinergic receptors. Mol Immunol 22: 731–739
Lafrado LJ, Quintana DA, Javadian MA, Kelliher JC (1989) Polymorphonuclear leukocyte function in HIV-1 infected chimpanzees. Immunol Immunopathol 21: 3–12
Lai MZ, Düzgünes N, Szoka C (1985) Effects of replacement of the hydroxyl group of cholesterol and tocopherol on the thermotropic behavior of phospholipid membranes. Biochemistry 24: 1646–1653
Laliberte JF, Sun IL, Crane FL, Clarke MJ (1987) Ruthenium amine complexes as electron acceptors for growth stimulation by plasma membrane electron transport. J Bioenerg Biomembr 19: 69–81
Lampe MA, Williams ML, Elias PM (1983) Human epidermal lipids: characterization and modulations during differentiation. J Lipid Res 24: 131–142
Lancaster-Smith M, Kumar P, Clark ML, Marks R, Johnson GD (1975) Antireticulin antibodies in dermatitis herpetiformis and coeliac disease. Br J Dermatol 92: 37–42
Lands WEM (1982) Control of prostaglandin biosynthesis. Prog Lipid Res 20: 875–883

Lands WEM (1984) Radicals and peroxides modulate the enzymic synthesis of eicosanoids from polyunsaturated fatty acids. In: Thaler-Dao H (ed) Iconasoids and cancer. Raven, New York, pp 41–48

Lands WEM (1985) Interactions of lipid hydroperoxides with eicosanoid biosynthesis. J Free Radic Biol Med 1: 97–101

Lands WEM (1989) n-3 Fatty acids as precursors for active metabolic substances: dissonance between expected and observed events. J Int Med 225 [Suppl 1]: 11–20

Lands WEM, Marshall PJ (1985) Peroxide stimulators of prostaglandin biosynthesis. In: Brooks PM, York JR (eds) Rheumatology. Elsevier, Amsterdam, pp 47–50

Lands WEM, Pendleton RB (1989) n-3 Fatty acids and hydroperoxide activation of fatty acid oxygenases. In: Simic MG, Taylor KA, Ward JF, von Sonntag C (eds) Oxygen radicals in biology and medicine. Plenum, New York, pp 675–681

Lands WEM, Marshall PJ, Kulmacz RJ (1985) Hydroperoxide availability in the regulation of the arachidonate cascade. In: Hayaishi O, Yamamoto S (eds) Advances in prostaglandin, thromboxane, and leukotriene research, vol 15. Raven, New York, pp 233–235

Landthaler M (1984a) Behandlung von hypertrophen Narben und Keloiden mit Orgotein. Hautarzt 35: 377–378

Landthaler M (1984b) Behandlung der Induratio penis plastica. Hautarzt 35: 327–328

Lange B, Meffert H, Böhm F (1980) Chemilumineszenzuntersuchung der Bindung von 8-Methoxypsoralen an Protein und Lymphocytenoberflächen. Dermatol Monatsschr 166: 599–603

Larussa FM, Rusciani L, De Sole P, De Simone C, Venier A, Serri F (1986) Chemiluminescence: an index of neutrophil activity in cutaneous disease. J Invest Dermatol 87: 152

Lash LH, Jones DP (1986) Purification and properties of the membranal thiol oxidase from porcine kidney. Arch Biochem Biophys 247: 120–130

Lash LH, Hagen TM, Jones DP (1986) Exogenous glutathione protects intestinal epithelial cells from oxidative injury. Proc Natl Acad Sci USA 83: 4641–4645

Lassus A, Reunala T, Idanpaa-Heikkila J, Juvakoski T, Salo O (1981) PUVA treatment and skin cancer: a follow up study. Acta Derm Venereol (Stockh) 61: 141–145

Last JA, Greenberg DB (1980) Ozone induced alterations in collagen metabolism of rat lungs. II. Long term exposures. Toxicol Appl Pharmacol 55: 108–114

Lau AS, Livesey JF (1989) Endotoxin induction of tumor necrosis factor is enhanced by acid label interferon-alpha in acquired immundeficiency syndrome. J Clin Invest 84: 738–743

Laubscher TA, Harris N, Chalton DO (1991) The mitogenic effects of sera from psoriatic subjects on normal dermal fibroblasts: an absence of correlation with the clinical activity of psoriasis. Br J Dermatol 124: 163–167

Lavie G, Zucker-Franklin D, Franklin EC (1978) Degradation of serum amyloid A protein surface associated enzymes of human blood monocytes. J Exp Med 148: 1020–1031

Lavker RM (1979) Structural alterations in exposed and unexposed aged skin. J Invest Dermatol 73: 59–66

Lavker RM, Grove GI, Kligman AM (1981) The atrophogenic effect of crude coal tar on human epidermis. Br J Dermatol 105: 77–82

Lawrence CM, Shuster S (1987) Effect of arachidonic acid on anthralin inflammation. Br J Clin Pharmacol 24: 125–131

Lawrence R, Burk R (1978) Species, tissue and subcellular distribution of non Se-dependent glutathione peroxidase activity. J Nutr 108: 211–215

Lebedev OL, Kazarnoskii SN (1959) Catalytic oxidation of aliphatic amines with hydrogen peroxide. Tr Khim Tekhnol Gork: 649–656 (in Russian)

Lee JAH (1989) The relationship between malignant melanoma of skin and exposure to sunlight. Photochem Photobiol 50: 493–496

Leffell DJ, Stetz ML (1988) In-vivo fluorescence of human skin. Arch Dermatol 124: 1514–1518

Legrand-Poels S, Vaira D, Pincemail J, van de Vorst A, Piette J (1990) Activation of human immunodeficiency virus type I by oxidative stress. AIDS Res Hum Retroviruses 6: 1389–1397

Lehnert H (1990) Eosinophilie-Myalgie-Syndrom und Einnahme L-Tryptophan-haltiger Arzneimittel. Dtsch Ärztebl 87: B 1474–1476
Leh F, Mudd JB (1974) Reaction of peroxylacetyl nitrate with cysteine, cystine, methionine, lipoic acid, papain and lysozym. Arch Biochem Biophys 161: 216–221
Leibsohn E, Appel B, Ullrick WC, Tye MJ (1958) Respiration of human skin. J Invest Dermatol 30: 1–8
Lengfelder E (1984) Can anti-inflammatory drugs act as scavengers of oxygen radicals? Agents Action 15: 56
Leonhardi G (1961) Die Beeinflussung von Atmung und Glycolyse der Haut durch verschiedene Pharmaka. Arch Klin Exp Dermatol 213: 839–845
Leranth G, Laszlo F (1936) Dermatologische Bedeutung des „E" Vitamins. Orv Hetil 778–779
Leroy EC (1982) Pathogenesis of scleroderma (systemic sclerosis). J Invest Dermatol 79 [Suppl 1]: 87–89
Leroy EC, Kahaleh MB, Mercuric S (1983) A fibroblast mitogen present in scleroderma but not control sera. Rheumatol Int 3: 35–38
Lever WF (1987) Hautentzündung und Hypereosinophilen. Z Hautkr 62: 766–772
Levine M, Hartzell W (1987) Ascorbic acid: the concept of optimum requirement. Ann NY Acad Sci 498: 424–444
Levine SA, Kidd PM (1985 a) Adaptation to oxidative stress. Pathway to degenerative disease. In: Levine SA, Kidd PM (eds) Antioxidant adaptation. Biocurrents Division Allergy Research Group, San Leonardo, pp 171–218
Levine SA, Kidd PM (1985 b) Biochemical pathologies initiated by free radical oxidant compounds in the aetiology of food hypersensitivity disease. Int Clin Nutr Rev 5: 5–24
Levine SA, Reinhardt JH (1983) Biochemical pathology initiated by free radicals, oxidant chemicals, and therapeutic drugs in the etiology of chemical hypersensitivity disease. Orthomol Psych 12: 166–183
Levinsky RJ, Lehner T (1978) Circulating soluble immune complexes in recurrent oral ulcerations and Behcet's syndrome. Clin Exp Immunol 32: 193–198
Lewin G, Popov I (1989) Zum Verhalten der antioxidativen Kapazität des Plasmas nach UV-Blutbestrahlung. 1. In-vitro Studie. Z Klin Med 44: 1599–1602
Li ASW, Chignell CF, Hall RD (1987 a) Cutaneous phototoxicity of tetracyclin antibiotics: generation of free radicals and singlet oxygen during photolysis as measured by spin trapping and the phosphorescence of singlet molecular oxygen. Photochem Photobiol 46: 379–382
Li ASW, Roethling P, Cummings KB, Chignell CF (1987 b) O2-photogenerated from aqueous solutions of tetracycline antibiotics (pH 7.3) as evidenced by DMPO spin trapping and cytochrome c reduction. Biochem Biophys Res Commun 146: 1191–1195
Likhtenstein GI (ed) (1976) Spin labeling methods in molecular biology. Wiley, New York
Lill PH (1983) Latent period and antigenicity of murine tumors induced in C3H mice by short wavelength ultraviolet radiation. J Invest Dermatol 81: 342–346
Lim HW, Gigli I (1981) Role of complement in porphyrin induced photosensitivity. J Invest Dermatol 76: 4–9
Lim HW, Perez HD, Poh-Fitzpatrick M, Goldstein IM, Gigli I (1981) Generation of chemotactic activity in serum from patients with erythropoietic protophorphyria and porphyria cutanea tarda. N Engl J Med 304: 212–216
Lind C, Hochstein P, Ernster L (1982) DT-diaphorase as a quinone reductase: a cellular control device against semiquinone and superoxide radical formation. Arch Biochem Biophys 216: 178–185
Lindner E (1958) Therapeutische Bedeutung des Vitamin E bei Störungen der Spermiogenese. Int Z Vitaminforsch 29: 33–40
Link EM, Riley PA (1988) Role of hydrogen peroxide in the cytotoxicity of the xanthine/xanthine oxidase system. Biochem J 249: 391–399
Lipkin G, Herrman F, Mandol L (1962) Studies on serum copper: I. The copper content of blood and serum in patients with psoriasis. J Invest Dermatol 39: 534–546

Lippman RD (1985) Rapid in-vivo quantification and comparison of hydroperoxides and oxidized collagen in aging mice, rabbits and man. Exp Gerontol 20: 1–5

Lo WB, Black HS (1973) Inhibition of carcinogen formation in skin irradiated with ultraviolet light. Nature 246: 489–491

Löblich HJ (1961) Elektronenmikroskopische Untersuchungen über die Entwicklung des Staubgranuloms. Beitr Silikoseforschung 71: 33–49

Loevenhart AS (1905) Benzoylperoxyd, ein neues therapeutisches Agens. Ther Monatsschr 19: 426

Logani MK, Solanki V, Slaga TJ (1982) Effect of tumor promotors on lipid peroxidation in mouse skin. Carcinogenesis 11: 1303–1306

Loginov AS, Nilova TV, Bendikov EA, Petrakov AV (1989) Pharmakokinetik der Liponsäurepräparate und ihr Einfluß auf die ATP Synthese sowie die Prozesse der mikrosomalen und zytosolen Oxidation in den Hepatocyten bei der Leberschädigung am Menschen. Farmakol Toksikol 52: 78–82 (in Russian)

Lohel H (1951) Vitamin E in der Dermatologie. Dtsch Gesundheitswes 6: 984–987

Lohmann W, Neubacher H (1985) Stable tissue free radicals. Methods Enzymol 104: 451–456

Lorincz AL, Pearson RW (1962) Sulfapyridine and sulfone type drugs in dermatology. Arch Dermatol 85: 42–52

Loven DP (1988) A role for reduced oxygen species in heat induced cell killing and the induction of thermotolerance. Med Hypotheses 26: 39–50

Löw H, Crane FL (1978) Redox function in plasma membranes. Biochim Biophys Acta 515: 141–161

Lowe NJ, Wortzman MS, Breeding J (1981) Efficiacy of a new coal tar extract and coal tar shampoos by epidermal DNA synthesis suppression assay. J Invest Dermatol 76: 315–319

Lowe NJ, Breeding J, Wortzman MS (1982) The pharmacological variability of crude coal tar. Br J Dermatol 107: 475–480

Lowitz BB (1975) Streaking with bleomycin. N Engl J Med 12: 1300–1301

Lowrey J (1976) An unusual diet-derived inflammatory dermatosis in a Dalmatian dog responds to orgotein. Vet Med 71: 289–295

Lowy RO, Baker DG (1972) Protection against local irradiation injury to the skin by locally and systemically applied drugs. Radiology 105: 425–428

Lukiewicz S (1972) The biological role of melanin. I. New concepts and methodical approaches. Folia Histochem Cytochem 10: 93–108

Lunec J, Dormandy TL (1979) Fluorescent lipid peroxidation products in synovial fluid. Clin Sci 56: 53–59

Lynch V, Smith HW, Marshall EK (1918) On dichlorethylsulfide (mustard gas). I. The systemic effects and mechanism of action. J Pharmacol Exp Ther 12: 265–289

Machtey I, Quaknine L (1978) Tocopherol in osteoarthritis: a controlled pilot study. J Am Geriatr Soc 26: 328–330

Maderson PFA (1972) When? Why? and How?: some speculations on the evolution of the vertebrate integument. Am Zool 12: 159–171

Maguire JJ, Wilson DS, Packer L (1989) Mitochondrial electron transport linked tocopheroxyl radical reduction. J Biol Chem 264: 21462–21465

Majewski S, Tigalonowa M, Jablonska S (1987) Serum samples from patients with active psoriasis enhance lymphocyte induced angiogenesis and modulate endothelial cell proliferation. Arch Dermatol 123: 221–225

Mahrle G, Orfanos CE (1977) The plasma unit membrane. Br J Dermatol 96: 215–223

Maisuradze VN, Platonov AG, Gudz TI, Goncharenko EN, Kudriashov IUB (1987) Effect of ultraviolet rays on lipid peroxidation and various factors of its regulation in the rat skin. Biol Nauki 5: 31–35 (in Russian)

Makino K, Mossoba MM, Riesz P (1982) Chemical effects of ultrasound on aqueous solutions. Evidence for •O and •H by spin trapping. J Am Chem Soc 104: 3537–3539

Makino K, Mossoba MM, Riesz P (1983a) Formation of •OH and •H in aqueous solutions by ultrasound using clinical equipment. Radiat Res 96: 416–421

Makino K, Mossoba MM, Riesz P (1983 b) Chemical effects of ultrasound on aqueous solutions. Formation of hydroxyl radicals and hydrogen atoms. J Phys Chem 87: 1369–1377
Malkinson FD (1981) Some principles of radiobiology: a selective review. J Invest Dermatol 76: 32–38
Malkinson FD, Keane JT (1981) Radiobiology of the skin: review of some effects on epidermis and hair. J Invest Dermatol 77: 133–138
Malkinson FD, Keane JT (1983) Radiobiology of the skin. In: Goldsmith LA (ed) Biochemistry and physioplogy of the skin, vol 2. Oxford University Press, New York, pp 769–814
Malmgren R, Zetterström O, Theorell H, de Wahl K (1986) Lowered glutathione peroxidase activity in asthmatic patients with food- and aspirin-intolerance. Allergy 41: 43–45
Malviya AN, Anglard P (1986) Modulation of cytosolic protein kinase c activity by ferricyanide: priming event seems transmembrane redox signalling. FEBS Lett 200: 265–270
Maly FE (1990) The B-lymphocyte: a newly recognized source of reactive oxygen species with immunoregulatory potential. Free Radic Res Commun 8: 143–148
Maly FE, Kapp A, De Weck AL (1987) A novel interleukin stimulating free radical production by granulocytes. Free Radic Res Commun 3: 57–67
Manabe M, Ogawa H (1987) In-vitro damage of basal lamina associated anionic sites by hydroxyl radical. J Invest Dermatol 88: 657–660
Mandell GL (1975) Catalase, superoxide dismutase, and virulence of Staphylococcus aureus: in-vitro and in-vivo studies with emphasis on staphylococcal-leucocyte interaction. J Clin Invest 55: 561–566
Mann T, Keilin D (1939) Haemocuprein and hepatocuprein copper protein compounds of blood and liver in mammals. Proc R Soc [Biol] 126: 303–315
Mann T, Jones R, Sherins R (1980) Oxygen damage, lipid peroxidation and motility of spermatozoa. In: Steinberger A, Steinberger E (eds) Testicular development, structure and function. Raven, New York, pp 497–501
Mannaioni PF, Masini E (1988) The release of histamine by free radicals. Free Radic Biol Med 5: 177–197
Manson PN, Anthenelli RM, Michael MA, Im MJ, Bulkley GB, Hoopes JE (1983) The role of oxygen free radicals in ischemic injury in island skin flaps. Ann Surg 198: 87–90
Manson PN, Narayan KK, Im MJ, Bulkley GB, Hoopes JE (1986) Improved survival in free skin flap transfers in rats. Surgery 99: 211–214
Manson PN, Jesudass R, Marzella L, Bulkley GB, Im MJ, Narayan KK (1991) Evidence for an early free radical-mediated reperfusion injury in frostbite. Free Radic Biol Med 10: 7–11
Marazzini Z (1955) Induratio penis plastica e vitamina E. Urologia (Treviso) 22: 352–355
Maria CS, Machado A (1986) Age and sex related differences in some rat renal NADPH consuming detoxification enzymes. Arch Gerontol Geriatr 5: 235–247
Maria CS, Machado A (1987) Effects of development and ageing on pulmonary NADPH cytochrome c reductase, glutathione peroxidase, glutathione reductase and thioredoxin reductase activities in male and female rats. Mech Ageing Dev 37: 183–195
Mariqu HR (1981) Vinyl chloride disease. International conference on progressive systemic sclerosis, Austin, 9 Oct 1981
Maricq HR (1985) Vinyl chlorid disease. In: Black CM, Myers AR (eds) Current topics in rheumatology: systemic sclerosis (scleroderma). Gower, New York, pp 105–113
Marklund S (1980) Distribution of CuZn superoxide dismutase and Mn superoxide dismutase in human tissues and extracellular fluids. Acta Physiol Scand Suppl 492: 19–23
Marklund S, Westmann NG, Lundgren E, Roos G (1982) Copper- and zinc-containing superoxide dismutase, manganese-containing superoxide dismutase, catalase, and glutathione peroxidase in normal and neoplastic human cell lines and normal human tissues. Cancer Res 42: 1955–1961
Markowitz SS, McDonald CJ, Fethiere W (1972) Occupational acroosteolysis. Arch Dermatol 106: 219–233
Marks F, Fürstenberger G (1985) Tumor promotion in skin: Are active oxygen species involved? In: Sies H (ed) Oxidative stress. Academic, London, pp 437–475
Marlhens F, Nicole A, Sinet PM (1985) Lowered level of translatable messenger RNAs for

manganese superoxide dismutase in human fibroblasts transformed by SV 40. Biochem Biophys Res Commun 129: 300–305
Marnett LJ, Bienkowski MJ (1977) Nonenzymatic reduction of prostaglandin H by lipoic acid. Biochemistry 16: 4303–4307
Marshall PJ, Lands WEM (1986) In-vitro formation and activators for prostaglandin synthesis by neutrophils and macrophages from humans and guinea pigs. J Lab Clin Med 108: 525–534
Marshall EK, Lynch V, Smith H (1918) On dichlorethylsulphide (mustard gas). II Variations in susceptibility of the skin to dichlorethylsulphide. J Pharmacol Exp Ther 12: 291–301
Marsico AR, Eaglstein WH (1973) Role of long wave ultraviolet light in the Goeckerman treatment. Arch Dermatol 108: 48–49
Marsico AR, Eaglstein WH, Weinstein GD (1976) Ultraviolet light and tar in the Goeckerman treatment of psoriasis. Arch Dermatol 112: 1249–1250
Martin HAF (1952) Über die Behandlung des Pruritus und der Kraurosis vulvae mit Vitamin E. Med Klin: 1603–1605
Martin JP, Logsdon N (1987) Oxygen radicals mediate cell inactivation by acridine dyes, fluorescein, and lucifer yellow CH. Photochem Photobiol 46: 45–53
Martin W, Loschen G, Günzler WA, Flohe L (1985) Superoxide dismutase inhibits LTB 4 induced leukotaxis. Agents Actions 16: 48–49
Martini MC (1984) Role des microemulsions dans l'absorption percutaneous de l'alphatocopherole. J Pharm Belg 39: 348–354
Marzulli FN, Maibach HI (1983) Dermatotoxicology, 2nd edn. Hemisphere, Washington
Mascio PD, Kaiser S, Devasagayam TPA, Murphy ME, Sies H (1990) Carotenoids, tocopherols and thiols as biological singlet oxygen quenchers. Free Rad Biol Med 9 (Suppl 1): 75
Mason HS, Ingram DJE, Allen B (1960) The free radical property of melanins. Arch Biochem Biophys 86: 225–230
Mason RP (1982) Free radical intermediates in the metabolism of toxic chemicals. Free Radic Biol 5: 161–222
Mason RP, Chignell CF (1982) Free radicals in pharmacology and toxicology – selected topics. Pharmacol Rev 33: 189–211
Massey JB, She HS, Pownall HJ (1982) Interaction of vitamin E with saturated phospholipid bilayers. Biochem Biophys Res Commun 106: 842–847
Massey V, Strickland S, Mayhew SG, Howell LG, Engel PC, Matthews RG, Schuman M, Sullivan PA (1969) The production of superoxide anion radicals in the reaction of reduced flavins and flavo-proteins with molecular oxygen. Biochem Biophys Res Commun 36: 891–897
Mata JM, Assad R, Peterkosfsky B (1981) An intramembranous reductant which participates in the proline hydroxylation reaction with intracisternal prolylhydroxylase and unhydroxylated procollagen in isolated microsomes from L-929 cells. Arch Biochem Biophys 206: 93–104
Mathews MM (1964) Protective effect of beta-carotene against lethal photosensitization by hematoporphyrin. Nature 203: 1092
Mathewson NR, Wong PS, Schuyler M, Travis J (1981) Interaction of human alpha-1-proteinase inhibitor with neutrophil myeloperoxidase. Biochemistry 20: 331–336
Mathews-Roth MM (1982) Antitumor activity of β-carotene, canthaxanthin and phytoene. Oncology 39: 33–37
Mathews-Roth MM (1983) Carotenoid pigment administration and delay in development of UV-B induced tumors. Photochem Photobiol 37: 509–511
Mathews-Roth MM (1986) Carotenoids quench evolution of excited species in epidermis exposed to UV-B (290–320 nm) light. Photochem Photobiol 43: 91–93
Mathews-Roth MM (1987) Photoprotection by carotenoids. Fed Proc 46: 1890–1893
Mathews-Roth MM, Krinsky N (1984) Effect of dietary fat level on UVB induced skin tumors and anti-tumor actions of β-carotene. Photochem Photobiol 4: 671–673
Mathews-Roth MM, Krinsky NI (1985) Carotenoid dose level and protection against UV-B induced skin tumors. Photochem Photobiol 42: 35–38

Mathews-Roth MM, Pathak MA, Fitzpatrick TB, Harber LC, Kass EH (1970) β-carotene as a photoprotective agent in erythropoietic protoporphyria. N Engl J Med 282: 1231–1234

Mathews-Roth MM, Pathak MA, Parrish J, Fitzpatrick TB, Kass EH, Toda K, Clemens W (1972) A clinical trial on the effects of oral beta-carotene on the responses of human skin to solar radiation. J Invest Dermatol 59: 349–353

Mathews-Roth MM, Pathak A, Fitzpatrick TB, Harber CC, Kass EH (1974) Beta carotene as an oral photoprotective agent in erythropoietic protoporphyria. JAMA 228: 1004–1008

Mathews-Roth MM, Hummel D, Crean C (1977) The carotenoid content of various organs of animals administered large amounts of beta-carotene. Nutr Rep Int 16: 419–423

Mathias CGT (1988) Occupational dermatoses. J Am Acad Dermatol 19: 1107–1114

Matkovics B, Novak R, Hanh HD, Szabo L, Varga SI (1977) A comparative study of some more important experimental animal peroxide metabolism enzymes. Comp Biochem Physiol 56B: 397–402

Matoltsy AG (1986) Structure and function of the mammalian epidermis. In: Bereiter-Hahn J, Matoltsy AG, Richards KS (eds) Biology of the integument, vol 2: Vertebrates. Springer, Berlin Heidelberg New York, pp 255–271

Matoltsy AG, Parakkal PF (1967) Keratinization. In: Zelickson AS (ed) Ultrastructure of normal and abnormal skin. Lea and Febiger, Philadelphia, pp 76–104

Matsumara G, Pigman W (1965) Catalytic role of copper and iron ions in the depolymerization of hyaluronic acid by ascorbic acid. Arch Biochem Biophys 110: 526–533

Matsumara N, Mizushima Y (1975) Leukocyte movement and colchicine treatment in Behcet's disease. Lancet ii: 813

Matsuo I, Fujita H, Hayakawa K, Ohkido M (1986) Lipid peroxidative potency of photosensitized thiazide diuretics. J Invest Dermatol 87: 637–641

Matuda S (1982) Reduction of cytochrome c by NADPH, lipoic acid and lipoamide dehydrogenase and effect of copper ions. Acta Med Univ Kagoshima 24: 71–79

Mayeno AN, Curran AJ, Roberts RL, Foote CS (1989) Eosinophils preferentially use bromide to generate halogenating agents. Arch Biochem Biophys 264: 5660–5668

Mayer H, Bollag W, Hänni R, Rüegg R (1978) Retinoids, a new class of compounds with prophylactic and therapeutic activities in oncology and dermatology. Experientia 34: 1105–1119

Maytin EV, Wimberly JM, Anderson RR (1990) Thermotolerance and the heat shock response in normal human keratinocytes in culture. J Invest Dermatol 95: 635–642

Mazur A, Green S, Saha A, Carleton A (1958) Mechanism of release of ferritin iron in-vivo by anthine oxidase. J Clin Invest 37: 1809–1817

McCann J, Choi E, Yamasaki E, Ames BN (1975) Detection of carcinogens as mutagens in the salmonella/microsome test: assay of 300 chemicals. Proc Natl Acad Sci USA 72: 5135–5139

McCarty DJ (1979) Short term drug control of crystal induced inflammation. In: Vane JR, Ferreira SH (eds) Anti-inflammatory drugs. Springer, Berlin Heidelberg New York, pp 92–107 (Handbook of experimental pharmacology, vol 50, pt 2)

McCay PB (1985) Vitamin E: interactions with free radicals and ascorbate. Annu Rev Nutr: 323–340

McCay PB, Fong KL, Lai EK, King M (1978) Possible role of vitamin E as a free radical scavenger and singlet oxygen quencher in biological system which initiate radical mediated reactions. In: de Duve C, Hayashi O (eds) Tocopherol, oxygen and biomembranes. Elsevier/North Holland, New York, pp 41–57

McCay PB, Lai EK, Powell SR, Breugmann G (1986) Vitamin E functions as an electron shuttle for glutathione dependent "free radical reductase" activity in biological membranes. Fed Proc 45: 451

McChesney EW, Fitch CD (1984) Antimalarial drugs in current use. In: Peters W, Richards WHG (eds) Antimalarial drugs, part 2, Current antimalarials and new drug developments. Springer, Berlin Heidelberg New York, pp 3–60 (Handbook of experimental pharmacology, vol 68)

McConnel HM, McFarland BG (1970) Physics and chemistry of spin labels. Q Rev Biophys 3: 191–214
McCord JM (1985) Oxygen derived free radicals in postischemic tissue injury. N Engl J Med 312: 159–163
McCord JM (1986) Superoxide radical: a likely link between reperfusion injury and inflammation. Adv Free Radic Biol Med 2: 325–345
McCord JM, Fridovich I (1969) Superoxide dismutase – an enzymic function for erythrocuprein (hemocuprein). J Biol Chem 244: 6049–6055
McCord JM, Fridovich I (1973) Production of O2- in photolyzed water demonstrated through the use of superoxide dismutase. Photochem Photobiol 17: 115–121
McCord JM, Wong K, Stokes SH, Petrone WF, English D (1980) Superoxide and inflammation: a mechanism for the anti-inflammatory activity of superoxide dismutase. Acta Physiol Scand 492 [Suppl]: 25–30
McCormick JP, Fisher JR, Pachlatko JP, Eisenstark A (1976) Characterization of a cell lethal tryptophan photooxidation product: hydrogen peroxide. Science 191: 468–469
McCormick JR, Harkin MM, Johnson KJ, Ward PA (1981) Suppression by superoxide dismutase of immune complex induced pulmonary alveolitis and dermal inflammation. Am J Pathol 102: 55–61
McCue JP (1979) Changes in oxygen radical scavenging by human blood cell lysates concurrent with viral infections. Exp Hematol 7: 361–368
Mc Cullough J, Weinstein G, Eaglestein WH (1983) Development of topical HPD formulation. Characterisation of photosensitizing effect in-vivo. J Invest Dermatol 81: 928–533
Mc Cullough JL, Weinstein GD, Douglas JL, Berns MW (1987) Photosensitizers in dermatology. Photochem Photobiol 46: 77–82
McDonald RA, Robbins SL, Mallory GK (1958) Dermal fibrosis following subcutaneous injections of serotonin creatinine sulfate. Proc Soc Exp Biol Med 97: 334–338
McMurray W, Dormandy TL (1974) Lipid autoxidation in human skeletal muscle. Clin Chim Acta 52: 105–114
Medeiros MHG, Wefers H, Sies H (1987) Generation of excited species catalyzed by horseraddish peroxidase or hemin in the presence of reduced glutathione and H_2O_2. Free Radic Biol Med 3: 107–110
Meffert H, Lohrisch I (1971 a) Hemmung der Atmung von Epidermiszellen und Fibroblasten durch Produkte der Lipoperoxidation. Dermatol Monatsschr 157: 793–801
Meffert H, Lohrisch L (1971 b) Untersuchung der Lipoperoxydation menschlicher Haut invitro. Dermatol Monatsschr 157: 338–344
Meffert H, Reich P (1969 a) Beeinflussung der Lipoperoxide der menschlichen Hautoberfläche durch ultraviolette Strahlung in-vitro und in-vivo. Dermatol Monatsschr 155: 948–954
Meffert H, Reich P (1969 b) Lipoperoxide und Dithranol-(Cignolin-) Effekt bei Psoriasis vulgaris. Dermatol Wochenschr 155: 157–161
Meffert H, Reichmann G (1972) Lipidperoxidation an der menschlichen Hautoberfläche durch erythemwirksame UV Strahlung. Acta Biol Med Ger 28: 667–673
Meffert H, Waldmann U (1970) Beschleunigung der UV-provozierten Lipidperoxidation des menschlichen Oberflächenfettes durch antipsoriatisch wirksame Substanzen. Dermatol Wochenschr 156: 771–775
Meffert H, Dressler C, Meffert B (1972) UV provozierte Lipidperoxidation in Epidermis, Korium und Subkutis des Menschen in-vitro. Acta Biol Med Ger 29: 667–675
Meffert H, Diezel W, Sönnichsen N (1976 a) Stable lipid peroxidation products in human skin: detection, ultraviolet light induced increase, pathogenic importance. Experientia 32: 1397–1398
Meffert H, Diezel W, Günther W, Sönnichsen N (1976 b) Fotochemotherapie der Psoriasis mit 8-Methoxypsoralen und UVA II. Bindung des Fotosensibilisators an Protein. Dermatol Monatsschr 162: 887–892
Meffert H, Barthelmes H, Metz D, Sönnichsen N (1977) Fotochemotherapie der Psoriasis mit 8-Methoxypsoralen und UVA III. Fluoreszenzmikroskopische Untersuchungen

der Bindung und des Transportes von 8-MOP in der menschlichen Haut. Dermatol Monatsschr 163: 619–627

Meffert H, Böhm F, Sönnichsen N (1980) Zerstörung peripherer Lymphocyten durch 8-MOP + UVA. Dermatol Monatsschr 166: 244–246

Meffert H, Böhm F, Röder B (1981) Mechanismus der Membrandestruktion durch 8-MOP plus UVA. Dermatol Monatsschr 167: 620–621

Meffert H, Böhm F, Röder B, Sönnichsen N (1982) Ist angeregter Singulettsauerstoff am PUVA Therapie Effekt beteiligt? Dermatol Monatsschr 168: 387–393

Meffert H, Pres H, Diezel W, Sönnichsen N (1989) Antipsoriatische und phototoxische Wirksamkeit von Hämatoporphyrin-Derivat nach topischer Aplikation und Bestrahlung mit sichtbarem Licht. Dermatol Monatsschr 175: 28–34

Mehlhorn RJ, Cole G (1985) The free radical theory of aging: a critical review. Adv Free Radic Biol Med 1: 165–223

Meier B, Sies H (1990) Superoxide release in fibroblasts by an NADPH oxidase is related to cell cycle and proliferation. Free Radic Biol Med 9 (Suppl 1): 31

Meier B, Radeke H, Selle S, Sies H, Younes M, Resch K, Habermehl GG (1988) Reactive oxygen species generation by human fibroblasts. Society for Free Radical Research, winter meeting on free radicals in medicine, Current status of antioxidant therapy, Paris, 9–10 Dec 1988

Meier B, Radecke HH, Selle S, Younes M, Sies H, Resch H, Habermehl GG (1989) Human fibroblasts release reactive oxygen species in response to interleukin-1 or tumor necrosis factor. Biochem J 263: 324–330

Meier B, Radeke HH, Selle S, Raspe HH, Sies H, Resch K, Habermehl GG (1990) Human fibroblasts release reactive oxygen species in response to treatment with synovial fluids from patients suffering from arthritis. Free Radic Res Commun 8: 149–160

Meigel W (1987) Einteilung und Klinik des Lupus erythematodes. In: Holzmann H, Altmeyer P, Marsch WC, Vogel HG (eds) Dermatologie und Rheuma. Springer, Berlin Heidelberg New York, pp 269–277

Melikian V, Laverson S, Zawacki B (1987) Oxygen derived free radical inhibition in the healing of experimental zone of stasis burns. J Trauma 27: 151–154

Melkumian AS, Tumanian EL, Aghajanov MI (1978) Histologic estimation of regenerative ability of skin in treatment of burn traumas with the ointment, prepared on the basis of vitamin E. Zh Eksp Klin Med 18: 52–53 (in Russian)

Mellors A, Tappel AL (1966) The inhibition of mitochondrial peroxidation by ubiquinone and ubiquinol. J Biol Chem 241: 4353–4356

Menon GK, Placzek D, Hincenbergs M, Williams ML (1989) Lovastatin induces peroxisomes in cultured keratinocytes. J Invest Dermatol 92: 480

Menon IA, Haberman HF (1977) Mechanisms of action of melanins. Br J Dermatol 97: 109–112

Menzel I (1977) Induratio penis plastica. Z Allgemeinmed 53: 324–326

Mercurio SD, Combs GF (1985) Drug induced changes in selenium dependent glutathione peroxidase activity in the chick. J Nutr 115: 1459–1470

Meredith MJ, Dodson ML (1987) Impaired glutathione biosynthesis in cultured human ataxia-teleangiectasia cells. Canc Res 47: 4576–4581

Merk HF (1987) Die Bedeutung der Sauerstoff Metabolite für die Entzündungsreaktion. In: Holzmann H, Altmeyer P, Marsch WC, Vogel HG (eds) Dermatologie und Rheuma. Springer, Berlin Heidelberg New York, pp 51–56

Merk HF, Khan WA, Bickers DR, Mukhtar H (1988) Induction of epidermal and liver NAD(P)H: quinone reductase by anthralin. J Invest Dermatol 90: 248

Merville MP, Decuyper J, Lopez M, Piette J, Van de Vorst A (1984) Phototoxic potentialities of tartrazine: screening test. Photochem Photobiol 40: 221–226

Metschnikoff E (1883) Untersuchungen über die intrazelluläre Verdauung bei wirbellosen Tieren. Arb Zool Inst Univ Wien 5: 141–168

Meyer-Betz F (1913) Untersuchungen über die biologische (photodynamische) Wirkung des Hämatoporphyrins und anderer Derivate des Blut- und Gallenfarbstoffs. Dtsch Arch Klin Med 112: 476–503

Michaelson G, Edquist LV (1984) Erythrocyte glutathione peroxidase activity in acne vulgaris and the effect of selenium and vitamin E treatment. Acta Derm Venereol (Stockh) 64: 9–14

Michaelson G, Ljunghall K (1990) Patients with dermatitis herpetiformis, acne, psoriasis and Darier's disease have low epidermal zinc concentrations. Acta Derm Venereol (Stockh) 70: 304–308

Michelson AM (1982) Oxygen radicals. Agents Actions 11 [Suppl]: 179–201

Michelson AM (1986a) Pro-oxidant action of superoxide dismutase in the autoxidation of rifamycin SV. Free Radic Res Commun 1: 185–188

Michelson AM (1986b) Free radicals and disease: treatment and clinical application with superoxide dismutase. In: Johnson JE, Walford R, Harman D, Miquel J (eds) Free radicals, aging and degenerative diseases. Liss, New York, pp 263–291

Michelson JD, Schmidt JD, Dresden MH, Duncan C (1974) Vitamin E treatment of epidermolysis bullosa. Arch Dermatol 109: 67–69

Mielens ZE, Rozitis J, Sansone VJ (1968) The effect of oral iodides on inflammation. Tex Rep Biol Med 26: 117–122

Milbradt R (1966) Neuere Erkenntnisse in der Ätiologie, Therapie und Prophylaxe männlicher Fertilitätsstörungen. Hautarzt 17: 97–100

Miller AW, McDonald TJ (1984) Dermal lipofuscinosis associated with amiodarone therapy. Arch Dermatol 12: 646–649

Miller DM, Aust SD (1989) Studies of ascorbate dependent iron catalyzed lipid peroxidation. Arch Biochem Biophys 271: 113–119

Miller SJ, Aly R, Shinefeld HR, Elias P (1988) In-vitro and in-vivo antistaphlococcal activity of human stratum corneum lipids. Arch Dermatol 124: 209–215

Millis AJT, Pious DA (1973) Oxidative phosphorylation in mitochondria isolated from human fibroblasts. Biochim Biophys Acta 292: 73–77

Mills EL, Debets-Ossenkopp Y, Verbrugh H, Verhoef J (1981) Initiation of the respiratory burst of human neutrophils by influenza virus. Infect Immun 32: 1200–1205

Mills O, Porte M, Kligman AM (1978) Enhancement of comedogenic substances by UV radiation. Br J Dermatol 98: 145–150

Milton SH (1972) Experimental studies on island flaps. II. Ischemia and delays. Plast Reconstr Surg 49: 444–447

Mino M, Kasugai O, Nagita A (1985) Relationship between red blood cell and liver tocopherol concentration after administration and depletion of vitamin E. Int J Vitamin Nutr Res 55: 7–51

Mirabelli F, Salis A, Vairetti M, Bellomo G, Thor H, Orrenis S (1989) Cytoskeletal alterations in human platelets exposed to oxidative stress are mediated by oxidative and Ca^{2+} dependent mechanisms. Arch Biochem Biophys 270: 478–488

Mirzoev IUA, Akimov VG, Grebeniuk VN, Polianskaia NP (1985) Role of sulfhydryl groups in the pathogenesis of Duhring's dermatitis herpetiformis. Vestn Dermatol Venerol 10: 24–27 (in Russian)

Mirzoev IUA, Lashmanova AP, Akimov VG (1986) Relation between the clinical course of Duhring's herpetiformis dermatitis and the degree of free radical lipid peroxidation. Vestn Dermatol Venerol 9: 19–22 (in Russian)

Mirzoev YA, Grebenyuk VN, Lashmanova AP, Polyanskaya NP (1988) Combined therapy, including antioxidants, of Duhring's dermatitis herpetiformis. Vestn Dermatol Venereol 2: 56–59 (in Russian)

Mitchell RL, Anderson IC (1965) Catalase photoinactivation. Science 150: 74

Mittal CK, Murad F (1977a) Properties and oxidative regulation of guanylate cyclase. J Cyclic Nucleotide Res 3: 381–391

Mittal CK, Murad F (1977b) Activation of guanylate cyclase by superoxide dismutase and hydroxyl radical: a physiological regulator of cyclic guanosine-3′,5′-monophosphate formation. Proc Natl Acad Sci USA 74: 4360–4364

Miyachi Y, Niwa Y (1982) Effects of potassium iodide, colchicine and dapsone on the generation of polymorphonuclear leukocyte derived oxygen intermediates. Br J Dermatol 107: 209–214

Miyachi Y, Niwa Y (1983) Effects of psoriatic sera on the generation of oxygen intermediates by normal polymorphonuclear leukocytes. Arch Dermatol Res 275: 23–26

Miyachi Y, Ozaki M, Uchida K, Niwa Y (1982a) Effect of thalidomide on the generation of oxygen intermediates by zymosan stimulated normal polymorphonuclear leukocytes. Arch Dermatol Res 274: 363–367

Miyachi Y, Yanase K, Imamura S, Niwa Y (1982b) Increased hydroxyl radical generation by normal polymorphonuclear leukocytes incubated in sera from patients with leukocytoclastic vasculitis. Arch Dermatol Res 274: 65–71

Miyachi Y, Horio T, Imamura S (1983) Sunburn cell formation is prevented by scavenging oxygen intermediates. Clin Exp Dermatol 8: 305–310

Miyachi Y, Uchida K, Komura J, Asada Y, Niwa Y (1985) Auto oxidative damage in cement dermatitis. Arch Dermatol Res 277: 288–292

Miyachi Y, Imamura S, Niwa Y, Tokura Y, Takigawa M (1986a) Mechanisms of contact photosensitivity in mice. VI. Oxygen intermediates are involved in contact photosensitization but not in ordinary contact sensitization. J Invest Dermatol 86: 26–28

Miyachi Y, Imamura S, Niwa Y (1986b) Erythema multiforme: a possible pathogenetic role of increased reactive oxygen species. J Clin Lab Immunol 19: 11–14

Miyachi Y, Imamura S, Niwa Y (1986c) Anti-oxidant action of metronidazole: a possible mechanism of action in rosacea. Br J Dermatol 114: 231–234

Miyachi Y, Yoshioka A, Imamura S, Niwa Y (1986d) Effect of antibiotics on the generation of reactive oxygen species. J Invest Dermatol 86: 449–453

Miyachi Y, Imamura S, Niwa Y (1987a) Decerased skin superoxide dismutase activity by a single exposure to ultraviolet radiation is reduced by liposomal superoxide dismutase pretreatment. J Invest Dermatol 89: 111–112

Miyachi Y, Yoshioka A, Imamura S, Niwa Y (1987b) Polymorphonuclear leucocyte derived reactive oxygen species in inflammatory skin diseases. In: Hayaishi O, Imamura S, Miyachi Y (eds) The biological role of reactive oxygen species in skin. Elsevier, New York, pp 135–140

Miyachi Y, Yoshioka A, Imamura S, Niwa Y (1987c) Anti-oxidant effects of gold compounds. Br J Dermatol 116: 39–46

Miyachi Y, Yoshioka A, Imamura S, Niwa Y (1988) The cumulative effect of continual oxidative stress to the skin and cutaneous aging. In: Kligman AM, Takase Y (eds) Cutaneous Aging. University of Tokyo Press, Tokyo, pp 435–447

Miyachi Y, Hamanaka H, Suzuki S, Yoshioka N, Imamura S (1989) Reactive oxygen species linking to ischemia/reperfusion injury and its reasonable prevention using skin flap models. J Invest Dermatol 92: 482

Miyauchi S, Miki Y (1983) Normal human skin echogram. Arch Dermatol Res 275: 345–349

Mizuno N, Esaki K (1987) Effect of PUVA on the chemotactic activity of polymorphonuclear neutrophils to C5a Des Arg. In: Hayaishi O, Imamura S, Miyachi Y (eds) The biological role of reactive oxygen species in skin. Elsevier, New York, pp 65–69

Moan J (1986) Porphyrin photosensitization and phototherapy. Photochem Photobiol 43: 681–690

Mobilio S, Palumbo M, Pathak MA (1990) The significance of singlet oxygen and DNA cross-links in skin photosensitization by tricyclic and tetracyclic psoralens. J Invest Dermatol 94: 556

Möhn E (ed) (1984) System und Phylogenie der Lebewesen. Schweizerbart, Stuttgart (Physikalische, chemische und biologische Evolution, vol 1)

Molin L (1975) Clofazimine enhanced phagocytosis in pustulosis palmaris et plantaris. Acta Derm Venereol (Stockh) 55: 151–153

Molin L, Stendahl O (1980) The effects of salazopyridine and its active components on human polymorphonuclear leucocyte function in relation to ulcerative colitis. Acta Med Scand 206: 451–457

Möller H, Potokar M, Wallat S (1987) Vitamin E als kosmetischer Wirkstoff, Parfüm Kosmet 68: 688–694

Monaghan BR, Schmitt FO (1932) The effects of carotene and of vitamin A on the oxidation of linoleic acid. J Biol Chem 96: 387–395

Monboisse JC, Braquet P, Randoux A, Borel JP (1983) Non-enzymatic degradation of acid soluble calf skin collagen by superoxide ion: protective effect of flavonoids. Biochem Pharmacol 32: 53–58

Monboisse JC, Braquet P, Borel JP (1984) Oxygen free radicals as mediators of collagen breakage. Agents Action 15: 49–50

Moneret-Vautrin DA (1985) Intolerance to Aspirin: a role for free radicals? Lancet i: 929

Moneret-Vautrin DA, Belanyi P, Martin R (1986) Role des radicaux libres dans l'urticaire chronique. Presse Med 4: 164

Monteiro H, Abdalla DSP, Augusto O, Bechara EJH (1989) Free radical generation during d-aminolevulinic acid autoxidation: induction by hemoglobin and connections with porphyrinpathies. Arch Biochem Biophys 271: 206–216

Montsko T, Tigyi A, Zsoldos T (1984) Morphological and histochemical study of membrane damages occuring in the early phase of experimental silicosis. Oxygen free radicals and tissue damage, scientific session, Pécs, 10–11 Jan 1984

Morgan J (1951) A note on the treatment of Lupus erythematosus with Vitamin E. Br J Dermatol 63: 224–227

Morison WL (1981) Photoimmunolgy. J Invest Dermatol 77: 71–76

Morison WL (1989) Effects of ultraviolet radiation on the immune system in humans. Photochem Photobiol 50: 515–524

Morlier P, Dubertret L, Melo TSE, Salet C, Fosse M, Santus R (1985) The effects of anthralin (dithranol) on mitochondria. Br J Dermatol 112: 509–515

Morrison NE, Marley GM (1976) The mode of action of clofazimine. DNA binding studies. Int J Leprosy 44: 133–134

Motoyoshi K (1983) Enhanced comedo formation in rabbit ear skin by squalene and oleic acid peroxides. Br J Dermatol 109: 191–198

Motten AG, Chignell CF (1983) Spectroscopic studies of cutaneous photosensitizing agents. III. Spin trapping of photolysis products from sulfanilamides analogs. Photochem Photobiol 37: 17–26

Motten AG, Buettner GR, Chignell CF (1985) Spectroscopic studies of cutaneous photosensitizing agents – VIII. A spin trapping study of light induced free radicals from chlorpromazine and promazine. Photochem Photobiol 42: 9–15

Mountz JD, Downs-Minor MB, Turner R, Tohams MB, Richaros F, Pisko E (1983) Bleomycin induced cutaneous toxicity in the rat: analysis for histopathology and ultrastructure compared with progressive systemic sclerosis (scleroderma). Br J Dermatol 108: 679–686

Moynahan EJ (1974) Acrodermatitis enterohepatica: a lethal inherited human zinc deficiency disorder. Lancet i: 399–400

Msika P, Cesarini JP, Poelman MC (1988) Antioxidants and agrressions from ultraviolet radiation in man. Society for Free Radical Research, winter meeting on free radicals in medicine, Current status of antioxidant therapy, Paris, 9–10 Dec 1988

Muglia JJ, Tonnesen MG, Osborn RL, Kissinger RM, Norris DA (1986) Ineffective antioxidant defense in human melanocytes. J Invest Dermatol 86: 496

Müller A, Bonne C (1988) Free radical scavengers in prevention of cutaneous aging. Society for Free Radical Research, winter meeting on free radicals in medicine, Current status of antioxidant therapy, Paris, 9–10 Dec 1988

Müller K, Mayer KK, Wiegrebe W (1986a) 102 – Oxidation of dithranol to chrysazin. Arch Pharm (Weinheim) 319: 1009–1018

Müller K, Eibler E, Mayer KK, Wiegrebe W (1986b) Dithranol, singlett oxygen and unsaturated fatty acids. Arch Pharm (Weinheim) 319: 2–9

Müller K, Wiegrebe W, Younes M (1987) Dithranol, active oxygen species and lipid peroxidation in-vivo. Arch Pharm (Weinheim) 320: 59–66

Mullin CH, Frings G, Abel J, Kind PK, Goerz G (1987) Specific induction of metallothionein in hairless mouse skin by zinc and dexamethasone. J Invest Dermatol 89: 164–166

Munday R (1989) Toxicity of thiols and disulphides: involvement of free radical species. Free Radic Biol Med 7: 659–673

Murad S, Grive D, Lindberg KA, Reynolds G, Sivarajah A, Pinell SR (1981a) Regulation of collagen synthesis by ascorbic acid. Proc Natl Acad Sci USA 78: 2879–2882

Murad S, Sivarajah A, Pinell SR (1981 b) Regulation of prolyl and lysyl hydroxylase activities in cultured cells by ascorbic acid. Biochem Biophys Res Commun 101: 868–875

Murant RS, Gibson SL, Hilf R (1987) Photosensitizing effects of Photofrin II on the site selected mitochondrial enzymes adenylate kinase and monoamine oxidase. Cancer Res 47: 4323–4328

Murina MA, Roschupkin DI (1983) Decrease in erythrocyte aggregation and peroxidation of their membraneous lipids. Biofizika 28: 716 (in Russian)

Murina MA, Roschupkin DI (1984) Effect of ultraviolet radiation on aggregation of human erythrocytes. Photobiochem Photobiophys 7: 59–65

Murray AW, Solanki V, Verma AK (1977) Acculumation of cyclic adenosine 3′,5′-monophosphate in adult and newborn mouse skin: responses to ischemia and isoproterenol. J Invest Dermatol 68: 125–127

Murrell GAC, Pilowsky E (1990) Treatment of Dupuytren's contracture with allopurinol. Antioxidants and degenerative diseases, satellite meeting of the Oxygen Society and the International Society for Free Radical Research, Berkeley, CA, 26–27 Jan 1990

Murrell GAC, Francis MJO, Bromley L (1990 a) Modulation of fibroblast proliferation by oxygen free radicals. Free Rad Biol Med 9 (Suppl 1): 109

Murrell GAC, Francis MJO, Bromley L (1990 b) Free radicals in dupuytren's contracture. Free Radic Biol Med 9 (Suppl 1): 188

Murty HS, Pinelli A, Nair RP, Mendeloff AI (1969) Porphyria cutanea tarda: Therapeutic response to vitamin E. Clin Res 17: 474

Musajo J, Rodighiero G (1962) The skin photosensitizing furocoumarins. Experientia 18: 153–161

Musajo L, Rodighiero G (1972) Mode of photosensitizing action of furocoumarins. In: Giese A (ed) Photophysiology, vol 7. Academic, London, pp 115–147

Musk P, Parsons PG (1987) Resistance of pigmented human cells to killing by sunlight and oxygen radicals. Photochem Photobiol 46: 489–494

Mustajoki P (1972) Vitamin E in porphyria. JAMA 221: 714–715

Mustakallio KK, Martinmaa J, Vilvala R, Halmekoski J (1984) Free radicals and the treatment of psoriasis with special reference to dithranol. Med Biol 62: 155–158

Muus P, Bonta IL, Oudstenden SA (1979) Plasma levels of malondialdehyde. A product of cyclooxygenase dependent and independent lipid peroxidation in rheumatoid arthritis: a correlation with disease activity. Prostagland Med 2: 63–70

Nadiger HA (1980) Role of vitamin E in the aetiology of phrynoderma (follicular hyperkeratosis) and its interrelationship with B-complex vitamins. Br J Nutr 211–214

Nagasawa H, Little JB (1979) Effect of tumor promotors, protease inhibitors and repair processes on X-ray induced sister chromatid exchanges in mouse cells. Proc Natl Acad Sci USA 76: 1943–1947

Nair PP, Mezey E, Murty HS, Quartner J, Mendeloff AI (1971) Vitamin E and porphyrin metabolism in man. Arch Intern Med 128: 411–415

Nakagawara A, Nathan CF, Cohn ZA (1981) Hydrogen peroxide metabolism in human monocytes during differentiation in-vitro. J Clin Invest 68: 1243–1252

Nakamura Y, Colburn NH, Gindhart TD (1985) Role of reactive oxygen in tumor promotion: implication of superoxide anion in promotion of neoplastic transformation in JB-6 cells by TPA. Carcinogenesis 6: 229–235

Namigata A (1960) A study on the vitamin E and skin. Jpn J Dermatol 70: 50–529 (in Japanese)

Naqui A, Chance B, Cadenas E (1986) Reactive oxygen intermediates in biochemistry. Annu Rev Biochem 55: 137–166

Nathan CF, Murray HW, Cohn ZA (1980) The macrophage as an effector cell. N Engl J Med 303: 622–626

Nazzaro-Porro M (1987) Azelaic acid. J Am Acad Dermatol 17: 1033–1041

Nazzaro-Porro M, Picardo M, Finotti E, Passi S (1986) Comparative studies on skin lipid peroxidation induced by physical and biological agents. J Invest Dermatol 89: 320

Neldner KH, Hambidge KM (1975) Zinc therapy of acrodermatitis enterohepatica. N Engl J Med 292: 879–882

Nemeth AJ, Eaglstein WH, Falanga V (1989a) Clinical parameters and transcutaneous oxygen measurements for the prognosis of venous ulcers. J Am Acad Dermatol 20: 186–190
Nemeth AJ, Falanga V, Alstadt SP, Eaglstein WH (1989b) Ulcerated edematous limbs: effect of edema removal on transcutaneous oxygen measurement. J Am Acad Dermatol 20: 191–197
Neuman E (1934) Potassium superoxide and the three electron bond. J Chem Phys 2: 31–33
Nicolaides N, Apon JMB (1977) The saturated methyl branched fatty acids of adult human skin surface lipids. Biomed Mass Spectrom 4: 337–342
Nicolaides N, Fu HC, Rice GR (1968) The skin surface lipids of man compared with those of eigtheen species of animals. J Invest Dermatol 51: 83–89
Niinikoski J, Hunt TK, Dunphy JE (1972) Oxygen supply in healing tissue. Am J Surg 123: 247–252
Niki E, Tsuchiya J, Tanimura R, Kamiya Y (1982) Regeneration of vitamin E from a-chromanoxyl radical by glutathione and vitamin C. Chem Lett: 789–792
Niki E, Kawakami A, Yamamoto Y, Kamiya Y (1985) Oxidation of Lipids. VIII. Synergistic inhibition of oxidation of phosphatidylcholine liposome in aqueous dispersion by vitamin E and vitamin C. Bull Chem Soc Jpn 58: 1971–1975
Nikolowski W (1950) Die Bedeutung des Vitamin E für die Behandlung männlicher Fertilitätsstörungen. Ther Ggw 89: 29–332
Nikolowski W (1959) Die Indikation für die Behandlung mit Vitamin E in der Dermatologie. Therapiewoche 10: 136–137
Nikolowski W (1960) Vitamin E und seine Indikation für die Behandlung von Hautkrankheiten. Med Klin 55: 415–418
Nikolowski W (1962) Inwieweit kann sich die Vitamin E-Therapie der Subfertilität des Mannes auf gesicherte Ergebnisse stützen? Dermatol Wochenschr 145: 185–192
Nikolowski W (1970) Beschwerden während der sogenannten Wechseljahre des Mannes und Möglichkeiten ihrer Behandlung. Z Gerontol 3: 404–407
Nikolowski W (1974) Das sogenannte climacterium virile. Deutsches Ärzteblatt, Heft 5, 31. Januar: 299–302
Nikolowski W (1985) Vitamin E in der Dermatologie. Apoth Z 7: 53–56
Nishi J, Ogura R, Sugiyama M, Hidaka T, Kohno M (1991) Involvement of active oxygen in lipid peroxide radical reaction of epidermal homogenate following ultraviolet light exposure. J Invest Dermatol 97: 115–119
Nishida KI, Takagi M, Yano K (1981) Escherichia coli inactivation by superoxide anion and hydrogen peroxide. J Gen Microbiol 27: 447–485
Nishigaki I, Hagihara M, Hiramatsu M, Izawa Y, Yagi K (1980) Effect of thermal injury on lipid peroxide levels of rat. Biochem Med 24: 185–189
Nishigori C, Miyachi Y, Imamura S, Takebe H (1989) Reduced superoxide dismutase activity in xeroderma pigmentosum fibroblasts. J Invest Dermatol 92: 491
Nishikimi M (1975) A function of tetrahydropteridine as cofactor for indolamine 2,3 dioxygenase. Biochem Biophys Res Commun 63: 92–98
Nishikimi N, Yamada H, Yagi K (1982) Oxidation products of a-tocopherol in its reaction with superoxide. In: Nozaki M, Yamamoto S, Ishimura Y, Coon MJ, Ernster L, Estabrook RW (eds) Oxygenases and oxygen metabolism. Academic, New York, pp 605–609
Nishizuka Y (1984) The role of protein kinase C in cell surface signal transduction and tumor promotion. Nature 308: 693–698
Niwa Y (1987a) Oxygen radical related inflammatory disease. In: Hayaishi O, Imamura S, Miyachi Y (eds) The biological role of reactive oxygen species in skin. Elsevier, New York, pp 151–161
Niwa Y (1987b) Superoxide dismutase. Biomed Ther (in Japanese) 19: 30–34
Niwa Y (1989) Lipid peroxides and superoxide dismutase (SOD) induction in skin inflammatory disease, and treatment with SOD preparations. Dermatologica 179 (Suppl 1): 101–106
Niwa Y, Sohmiya K (1984) Enhanced neutrophilic functions in mucocutaneous lymph node

syndrome, with special reference to the possible role of increased oxygen intermediate generation in the pathogenesis of coronary thrombarteritis. J Pediatr 104: 56–60
Niwa Y, Miyake S, Sakane T, Shingu M, Yokoyama M (1982) Auto oxidative damage in Behcet's disease – endothelial cell damage following the elevated oxygen radicals generated by stimulated neutrophils. Clin Exp Immunol 49: 247–255
Niwa Y, Sakane T, Miyachi Y, Kanoh T, Somiya K (1984 a) Decrease in generation of reactive oxygen species by neutrophils from patients with infectious mononucleosis: role of suppressor T Lymphocytes. Blood 6: 994–999
Niwa Y, Sakane T, Miyachi Y, Ozaki M (1984 b) Oxygen metabolism in phagocytes of leprotic patients: enhanced endogenous superoxide dismutase activity and hydroxyl radical generation by clofazimine. J Clin Microbiol: 837–842
Niwa Y, Sakane T, Miyachi Y (1984 c) Dissociation of the inhibitory effect of dapsone on the generation of oxygen intermediates in comparison with that of colchicine and various scavengers. Biochem Pharmacol 33: 2355–2360
Niwa Y, Sakane T, Shingu M, Miyachi Y (1985 a) Role of stimulated neutrophils from patients with systemic lupus erythemtosus in tissue injury, with special reference to serum factors and increased active oxygen species generated by neutrophils. Inflammation 9: 163–172
Niwa Y, Sakane T, Shingu M, Yanagida I, Komura J, Miyachi Y (1985 b) Neutrophil generated active oxygens in linear IgA bullous dermatosis. Arch Dermatol 121: 73–78
Niwa Y, Sakane T, Somiya K, Miyachi Y (1985 c) Decreased oxygen radical generation by neutrophils from patients with measles presumably owing to activation of suppressor T lymphocytes. J Clin Microbiol 21: 318–322
Niwa Y, Somiya K, Michelson AM, Puget K (1985 d) Effect of liposomal encapsulated superoxide dismutase on active oxygen related human disorders. A preliminary study. Free Radic Res Commun 1: 137–153
Niwa Y, Kanoh T, Sakane T, Soh H, Kawai S, Miyachi Y (1987 a) Detection of enhanced lipid peroxide levels in patients with inflammatory skin diseases. J Clin Biochem Nutr 2: 245–251
Niwa Y, Kanoh T, Sakane T, Soh H, Kawai S, Miyachi Y (1987 b) The ratio of lipoperoxides to superoxide dismutase activity in the skin lesions of patients with severe skin diseases: an accurate prognostic indicator. Life Sci 40: 921–927
Niwa Y, Kasama T, Kawai S, Komura J, Sakane T, Kanoh T, Miyachi Y (1988) The effect of aging on cutaneous lipid peroxide levels and superoxide dismutase activity in guinea pigs and patients with burns. Life Sci 42: 351–356
Niwa Y, Ishimoto K, Kanoh T (1990) Induction of superoxide dismutase in leukocytes by paraquat: correlation with age and possible predictor of longevity. Blood 76 (4): 1–7
Nobbs P, Barr RD (1983) Soft tissue injury caused by antineoplastic drugs is inhibited by topical dimethyl sulphoxide and alpha tocopherol. Br J Cancer 48: 873–876
Nohl H (1989) 2,3,7,8-Tetrachlorodibenzo-p-dioxin induced oxidative stress and its consequences for energy supply to the cell. In: Hayaishi M, Niki E, Kondo M, Yoshikawa T (eds) Medical, biochemical and chemical aspects of free radicals. Elsevier, Amsterdam, pp 117–123
Nohl H (1990) Generation of activated oxygen species as side products of cell respiration. Free Radic Biol Med 9 (Suppl 1): 27
Nohl H, Hegner D, Summer KH (1981) The mechanism of toxic action of hyperbaric oxygenation on the mitochondria of rat heart cells. Biochem Pharmacol 30: 1753–1757
Nohl H, Jordan W, Youngman RJ (1986) Quinones in biology: functions in electron transfer and oxygen activation. Adv Free Radic Biol Med 2: 211–279
Nohl H, Silva DD, Summer KH (1989) 2,3,7,8-Tetrachlorodibenzo-*p*-dioxin induces oxygen activation associated with cell respiration. Free Radic Biol Med 6: 369–374
Nomura H, Hachisuka H, Mori O, Sakamoto F, Sasai Y (1985) Cytofluorometric study of thiol and disulfide groups in guinea pig epidermis. Acta Histochem 76: 213–218
Nonhebel DC, Tedder JM, Walton JC (eds) (1979) Radicals. Cambridge University Press, Cambridge
Noonan PK, Wester RC (1983) Cutaneous biotransformation and some pharmacological

and toxicological implications. In: Marzulli FN, Maibach HI (eds) Dermatotoxicology. Hemisphere, Washington, pp 71–90

Nordlund JJ, Klaus SN, Mathews-Roth MM, Pathak MA (1973) New therapy for polymorphous light eruptions. Arch Dermatol 108: 710–718

Norins AL (1962) Free radical formation in the skin following exposure to ultraviolet light. J Invest Dermatol 39: 445–448

Norris D (1985) Zinc and cutaneous inflammation. Arch Dermatol 121: 985–989

Norton L (1985) Further observations on the yellow nail syndrom with therapeutic effects of oral α-tocopherol. Cutis 457–462

Nye AC, Rosen GM, Gabrielson EW, Keana JFW, Prabhu VS (1987) Diffusion of singlet oxygen into bronchial epithelial cells. Biochim Biophys Acta 928: 1–7

Oberley LW, Buettner GR (1979) Role of superoxide dismutase in cancer: a review. Cancer Res 39: 1141–1149

Oberley LW, Bize IB, Shau SK, Chan Leuthauser SWH, Gruber HE (1978) Superoxide dismutase activity of normal murine liver, regenerating liver and H6 hepatoma. J Natl Cancer Inst 61: 375–379

Oberley LW, McCormick ML, Sierra-Rivera E, Kasemset-St Clair D (1989) Manganese superoxide dismutase in normal and transformed human embryonic lung fibroblasts. Free Radic Biol Med 6: 379–384

O'Brien PJ (1984) Multiple mechanism of metabolic activation of aromatic amine carcinogens. Free Radic Biol 6: 289–322

Ochsendorf FR, Fuchs J, Milbradt R (1991) Reaktive Oxidantien, Lipidperoxidation und Antioxidantien: Übersicht über ihre Bedeutung für die Spermatozoenfunktion. Dreiländertagung Fertilität und Sterilität, Goslar, May 29–June 1, 1991

O'Connell JF, Klein-Szanto AJP, DiGiovanni DM, Fries JW, Slaga TJ (1986) Enhanced malignant progression of mouse skin tumors by the free radical generator benzoyl peroxide. Cancer Res 46: 2863–2865

Oda T, Akaike T, Hamamoto T, Suzuki F, Hirano T, Maeda H (1989) Oxygen radicals in influenza induced pathogenesis and treatment with pyran polymer conjugated SOD. Science 244: 974–976

Ogawa H, Taneda A, Kanaoka Y, Sekine T (1979) The histochemical distribution of protein bound sulfhydryl groups in human epidermis by the new staining method. J Histochem Cytochem 27: 942–946

Oginsky EL, Green GS, Griffith DG, Fowlks WL (1959) Lethal photosensitization of bacteria with 8-methoxypsoralen to long wavelength ultraviolet radiation. J Bacteriol 78: 821–833

Ogura R (1981) Cellular damage of epidermis exposed to ultraviolet light. Lipid peroxide formation in mitochondria and effect of antioxidants. Kurume Med J 45: 279–301 (in Japanese)

Ogura RM, Knox JM (1967) The effect of single exposure of ultraviolet light on the sulhydryl and disulfide of rat epidermis. Jpn J Dermatol 77: 17–21 (in Japanese)

Ogura R, Knox JM, Griffin AC, Kusuhara M (1962) The concentration of sulfhydryl and disulfide in human epidermis, hair and nail. J Invest Dermatol 38: 69–75

Ogura R, Sugiyama M, Sakanashi T, Hidaka T (1987) Role of oxygen in lipid peroxide of skin exposed to ultraviolet light. In: Hayaishi O, Imamura S, Miyachi Y (eds) The biological role of reactive oxygen species in skin. Elsevier, New York, pp 55–63

Ogura R, Ueta H, Sugiyama M, Hidaka T (1990) Distribution of superoxide dismutase activity in the epidermis: Measurement with electron spin resonance spin trapping. J Invest Dermatol 94: 227–229

Ogura R, Sugiyama M, Nishi J, Haramaki N (1991) Mechanism of lipid radical formation following exposure of epidermal homogenate to ultraviolet light. J Invest Dermatol 97: 1044–1047

Oh SH, Deagen JT, Whanger PD, Weswig PH (1978) Biological function of metallothionein. V. Its function in rats by various stresses. Am J Physiol 234: E282–E285

Ohkawa Y, Iwata K, Shibuya H, Fujita H (1984) A rapid screening method for screening skin tumor-promotors using mouse peritoneal macrophages in-vitro. Cancer Lett 21: 257–260

Ohki K, Takamura T, Nozawa Y (1984) Effect of a-tocopherol on lipid peroxidation and acyl chain mobility of liver microsomes from vitamin E deficient rat. J Nutr Sci Vitaminol (Tokyo) 30: 221–234

Ohkuma N, Iizka H, Mizumoto T, Ohkawara A (1987) Superoxide dismutase in epidermis: its relation to keratinocyte proliferation. In: Hayaishi O, Imamura S, Miyachi Y (eds) The biological role of reactive oxygen species in skin. Elsevier, New York, pp 231–237

Ohmori H, Komoriya K, Azuma A, Kurozumi S, Oto YH (1979) Xanthine oxidase induced histamine release from isolated rat peritoneal mast cells: involvement of hydrogen peroxide. Biochem Pharmacol 28: 333–334

Ohmori H, Yamamoto I, Akagi M, Tasaka K (1980) Properties of hydrogen peroxide induced histamine release from mast cells. Biochem Pharmacol 29: 741–745

Ohmori H, Yamauchi T, Yamamoto I (1986a) Augmentation of the antibody response by lipoic acid in mice. II Restoration of the antibody response in immunosuppressed mice. Jpn J Pharmacol 42: 275–280

Ohmori H, Yamauchi T, Yamamoto I (1986b) Augmentation of the antibody response by lipoic acid in mice. I Analysis of the mode of action in an in-vitro culture system. Jpn J Pharmacol 42: 135–140

Ohsawa K, Watanabe T, Matsukawa R, Yoshimura Y, Imaeda K (1984) The possible role of squalene and its peroxide of the sebum in the ocurrence of sunburn and protection from the damage caused by UV iradiation. J Toxicol Sci 9: 151–159

Ohzawa S, Arai M, Takeda Y (1984) Vitamin E in topical preparations. Nippon Koshohin Kakakkaishi (BJJAEV) 8: 18–27 (in Japanese)

Okazaki I, Maruyama K (1974) Collagenase activity in experimental hepatic fibrosis. Nature 252: 49–50

Olcovich HS, Mattill HA (1930) The unsaponifiable lipids of lettuce. I. Carotene. Proc Soc Exp Biol Med 28: 105–117

Oliver EA, Schwartz L, Warren LH (1939) Occupational leukoderma. JAMA 113: 927–928

Oliver EA, Schwartz L, Warren LH (1940) Occupational leukoderma. Arch Dermatol Syphilol 42: 993–1014

Omar RA, Yano S, Kikkawa Y (1987) Antioxidant enzymes and survival of normal and Simian virus 40 transformed mouse embryo cells after hyperthermia. Cancer Res 47: 3473–3476

Oradei A, Lippa S, Littarru GP, Perosino E, Rusciani L (1990) Evaluation of CoQ10 content in exposed and unexposed human skin. 6th International symposium on biomedical and clinical aspects of coenzyme Q, 22–24 Jan 1990

Orfanos CE (1980) Oral retinoids – present status. Br J Dermatol 103: 473–482

Oski FA (1980) Vitamin E – a radical defense. N Engl J Med 303: 454–455

Oya Y, Tonomura A, Yamamoto K (1987) The biological activity of hydrogen peroxide. III. Induction of Epstein Barr virus via indirect action, as compared with TPA and teleocidin. Int J Cancer 40: 69–73

Oyanagui Y (1981) Steroid like antiinflammatory effect of superoxide dismutase in serotonin-, histamine- and kinin-induced edemata of mice: existence of vascular permeability regulating protein(s). Biochem Pharmacol 30: 1791–1798

Ozaki Y, Ohashi T, Niwa Y (1987) Effects of lipoxygenase inhibitors on neutrophil function: reevaluation of the mechanism of their inhibitory effects and the role of leukotriene B4. In: Hayaishi O, Imamura S, Miyachi Y (eds) The biological role of reactive oxygen species in skin. Elsevier, New York, pp 179–188

Pace WE (1976) Treatment and cutaneous ulcers with benzoyl peroxide. Can Med Assoc J 115: 1101–1106

Packer L (1991) Vitamin E – biological activity and health benefits: summing-up overview. German International Workshop on α-tocopherol. Kluvensiek, Germany, Jan 31–Feb 5, 1991

Packer JE, Slater TF, Willson RI (1979) Direct observation of a free radical interaction between vitamin E and vitamin C. Nature 278: 737–738

Packer L, Fuehr K (1977) Low oxygen concentration extends the lifespan of cultured human diploid cells. Nature 267: 423–425

Packer L, Deamer DW, Heath RL (1966) Regulation and deterioration of structure in membranes. In: Marner WL, Cowgill GR (eds) The art of predictive medicine. Thomas, Springfield, pp 77–119

Packer L, Maguire JJ, Mehlhorn RJ, Serbinova E, Kagan VE (1989) Mitochondrial and microsomal membranes have a free radical reductase activity that prevents chromanoxyl radical accumulation. Biochem Biophys Res Commun 159: 229–235

Packer L, Fuchs J, Leipala J, Witt E, Marenus DM, Smith WP (1990) UV effects on skin antioxidants. J Invest Dermatol 94: 563

Palestine RF, Millns JL, Spigel GT (1980) Skin manifestations of pentazocin abusus. J Am Acad Dermatol 2: 47–55

Palmer RMJ, Ferrige AG, Moncada S (1987) Nitric oxide release accounts for the biological activity of endothelium-derived relaxing factor. Nature 327: 524–526

Pambor M, Schmidt W, Wiesner M, Jahr U (1985) Induratio penis plastica – Ergebnisse nach kombinierter Behandlung mit Röntgenbestrahlung und Tokopherol. Z Klin Welt 40: 1425–1427

Pan SF, Rodnan GP, Deutsch M, Wald N (1975) Chromosomal abnormalities in progressive systemic sclerosis (scleroderma) with consideration of radiation effect. J Lab Clin Med 86: 300–308

Panganamala RV, Cornwell DG (1982) The effects of vitamin E on arachidonic acid metabolism. Ann NY Acad Sci 393: 376–390

Papadopulos-Eleopolus E (1988) Reappraisal of AIDS – is the oxidation induced by the risk factors the primary cause? Med Hypotheses 25: 151–162

Parks DA, Granger DN (1986) Xanthine oxidase: biochemistry, distribution and physiology. Acta Physiol Scand 548 [Suppl]: 87–99

Parrish JA (1981) Phototherapy and photochemotherapy of skin diseases. J Invest Dermatol 77: 167–171

Parrish JA, Fitzpatrick TB, Tannenbaum L, Pathak MA (1974) Photochemotherapy of psoriasis with oral methoxsalen and longwave ultraviolet light. N Engl J Med 291: 1207–1211

Pascher G, Röckl H (1960) Quantitative Bestimmung einiger Vitamine im Wasserlöslichen der menschlichen Hornschicht hautgesunder Versuchspersonen. Arch Klin Exp Dermatol 210: 485–495

Pascher F, Sawicky HH, Mabel MD, Silverberg MG, Braitman M, Kanof NB (1951) Therapeutic assays of the skin and cancer unit of the New York University hospitals. Assay V. Tocopherols (Vitamin E) for discoid lupus erythematosus and other dermatoses. J Invest Dermatol 17: 261–263

Pätel M, Schaarschmidt B, Reichert U (1981) Calorimetric and manometric measurements on human skin fibroblasts in culture. Br J Dermatol 105 (Suppl 20): 60–61

Pathak MA, Carraro C (1987 a) Characterization of superoxide dismutase from mammalian skin epidermis. J Invest Dermatol 88: 511

Pathak MA, Carraro C (1987 b) Reactive oxygen species in cutaneous photosensitivity reactions in porphyrias and PUVA photochemotherapy and in melanin pigmentation. In: Hayaishi O, Imamura S, Miyachi Y (eds) The biological role of reactive oxygen species in skin. Elsevier, New York, pp 75–94

Pathak MA, Joshi PC (1983) The nature and molecular basis of cutaneous photosensitivity reactions to psoralens and coal tar. J Invest Dermatol 80: 66 s–74 s

Pathak MA, Stratton K (1968) Free radicals in human skin before and after exposure to light. Arch Biochem Biophys 123: 468–476

Pathak MA, Allen B, Ingram DIE, Fellman JH (1961) Photosensitization and the effect of ultraviolet radiation on the production of unpaired electrons in the presence of furocoumarins (psoralens). Biochim Biophys Acta 54: 506–515

Pathak MA, Riley FC, Fitzpatrick TB (1962) Melanogenesis in human skin following exposure to long wave ultraviolet and visible light. J Invest Dermatol 39: 435–443

Pathak MA, Kramer DM, Fitzpatrick TB (1974) Photobiology and photochemistry of furocoumarins (psoralens). In: Fitzpatrick TB (ed) Sunlight and man. University of Tokyo Press, Tokyo, pp 335–368

Pathak MA, Carbonare MD, Mobilio S (1990) The role of reactive oxygen (1O2, O2-, OH) in collagen cross linking and photoaging. J Invest Dermatol 94: 564
Patriarca P, Zatti M, Cramer R, Rossi F (1970) Stimulation of the respiration of polymorphonuclear leukocytes by phospholipase C. Life Sci 9: 841–849
Payne PA (1985) Applications of ultrasound in dermatology. Bioeng Skin 1: 293–320
Peak JG, Peak MJ, Tuveson RW (1983) Ultraviolet action spectra for aerobic and anaerobic inactivation of Escherichia coli strains specifically sensitive and resistant to near ultraviolet radiations. Photochem Photobiol 38: 541–543
Peak MJ, Peak JG, Moehring MP, Webb RB (1984) Ultraviolet action spectrum for DNA dimer induction, lethality, and mutagenesis in Escherichia coli with emphasis on the UVB region. Photochem Photobiol 40: 613–620
Peak MJ, Peak JG, Jones CA (1985) Different (direct and indirect) mechanisms for the induction of DNA-protein crosslinks in human cells by far- and near-ultraviolet radiations (290 and 405 nm). Photochem Photobiol 42: 141–146
Peller OG (1982) The pathogenesis of systemic lupus erythematosus. Med Hypotheses 8: 643–647
Pence BC, Naylor MF (1990) Effects of single-dose ultraviolet radiation on skin superoxide dismutase, catalase, and xanthine oxidase in hairless mice. J Invest Dermatol 95: 213–216
Penney KB, Smith CJ, Allen JC (1984) Depigmenting action of hydroquinone depends on disruption of fundamental cell processes. J Invest Dermatol 82: 308–310
Penneys NS (1979) Gold therapy: dermatologic uses and toxicities. J Am Acad Dermatol 1: 315–320
Penneys NS, Eaglstein WH, Frost P (1976) Management of pemphigus with gold compounds. Arch Dermatol 112: 185–187
Pentland AP, Marcelo CL, Jordan MA, Voorhees JJ (1986) Effect of gas tension on epidermal keratinocytes DNA synthesis and prostaglandin production. J Invest Dermatol 86: 177–180
Perchellet JP, Perchellet EM (1989) Antioxidants and multistage carcinogenesis in mouse skin. Free Radic Biol Med 7: 377–408
Perchellet JP, Owen MD, Posey TD, Orten DK, Schneider BA (1985 a) Inhibitory effects of glutathione level raising agents and *d-α*-tocopherol on ornithine decarboxylase induction and mouse skin tumor promotion by 12-O-tetradecanoylphorbol-13-acetate. Carcinogenesis 6: 567–573
Perchellet JP, Perchellet EM, Orten DK, Schneider BA (1985 b) Inhibition of the effects of 12-O-tetradecanoylphorbol-13-myristate on mouse epidermal glutathione peroxidase and ornithine decarboxylase activities by glutathione level raising agents and selenium containing compounds. Cancer Lett 26: 283–293
Perchellet JP, Perchellet EM, Abney NL, Zirnstein JA, Belman S (1986 a) Effects of garlic and onion oils on glutathione peroxidase activity, the ratio of reduced/oxidized glutathione and ornithine decarboxylase induction in isolated mouse epidermal cells treated with tumor promotors. Cancer Biochem Biophys 8: 299–312
Perchellet JP, Perchellet EM, Orten DK, Schneider BA (1986 b) Decreased ratio of reduced/oxidized glutathione in mouse epidermal cells treated with tumor promotors. Carcinogenesis 7: 503–506
Perchellet JP, Abney NL, Thomas RM, Guislain YL, Perchellet EM (1987) Effects of combined treatments with selenium, glutathione, and vitamin E on glutathione peroxidase activity, ornithine decarboxylase induction and complete and multistage carcinogenesis in mouse skin. Cancer Res 47: 477–485
Pereira OM, Smith JR, Packer L (1976) Photosensitization of human diploid cell cultures by intracellular flavins and protection by antioxidants. Photochem Photobiol 24: 237–242
Perez HD, Goldstein JM (1980) Generation of a chemotactic lipid from arachidonic acid by exposure to a superoxide generating system. Fed Proc 39: 1170–1176
Perisho K, Wertz PW, Madison KC, Stewart ME, Downing DT (1988) Fatty acids of acylceramides from comedones and from the skin surface of acne patients and control subjects. J Invest Dermatol 90: 350–353

Perrot H, Ortonne JP (1978) Hyperpigmentation after bleomycin therapy. Arch Dermatol Res 261: 245–252

Person JR, Rogers RS (1977) Bullous pemphigoid responding to sulphapyridine and the sulphones. Arch Dermatol 113: 610–616

Peruzzi J, Corsi M (1957) Prove sperimentali sulle proprieta della cisteamine verso le rleazioni; cutanee da irradiazione. Radioter Radiobiol Fis Med 12: 61–65

Peskin AV, Koen YM, Zbarsky IB, Konstantinov AA (1977) Superoxide dismutase and glutathione peroxidase activities in tumors. FEBS Lett 78: 41–45

Peterhans E, Grob M, Bürge TH, Zanoni R (1987) Virus induced formation of reactive oxygen intermediates in phagocytic cells. Free Radic Res Commun 3: 390–446

Peterhans E, Jungi TW, Stocker R (1988) Autotoxicity and reactive oxygen in viral disease. In: Oxy-radicals in molecular biology and pathology. Liss, New York, pp 543–562

Petkau A (1980) Radiation carcinogenesis from a membrane perspective. Acta Physiol Scand 492 [Suppl]: 81–99

Petkau A, Monasterski LG, Kelly K, Friesen HG (1977) Modification of superoxide dismutase in rat mammary carcinoma. Res Commun Chem Pathol Pharmacol 17: 125–132

Petkau A, Chelack WS, Palamar E, Gerrard J (1987) In-vitro response of human bone marrow progenitor cells to superoxide radicals. Res Commun Chem Pathol Pharmacol 57: 107–116

Petkovich M, Brand NJ, Krust A, Chambon P (1987) A human retinoic acid receptor which belongs to the family of nuclear receptors. Nature 330: 444–450

Peto R, Doll R, Buckley JD, Sporn MB (1981) Can dietary beta-carotene materially reduce human cancer rates? Nature 290: 201–208

Petrone WF, English DK, Wong K, McCord JM (1980) Free radicals and inflammation: superoxide dependent activation of a neutrophil chemotactic factor in plasma. Proc Natl Acad Sci USA 77: 1159–1163

Petrozzi JW, Barton JO, Kaidbey KK, Kligman AM (1978) Updating the Goeckerman regimen for psoriasis. Br J Dermatol 98: 437–444

Peyron JL, Meynadier J (1986) The pharmacological basis for the treatment of photodermatoses. Biochimie 68: 899–904

Pfau RG, Hood AF, Morison WL (1986) Photoageing: the role of UVB, solar-simulated UVB, visible and psoralen UVA radiation. Br J Dermatol 114: 319–327

Phelps RG, Fleischmajer R (1988) Clinical, pathologic, and immunologic manifestations of toxic oil syndrom. J Am Acad Dermatol 18: 313–324

Phillips BJ, James TEB, Anderson D (1984) Genetic damage in CHO cells exposed to enzymatically generated active oxygen species. Mutat Res 126: 265–271

Picardo M, De Luca C, Zompetta C, Passi S, Nazzaro-Porro M (1989) Skin surface lipoperoxides and immune response. J Invest Dermatol 92: 500

Pierce AW (1954) Carotene and vitamin A in human fat. Med J Aust 1: 589–592

Piette J, Decuyper J, Van der Vorst A (1986) DNA alterations photosensitized by tetracycline and some of its derivatives. J Invest Dermatol 86: 653–658

Piette J, Legrand-Poels S, Vaira D, Rentier B, van de Vorst A (1990) Reactivation of HIV-1 after an oxidative stress. Free Radic Biol Med 9 (Suppl 1): 5

Piloty O, Graf Schwerin B (1901) Über die Existenz von Derivaten des vierwertigen Stickstoffs. Ber Dtsch Chem Ges 34 (II): 2354–2367

Pirie A, Dilley KJ (1974) Photo-oxidation of *N*-formylkynurenine and tryptophan peptides by sunlight or simulated sunlight. Photochem Photobiol 19: 115–118

Plaa GL, Witschi H (1976) Chemicals, drugs and lipid peroxidation. Annu Rev Pharmacol Toxicol 16: 125–141

Pliquett F, Ermolayev AV, Sarycheva IK, Evstigneeva RP (1984) PUVA induced erythema and changes in mechanoelectrical properties of skin. Inhibition by tocopherols. Arch Dermatol Res 276: 12–16

Plonka A, Metodiewa D (1979) ESR evidence of superoxide radical dismutation by human ceruloplasmin. Biochem Biophys Res Commun 95: 978–984

Poensin D, Beani JC, Berthod F, Perrillat Y, Monnier C, Mezin P, Reymond JL, Amblard P (1990) Un nouveau cas de syndrome eosinophilie-myalgies associé a la prise de *L*-tryptophane. Nouv Dermatol 9: 754–756

Pohl R, Philpot R, Fouts J (1976) Cytochrome P-450 content and mixed function oxidase activity in microsomes isolated from mouse skin. Drug Metab Dispos 4: 442–450

Polkanov VS, Bochkarev IUM, Shmeleva LT, Kipper SN (1987) Lipid peroxidation and the blood antioxidant activity in psoriasis. Vestn Dermatol Venerol 7: 42–46 (in Russian)

Polla BS (1990) Heat (shock) and the skin. Dermatologica 180: 113–117

Pollitt N (1975) β-Carotene and the photodermatoses. Br J Dermatol 93: 721–724

Pooler JP (1989) Photooxidation of cell memebranes using eosin derivatives that locate in lipid or protein to study the role of diffusible intermediates. Photochem Photobiol 50: 55–68

Poot M, Hoehn H (1989) Cell kinetic evidence suggests elevated oxidative stress in cultured cells of Bloom's syndrome. Free Radic Res Commun 7: 179–187

Poot M, Hoehn H, Rabinovitch PS (1988) Is Bloom's syndrome an inherited disease of elevated generation of free radicals? International Symposium on free radicals in medicine, Vienna, 8–12 Nov 1988 (abstr POS-61)

Popov I, Lewin G (1989) Psoriasis – deregulation in the antioxidant system of the organism? International conference on regulation of free radical reactions, Varna, 13–16 Sept 1989

Popov I, Lewin G, Schade R, Göhler K, von Baehr R (1988a) Die Beteiligung des antioxidativen Systems des Organismus an der Akutphasereaktion. 18th Annual Meeting of the Biochemical Society of the GDR, Neubrandenburg, 13–15 Jan 1988

Popov I, Lewin G, Matthes G, Meffert H, Scherf HP, Hengst P, Conradi E, Heilmann HH, Zimmermann I, von Baehr R (1988b) Zur Bedeutung der antioxidativen Kapazität des Blutplasmas. Z Klin Med 43: 1663–1666

Popov I, Lewin G, Scherf HP, Meffert H (1989) Zum Verhalten der antioxidativen Kapazität des Plasmas nach UV-Bestrahlung. 2. Untersuchungen an Patienten mit Psoriasis und peripherer arterieller Verschlußkrankheiten. Z Klin Med 44: 1857–1860

Popov I, Müller GM, Miehe M, Lewin G, von Baehr R (1990) Die Beziehung zwischen dem antioxidativen Potential des psoriatischen Blutplasmas und der Reaktivität von Granulocyten. Dermatol Monatsschr 176: 43–48

Potapenko AY, Abijev GA, Pliquett F (1980) Inhibiton of erythema of the skin photosensitized with 8-methoxypsoralene by a-tocopherol. Bull Exp Biol Med 89: 611–615

Potapenko AJ, Piszov MJ, Abijev GA, Pliquett F (1983) α-Tokopherol, ein Inhibitor von durch UV-Strahlung induzierten Veränderungen mechanoelektrischer Hauteigenschaften. Dermatol Monatsschr 169: 300–304

Potapenko AY, Abijev GA, Pitsov MY, Roschchupkin DI, Vladimirov YA, Pliquett F, Ermolayev AV, Sarycheva IK, Evstigneeva RP (1984) PUVA induced erythema and changes in mechanoelectrical properties of skin. Inhibition by tocopherols. Arch Dermatol Res 276: 12–16

Potokar M, Holtman W, Werner-Busse A (1988) Protective effect of vitamin E against ultraviolet light: comparative testing of natural tocopherol homologues on the skin of hairless mice. Society for Free Radical Research, winter meeting on free radicals in medicine, current status of antioxidant therapy, Paris, 9–10 Dec 1988

Poulin Y, Perry HO, Muller SA (1984) Pemphigus vulgaris: results of treatment with gold as a steroid sparing agent in a series of thirteen patients. J Am Acad Dermatol 11: 851–857

Pres H, Meffert H, Sönnichsen N (1989) Photodynamische Therapie der Psoriasis palmaris et plantaris mit topisch appliziertem Hämatoporphyrin Derivat und sichtbarem Licht. Dermatol Monatsschr 175: 745–750

Price JM, Brown JJ, Rukavina JG, Mendelson C, Johnson SAM (1957) Scleroderma (acrosclerosis). II Tryptophan metabolism before and during treatment by chelation (EDTA). J Invest Dermatol 29: 289–298

Priestly GC, Adams LW (1985) Mitogenic effects of sera from normal and psoriatic subjects on human skin fibroblasts. Arch Dermatol Res 277: 13–15

Probskina MP, Matyuskin AI (1982) Effect of hormones on lipid peroxidation in the heart and liver. Farmakol Toksikol 45: 24–26

Proctor PH, Reynolds ES (1984) Free radicals an disease in man. Physiol Chem Phys Med NMR 16: 175–195

Prosser CL (ed) (1986) Adaptational biology. Molecules to organisms. Wiley, New York

Pryor WA (1975) The role of free radical reactions in biological systems. Free Radic Biol 1: 1–49
Pryor WA (1982) Free radical biology: xenobiotics, cancer and aging. Ann NY Acad Sci: 1–22
Pryor WA (1986) Oxygen radicals and related species: their formation, lifetimes, and reactions. Annu Rev Physiol 48: 657–667
Pugliese PT, Lampley CB (1985) Biochemical assessment of the anti-aging effects of cosmetic products. J Appl Cosmet 3: 129–138
Pugliese PT, Saylor GB, Salter DC (1983) The effects of dl-alpha-tocopherol acetate on ultraviolet light induced epidermal ornithine decarboxylase activity. Annual meeting of the SCC, New York, 1–2 Dec 1983
Puhvel SM, Connor JM, Sakamoto M (1990) Vitamin A deficiency and dioxin induced cutaneous toxicity in murine skin. J Invest Dermatol 94: 568
Puig-Parallada P, Planas JM (1978) Synovial fluid degradation induced by free radicals. In-vitro action of several free radical scavengers and antiinflammatory drugs. Biochem Pharmacol 27: 535–537
Pult H (1954) Erfahrungen bei der Behandlung der Kollagenosen mit Vitamin E. Dtsch Med Wochenschr 471–483
Punnonen K, Jansen CT, Puntala A, Ahotupa M (1991 a) Effect of in-vitro UVA irradiation and PUVA treatment on membrane fatty acids and activities of antioxidant enzymes in human keratinocytes. J Invest Dermatol 96: 255–259
Punnonen K, Puntala A, Jansen CT, Ahotupa M (1991 b) UVB irradiation induces lipid peroxidation and reduces antioxidant enzyme activities in human keratinocytes in itro. Acta Derm Venereol (Stockholm) 71: 239–273
Puppo A, Halliwell B (1988) Formation of hydroxyl radicals from hydrogen peroxide in the presence of iron. Biochem J 249: 185–190
Puschmann M (1982) Klinisch experimentelle Untersuchungen zum Wirkungsnachweis von Benzoylperoxid. Hautarzt 33: 257–265
Quinlan GJ, Gutteridge MC (1988) Hydroxyl radical generation by the tetracycline antibiotics with free radical damage to DNA, lipids and carbohydrate in the presence of iron and copper salts. Free Radic Biol Med 5: 341–348
Raab O (1900) Über die Wirkung fluorescierender Stoffe auf Infusorien. Z Biol 39: 524–546
Raab W (1979) Die Wirkung von langwelligem Ultraviolettlicht (UV-A) und von mittelwelliger Ultraviolettstrahlung (UV-B) auf die menschliche Haut. Ein kritischer Vergleich. Z Hautkr 55: 497–513
Raab W, Patermann F (1961) Die Wirkung externer Antipsoriatika auf die Zellatmung. Arch Klin Exp Dermatol 226: 153–166
Raes M, Michiels C, Remacle J (1987) Comparative study of the enzymatic defense systems against oxygen derived free radicals: the key role of glutathione peroxidase. Free Radic Biol Med 3: 3–7
Raghow R, Lurie S, Seyer JM, Kang AH (1985) Profiles of steady state levels of messenger RNA's coding for type I procollagen, elastin and fibronectin in hamster lungs undergoing bleomycin induced interstitial pulmonary fibrosis. J Clin Invest 76: 1733–1739
Rajewsky B, Hobitz H, Harder D (1959) Biologische Grundlagen der Röntgentherapie. In: Marchionini A, Schirren CG (eds) Strahlentherapie von Hautkrankheiten. Springer, Berlin Göttingen Heidelberg, pp 86–180 (Handbuch der Haut- und Geschlechtskrankheiten, vol 5, pt 2)
Ramasarma T (1985) Natural occurrence and distribution of coenzyme Q. In: Lenaz G (ed) Coenzyme Q. Wiley, London, pp 67–81
Ramboer C, Piessens F, DeGroote J (1972) Serum xanthine oxidase and liver disease. Digestion 7: 183–195
Rampen FHJ, Fleuren E (1987) Melanoma of the skin is not caused by ultraviolet radiation but by a chemical xenobiotic. Med Hypotheses 22: 341–346
Ranadive NS, Menon IA (1986) Role of reactive oxygen species and free radicals from melanins in photoinduced cutaneous inflammation. Pathol Immunopathol Res 5: 118–139
Ranadive NS, Shirwadkar S, Persad S, Menon A (1986) Effects of melanin induced free radicals on the isolated rat peritoneal mast cells. J Invest Dermatol 86: 303–307

Rao NA, Sevanian A, Fernandez MAS, Romera JL, Faure JP, de Kozak Y, Till GO, Marak GE (1987) Role of oxygen radicals in experimental allergic uveitis. Invest Ophthalmol Vis Sci 28: 886–892
Rapaport MJ (1982) Depigmentation with cerium oxide. Contact Dermatitis 8: 282–283
Rasbridge MR, Scott GL (1973) The haemolytic action of dapsone: the effect on red cell glycolysis. Br J Haematol 24: 169–181
Rassat A (1971) Application of electron spin resonance to conformational analysis. Pure Appl Chem 25: 623–629
Rave O, Wagner H, Junge-Hülsing G, Hauss WH (1971) Zur Wirkung von Vitamin E auf den Mesenchymstoffwechsel. Z Rheumaforsch 30: 266–274
Reddy CC, Scholz RW, Thomas CE, Massaro EJ (1982) Vitamin E dependent reduced glutathione inhibition of rat liver microsomal lipid peroxidation. Life Sci 31: 571–576
Reed BR, Clark RAF (1985) Cutaneous tissue repair: Practical implications of current knowledge. II. J Am Acad Dermatol 13: 919–941
Reed CJ, Douglas KT (1989) Single strand cleavage of DNA by Cu (II) and thiols: a powerful chemical DNA cleaving system. Biochem Biophys Res Commun 162: 1111–1117
Rees SJ, Slater T (1988) Tetracycline – an iron chelator for lipid peroxidation. International symposium on free radicals in medicine, Vienna, 8–12 Nov 1988 (abstr POS-53)
Reffy F (1936) Über die Bedeutung der Vitamin in der Wundheilung. Verh 9. Internat Kong Dermat 2: 528–530
Reichert U (1986) Skin toxicity and cellular metabolism: in-vitro models. Br J Dermatol 115 (Suppl 31): 108–116
Reichert U, Schmidt R, Schroot B (1985) The mode of action of anthralin: antirespiratory activity and subcellular distribution in cultured human keratinocytes. J Invest Dermatol 84: 302
Reifferscheid M, Matis P (1951) Das Vitamin E in der Behandlung von Durchblutungsschäden, Dupuytren-Kontraktur und Thrombosen. Med Welt: 1168–1172
Reiners JJ, Rupp T (1989) Conversion of xanthine dehydrogenase to xanthine oxidase occurs during keratinocyte differentiation: modulation by 12-O-tetradecanoylphorbol-13-acetate. J Invest Dermatol 93: 132–135
Reinl W (1957) Sklerodermie durch Trichlorethylen Einwirkung. Bull Hyg 32: 678–679
Reiss U, Gershon D (1976) Comparison of cytoplasmic superoxide dismutase in liver, heart and brain of aging rats and mice. Biochem Biophys Res Commun 73: 255–262
Requena L, Requena C, Sanchez M, Jaqueti G, Aguilar A, Sanchez-Yus E, Hernandez-Moro B (1988) Chemical warfare: cutaneous lesions from mustard gas. J Am Acad Dermatol 19: 529–536
Rettie AE, Williams FM, Rawlins MD (1986) Substrate specificity of the mouse skin mixed function oxidase system. Xenobiotika 16: 205–211
Reusch MK, Meager K, Leadon SA, Hanawalt PC (1988) Comparative removal of pyrimidine dimers from human epidermal keratinocytes in-vivo and in-vitro. J Invest Dermatol 91: 349–352
Reyftmann JP, Kohen E, Morliere P, Santus R, Kohen C, Mangel WF, Dubertret L, Hirschberg JG (1986) A microspectrofluorimetric study of porphyrin-photosensitized single living cells. I. Membrane alterations. Photochem Photobiol 44: 461–469
Reyftmann JP, Santus R, Maziere JC, Morliere P, Salmon S, Candide C, Maziere C, Haigle J (1990) Sensitivity of tryptophan and related compounds to oxidation induced by lipid autoperoxidation. Application to human serum low- and high-density lipoproteins. Biochim Biophys Acta 1042: 159–167
Riesz P, Berdahl D, Christman CL (1985) Free radical generation by ultrasound in aqueous and nonaqueous solutions. Environ Health Perspect 64: 233–252
Rietschel L (1963) Die Bestimmung des Vitamin A and E im Serum von Hautkranken. Dermatol Wochenschr 25: 635–638
Riley PA (1969a) Hydroxyanisole depigmentation: in-vitro studies. J Pathol 97: 193–206
Riley PA (1969b) Hydroxyanisole depigmentation: in-vivo studies. J Pathol 97: 185–191
Riley PA (1970) Mechanism of pigment cell toxicity produced by hydroxyanisol. J Pathol 101: 163–169

Riley PA (1985) Radicals and melanomas. Phil Trans R Soc Lond [B] 311: 679–689
Riley PA, Sawyer B, Wolff MA (1975) The melanocytotoxic action of 4-hydroxyanisole. J Invest Dermatol 64: 86–89
Riley PA, Cooksey CJ, Land EJ (1989) Free radicals and melanomas. In: Hayaishi O, Niki E, Kondo M, Yoshokawa T (eds) Medical biochemical and chemical aspects of free radicals. Elsevier, Amsterdam, pp 1485–1490
Riordan JF, Vallee BL (1976) Structure and function of metalloenzymes. In: Prasad AS (ed) Trace elements in human health and disease, vol 1. Academic, New York, chap 14
Risse O (1929) Über die Röntgenphotolyse des Hydroperoxyds. Z Phys Chem [Abt A] 140: 133
Rister M, Baehner RL (1976) The alteration of superoxide dismutase, catalase, glutathione peroxidase and NAD(P)H cytochrome c reductase in guinea pig polymorphonuclear leukocytes and alveolar macrophages during hyperoxia. J Clin Invest 58: 1174–1184
Rister M, Baumeister K, Gravest O, Gladtke E (1978) Superoxide dismutase deficiency in rheumatoid arthritis. Lancet i: 1094
Rister M, Mitrenga D, Schneider A (1984) New aspects of the inflammatory process in systemic lupus erythematosus. Agents Action 15: 44–45
Robak J, Gryglewski RJ (1988) Flavonoids are scavengers of superoxide anion. Biochem Pharmacol 37: 837–841
Roberfroid MB, Viehe HG, Remacle J (1987) Free radicals in drug research. Adv Drug Res 16: 1–84
Roberts AB, Sporn MB (1984) Cellular biology and biochemistry of the retinoids. In: Sporn MB, Roberts AB, Goodman DS (eds) The retinoids, vol 2. Academic, Orlando, pp 210–287
Robertson FM, Laskin DL, Laskin JD (1988) Production of reactive oxygen intermediates in mouse skin following topical application of the tumor promotor 12-O-tetradecanoyl phorbol-13-acetate (PMA). Proc Am Assoc Cancer Res 29: 155
Robins SP, Shimokomaki M, Bailey AJ (1973) Age related changes in the reducible components of intact bovine collagen fibers. Biochem J 131: 771–780
Rodnan GP, Medsger TA (1981) Pentazocine induced fibrosis. International conference on progressive systemic sclerosis, Austin, TX, Oct 1981
Rodnan GP, Benedek TG, Medsger TA, Cammarata RJ (1967) The association of progressive systemic sclerosis (scleroderma) with coal miners' pneumokoniosis an other forms of silicosis. Ann Intern Med 66: 323–334
Roederer M, Staal FJT, Raju PA, Ela SW, Herzenberg LA, Herzenberg LA (1990) Cytokine-stimulated human immunodeficiency virus replication is inhibited by *N*-acetyl-L-cysteine. Proc Natl Acad Sci USA 87: 4884–4888
Roelfzema H, Bergers M, Van Erp PEJ, Gommans JM, Mier PD (1981) Studies on the plasma membrane of normal and psoriatic keratinocytes. 3. Uptake of labelled sugar and their incorporation into glycoconjugates. Br J Dermatol 104: 635–640
Roenigk HH Jr, Caro WA (1981) Skin cancer in the PUVA-48 co-operative study. J Am Acad Dermatol 4: 319–324
Roisman FR, Walz DT, Finkelstein AE (1983) Superoxide radical production by human leukocytes exposed to immune complexes: inhibitory action of gold compounds. Inflammation 7: 355–362
Rony HR, Scheff GJ, Cohen DM, Rennagel WR (1958) Sulfhydryl oxidase activity in skin homogenates. J Invest Dermatol 30: 43–50
Roos D, Bragt PC (1984) Workshop session summary: is phagocyte radical production relevant to inflammation? Agents Action 15: 57–58
Roos DJ, Hamon-Müller WT, Weening RJ (1976) Effect of cytochalasin B on the oxidative metabolism of human peripheral blood granulocytes. Biochem Biophys Res Commun 68: 71–77
Roos D, Weenig RS, Wyss SR, Aebi HE (1980) Protection of human neutrophils by endogenous catalase: studies with catalase deficient individuals. J Clin Invest 65: 1515–1522
Root RK, Metcalf JA (1977) H_2O_2 release from human granulocytes during phagocytosis:

relationship to superoxide anion formation and cellular catabolism of H_2O_2: studies with normal and cytochalasin B treated cells. J Clin Invest 60: 1266–1279
Roschchupkin DI, Pitsov MY, Potapenko AY (1979) Inhibition of ultraviolet light-induced erythema by antioxidants. Arch Dermatol Res 266: 91–94
Roschchupkin DI, Anosov AK, Potapenko AY (1983) Antiaggregation action of ultraviolet irradiation on platelet rich plasma in the presence of antioxidants. Photobiochem Photophys 5: 307–313
Rose DP (1972) Aspects of tryptophan metabolism in health and disease: a review. J Clin Pathol 25: 17–25
Rosenberg HR, Culik R (1959) Effect of a-lipoc acid on vitamin C and vitamin E deficiencies. Arch Biochem Biophys 80: 86–93
Rosett T, Ohkido M, Smith JG, Yardley H (1967) Studies in the biochemistry of skin. IV. Some properties of mitochondria isolated from the epidermis of the adult rat. J Invest Dermatol 48: 67–78
Ross D, Moldeus P (1985) Generation of reactive species and fate of thiols during peroxidase catalyzed metabolic activation of aromatic amines and phenols. Environ Health Perspect 64: 253–257
Ross D, Albano E, Nilsson U, Moldeus P (1984) Thiyl radicals – formation during peroxidase catalyzed metabolism of acetaminophen in the presence of thiols. Biochem Biophys Res Commun 125: 109–115
Ross D, Cotgreave I, Moldeus P (1985) The interaction of reduced glutathione with active oxygen species generated by xanthine oxidase catalyzed metabolism of xanthine. Biochim Biophys Acta 841: 278–282
Röth E, Török B, Zsoldos T, Matkovics B (1985) Lipid peroxidation and scavenger mechanism in experimentally induced heart infarcts. Basic Res Cardiol 80: 530–536
Roth NG, Halvorson HO (1952) Effect of oxidative rancidity in unsaturated fatty acids on germination of bacterial spores. J Bacteriol 63: 429–434
Rothman S, Krysa HF, Smiljanic AM (1946) Inhibitory action of human epidermis on melanin formation. Proc Soc Exp Biol Med 62: 208–212
Roumm AD, Medsger TA (1985) Cancer and systemic sclerosis. Arthritis Rheum 28: 1336–1340
Rous P, Kidd JG (1941) Conditional neoplasms and subthreshold neoplastic states. A study of tar tumors in rabbits. J Exp Med 73: 365–389
Rousseau-Richard C, Richard C, Martin R (1988) Kinetics of bimolecular decay of a-tocopherol free radicals studied by ESR. FEBS Lett 233: 307–310
Roux-Lombard P, Modoux C, Cruchaud A, Dayer JM (1989) Purified blood monocytes from HIV 1 infected patients produce high levels of TNF alpha and IL-1. Clin Immunol Immunopathol 50: 374–384
Rowell NR (1986) Lupus erythematosus, scleroderma and dermatomyositis. The "collagen" or "connective tissue" diseases. In: Rook A, Wilkinson DS, Ebling FJG, Champion RH, Burton JL (eds) Textbook of dermatology, vol 2. Blackwell, Oxford, pp 1281–1392
Rowley AH, Shulman ST (1987) Kawasaki disease: new etiologic clues and advances in therapy. Pediatr Dermatol 4: 134–138
Rowley D, Gutteridge JMC, Blake D, Farr M, Halliwell B (1984) Lipid peroxidation in rheumatoid arthritis: thiobarbituric acid reactive material and catalytic iron salts in synovial fluid from rheumatoid patients. Clin Sci 66: 691–700
Roza LGP, Van der Schans P, Lohman PH (1985) The induction and repair of DNA damage and its influence on cell death in primary human fibroblasts exposed to UV-A or UV-C irradiation. Mutat Res 146: 89–98
Rozantsev EG (ed) (1970) Free nitroxyl radicals. Plenum, New York
Rozell B, Hansson HA (1985) Immunohistochemical localization of thioredoxin and thioredoxin reductase in adult rats. Eur J Cell Biol 38: 79–86
Rubanyi GM (1988) Vascular effects of oxygen-derived free radicals. Free Radic Biol Med 4: 107–120
Rubin P, Casarett GW (1968) Skin and adnexa. In: Clinical radiation pathology, vol 1. Saunders, Philadelphia, pp 62–119

Rukavina B, Mohar N (1979) An approach of ultrasound diagnostic techniques of the skin and subcutaneous tissue. Dermatologica 158: 81–92

Rupec M, Vakilzadeh F, Brühl R (1969) Die Häufigkeitsverteilung der Mitochondrien in der normalen Meerschweinchenepidermis. (Eine quantitative elektronenmikroskopische Untersuchung). Arch Klin Exp Dermatol 235: 228–233

Rush PJ, Bell MJ, Fam AG (1984) Toxic oil syndrome (spanish oil disease) and chemically induced scleroderma like conditions. J Rheumatol 11: 262–264

Sa E Melo T, Dubertret L, Prognon P, Gond A, Mahuzier G, Santus R (1983) Physicochemical properties and stability of anthralin in model systems and human skin. J Invest Dermatol 80: 1–6

Sacks T, Moldow CF, Craddock PR, Boxers TK, Jacob HS (1978) Oxygen radicals mediate endothelial cell damage by complement stimulated granulocytes: an in-vitro model of immune vascular damage. J Clin Invest 61: 1161–1167

Sadrazadeh SMH, Graf E, Panter SS, Hallaway PE, Eaton JW (1984) Hemoglobin. A biological Fenton reagent. J Biol Chem 259: 14354–14356

Sahu SK, Oberley LW, Stevens RH, Riley EE (1977) Superoxide dismutase of Ehrlich ascites tumor cells. J Natl Cancer Inst 58: 1125–1128

Sahut A, Aiache S, Andermann G, De Burlet G, Aiache JM (1985) Topical and bioavailability of benzoyl peroxide. Int J Cosmet Sci 7: 61–69

Saihan EM, Burton JL, Heaton KW (1978) A new syndrom with pigmentation, scleroderma, gynaecomastia, Raynaud's phenomenon and peripheral neuropathy. Br J Dermatol 99: 437–440

Saint-Leger D, Bague A, Cohen E, Chivot M (1986) A possible role for squalene in the pathogenesis of acne. I. In-vitro study of squalene oxidation. Br J Dermatol 114: 535–542

Salet C, Passarella S, Quagliariello E (1987) Effects of selective irradiation on mammalian mitochondria. Photochem Photobiol 45: 433–438

Salminen A, Saari P, Kihlstrom M (1988) Age- and sex-related difference in lipid peroxidation of mouse cardiac and skeletal muscle. Comp Biochem Physiol 89B: 695–698

Salomon L, Stubbs DW (1961) Some aspects of the metabolism of ascorbic acid in rats. Ann NY Acad Sci 92: 128–140

Sammartano LJ, Tueson RW, Davenport R (1986) Control of sensitivity to inactivation by broad spectrum near UV radiation by Escherichia coli katF locus. J Bacteriol 168: 12–21

Sandberg S (1981) Protoporphyrin induced photodamage to mitchondria and lysosomes from rat liver. Clin Chim Acta 111: 55–60

Sandberg S, Romslo I (1981) Phototoxicity of protoporphyrin as related to its subcellular localization in mice livers after short term feeding with griseofulvin. Biochem J 198: 67–71

Santus R, Reyftmann JP (1986) Photosensitization of membrane components. Biochimie 68: 843–848

Saran M, Bors W (1989) Oxygen radicals acting as chemical messengers: a hypothesis. Free Radic Res Commun 7: 3–6

Sarna T, Sealy RC (1984a) Photoinduced oxygen consumption in melanin systems. Action spectra and quantum yields for eumelanin and synthetic melanin. Photochem Photobiol 39: 69–74

Sarna T, Sealy RC (1984b) Free radicals from eumelanins: quantum yields and wavelength dependence. Arch Biochem Biophys 232: 574–578

Sarna T, Menon IA, Sealy RC (1984c) Photoinduced oxygen consumption in melanin systems-II. Action spectra and quantum yields for pheomelanins. Photochem Photobiol 39: 805–809

Sarna T, Menon IA, Sealy RC (1985a) Photosensitization of melanins: a comparative study. Photochem Photobiol 42: 529–532

Sarna T, Korytowski W, Sealy RC (1985b) Nitroxides as redox probes of melanins: dark induced and photoinduced changes in redox equilibria. Arch Biochem Biophys 239: 226–233

Sarna T, Pilas B, Land EJ, Truscott TG (1986) Interaction of radicals from water radiolysis with melanin. Biochim Biophys Acta 883: 162–167

Sasaki R, Kurokawa T, Tero-Kubota S (1983) Ascorbate radical and ascorbic acid level in human serum and age. J Gerontol 38: 26–30
Sato T, Nishikawa M, Higuchi C, Omata M, Sanaka T, Sugino N (1989) Effect of SOD on lupus nephritis. In: Hayaishi O, Niki E, Kondo M, Yoshokawa T (eds) Medical, biochemical and chemical aspects of free radicals. Elsevier, Amsterdam, pp 735–738
Sawicki HH (1950) Therapy of lupus erythematosus. Arch Dermatol Syphilol 61 [Suppl]: 906–908
Scarpa M, Rigo A, Maiorino M, Ursini F, Gregolin C (1984) Formation of a-tocopherol radical and recycling of a-tocopherol by ascorbate during peroxidation of phosphatidylcholine liposomes. Biochim Biophys Acta 801: 215–219
Schade W (1952) Zur Therapie der Psoriasis vulgaris. Hautarzt 3: 373–374
Schaich KM, Borg DC (1988) Fenton reactions in lipid phases. Lipids 23: 570–579
Schalkwjik J, van den Berg WB, van de Putte LBA, Joosten LAB (1986) An experimental model for hydrogen peroxide induced tissue damage. Arthritis Rheum 29: 532–538
Schallreuter KU (1988) Freie Radikale an der Oberfläche der lebenden Haut. Z Phys Med Baln Med Klim 17: 283
Schallreuter KU, Pittelkow MR (1987) Anthralin inhibits elevated levels of thioredoxin reductase in psoriasis. Arch Dermatol 123: 1494–1498
Schallreuter KU, Witkop CJ (1988) Thioredoxin reductase activity in Hermansky-Pudlak syndrome: a method for identification of putative heterozygotes. J Invest Dermatol 90: 372–377
Schallreuter KU, Wood JM (1986) The role of thioredoxin reductase in the reduction of free radicals at the surface of the epidermis. Biochem Biophys Res Commun 136: 630–637
Schallreuter KU, Wood JM (1987) Azelaic acid as a competitive inhibitor of thioredoxin reductase in human melanoma cells. Cancer Lett 36: 297–305
Schallreuter KU, Wood JM (1989) Free radical reduction in the human epidermis. Free Radic Biol Med 6: 519–532
Schallreuter KU, Wood JW (1990) A possible mechanism of action for azelaic acid in the human epidermis. Arch Dermatol Res 282: 168–171
Schallreuter KU, Pittelkow MR, Wood JM (1986a) Free radical reduction by thioredoxin reductase at the surface of normal and vitiliginous human keratinocytes. J Invest Dermatol 87: 728–732
Schallreuter KU, Pittelkow MR, Gleason FK, Wood JM (1986b) The role of calcium in the regulation of free radical reduction by thioredoxin reductase at the surface of the skin. J Inorg Biochem 28: 227–238
Schallreuter KU, Hordinsky MK, Wood JM (1987) Thioredoxin reductase. Arch Dermatol 123: 615–619
Schallreuter KU, Sundaram C, Köster W (1989) The effect of UVA and UVB on the activity of thioredoxin reductase in guinea pig and human skin. J Invest Dermatol 92: 513
Schallreuter KU, Pittelkow MR, Wood JM (1990a) The regulation of melanogenesis by the redox status of the epidermis. Arch Dermatol Res 281: 560–561
Schallreuter KU, Pittelkow MR, Wood JM (1991) Defects in antioxidant defense and calcium transport in the epidermis of xeroderma pigmentosum patients. Arch Dermatol Res 283: 449–455
Schaltegger A, Krebs A (1984) Oxidation behaviour of dithranol. Arch Dermatol Res 276: 274–275
Schalwijk J, van den Berg WB, van de Putte LBA, Joosten LAB (1986) An experimental model for hydrogen peroxide induced tissue damage. Arthritis Rheum 29: 532–538
Schauder S, Mahrle G (1982) Kombinierte Einstundentherapie der Psoriasis mit Anthralin und UV Licht. Hautarzt 33: 206–209
Schiltz JR, Michel B, Papay R (1979) Appearance of pemphigus acantholytic factor in human skin cultured with pemphigus antibody. J Invest Dermatol 73: 575–581
Schirren SM (1971) Skin lesions caused by trichlorethylen in a metalworking plant. Berufsdermatosen 19: 240–254
Schmidt R, Pautrat G, Michel S, Cavey M, Gazith J, Dalbiez C, Reichert U (1985) High

yield purification of plasma membranes from transformed human keratinocytes in culture. J Invest Dermatol 85: 50–53

Schmidt RJ, Khan L, Chung LY (1990) Are free radicals and not quinones the haptenic species derived from urushiols and other contact allergenic mono- and dihydric alkylbenzenes? The significance of NADH, glutathione, and redox cycling in the skin. Arch Dermatol Res 282: 56–64

Schmidt RJ, Chung LY (1990) Oxidative stress occurs in skin in the preimmunologic phase of the cell mediated immune response to allergenic 2,4-dinitrohalobenzenes. Free Radic Biol Med 9 (Suppl 1): 147

Scholich H, Murphy ME, Sies H (1989) Antioxidant activity of dihydrolipoate against microsomal lipid peroxidation and its dependence on alpha-tocopherol. Biochim Biophys Acta 1101: 256–261

Scholz RW, Todhunter DA, Cook LS (1981 a) Selenium content and glutathione peroxidase activity in tissues of young cattle fed supplemented whole milk diets. Am J Vet Res 42: 1718-1723

Scholz RW, Cook LS, Todhunter DA (1981 b) Distribution of selenium dependent and non-selenium dependent glutathione peroxidase activity in tissues of young cattle. Am J Vet Res 42: 1724–1729

Schöneich C, Asmus KD, Dillinger U, von Bruchhausen F (1989) Thiyl radical attack on polyunsaturated fatty acids: a possible route to lipid peroxidation. Biochem Biophys Res Commun 161: 113–120

Schonhofer P, Peters H, Karzel K, Dinnendahl V, Westhofen P (1974) Influence of antiphlogistic drugs on prostaglandin E stimulated cyclic AMP levels and glycosaminoglycan synthesis in fibroblast tissue cultures. Pol J Pharmacol 26: 51–60

Schoonderwoerd SA, Beijersbergen van Henegouwen GMJ, Van Belkum S (1989) In-vivo photodegradation of chlorpromazin. Photochem Photobiol 50: 659–664

Schothorst AA, De Haas CAC, Suurmond D (1980) Photochemical damage to skin fibroblasts caused by protoporphyrin and violet light. Arch Dermatol Res 268: 31–42

Schothorst AA, Ennings IC, Schouten R, Zeeland A, Filon R, Simons JWIM (1985) Survival and induction of endonuclease sensitive sites (E.S.S) after irradiation with monochromatic ultraviolet light of human fibroblasts and keratinocytes. Photochem Photobiol 41: 92S

Schreiber J, Foureman GL, Hughes MF, Mason RP, Eling TE (1989) Detection of glutathione thiyl free radical catalyzed by prostaglandin H synthase present in keratinocytes. J Biol Chem 264: 7936–7943

Schröder JM, Christophers E (1985) Transient absence of C5a specific neutrophil function in inflammatory disorders of the skin. J Invest Dermatol 85: 194–198

Schröder JM, Kosfeld U, Christophers E (1985) Multifunctional inhibition by anthralin in nonstimulated and chemotactic factor stimulated human neutrophils. J Invest Dermatol 85: 30–34

Schroot B, Brown C (1986) Free radicals in skin exposed to dithranol and its derivatives. Arzneimittelforschung/Drug Res 36: 1253–1255

Schulz EJ, Whiting DA (1976) Treatment of erythema nodosum and nodular vasculitis with potassium iodide. Br J Dermatol 94: 75–78

Schwartz E, Cruickshank FA, Perlish JS, Fleischmajer R (1989) Alterations in dermal collagen in ultraviolet irradiated hairless mice. J Invest Dermatol 93: 142–146

Schwarz M, Peres G, Kunz W, Fürstenberger G, Kittstein W, Marks F (1984) On the role of superoxide anion radicals in skin tumor promotion. Carcinogenesis 5: 1663–1670

Schwarz RI, Kleinman P, Owens N (1987) Ascorbate can act as an inducer of the collagen pathway because most steps are tightly coupled. Ann NY Acad Sci 498: 172–185

Schwegler F, Holzmann H (1990) Ratschläge für Kranke mit progressiver systemischer Sklerodermie. Dtsch Dermatol 38: 958–960

Schweizer J, Loehrke H, Edler L, Goertler K (1987) Benzoyl peroxide promotes the formation of melanotic tumors in the skin of 7,12-dimethylbenz[*a*]anthracene initiated Syrian golden hamsters.

Scott G (1985) Antioxidants in-vitro and in-vivo. Chem Br: 648–653

Scott JA, Rabito CA (1988) Oxygen radicals and plasma membrane potential. Free Radic Biol Med 5: 237–246

Scott JA, Fischman AJ, Khaw BA, Homcy CJ, Rabito CA (1987) Free radical mediated membrane depolarization in renal and cardiac cells. Biochim Biophys Acta 899: 76–82

Scott JA, Fischmann AJ, Homcy CJ, Fallow JT, Khaw BA, Peto CA, Rabito CA (1989) Morphologic and functional correlates of plasma membrane injury during oxidant exposure. Free Radic Biol Med 6: 361–367

Scudder PR, Stocks J, Dormandy TL (1976) The relationship between superoxide dismutase activity and erythrocyte copper levels in normal subjects and in patients with rheumatoid arthritis. Clin Chim Acta 69: 397–403

Sealy RC (1984) Radicals in melanin biochemistry. Methods Enzymol 105: 479–483

Sealy RC, Felix CC, Hyde JS, Swartz HM (1980) Structure and reactivity of melanins: influence of free radicals and metal ions. Free Radic Biol 4: 209–259

Sealy RC, Sarna T, Wanner EJ, Reszka K (1984) Photosensitization of melanin: an electron spin resonance study of sensitized radical production and oxygen consumption. Photochem Photobiol 39: 453–459

Sedgwick JB, Bergstresser PR, Hurd ER (1980) Increased granulocyte adherence in psoriasis and psoriatic arthritis. J Invest Dermatol 74: 81–84

Sedwick JB, Bergstresser PR, Hurd ER (1981) Increased superoxide generation by normal granulocytes incubated in sera from patients with psoriasis. J Invest Dermatol 76: 158–163

Segal A, Katz C, Van Duuren BL (1971) Structure and tumor promoting activity of anthralin (1.8-dihydroxy -9-anthrone) and related compounds. J Med Chem 14: 1152–1154

Sehgal VN, Vadiraj SN, Rege VL, Beohar PC (1972) Dystrophic epidermolysis bullosa in a family. Dermatologica 144: 27–34

Senff H, Köllner A, Engelmann L, Woort-Menker M, Mensing H, Kunze J (1990) L-Tryptophan-induziertes Eosinophilie-Myalgie-Syndrom unter dem Bild einer diffusen Fasciitis mit Eosinophilie. Hautarzt 41: 578–582

Serhan CN, Radin A, Smolen JE, Korchak H, Samuelson B, Weissmann G (1982) Leukotriene B4 is a complete secretagogue in human neutrophils: a kinetic analysis. Biochem Biophys Res Commun 107: 1006–1012

Serri F, Bartoli GM, Seccia A, Borello S, Galeotti T (1979) Age related mitochondrial lipoperoxidation in human skin. J Invest Dermatol 73: 123–125

Sestili P, Piedimonte G, Flaminio C, Cantoni O (1986) Induction of DNA breakage and suppression of DNA synthesis by the OH radical generated in a Fenton like reaction. Biochem Int 12: 493–501

Shaham M, Becker Y, Cohen M (1980) A diffusable clastogenic factor in ataxia telangiectasia. Cytogenet Cell Genet 27: 155–161

Shalita A (1974) Genesis of free fatty acids. J Invest Dermatol 62: 332–336

Shamma'a MH, Nasrallah SM, Al-Khalidi UAS (1973) Serum xanthine oxidase: an experience with 2000 patients. Dig Dis 18: 15–22

Shanahan WR, Korn JH (1982) Cytotoxic activity of sera from scleroderma and other connective tissue disease. Rheumatology 25: 1391–1395

Sharkey P, Eedy DJ, Burrows D, McCaigue MD, Bell AL (1991) A possible role for superoxide production in the pathogenesis of contact dermatitis. Acta Derm Venereol (Stockh) 71: 156–159

Shea CR, Wimberly J, Hasan T (1986) Mitochondrial phototoxicity sensitized by doxycycline in cultured human carcinoma cells. J Invest Dermatol 87: 338–342

Shekelle RB, Liu S, Raynor WJ, Lepper M, Maliza C, Rossof AH (1981) Dietary vitamin A and risk of cancer in the western electric study. Lancet ii: 1185–1189

Shelley WB (1974) p-Cresol: cause of ink-induced hair depigmentation in mice. Br J Dermatol 90: 169–174

Sherer GK, Jackson BB, Leroy EC (1981) Chromosome breakage and sister chromatide exchange frequencies in scleroderma. Arthritis Rheum 24: 1409–1413

Shimizu K, Kasai R, Yamamuro T, Hosokawa M, Takeshita S, Takeda T (1981) Amyloid deposition in articular structures of AR senescent mice. Arthritis Rheum 24: 1540–1544

Shingu M (1987) Complement activation and vascular damage by hydrogen peroxide. In: Hayaishi O, Imamura S, Miyachi Y (eds) The biological role of reactive oxygen species in skin. Elsevier, New York, pp 189–195

Shingu M, Oribe M, Todoroki T, Tatsukawa K, Tomo-Oka K, Yasuda M, Nobunaga M (1983) Serum factors from patients with systemic lupus erythematosus enhancing superoxide generation by normal neutrophils. J Invest Dermatol 81: 212–215

Shinskii GE, Telegina KA, Shephovtsova VV (1962) Experience with vitamin E in the treatment of erythematosus. Vestn Dermatol Venereol 36: 64–68 (in Russian)

Shirahama T, Cohen AS (1974) Blockage of amyloid induction by colchicine in an animal model. J Exp Med 140: 1102–1105

Shiraishi N, Aona K, Utsumi K (1983) Increased metallothionein content in rat liver induced by X irradiation and exposure to high oxygen tension. Radiat Res 95: 298–302

Shiratori T (1974) Uptake, storage and excretion of chylomicra-bound 3H-alpha-tocopherol by the skin of the rat. Life Sci 14: 929–935

Shulman ST, Rowley AH (1986) Does Kawasaki disease have a retroviral aetiology? Lancet ii: 545–546

Shupack JL (1976) The skin as a target organ for systemic agents. In: Drill VA, Lazar P (eds) Cutaneous toxicity. Academic, New York, pp 43–52

Siedentopf H, Krüger A (1949) Die Wirkung hoher Vitamin E (alpha-Tocopherol) Gaben auf Gefäßerkrankungen, im besonderen auf das Ulcus cruris. Med Klin 1060–1062

Sies H, Murhy ME (1991) Role of tocopherol in the protection of biological systems against oxidative damage. J Photochem Photobiol B: Biol., 8: 211–224

Sies H (ed) (1991) Oxidative stress: introduction. In: Oxidative stress oxidants and antioxidants. Academic, London, pp XV–XXII

Sies H, Cadenas E (1983) Biological basis of detoxification of oxygen free radicals. In: Caldwell J, Jacoby WB (eds) Biological basis of detoxification. Academic, New York, pp 181–211

Sies H, Mehlhorn R (1986) Mutagenicity of nitroxide free radicals. Arch Biochem Biophys 251: 393–396

Sies H, Wendel A, Bors W (1982) Metabolism of organic hydroperoxides. In: Caldwell J, Jacoby W (eds) Metabolic basis of detoxication. Academic, New York, pp 307–321

Silver IA (1972) Oxygen tension and epithelialization. In: Maibach HI, Rovee DT (eds) Epidermal wound healing. Yearbook, Chicago, pp 291–305

Silvers SH (1950) Chronic lupus erythematosus successfully treated with vitamin E. Arch Dermatol Syphilol 62 [Suppl]: 163

Silverstein JL, Steen VD, Medsger TA, Falanga V (1988) Cutaneous hypoxia in patients with systemic sclerosis (scleroderma). Arch Dermatol 124: 1379–1382

Simic MG (1981) Vitamin E radicals. In: Rodger MA, Powers EL (eds) Oxygen and oxyradicals in chemistry and biology. Academic, New York, pp 109–118

Simon RH, Scoggin CH, Patterson D (1981) Hydrogen peroxide causes the fatal injury to human fibroblasts exposed to oxygen radicals. J Biol Chem 256: 7181–7186

Sinet PM (1982) Metabolism of oxygen derivatives in Down syndrome. Ann NY Acad Sci 396: 83–94

Singer KH, Sawka NJ, Samowitz HR, Lazzarus GS (1980) Proteinase activation: a mechanism for cellular dyshesion in pemphigus. J Invest Dermatol 74: 363–367

Singh N, Clausen J (1980) Different tissue responses of mixed function oxidases and detoxifying enzymes to aflatoxin B1 administration in the rat. Br J Exp Pathol 61: 611–616

Slaga TJ (1984) Multistage skin carcinogenesis. a useful model for the study of the chemoprevention of cancer. Acta Pharmacol Toxicol 55 (Suppl II): 107–125

Slaga TJ, Bracken WM (1977) The effects of antioxidants on skin tumor initiation and aryl hydrocarbon hydroxylase. Cancer Res 37: 1631–1635

Slaga TJ, Fischer SM, Nelson K, Gleason GL (1980) Studies on the mechanism of skin tumor promotion: evidence for several stages in promotion. Proc Natl Acad Sci USA 77: 3659–3663

Slaga TJ, Fischer SM, Weeks CE, Klein-Szanto AJP (1981 a) Cellular and biochemical mechanisms of mouse skin tumor promotors. Rev Biochem Toxicol 3: 231–281

Slaga TJ, Klein-Szanto AJP, Triplett LL, Yotti LP, Trosko JE (1981 b) Skin tumor promoting activity of benzoyl peroxide, a widely used free radical generating compound. Science 213: 1023–1025
Slaga TJ, Fischer SM, Weeks CE, Klein-Szanto AJP, Reinders J (1982) Studies on the mechanism involved in multistage carcinogenesis in mouse skin. J Cell Biochem 18: 99–119
Slater TF, Riley PA (1966) Photosensitization and lysosomal damage. Nature 209: 151–154
Smith CV, Anderson RE (1987) Methods for determination of lipid peroxidation in biological samples. Free Radic Biol Med 3: 341–344
Smith EB, Michener WM (1973) Vitamin E treatment of dermolytic bullous dermatosis. Arch Dermatol 108: 254–256
Smith JL (1897) The influence of pathological conditions on active absorption of oxygen by the lungs. J Physiol 22: 307–308
Smith JL (1899) The pathological effects due to increase of oxygen tension in the air breathed. J Physiol 24: 19–35
Smith LL (1985) Paraquat toxicity. Phil Trans R Soc Lond [B] 311: 647–657
Snell RS (1964) Monobenzylether of hydroquinone. Arch Dermatol 90: 63–70
Snyder DS (1975) Cutaneous effect of topical indomethacin, an inhibitor of prostaglandin synthesis, on UV-damaged skin. J Invest Dermatol 64: 322–325
Snyderman R (1983) Pharmacological manipulation of leucocyte chemotaxis. Am J Med: 10–18
Sobel JD, Haiman S, Obedeanu N, Meshulam T, Merzbach D (1977) Polymorphonuclear leucocyte function in Behcet's disease. J Clin Pathol 30: 250–253
Society of the Plastics Industry (1982) Commercial organic peroxide toxicological data. SPI Bulletin. SPI, New York
Sohal RS (1986) Relationship between oxygen metabolism, aging and development. Adv Free Radic Biol Med 2: 117–160
Solanki V, Murray AW (1982 a) The effect of ischemia on cyclic adenosine 3′,5′-monophosphate accumulation in mouse skin. J Invest Dermatol 78: 48–51
Solanki V, Murray AW (1982 b) Decreased accumulation of cyclic adenosine 3′,5′-monophosphate in "ischemic" skin after 12-O-tetradecanoyl-phorbol-13-acetate treatment. J Invest Dermatol 78: 264–266
Solanki V, Rana RS, Slaga TJ (1982) Diminution of mouse epidermal superoxide dismutase and catalase activities by tumor promotors. Carcinogenesis 2: 1141–1146
Solberg CO, Kalager T, Hill HR, Glette J (1982) Polymorphonuclear leukocyte function in bacterial and viral infections. Scand J Infect Dis 14: 11–18
Solic F (1977) The influence of some physical parameters on the quantity of free radicals in skeletal muscles. Period Biol 79: 47–54
Solic F (1979) The investigation of free radicals in stimulated muscles at different temperatures. Period Biol 81: 547–551
Sollberg S, Holzmann H (1987) Genetik der Sklerodermien. In: Holzmann H, Altmeyer P, Marsch WC, Vogel HG (eds) Dermatologie und Rheuma. Springer, Berlin Heidelberg New York, pp 241–252
Sollmann T (1918 a) Dichlorethylsulfide ("mustard gas"). I The influence of solvents, adsorbents and chemical antidotes on the severity of the human skin lesions. J Pharmacol Exp Ther 12: 303–317
Sollmann T (1918 b) Dichlorethylsulfide ("mustard gas"). II The question of induced hypersusceptibility of the skin. J Pharm Exp Ther 12: 319–321
Soloshenko EN (1989) The role of lipid peroxidation disorders in the pathogenesis of drug disease. Vestn Dermatol Venerol 11: 4–8 (in Russian)
Somani SM, Babu SR (1989) Toxocodynamics of sulfur mustard. Int J Clin Pharmacol Ther Toxicol 27: 419–435
Somiya K, Niwa Y, Fukami S, Puget K, Michelson AM (1987) Treatment with liposomal superoxide dismutase of patients with Kawasaki disease. In: Hayaishi O, Imamura S, Miyachi Y (eds) The biological role of reactive oxygen species in Skin. Elsevier, New York, pp 211–216

Song PA, Tapley JJ Jr (1979) Photochemistry and photobiology of psoralens. Photochem Photobiol 29: 1177–1197
Sönneborg A, Carlin G, Åkerlund B, Jarstrand C (1988) Increased production of malondialdehyde in patients with HIV infection. Scand J Infect Dis 20: 287–290
Sorensen JRJ (1984) Bis (3,5-diisopropylsalicylate) copper II, a potent radioprotectant with superoxide dismutase mimetic activity. J Med Chem 27: 1747–1749
Sparrow GP (1977) A connective tissue disorder similar to vinyl chloride disease in a patient exposed to perchlorethylene. Clin Exp Dermatol 2: 17–22
Spector A, Yan GZ, Huang RRC, McDermott MJ, Gascoyne PRC, Pigiet V (1988 a) The effect of H_2O_2 upon thioredoxin enriched lens epithelial cells. J Biol Chem 263: 4984–4990
Spector A, Huang RRC, Yan GZ, Wang RR (1988 b) Thioredoxin fragment 31–36 is reduced by dihydrolipoamide and reduces oxidized protein. Biochem Biophys Res Commun 150: 156–162
Speier C, Newburger PE (1986) Changes in superoxide dismutase, catalase and the glutathione cycle during induced myeloid differentiation. Arch Biochem Biophys 251: 551–557
Spencer MC (1961) Hydroquinone bleaching. Arch Dermatol 84: 131–134
Spengler J (1989) Toxikologische Prüfung und gesundheitliche Bewertung oxidativer Haarfärbemittel. Ärztl Kosmet 19: 41–53
Spiera H (1988) Scleroderma after silicone augmentation mammaplasty. JAMA 260: 236–238
Spisani S, Marangoni C, Traniello S (1984) Effect of anti-inflammatory drugs on leukocyte superoxide production by soluble and particulate stimuli. Biochem Int 9: 1–8
Sporn MB, Newton DL (1979) Chemoprevention of cancer with retinoids. Fed Proc 38: 2528–2534
Spurzem JR, Lockey JE (1984) Toxic oil syndrome. Arch Intern Med 144: 249–250
Staberg B, Wulf HC, Klemp P, Poulsen T, Brodthagen H (1983 a) The carcinogenic effect of UVA irradiation. J Invest Dermatol 81: 517–519
Staberg B, Wulf HC, Poulsen T, Klemp P, Brodthagen H (1983 b) Carcinogenic effect of sequential artificial sunlight and UVA irradiation in hairless mice. Arch Dermatol 119: 641–643
Stacey NH, Klaassen CD (1982) Effects of phospholipase A2 inhibitors on diethyl-maleate induced lipid peroxidation and cellular injury in isolated rat hepatocytes. J Toxicol Environ Health 9: 439–450
Stachow A, Jablonska S, Skiendzielewska A (1979) Biogenic amines derived from tryptophan in systemic and cutaneous scleroderma. Acta Derm Venereol (Stockh) 59: 1–5
Stadtmann ER (1990) Protein damage and repair. Free Radic Biol Med 9 (Suppl 1): 82
Steigleder GK (1952) Histochemische Untersuchungen im psoriatischen Herd über Oxydation, Reduktion und Lipidstoffwechsel. Arch Dermatol Syph 194: 296–307
Stary FE, Jindal SL, Murray RW (1975) Oxidation of α-lipoic acid. J Org Chem 40: 58
Stendahl O, Molin L, Dahlgren C (1978) The inhibition of polymorphonuclear leukocyte cytotoxicity by dapsone. A possible mechanism in the treatment of dermatitis herpetiformis. J Clin Invest 62: 214–220
Sterenborg HJCM, Van der Leun JC (1990) Tumorigenesis by long wavelenght UV-A source. Photochem Photobiol 51: 325–330
Stern RS, Lange R (1988) Non-melanoma skin cancer occuring in patients treated with PUVA five to ten years after first treatment. J Invest Dermatol 91: 120–124
Stern RS, Gange RW, Parrish JA, Tang SV, Arndt KA (1986) Contribution of topical tar oil to ultraviolet B phototherapy for psoriasis. J Am Acad Dermatol 14: 742–747
Sternberg EM, Van Woert MH, Young SN, Magnussen I, Baker H, Gauthier S, Osterland CK (1980) Development of a sleroderma like illness during therapy with L-5-hydroxytryptophan and carbidopa. N Engl J Med 303: 782–787
Sterry W (1987) Kolchizin in der Dermatologie. Z Hautkr 62: 697–704
Sterzi G (1950) La Vitamina E, con speciale riferimento alla dermatologia. Arch Ital Dermatol Sifilogr Venereol 23: 257–285

Stevens B, Perez SR, Small RD (1974) The photoperoxidation of unsaturated organic molecules – IX. Lipoic acid inhibition of rubrene autoperoxidation. Photochem Photobiol 19: 315–316

Stevens JB, Autor AP (1980) Proposed mechanism for neonatal rat tolerance to normobaric hyperoxia. Fed Proc 39: 3138–3143

Stevenson KJ, Pratt KJ, Carles C, Carne TJ, Danson MJ (1988) Detection of cellular lipoic acid: a modified gas chromatograph – mass spectrometric procedure. Fed Am Soc Exp Biol J 2: 8554

Stich HF, Hornby AP, Dunn BP (1986) Beta-carotene levels in exfoliated human mucosa cells following its oral administration. Cancer Lett 30: 133–141

Stillwell W, Bryant L (1983) Membrane permeability changes with vitamin A/vitamin E mixed bilayers. Biochim Biophys Acta 731: 483–486

Stocker R, Peterhans E (1989) Antioxidant properties of conjugated bilirubin and biliverdin: biologically relevant scavenging of hypochlorous acid. Free Radic Res Commun 6: 57–66

Stocker R, Yamamoto Y, McDonagh AF, Glazer AN, Ames BN (1987) Bilirubin is an antioxidant of possible physiological importance. Science 235: 1043–1046

Stocks J, Gutterridge JM, Sharp RJ, Dormandy TL (1974) The inhibition of lipid autoxidation by human serum and its relation to serum proteins and α-tocopherol. Clin Sci Mol Med 47: 223–232

Stöhr L, Altmeyer AP, Sessler MJ, Scharrer I, Helm EB, Holzmann H (1984) Chemiluminescenzmessung bei AIDS-, Lymphadenopathie- und Hämophiliepatienten. Z Hautkr 60: 1214–1223

Stohs SJ (1990) Oxidative stress induced by 2,3,7,8-tetrachlorodibenzo p-dioxin (TCDD). Free Rad Biol Med 9: 79–90

Stoien JD, Wang RJ (1974) Effect of near ultraviolet and visible light on mammalian cells in culture. II. Accumulation of toxic photoproducts in tissue culture medium by black light. Proc Natl Acad Sci USA 71: 3961–3965

Stokinger HE (1965) Ozone toxicoloty: a review of research and industrial experience, 1954–1964. Arch Environ Health 10: 719–731

Stokinger HE, Wagner WD, Wright PC (1956) Studies on ozone toxicity. I. Potentiating effects of exercise and tolerance development. Arch Ind Health 14: 158–162

Stone OJ (1971) What are the non-endocrine effects of iodides? Med Times 99: 143–151

Stone OJ (1988) Lupus erythematosus and rheumatoid arthritis – groups of defects in microdebridement (Polygenetic defects exceeding the fault tolerance threshold – a consequence of natural defense mechanism). Med Hypotheses 27: 327–332

Stone OJ (1991) Autoimmunity as a secondary phenomenon in scleroderma (and so called human adjuvant disease). Med Hypotheses 34: 127–130

Stone OJ, Willis CJ (1967) Iodide enhancement of inflammation – experimental with clinical correlation. Tex Rep Biol Med 25: 205–210

Stossel TP, Mason RJ, Smith AL (1974) Lipid peroxidation by human blood phagocytes. J Clin Invest 54: 638–644

Stout OM (1951) Pseudoxanthoma elasticum with retinal angioid streaking decidedly improved on tocopherol therapy. Arch Dermatol Syphil 63: 510–511

Stoyanovsky DA, Kagan VE, Packer L (1989) Iron binding to α-tocopherol containing phospholipd liposomes. Biochem Biophys Res Commun 160: 834–838

Straight RC, Spikes JD (1985) Photosensitized oxidation of biomolecules. In: Frimer AA (ed) Singlet oxygen, vol 4. CRC, Boca Raton, pp 91–143

Stratigos J, Kasimatis B, Panas E, Capetanakis J (1975) Biochimie du psoriasis: le cuivre et la ceruleoplasmine dans le serum des psoriasiques. Ann Dermatol Venerol 103: 584–587

Stratton K, Pathak MA (1968) Photoenhancement of the electron spin resonance signal from melanins. Arch Biochem Biophys 123: 477–483

Strelkoy RB, Chizhov A, Tokarey O, Tolstykh VN, Kucherenko NG (1983) Radioprotective action of a gas hypoxic mixture GHM-10 in multiple irradiations of rat and human skin. Radiobiologiia 23: 358–362 (in Russian)

Strickland PT (1986) Photocarcinogenesis by near-ultraviolet (UVA) radiation in Sencar mice. J Invest Dermatol 87: 272–275
Stüttgen G (1955) Zur Atmung und Glycolyse der normalen und krankhaft veränderten Haut. Arch Klin Exp Dermatol 201: 507–520
Stüttgen G, Betzler H, Witte W (1957) Zur Erfassung des Vitamin E im Serum und Hautfett mit Phosphormolybdänsäure. Arzneimittelforschung/Drug Res 7: 407–408
Stüttgen G, Schaefer E (1974) Vitamine und Haut. In: Stüttgen G, Schaefer E (eds) Funktionelle Dermatologie. Springer, Berlin Heidelberg New York, pp 78–79
Sugioka Y, Suzuki M, Sugioka K, Nakano M (1987) A ferriprotoporphyrin IX-chloroquine complex promotes membrane phospholipid peroxidation. FEBS Lett 223: 251–254
Sugiura K, Ueda H, Hirano K, Adachi T (1985a) Studies on superoxide dismutase in human skin (2). Contents of superoxide dismutase and lipoperoxide in normal human skin. Jpn J Dermatol 95: 1541–1545 (in Japanese)
Sugiura K, Ueda H, Hirano K, Adachi T (1985b) Studies on superoxide dismutase in human skin (1). Contents of superoxide dismutase determined by enzyme immonoassay. Jpn J Dermatol 95: 1535–1539 (in Japanese)
Sugiura K, Abe M, Inasaka H, Ueda H, Hirano K, Adachi T (1986a) Studies on superoxide dismutase in human skin (4). Contents of superoxide dismutase and lipid peroxide in keloid, hypertrophic scar and scar. Jpn J Dermatol 96: 171–174 (in Japanese)
Sugiura K, Iida Y, Ohgoshi H, Ueda H, Hirano K, Adachi T (1986b) Studies on superoxide dismutase in human skin (3). Contents of superoxide dismutase and lipid peroxides in skin of patients with atopic dermatitis. Jpn J Dermatol 96: 165–170 (in Japanese)
Sulzberger MB, Witten VH (1952) The effect of topically applied compound F in selected dermatoses. J Invest Dermatol 19: 101–104
Sumimoto H, Takeshige K, Minakami S (1984) Superoxide production of human polymorphonuclear leukocytes stimulated by leukotriene B4. Biochim Biophys Acta 803: 271–277
Summerfield M, Tudhope GR (1978) Studies with primaquine in-vitro: superoxide radical formation and oxidation of haemoglobin. Br J Clin Pharmacol 6: 319–323
Summerhayes IC, Lampidis TJ, Bernal LSD, Nadakavukaren JJ, Nadakavukaren KK, Shepherd EL, Chen LB (1982) Unusual retention of rhodamine 123 by mitochondria in muscle and carcinoma cells. Proc Natl Acad Sci USA 79: 5292–5296
Summers RW, Maves BV, Reeves RD, Arjes LJ, Oberley LW (1989) Irradiation increases superoxide dismutase in rat intestinal smooth muscle. Free Radic Biol Med 6: 261–270
Sun IL, Crane FL, Grebing C, Löw H (1984) Properties of a transplasma membrane electron transport system in HeLa cells. J Bioenerg Biomembr 16: 583–595
Sun IL, Crane FL, Grebing C, Löw H (1985) Transmembrane redox in control of cell growth. Exp Cell Res 156: 528–536
Sun Y, Oberley LW (1989) The inhibition of catalase by glutathione. Free Radic Biol Med 7: 595–602
Sundaram C, Köster W, Schallreuter KU (1990) The effect of UV radiation and sun blockers on free radical defense and guinea pig epidermis. Arch Dermatol Res 282: 526–531
Sung C, Mirabelli CK, Badger AM (1984) Effect of gold compounds on phorbol myristic acetate (PMA) activated superoxide (O_2-) production by mouse peritoneal macrophages. J Rheumatol 11: 153–157
Suthanthiran M, Anderson ME, Sharma VK, Meister A (1990) Glutathione regulates activation-dependent DNA synthesis in highly purified normal human T lymphocytes stimulated via the CD2 and CD3 antigens. Proc Natl Acad Sci USA 87: 3343–3347
Sutherland BM, Rice M, Wagner EK (1975) Xeroderma pigmentosum cells contain low levels of photoreactivating enzyme. Proc Natl Acad Sci USA 72: 103–107
Sutherland BM, Harber LC, Kochevar IE (1980) Pyrimidine dimer formation and repair in human skin. Cancer Res 40: 3181–3185
Sutherland JC, Griffin KP (1981) Absorption spectrum of DNA for wavelenghts greater than 300 nm. Radiat Res 86: 399–409
Sutton HD (1985) Efficiency of chelated iron compounds as catalysts for the Haber Weiss reaction. J Free Radic Biol Med 1: 195–202
Svingen BA, Powis GP, Appel PL, Scot M (1981) Protection against adriamycin induced

skin necrosis in the rat by dimethyl sulfoxide and a-tocopherol. Cancer Res 41: 3395–3399
Swanbeck G, Wennersten G (1974) Photohemolytic activity of tryptophan and phenylalanine metabolites. Acta Derm Venereol (Stockh) 54: 99–104
Sweeney GD, Jones KG, Cole FM, Basford D, Krestynski F (1979) Iron deficiency prevents liver toxicity of 2,3,7,8-tetrachlorodibenzo-p-dioxin. Science 204: 332–335
Sweet RD (1948) Vitamin E in Collagenoses. Lancet ii: 310–311
Symons MCR (ed) (1990) Electron spin resonance, vol 12A. A specialist periodical report. Royal Society of Chemistry, London
Tabuenca JM (1981) Toxic-allergic syndrome caused by ingestion of rapseed oil denaturated with aniline. Lancet ii: 567–568
Taffe BG, Kensler TW (1986) Generation of free radicals from organic hydroperoxide tumor promotors by mouse epidermal cells. Pharmacologist 28: 175
Taffe BG, Takahashi N, Kensler TW, Mason RP (1987) Generation of free radicals from organic hydroperoxide tumor promoters in isolated mouse keratinocytes. J Biol Chem 262: 12143–12149
Takagi S, Wakabayashi T, Mizuno N (1989) Inhibition of oxidative phosphorylation in mitochondria is the key factor in demethylchlortetracycline (DMCT) phototoxic reaction. J Invest Dermatol 92: 526
Takahama U (1985) $O_2^{-\cdot}$ dependent and independent photooxidation of quercetin in the presence and absence of riboflavin and effects of ascorbate on the photooxidation. Photochem Photobiol 42: 89–91
Takahama U (1987) Oxidation products of kaempherol by superoxide anion radical. Plant Cell Physiol 28: 953–957
Takamori K, Thorpe JM, Goldsmith LA (1980) Skin sulfhydryl oxidase. Purification and some properties. Biochim Biophys Acta 615: 309–323
Takamori K, Ikeda S, Naito K, Ogawa H (1985) Proteases are responsible for blister formation in recessive dystrophic epidermolysis bullosa and epidermolysis bullosa simplex. Br J Dermatol 112: 533–538
Takayanagi R, Takeshige K, Minakamai S (1980) NADA- and NADPH-dependent lipid peroxidation in bovine heart submitochondrial particles. Biochem J 192: 853–860
Takeda T, Hosokawa M, Takeshita S, Irino M, Higuchi K, Matsushita T, Tomita Y, Yasuhira K, Hamamoto H, Shimizu K, Ishii M, Yamamuro T (1981) A new murine model of accelerated senescence. Mech Ageing Dev 17: 183–189
Takeuchi A, Kobayashi K, Mizushima Y (1978) Mechanism of hyperchemotaxis of neutrophils from patients with Behcet's disease: In: Behcet's Disease Research Committee of Japan. Studies on etiology, treatment and prevention of Behcet's disease. Japanese Ministry of Welfare, Tokyo
Tan CY, Statham B, Marks R, Payne PA (1982) Skin thickness measurement by pulsed ultrasound: its reproducibility, validation and variability. Br J Dermatol 106: 657–667
Tanaka T (1979) Skin damage and its prevention from lipoperoxide. Vitamins (Kyoto) 53: 577–586 (in Japanese)
Tanaka T, Hayakawa R (1986) Lipid peroxides in cosmetic products and their effect to irritate the skin. J Clin Biochem Nutr 1: 201–207
Tanenbaum L, Parrish JA (1975) Tar phototoxicity and phototherapy for psoriasis. Arch Dermatol 111: 467–470
Tanenier H, Jesionek A (1903) Therapeutische Versuche mit fluoreszierenden Stoffen. Münch Med Wochenschr 1: 2042–2044
Tankov IUP (1989) The effect of metabolites of the epidermal barrier on the development of phototoxic dermatitis. Vestn Dermatol Venerol 12: 18–20 (in Russian)
Tappel AL (1980) Measurement of and protection from in-vivo lipid peroxidation. Free Radic Biol 4: 2–47
Tarakhovsky AM, Umansky VJ, Shlyakhovenko VA, Balitsky KP (1985) Redox dependent activation of 5′-nucleosidase in rat liver plasma membranes. FEBS Lett 189: 338–340
Taren DL, Chvapil M, Weber CW (1987) Increasing the breaking strenght of wounds exposed to preoperative irradiation using vitamin E. Int J Vitam Nutr 57: 133–137

Tauber A, Babior B (1978) $O_2^{-}\cdot$ and host defense: the production and fate of $O_2^{-}\cdot$ in neutrophils. Photochem Photobiol 28: 701–709

Taniguchi S, Kono T, Mizuno N, Ishii M, Matsui-Yuasa I, Otani S, Hamada T (1991) Effects of butylated hydroxyanisol on ornithine decarboxylase activity and its gene expression induced by phorbol ester tumor promotor. J Invest Dermatol 96: 289–291

Taylor WG, Camalier RF (1982) Modulation of epithelial cells proliferation in culture by dissolved oxygen. J Cell Physiol 111: 21–27

Tegner E, Rorsman H, Rosengren E (1983) 5-S-cysteinyldopa and pigment response to UVA light. Acta Derm Venereol (Stockh) 63: 21–25

Terada H (1981) The interaction of highly active uncouplers with mitochondria. Biochim Biophys Acta 639: 225–242

Terakado M, Yamazaki M, Tsujimoto T, Nagashima K, Ogawa J, Fujita Y, Sugiya H, Sakai T, Furuyama S (1984) Lipid peroxidation as a possible cause of benzoyl peroxide toxicity in rabbit dental pulp – a microsomal lipid peroxidation in-vitro. J Dent Res 63: 901–905

Terao J, Yamauchi R, Murakami H, Matsushita S (1980) Inhibitory effects of tocopherols and β-carotene on singlet oxygen initiated photooxidation of methyl linoleat and soybean oil. J Food Process Preserv 4: 79–93

Test ST, Weiss SJ (1986) The generation of utilization of chlorinated oxidants by human neutrophils. Adv Free Radic Biol Med 2: 91–116

Texier L, Gauthier Y, Gauthier O, Delaunay M, Boineau D (1975) Hypodermite sclerodermiforme lombo-fessiere induite par injection de vitamine K1. Defense immunitaire conservee, evolution spontanement regressive. Bull Soc Fr Dermatol Syphiligr 82: 448–449

Tezuka T, Takahashi M (1987) Stratum corneum membrane proteins in newborn rat as scavengers of lipid peroxide. In: Hayaishi O, Imamura S, Miyachi Y (eds) The biological role of reactive oxygen species in skin. Elsevier, New York, pp 125–132

Thayer WS (1986) Role of catalase in metabolism of hydrogen peroxide by the perfused rat heart. FEBS Lett 202: 137–140

Theron A, Anderson R (1985) Investigation of the protective effects of the antioxidants ascorbate, cysteine and dapsone on the phagocyte mediated oxidative inactivation of human a1-protease inhibitor in-vitro. Am Rev Respir Dis 132: 1049–1054

Thomas CE, Morehouse LA, Aust SD (1985) Ferritin and superoxide dependent lipid peroxidation. J Biol Chem 260: 3275–3280

Thompson S, Slaga TJ (1976) Mouse epidermal aryl hydrocarbon hydroxylase. J Invest Dermatol 66: 108–111

Thor H, Smith MT, Hartzell P, Bellomo G, Jewell SA, Orrenius S (1982) The metabolism of menadione (2-methyl-1,4-naphtoquinone) by isolated hepatocytes. J Biol Chem 257: 12419–12425

Thornalley PJ (1985) Monosacharide autoxidation in health and disease. Environ Health Perspect 64: 297–307

Thornalley PJ, Vasak M (1985) Possible role for metallothionein in protection against radiation induced oxidative stress. Kinetics and mechanism of its reaction with superoxide and hydroxyl radicals. Biochim Biophys Acta 827: 36–44

Thürlimann W, Harms M (1982) Sklerodermiforme Veränderungen nach PUVA Therapie bei Vitiligo. Dermatologica 164: 305–313

Till GO, Hatherill JR, Tourtellotte WW, Lutz MJ, Ward PA (1985) Lipid peroxidation and acute lung injury after thermal trauma to skin. J Pathol 119: 376–384

Till GO, Guilds LS, Friedl HP, Ward PA (1988) Antioxidant treatment in experimental thermal injury. Society for Free Radical Research, winter meeting on free radicals in medicine; current status of antioxidant therapy, Paris, 9–10 Dec 1988

Tkac A (1975) Transition metal chelates and free radical redox processes. Int J Radiat Phys Chem 7: 457–474

Tkac A (1987) Investigation of hydrogen and electron transfer reactions of antioxidants by electron spin resonance. In: Scott G (ed) Developments in polymer stabilisation, vol 8. Elsevier, London, pp 61–178

Tkac A (1988) Alternating reactivity of free radicals coordinated to chelated transition metals and to hemoproteins. Collection Czech Chem Commun 53: 2429–2464

Toback AC, Kohn SR (1989) Manifesto of space medicine: the next dermatologic frontier. J Am Acad Dermatol 20: 489–495

Toda K, Miyachi Y, Nesumi N, Konishi J, Imamura S (1986) UVB/PUVA induced supression of human natural killer activity is reduced by superoxide dismutase and/or interleukin 2 in-vitro. J Invest Dermatol 86: 519–522

Tojo K, Lee AC (1987) Bioconversion of a provitamin to vitamins C and E in skin. J Soc Cosmet Chem 38: 333–339

Tolmasoff JM, Ono T, Cutler R (1980) Superoxide dismutase: correlation with life-span and specific metabolic rate in primate species. Proc Natl Acad Sci USA 77: 2777–2781

Tomita Y, Hariu A, Kato C, Seiji M (1984) Radical production during tyrosinase reaction, DOPA-melanin formation, and photoirradiation of DOPA-melanin. J Invest Dermatol 82: 573–576

Torel J, Cillard J, Cillard P (1986) Antioxidant activity of flavonoids and reactivity with peroxy radical. Phytochemistry 25: 383–385

Torinuki W, Miura T, Seiji M (1980) Lysosome destruction and lipoperoxide formation due to active oxygen generated from haematoporphyrin and UV irradiation. Br J Dermatol 102: 17–27

Törmä H, Valquist A (1984) Vitamin A uptake by human skin in-vitro. Arch Dermatol Res 276: 390–395

Tosca N, Stratigos JD (1988) Possible pathogenetic mechanisms in allergic cutaneous vasculitis. Int J Dermatol 27: 291–296

Trautinger F, Knobler RM, Granninger W, Macheiner W, Neumann R, Micksche M (1990a) The ability of neutrophils to release oxygen free radicals is reduced by extracorporal photochemotherapy (photopheresis). Arch Dermatol Res 281: 563 A

Trautinger F, Grunewald C, Trenz A, Pittermann W, Kokoschka EM (1990b) Influence of natural vitamin E and UV radiation on dermal collagen content in hairless mice. XVIII Jahrestagung der Arbeitsgemeinschaft Dermatologische Forschung. Mannheim, 9–11 November 1990

Trenam CW, Dabbagh AJ, Morris CJ, Blake DR (1991) Skin inflammation induced by reactive oxygen species (ROS): an in-vivo model. Br J Dermatol 125: 325–329

Triggs WJ, Willmore LJ (1984) In-vivo lipid peroxidation in rat brain following intracortical Fe2+ injection. J Neurochem 42: 976–980

Troll W, Wiesner R (1985) The role of oxygen radicals as a possible mechanism of tumor promotion. Annu Rev Pharmacol Toxicol 25: 509–528

Trozak DJ, Gould WM (1984) Cocaine abusus and connective tissue disease. J Am Acad Dermatol 10: 525

Trump BJ, Arstila AU (eds) (1975) Pathobiology of cell membranes, vol 1. Academic, New York

Trump BJ, Arstila AU (eds) (1980) Pathobiology of cell membranes, vol 2. Academic, New York

Tso POP, Caspary WJ, Lorentzen RJ (1980) The involvement of free radicals in chemical carcinogenesis. Free Radic Biol 4: 251–303

Tsuchiya J, Yamada T, Niki E, Kamiya Y (1985) Interaction of galvinoxyl radical with ascorbic acid, cysteine, and glutathione in homogenous solution and in aqueous dispersion. Bull Chem Soc Jpn 58: 326–330

Tsujimoto Y, Matsushima K, Miura T (1983) Benzoylperoxide dependent formation of peroxides in microsomal fraction of rabbit dental pulp. Nihon Univ J Oral Sci 9: 407–418

Tsujimoto Y, Nagashima K, Yamazaki M, Furuyama S (1988) Inhibition of benzoyl peroxide/Cu^{2+} dependent lipid peroxidation by manganese. Int J Biochem 20: 591–594

Turner SR, Campbell JA, Lynn WS (1975) Polymorphonuclear leucocyte chemotaxis toward oxidized lipid components of cell membranes. J Exp Med 141: 1437–1441

Turrens JF, Freeman BA, Crapo JD (1982a) Hyperoxia increases H_2O_2 release by lung mitochondria and microsomes. Arch Biochem Biophys 217: 411–421

Turrens JF, Freeman BA, Levitt JG, Crapo JD (1982b) The effect of hyperoxia on superoxide production by lung submitochondrial particles. Arch Biochem Biophys 217: 401–410

Tuschl H, Kovac R, Wolf A, Smolen JS (1984) SCE frequencies in lymphocytes of systemic lupus erythematosus patients. Mutat Res 128: 167–172

Tuveson RW, Sammartano LJ (1986) Sensitivity of hemA mutant Escherichia coli cells to inactivation by near ultraviolet light depends on the level of supplementation with delta-aminolevulinic acid. Photochem Photobiol 43: 621–626

Tyrrell RM, Pidoux M (1986) Endogenous glutathione protects human skin fibroblasts against the cytotoxic action of UVB, UVA and near visible radiations. Photochem Photobiol 42: 561–564

Tyrrell RM, Pidoux M (1988) Correlation between endogenous glutathione content and sensitivity of cultured human skin cells to radiation at defined wavelenghts in the solar ultraviolet range. Photochem Photobiol 47: 405–412

Tyrrell RM, Pidoux M (1989) Singlet oxygen involvement in the inactivation of cultured human fibroblasts by UVA (334 nm, 365 nm) and near visible (405 nm) radiations. Photochem Photobiol 49: 407–412

Tyrrell RM, Werfelli P, Moraes EC (1984) Lethal action of ultraviolet and visible (blue-violet) radiations at defined wavelenghts on human lymphoblastoid cells: action spectra and interaction sites. Photochem Photobiol 39: 183–189

Tyrrell RM, Keyse SM, Lautier D, Applegate L (1990) Inducible cellular defense against oxidative stress in cultured human cells. Free Radic Biol Med 9 (Suppl 1): 1

Ueda H, Hayakawa R, Izawa Y (1971) Significance of lipid peroxides in atopic dermatitis. Jpn J Dermatol 81: 736–741 (in Japanese)

Ueda K, Kobayashi S, Morita J, Komano T (1985) Site specific DNA damage caused by lipid peroxidation products. Biochim Biophys Acta 824: 341–348

Uito J, Jimenez S (1990) Fibrotic skin diseases. Arch Dermatol 126: 661–664

Unna PG (1916) Cignolin als Heilmittel der Psoriasis. Dermatol Wochenschr 7: 150–163

Unna PG, Golodetz L (1910) Zur Chemie der Haut VI. Dermatol Monatshefte 50: 451–460

Urano S, Yano K, Matsuo M (1988) Membrane stabilizing effect of vitamin E: effect of α-tocopherol and its model compounds on fluidity of lecithin liposomes. Biochem Biophys Res Commun 10: 469–475

Urbach F (1989) Potential effects of altered solar ultraviolet radiation on human skin cancer. Photochem Photobiol 50: 507–513

Vahlquist A, Lee JB, Michaelsson G, Rollman O (1982) Vitamin A in human skin: II concentrations of carotene, retinol and dehydroretinol in various components or normal skin. J Invest Dermatol 79: 94–97

Valerie K, Delers A, Bruck C, Thiriart C, Rosenberg H, Debouck C, Rosenberg M (1988) Activation of human immunodeficiency virus type 1 by DNA damage in human cells. Nature 333: 78–81

Van Baar HMJ, van de Kerkhof PCM, Schalkwijk J, Mier PD (1987 a) Cutaneous superoxide dismutase activity in psoriasis. Br J Dermatol 116: 462–463

Van Baar HMJ, van de Kerkhof PCM, Mier PD, Happle R (1987 b) Tetracyclines are potent scavengers of the superoxide radical. Br J Dermatol 117: 131–134

Van Balgooy JNA, Roberts E (1979) Superoxide dismutase in normal and malignant tissues in different species. Comp Biochem Physiol 62: 263–268

Van Duuren BL, Nelson N, Orris L, Palmes ED, Schmitt FL (1963) Carcinogenicity of epoxides, lactones, and peroxy compounds. J Natl Cancer Inst 31: 41–55

Van Dyke K, Van Dyke CV (1986) Cellular chemiluminescence associated with disease states. Methods Enzymol 133: 493–507

Vanhoutee PM (1989) Endothelium and control of vascular function. Hypertension 13: 658–667

Van Scott EJ, Flesch P (1954) Sulfhydryl groups and disulfide linkages in normal and pathological keratinization. Arch Dermatol Syphilol 70: 141–154

Van Weelden H, van der Leun JC (1985) Carcinogenesis by UV-A in pigmented and albino hairless mice. Photochem Photobiol 41: 109S

Van Weelden H, de Gruijl FR, van der Leun JC (1983) Tumors induced by UVA in mice. Photochem Photobiol 37: S79

Van Weelden H, Velthuis PJ, Baart de la Faille H (1989) Light induced skin lesions in lupus erythematosus. Arch Dermatol Res 281: 470–474

Varma SD, Srivastava VK, Richards RD (1982) Photoprotection in lens and cataract formation: prevention role of superoxide dismutase, catalase and vitamin C. Ophthalmic Res 14: 157–175

Varshavsky A (1981) Phorbol ester dramatically increase incidence of methotrexate resistant mouse cells: possible mechanisms and relevance to tumor promotion. Cell 25: 561–573

Vatassery GT, Brin MF, Fahn S, Kayden HJ, Traber MG (1988) Effect of high doses of dietary vitamin E on the concentrations of vitamin E in several brain regions, plasma, liver, and adipose tissue of rats. J Neurochem 51: 621–623

Vehaskari VM, Chang CTC, Stevens JK, Robson AM (1984) The effects of polycations on vascular permeability in the rat. A proposed role for charge sites. J Clin Invest 73: 1053–1061

Veltman G (1980) Klinische Befunde und arbeitsmedizinische Aspekte der Vinylchlorid Krankheit. Dermatol Monatsschr 166: 705–712

Verbenko EV, Ezhova MN, Varfolomeev VN (1986) Investigation of blood and tissue paramagnetic centeres in some dermatoses. Vestn Dermatol Venerol 2: 4–7 (in Russian)

Vesely DL, Rovere LE, Levey GS (1977) Activation of guanylate cyclase by streptozotocin and *l*-methyl-1-nitrosourea. Cancer Res 37: 28–31

Vesely D, Watson B, Levey GS (1979) Activation of liver guanylate cyclase by paraquat: possible role of superoxide anion. J Pharmacol Exp Ther 209: 162–164

Vidlakova M, Erazimova J, Horki J, Placer Z (1972) Relationship of serum antioxidative activity to tocopherol and serum inhibitor of lipid peroxidation. Clin Chim Acta 68: 61–66

Vladimirov YA, Olenev VI, Suslova TB, Cheremisina ZP (1980) Lipid peroxidation in mitochondrial membranes. Adv Lipid Res 17: 173–249

Vojvodic V, Milosavljevic Z, Boskovic B, Bojanic N (1985) The protective effect of different drugs in rats poisoned by sulfur and nitrogen mustards. Fundam Appl Toxicol 5: 160–168

Vollum DI (1971) Hypomelanosis from antioxidant in polyethylene film. Arch Dermatol 104: 70–72

Volodarsky LB (ed) (1988) Imidazolin nitroxides, vol 1. Academic, Orlando

Volodarsky LB (ed) (1989) Imidazolin nitroxides, vol 2. Academic, Orlando

Von Bettmann S (1901) Chlorakne eine besondere Form von professioneller Hauterkrankung. Dtsch Med Wochenschr 27: 437

Von Seebach A, Stahl KW (1987) Das Ulcus cruris venosum in der Dermatologischen Praxis: 1537 Therapieergebnisse mit der Oxoferin-Wundfeuchtbehandlung. Aktuel Dermatol 13: 142–145

Voorhees JJ (1983) Leukotrienes and other lipoxygenase products in the pathogenesis and therapy of psoriasis and other dermatoses. Arch Dermatol 119: 541–547

Vuillaume M, Decroix Y, Calvayrac R, Vallot R, Best-Belpomme M (1983) Xeroderma pigmentosum: H_2O_2 formation and catalytic activity during the evolution of the skin cells: application to an early diagnosis and to a tentative treatment. C R Acad Sci (Paris) 296: 845–850

Vuillaume M, Calvayrac R, Best-Belpomme M, Tarroux P, Hubert M, Decroix Y, Sarasin A (1986) Deficiency in the catalase activity of xeroderma pigmentosum cell and Simian virus 40 transformed human cell extracts. Cancer Res 46: 538–544

Wach F, Hein R, Adelmann-Grill BC, Krieg T (1987) Superoxide dismutase inhibits the chemotactic response of human skin fibroblasts. J Invest Dermatol 89: 445

Walder BK (1965) Solvents and scleroderma. Lancet ii: 436

Walder BK (1981) Solvent induced scleroderma. Melbourne meeting of the Royal Australian College of Dermatology, April 1981

Walder BK (1983) Do solvents cause scleroderma? Int J Dermatol 22: 157–158

Walrant P, Santus R (1974) Ultraviolet and *N*-formyl-kynurenine-sensitized photoinactivation of bovine carbonic anhydrase: an internal photodynamic effect. Photochem Photobiol 20: 455–460

Walter JF, Stoughton RB, De Quoy PR (1978) Suppression of epidermal DNA synthesis supression by ultraviolet light, coal tar and anthralin. Br J Dermatol 99: 89–96

Walz DT, Dimartino MJ, Griswold DE, Intoccia AP, Flanagan TL (1983) Biologic actions and pharmacokinetic studies of auranofin. Am J Med: 90–108

Walzer C, Frank E (1983) Ultrastructural demonstration of endogenous peroxidase activity in mammalian epidermis. Histochemistry 78: 491–501

Walzer C, Frank E (1986) Cytochemical expression of epidermal peroxidase and cytochrome oxidase activities in pathological skin conditions in man. Arch Dermatol Res 278: 460–464

Wanders RJA, Kos M, Meijer AJ, Schrakamp G, Heymans HSA, Tegelaers WHH, van den Bosch H, Schutgens RBH, Tager JM (1984) Activity of peroxisomal enzymes and intracellular distribution of catalase in Zellweger syndrome. Biochem Biophys Res Commun 123: 1054–1061

Wanders RJA, Strijland A, van Roermund CWT, van den Bosch H, Schutgens RBH, Tager JM, Schram AW (1987) Catalase in cultured skin fibroblasts from patients with the cerebro-hepato-renal (Zellweger) syndrome: normal maturation in peroxisome deficient cells. Biochim Biophys Acta 923: 478–482

Wang RJ (1975) Lethal effect of "daylight" fluorescent light on human cells in tissue culture medium. Photochem Photobiol 21: 373–375

Wang RJ (1976) Effect of room fluorescent light on the deterioration of tissue culture medium. In Vitro 12: 19–22

Wang RJ, Nixon BT, Dalton JM (1978) Identification of hydrogen peroxide as a photoproduct toxic to human cells in tissue culture medium irradiated with "daylight" fluorescent light. In Vitro 14: 715–722

Wantanowicz M, Panczeno-Kresowkska, Ziemlanski B, Kowalska M, Okolska G (1984) The effect of alpha-tocopherol and ascorbic acid on the serum lipid peroxide level in elderly people. Ann Nutr Metab 28: 186–191

Waravdekar VS, Saslaw LD, Jones WA, Kuhns JG (1965) Skin changes induced by UV irradiated linolenic acid extract. Arch Pathol 80: 91–95

Ward AM, Udnoon S, Watkins J, Walker AE, Darke CS (1976) Immunological mechanism in the pathogenesis of vinyl chloride disease. Br Med J 1: 936–938

Warren JS, Yabroff KR, Mandel DM, Johnson KJ, Ward PA (1990) Role of O_2- in neutrophil recriutment into sites of dermal and pulmonary vasculitis. Free Radic Biol Med 8: 163–172

Watabiki T, Ogawa K (1975) Electron microscopic studies of effects of vitamin E on changes in the mouse skin following ultraviolet irradiation. Vitamins (Kyoto) 49: 121–142 (in Japanese)

Watson RR, Yahya MD, Darban HR, Prabhala RH (1988) Enhanced survival by vitamin A supplementation during a retrovirus infection causing murine AIDS. Life Science 43: XIII–XVIIII

Wayner DDM, Burton GW, Ingold KU (1986) The antioxidant efficiency of vitamin C is concentration dependent. Biochim Biophys Acta 884: 119–123

Wayner DDM, Burton GW, Ingold KU, Barclay LRC, Locke SJ (1987) The relative contributions of vitamin E, urate, ascorbate and proteins to the total peroxyl radical trapping antioxidant activity of human blood plasma. Biochim Biophys Acta 924: 408–419

Webster GF, Alexander JC, McArthur WP, Leyden TJ (1984) Inhibiton of chemiluminescence in human neutrophils by dapsone. Br J Dermatol 110: 657–663

Weetman RM, Boxer LA, Brown MP, Mantich NM, Bachner RL (1980) In-vitro inhibition of granulopoiesis by 4-amino-4′-hydroxyaminodiphenyl sulfone. Br J Haematol 45: 361–372

Wefers H (1987) Singlet oxygen in biological systems. Bioelectrochem Bioenerg 18: 91–104

Weinraub Z, Brenner S, Krakowski A, Caspi E (1982) The water cushion in dermatological diagnosic ultrasound. Dermatologica 164: 357–359

Weiser H, Vecchi M (1982) Stereoisomers of a-tocopheryl acetate. II Biopotency of all eight stereoisomers, individually or in mixtures, as determined by rat resorption gestation test. Int J Vitam Nutr Res 52: 351–370

Weisman MH, Albert D, Mueller MR, Zvaifler NJ (1983) Gold therapy in patients with systemic lupus erythematosus. Am J Med 156–164

Weiss J (1944) Radiochemistry of aqueous solutions. Nature 153: 748–750

Weiss SJ (1986) Oxygen, ischemia and inflammation. Acta Physiol Scand Suppl 548: 9–37
Weiss SJ (1989) Tissue destruction by neutrophils. N Engl J Med 320: 365–376
Weiss SJ, Young J, Lobuglio AF, Slivka A, Nimeh NF (1981) Role of hydrogen peroxide in neutrophil mediated destruction of cultured endothelial cells. J Clin Invest 68: 714–721
Weitberg AB, Weitzmann SA, Destrempes M, Latt SA, Stossel TP (1983) Stimulated human phagocytes produce cytogenectic changes in cultured mammalian cells. N Engl J Med 308: 26–30
Weitzmann SA, Stossel TP (1981) Mutation caused by human phagocytes. Science 212: 546–547
Weitzmann SA, Stossel TP (1984) Phagocyte induced mutation in Chinese hamster ovary cells. Cancer Lett 22: 337–441
Weitzman SA, Turk PW, Rundell K (1990) Hydrogen peroxide causes amplification of viral and cellular genes. Free Radic Biol Med 9 (Suppl 1): 167
Welgus HG, Stricklin GP, Eisen AZ, Bauer EA, Cooney RV, Jeffrey JJ (1979) A specific inhibitor of vertebrate collagenase produced by human skin fibroblasts. J Biol Chem 25: 1938–1943
Wells RL, Han A (1984) Action spectra for killing and mutation of Chinese hamster cells exposed to mid- and near ultraviolet monochromatic light. Mutat Res 129: 251–258
Welsh AL (1954) Lupus erythematosus. Arch Dermatol 70: 181–198
Wennersten G, Brunk U (1977) Cellular aspects of phototoxic reactions induced by kynurenic acid. Acta Derm Venereol (Stockh) 57: 201–209
Wepierre J, Corroller M, Didry JR (1986) Distribution and dissociation of benzoylperoxide in cutaneous tissue after application on skin in the hairless rat. Int J Cosmet Sci 8: 97–104
Werninghaus K, Handjani RM, Gilchrest BA (1990) Protective effect of alpha-tocopherol in carrier liposomes on ultraviolet light mediated epidermal cell damage in-vitro. J Invest Dermatol 94: 590
Wertz PM, Downing DT (1984) Acylglucosylceramides of pig epidermis: structure determination. J Lipid Res 24: 753–761
Wertz PW, Miethke MC, Long SA, Strauss JS, Downing DT (1985) The composition of the ceramides from human stratum corneum and from comedones. J Invest Dermatol 84: 410–412
Westermarck T, Antila E, Nordberg UR, Szamotulska J, Atroshi F (1990) Beta-carotene supplementation in handicapped patients. Free Radic Biol Med 9 (Suppl 1): 80
Western A, van Camp JR, Benasson R, Land EJ, Kochevar IE (1987) Involvement of singlet oxygen in the phototoxicity mechanism for a metabolite of piroxicam. Photochem Photobiol 46: 469–475
Westman NG, Marklund SL (1981) Copper- and zinc-containing superoxide dismutase and manganese-containing superoxide dismutase in human tissues and human malignant tumors. Cancer Res 41: 2962–2966
Wheatley VR, James AT (1957) Studies of sebum. 7. The composition of the sebum of some common rodents. Biochem J 65: 36–42
Wheeler LA, Aswad A, Connor MJ, Lowe M (1986) Depletion of cutaneous glutathione and the induction of inflammation by 8-methoxypsoralen plus UVA radiation. J Invest Dermatol 87: 658–662
White AA, Crawford KM, Patt CS, Lad PJ (1976) Activation of soluble guanylate cyclase from rat lung by incubation or by hydrogen peroxide. J Biol Chem 251: 7304–7312
Whiteside C, Hassan H (1987) Induction and inactivation of catalase and superoxide dismutase of Echerichia coli by ozone. Arch Biochem Biophys 257: 464–471
Whiteside C, Hassan HM (1988) Role of oxyradicals in the inactivation of catalase by ozone. Free Radic Biol Med 5: 305–312
Wiebel FJ, Leutz JL, Gelboin HV (1975) Aryl hydrocarbon [benzo(*a*)pyrene] hydroxylase. A mixed function oxygenase in mouse skin. J Invest Dermatol 64: 184–190
Wiggington JM (1989) The potential role of serum ferritin in the pathogenesis of acquired immune deficiency syndrome (AIDS). Med Hypotheses 30: 65–70
Williams RJP (1985) The necessary and the desirable production of radicals in biology. Phil Trans R Soc Lond [B] 311: 593–603

Willis I, Menter JM, Whyte J (1981) The rapid induction of cancers in the hairless mouse utilizing the principle of photoaugmentation. J Invest Dermatol 76: 404–408
Wills ED (1966) Mechanism of lipid peroxide formation in animal tissues. Biochem J 99: 667–676
Wills ED, Rotblat J (1964) The formation of peroxides in tissue lipids and unsaturated fatty acids by irradiation. Int J Radiat Biol 8: 551–567
Wilson HD (1964) Treatment of epidermolysis bullosa dystrophica by alpha-tocopherol. Can Med Assoc J 90: 1315–1316
Wilson RL (1988) Zinc and iron in free radical pathology and cellular control. In: Mills CF (ed) Zinc in human biology. Springer, Berlin Heidelberg New York, pp 147–172
Witt EH, Motchnik PA, Han D, Ames B, Packer L (1991) Ultraviolet irradiation, destruction of lipophilic antioxidants, and formation of lipid hydroperoxides in skin of hairless mice. National Meeting of the Society for Investigative Dermatology, Seattle, May 1–3, 1991
Witz G, Goldstein BD, Amoruso M, Stone DS, Troll W (1980) Retinoid inhibition of superoxide anion radical production by human polymorphonuclear leucocytes stimulated with tumor promotors. Biochem Biophys Res Commun 97: 883–888
Wohlgemuth J, Klopstock E (1926) Die Fermente der Haut. V. Atmung und Glycolyse der Haut und ihre Beeinflussung durch Hormone. Biochem Z 175: 202–208
Wolbach SB, Howe PR (1925) Tissue changes following deprivation of fat soluble A vitamin. J Exp Med 42: 753–756
Wolbach SB, Howe PR (1928) Vitamin A deficiency in the guinea pig. Arch Pathol Lab Med 5: 239–245
Wolbach SB, Howe PR (1933) Epithelial repair in recovery from vitamin A deficiency. J Exp Med 57: 511
Wölbling RH, Fuchs J, Schäfer V, Milbradt R (1987) Zur Dithranoltherapie der seborrhoischen Dermatitis. Aktuell Dermatol 13: 301–305
Wölbling RH, Fuchs J, Milbradt R (1989) Low dose Dithranol bei seborrhoischer Dermatitis. Dtsch Dermatol 37: 70–72
Wolf C, Steiner A, Hönigsmann H (1988) Do oral carotenoids protect human skin against ultraviolet erythema, psoralen phototoxicity, and ultraviolet induced DNA damage? J Invest Dermatol 90: 55–57
Wolff SP, Jiang ZY, Hunt JV (1991) Protein glycation and oxidative stress in diabetes mellitus and ageing. Free Radic Biol Med 10: 339–352
Wong GHW, McHugh TM, Stiles DP, Goeddel DV (1990) HIV infected cells are mores susceptible than uninfected cells to killing by irradiation, heat or tumor necrosis factor. In: Antioxidants and degenerative disease, Satellite meeting of the Oxygen Society and the International Society for Free Radical Research, University of California Berkeley, CA, 26–27 Jan 1990
Wood DC, Wood J (1975) Pharmacologic and biochemical considerations of dimethyl sulfoxide. Ann NY Acad Sci 243: 7–19
Wooley DE, Roberts DR, Evanson JM (1976) Small molecular weight β1 serum protein which specifically inhibits human skin collagenase. Nature 261: 325–327
Wozel G (1987) Zum Wirkungsmechanismus von Dapsone bei chronisch entzündlichen Dermatosen. Dermatol Monatsschr 173: 569–576
Wozel G (1988) Zum Wirkungsmechanismus von Dapsone bei chronisch entzündlichen Dermatosen. Teil II. Beeinflussung der Lymphocytenfunktionen und der Entzündungsmechanismen. Dermatol Monatsschr 174: 1–9
Wozel G, Thiele B (1987) Urtikaria vaskulitis and sulphones. A case report. Z Hautkr 62: 407–409
Wright CD, Mülsch A, Busse R, Osswald H (1989) Generation of nitric oxide by human neutrophils. Biochem Biophys Res Commun 160: 813–819
Wu Z, Ahmad R, Armstrong DA (1984) Formation of lipoamide anion radicals by hydroxyl, formate and alcohol radicals at pH 6–9. Radiat Phys Chem 23: 251–257
Wuepper KD, Watson PA, Kazierowski JA (1980) Immune complexes in erythema multiforme and the Stevens Johnson syndrome. J Invest Dermatol 74: 368–371
Wulf K (1962) Vitamin E. In: Kimmig J (ed) Therapie der Haut- und Geschlechtskrank-

heiten. Springer, Berlin Göttingen Heidelberg, pp 447–469 (Handbuch der Haut- und Geschlechtskrankheiten, vol 5, pt 1)
Wulf K, Unna PJ, Willers M (1963) Experimentelle Untersuchungen über die photodynamische Wirksamkeit von Steinkohlenteerbestandteilen. Hautarzt 14: 292–297
Yagi K (ed) (1982) Lipid peroxides in biology and medicine. Academic, New York
Yagi K (1987) Lipid peroxides in the skin. In: Hayaishi O, Imamura S, Miyachi Y (eds) The biological role of reactive oxygen species in skin. Elsevier, New York, pp 109–116
Yagi K, Yamada H, Nishikimi M (1978) Oxidation of a-tocopherol with O_2-· In: de Duve C, Hayashi O (eds) Tocopherol, oxygen and biomembranes. Elsevier/North Holland, New York, pp 1–11
Yagi K, Yoshino K, Komura S, Kondo, Ohishi N (1988) Lipid peroxide levels in the senescence accelerated mouse. J Clin Biochem Nutr 5: 21–27
Yamada H, Takamori K, Ogawa H (1987) Localization and some properties of skin sulfhydryl oxidase. Arch Dermatol Res 279: 194–197
Yamada H, Suga Y, Takamori K, Ogawa H (1989) Characteristics of skin sulfhydryl oxidase – stochiometry of the enzymatic reaction. J Invest Dermatol 92: 544
Yamada K, Salopek T, Jimbow K, Ito S (1989) An extremely high content of pheomelanin in dysplastic melanocytic nevi. J Invest Dermatol 92: 544
Yamagiwa K, Ichikawa K (1915) Experimental study of the pathogenesis of carcinoma. Univ Tokyo 15: 295–344
Yamakage A, Ishikawa H (1982) Generalized morphea like scleroderma occuring in people exposed to organic solvents. Dermatologica 165: 186–193
Yamakage A, Ishikawa H, Saito Y, Hattori A (1980) Occupational scleroderma like disorder occuring in men engaged in polymerization of epoxy resins. Dermatologica 161: 33–44
Yamamoto A, Nagai T, Shimizu M, Shitara A, Sato Y, Inomata N (1982) Skin irritation and cytotoxicity by squalene peroxides. 16th international congress of dermatology. Tokyo, 1982
Yamamoto Y, Frei B, Ames BN (1990) Assay of lipid hydroperoxides using high performance liquid chromatography with isoluminol chemiluminescence detection. Methods Enzymol 186: 371–379
Yamazaki I, Mason HS, Piette L (1960) Identification, by electron paramagnetic resonance spectroscopy, of free radicals generated from substrates by peroxidase. J Biol Chem 235: 2444–2449
Yamazaki I, Tamamura M, Nakajima R, Nakamura M (1985) Physiological aspects of free radical reactions. Environ Health Perspect 64: 331–342
Yanai M, Maeda H, Hayashi K, Mori K, Tsuji Y (1987) Effects of liposomal SOD on human polymorphonuclear neutrophil functions. In: Hayaishi O, Imamura S, Miyachi Y (eds) The biological role of reactive oxygen species in skin. Elsevier, New York, pp 217–224
Yancey KB, Lawley TJ (1984) Circulating immune complexes: their immunochemistry, biology, and detection in selected dermatologic and systemic disease. J Am Acad Dermatol 10: 711–731
Yano S (1959) Experimental and clinical studies on vitamin E from the dermatological point of view. Sapporo Med J 16: 449–458 (in Japanese)
Yardley HJ (1983) Epidermal lipids. In: Goldsmith LA (eds) Biochemistry and physiology of the skin. Oxford University Press, New York, pp 363–381
Yen EHK, Sodek J, Melcher AH (1979) The effect of oxygen partial pressure on protein synthesis and collagen hydroxylation by mature peridontal tissues maintained in organ cultures. Biochem J 178: 605–612
Yohn JJ, Norris DA, Yrastorza DG, Buno IJ, Leff JA, Hake SS (1991) Disparate antioxidant enzyme activities in cultured human cutaneous fibroblasts, keratinocytes, and melanocytes. J Invest Dermatol 97: 405–409
Yokoo K, Hayakawa M, Sugiyama S, Ozawa T, Aoyama H, Izawa Y, Kondo T, Hayakawa Y (1986) A novel uncoupler of mitochondrial respiration, 9,10-epoxy-12-octadecenoate, exists in human burned epidermis. J Clin Biochem Nutr 1: 121–127
Yonaha M, Ohbayashi Y, Ichinose T, Sagai M (1982) Lipid peroxidation stimulated by mercuric chloride and its relation to toxicity. Chem Pharm Bull (Tokyo) 30: 1437–1442

Yoshino K, Matsuo, Ohkido M (1981) Antioxidant effects of human epidermal keratin of skin surface lipid peroxides. Jpn J Dermatol 91: 1175–1179 (in Japanese)
Yoshioka A, Miyachi Y, Imamura S, Niwa Y (1986) Anti-oxidant effect of retinoids on anti-inflammatory skin diseases. Arch Dermatol Res 278: 177–183
Yoshioka A, Miyachi Y, Imamura S, Niwa Y (1987a) The effect of the supernatants obtained from Sporotrix schenkii and Candida albicans on the generation of reactive oxygen species by polymorphonuclear leukocytes. Mycopathologia 100: 43–48
Yoshioka A, Miyachi Y, Imamura S (1987b) Mechanism of reactive oxygen species induced skin erythema and superoxide dismutase activities in guinea pigs. J Dermatol 14: 569–575
Youn JI, Gange RW, Maytum D, Parrish JA (1988) Effect of hypoxia on sunburn cell formation and inflammation induced by ultraviolet radiation, Photodermatology 5: 252–256
Youngman RJ, Wagner G, Kühne FW, Elstner EF (1985) Biochemical oxygen activation as the basis for the physiological action of tetrachlorodecaoxide (TCDO). Z Naturforsch 40c: 409–414
Youssef AR, Baron DN (1983) Leucocyte superoxide dismutase in rheumatoid arthritis. Ann Rheum Dis 42: 558–562
Yuan C, Mei Z, FengJu Q, Hua K, Wenan W, Tingzhi H, Weiming Z (1988) Burns and damage by lipid peroxidation. J Med Coll PLA 3: 55–59
Yusa T, Crapo JD, Freeman BA (1984) Hyperoxia enhances lung and liver nuclear superoxide generation. Biochim Biophys Acta 798: 167–174
Zackheim SH, Wolf P (1972) Serum copper in psoriasis and other dermatoses. J Invest Dermatol 58: 28–32
Zadorozhny BA, Adolf EV, Shevchenko VP (1973) Free radicals in the skin of patients with psoriasis and their relationship to oxidative-reductive processes. Vestn Dermatol Venerol 47: 25–28 (in Russian)
Zarafonetis CDJ, Lorber SH, Hanson SM (1958) Association of functioning carcinoid syndrome and scleroderma. Am J Med Sci 236: 1–14
Zavoisky E (1945) Paramagnetic relaxation of liquid solutions for perpendicular fields. J Phys (USSR) 9: 211–216
Ziboh VA, Chapkin RS (1987) Biological significance of polyunsaturated fatty acids in the skin. Arch Dermatol 123: 1686–1690
Ziboh VA, Lord JT (1979) Phospholipase A activity in skin. Modulators of arachidonic acid release of phosphatidyl choline. Biochem J 184: 283–290
Ziboh VA, Lord JT, Uematsu S, Blick G (1978) Activation of phospholipase A2 and increased release of prostaglandin precursor from skin by ultraviolet irradiation. J Invest Dermatol 70: 211
Ziegler DM (1985) Role of reversible oxidation-reduction of enzyme thiols-disulfides in metabolic regulation. Annu Rev Biochem 54: 305–329
Ziegler V, Pampel W, Zschunke E, Münzberger H, Mährlein W, Köpping H (1982) Kristalliner Quarz (eine) Ursache der progressiven Sklerodermie? Dermatol Monatsschr 168: 398–401
Ziegler V, Haustein UF, Mehlhorn J, Münzberger H, Rennau H (1986) Quarzinduzierte Sklerodermie. Sklerodermie-ähnliches Syndrom oder echte progressive Sklerodermie? Dermatol Monatsschr 172: 86–90
Zimmer G, Thürich T (1991) Membrane fluidity and vitamine E. In: Packer L, Fuchs J (eds) Vitamine E: biochemistry and clinical applications. Dekker, New York (in press)
Zimmer G, Beyersdorf F, Fuchs J (1985) Decay of structure and function of heart mitochondria during hypoxia and related stress and its treatment. Mol Physiol 8: 495–513
Zimmerman R, Cerutti P (1984) Active oxygen acts as a promoter of transformation in mouse embryo C3H/10T 1/2/C18 fibroblasts. Proc Natl Acad Sci USA 81: 2085–2087
Zweier JL, Flaherty JT, Weisfeldt ML (1987) Direct measurement of free radical generation following reperfusion of ischemic myocardium. Proc Natl Acad Sci USA 84: 1404–1407

Subject Index

T

MIX
Papier aus verantwortungsvollen Quellen
Paper from responsible sources
FSC® C105338

If you have any concerns about our products,
you can contact us on
ProductSafety@springernature.com

In case Publisher is established outside the EU,
the EU authorized representative is:
Springer Nature Customer Service Center GmbH
Europaplatz 3, 69115 Heidelberg, Germany

Printed by Libri Plureos GmbH
in Hamburg, Germany